(*Courtesy of Country Roads Board of Victoria*)

Breaking ground on the last section of sealed road between Sydney and Melbourne, 1966. The Prince's Highway in Australia's Alfred National Park.

This textbook is dedicated to

Nuala

who helped much
with 'heart an' hand'

HIGHWAYS

Volume 2
Highway Engineering

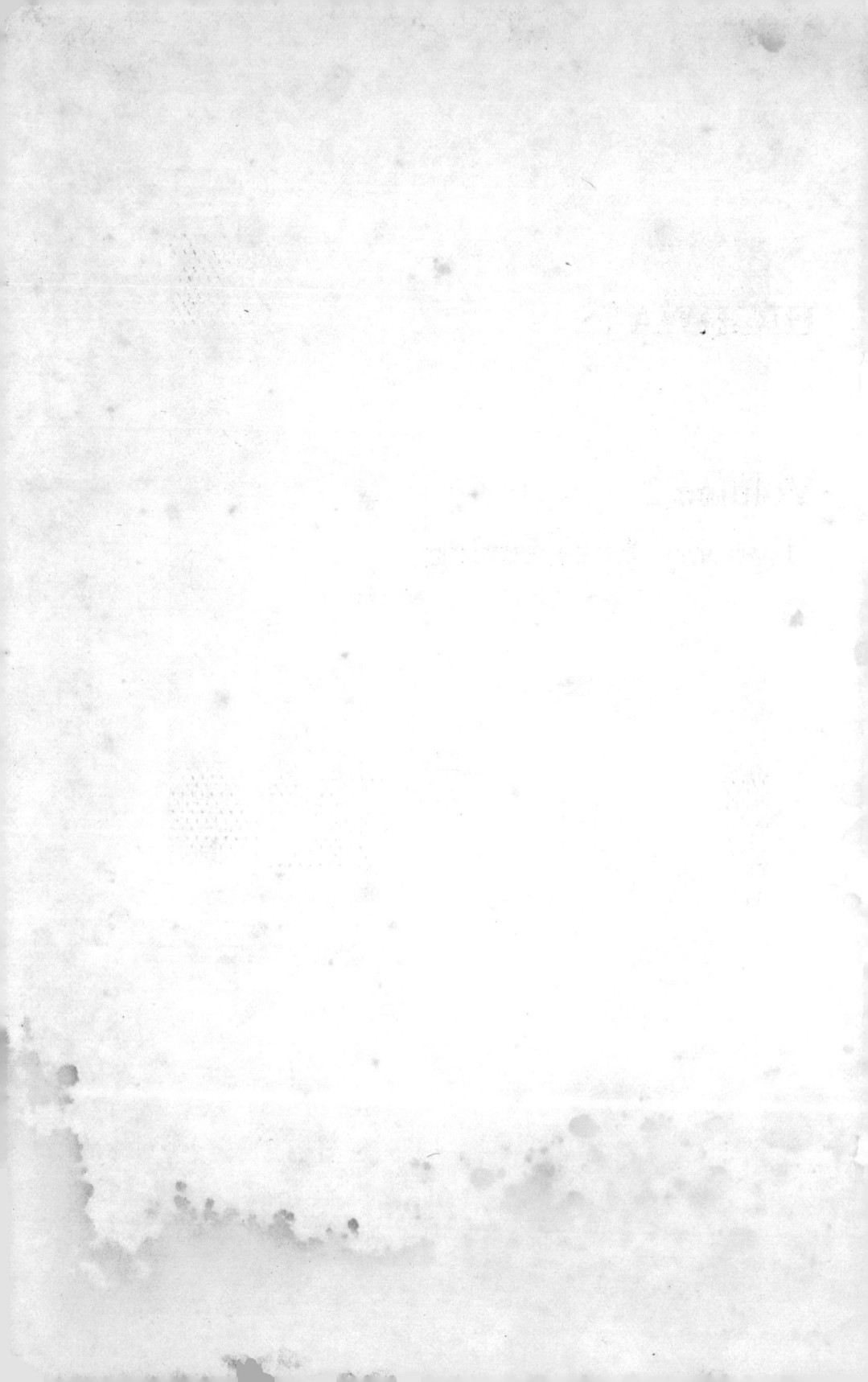

HIGHWAYS

Volume 2
Highway Engineering

C. A. O'Flaherty

B.E.(N.U.I.), M.S., Ph.D.(Iowa State), C.Eng., F.I.Mun.E., F.C.I.T.,
M.Inst.H.E., M.I.E.I.
*Formerly Professor of Transport Engineering and Director of the Institute for
Transport Studies, University of Leeds*

Edward Arnold

First published in 1967 by
Edward Arnold (Publishers) Ltd,
41 Bedford Square, London WC1B 3DQ

Second Edition 1974
Reprinted 1975, 1978, 1979

Boards edition ISBN 0 7131 3324 8
Paper edition   ISBN 0 7131 3325 6

Printed in Great Britain by Butler & Tanner Ltd, Frome and London

Preface to Second Edition

The primary aim of this work is as described for the first edition, to provide a basic textbook for the young engineer about to centre his career in highways, whether it be in their planning, design, construction, or operation. The approach, therefore, is similar — except that, for economic and handling reasons, it has been necessary to divide the material between two volumes. Thus, Volume 2: Highway Engineering covers the 'hard' side of highways, i.e. their physical location and structural design, as well as the materials used in their construction. Volume 1: Highways and Traffic contains the 'soft' material which will be of particular interest to traffic engineers. (The contents of Volume 1 are set out on p. x).

There are a number of other changes in the material in Volume 2. First, and most important, the text has been updated, thus every chapter has been modified in some way or another, particularly the chapters on Surface and Subsurface Moisture Control, Flexible Pavements, Bituminous Surfacings, and Rigid Pavements.

The opportunity has also been taken to 'metricate' the text. This has meant in certain instances that assumptions have been made, with results which may not be in accord with evolving practice, e.g. where sieve sizes are given in imperial units, they have been consistently changed to the nearest direct metric equivalent. While it is hoped that this volume will also be of use to the practising engineer, it must be kept in mind that the factual material in the text is primarily given for demonstration purposes; hence, the reader is urged to seek out the original material (references are given in most instances), or to consult the most up-to-date 'official' versions of the codes, standards, specifications, recommendations, etc, when engaged in actual design and field-work. Also, certain errors may have occurred during the metrication process, and the writer will welcome any being brought to his attention.

COLEMAN O'FLAHERTY

Institute for Transport Studies
and Department of Civil Engineering,
University of Leeds,
Leeds 2.

Feb. 12, 1973.

Acknowledgments

I am indebted to many colleagues for their help and encouragement during the preparation of this text. I am also grateful to the many organizations who gave me permission to reproduce diagrams and tables from their publications. (The number in the title of a table or figure indicates the reference at the end of the chapter where the source may be found.)

The names of the people and organizations who were most helpful are given in the 'Acknowledgments' section in Volume 1.

Contents

		Page
Preface to second edition		vii
Acknowledgments		viii
1	Highway location, physical surveys, plans and specifications	1
2	Surface and sub-surface moisture control	56
3	Pavement materials	90
4	Soil engineering for highways	154
5	Soil stabilization	221
6	Flexible pavements	269
7	Bituminous surfacings	352
8	Rigid pavements	392
	Index	445

Complementary to this volume

HIGHWAYS, Volume 1: HIGHWAYS AND TRAFFIC

1   The road in perspective
2   Highway legislation, classification, finance and administration
3   Highway planning
4   Parking
5   The economics of road improvements
6   Geometric design
7   Traffic management
8   Highway lighting
    Index

1 Highway location, physical surveys, plans and specifications

The making of a major highway is now acknowledged to require the combined skills of the planner, economist, surveyor and geologist, as well as the expected engineering ones. Before an attempt can be made to physically locate the highway improvement, data must be available regarding traffic desires and needs, the planning intentions within the area to be traversed, and estimates of the future physical characteristics of the highway itself. Location surveys involving geologic and photogrammetric skills provide the basic information for structural design, as well as the economic analyses which have a considerable influence on the final location of the highway. The coupling of advanced computing techniques with improved methods of obtaining data ensure that, when the final costings are done and the specifications written, the highway location selected will yield the highest social return on the transport investment, and reconcile as effectively as possible the conflicting interests of the various individuals and groups affected by the facility.

In essence, this whole textbook is concerned with as many of these considerations as is deemed practical. In this chapter, however, are outlined the methods of determining the *physical* characteristics which influence location and design. Integrated with these surveys are the plans and specifications resulting from them, so these are also described here.

PRINCIPLES OF HIGHWAY LOCATION

Before discussing the means by which the engineering data are obtained it is worthwhile considering some detailed guiding principles which should be kept in mind in selecting the location for a highway. The following outline is not in any particular order, nor indeed is it, or can it be, complete. In addition some of the elements tend to contradict one another; in practice the location is selected which represents the best compromise solution.

1. For the highway to serve its function of allowing convenient, continuous, free-flowing traffic operation, it should be located where it can best meet the major traffic desire lines and be as direct as possible.

2. Keep grades and curvature to the minimum necessary to satisfy the service requirements of the highway.

3. Avoid sudden changes in sight distance, especially near junctions.

4. Avoid having a sharp horizontal curve on or adjacent to a pronounced vertical curve.

5. In urban areas site the highway through undeveloped or blighted areas, along the edges of large parklands and, in general, away from highly-developed, expensive land areas.

6. In urban areas locate the highway as close as possible to the principal parking terminals.

7. In rural areas locate as much as possible of the new highway on existing ones, so as to minimize the use of farmland and reduce total initial and maintenance costs.

8. Locate along the edges of properties rather than through the middle, so as to cause the minimum interference to cultivation and avoid the need for subway construction.

9. Avoid the destruction or removal of man-made culture.

10. Keep the highway away from cemeteries, places of worship, hospitals, schools and playgrounds.

11. The effect of the proposed highway on existing or future utilities above, on or under the ground should be considered. It may be such as to warrant changes in order to avoid expensive relocation of these utilities.

12. Never have two roads intersecting near a bend or at the top or bottom of a hill.

13. In the case of a motorway, the need for an interchange with another road may dictate an alignment that will intersect the other highway at a place, at an angle and in terrain that will best permit the interchange to be constructed.

14. Avoid intersections at-grade with railway lines. If possible have the highway pass over the railway where it goes into a cutting.

15. Seek favourable sites for river crossings. Preferably these should be at right-angles to the stream centre line.

16. Do not have a bridge located on or adjacent to a highway curve.

17. Avoid the need for deep cuttings and expensive tunnel construction.

18. Avoid locations where rock is close to the surface as this will usually require at least some expensive excavation.

19. In hilly terrain, be aware of the possibilities of landslides.

20. To minimize drainage problems, select a location on high ground in contrast to one in a valley.

21. Avoid bogs, marshes and other low-lying lands subject to flooding.

22. Locate the highway on soil which will require the least pavement thickness above it.

23. Locate the highway adjacent to sources of pavement materials.

24. When the needs of all other factors have been satisfied, the best location is the one which results in the minimum total cost of earthworks. This means that the minimum quantities of excavation should be so balanced with the quantities of embankment as to require a minimum of haulage with little need for overhaul.

25. In hilly terrain the highway should cross ridges at their lowest points. (This usually results in cheaper construction as well as more economical vehicle operating costs.) Avoid creating severe breaks in the natural skyline.

26. In hilly country also, select a location subject to sunlight and avoid areas where snow and ice will accumulate.

27. Avoid the unnecessary and expensive destruction of wooded areas. If intrusion is unavoidable, the road should be on a curve where possible so as to preserve an unbroken background.

28. Avoid ground subject to mining subsidence.

29. Avoid placing the highway at right-angles to the natural drainage channels.

30. To relieve the monotony of driving on a long straight road it is an advantage to site it so as to give a view of some prominent feature ahead.

LOCATION SURVEYS IN RURAL AREAS

Structure of the location process

In general, the approach to selecting the route for a long highway can be described[1] as a 'hierarchically structured decision process'. This very logical approach is easiest described by referring to Fig. 1.1.

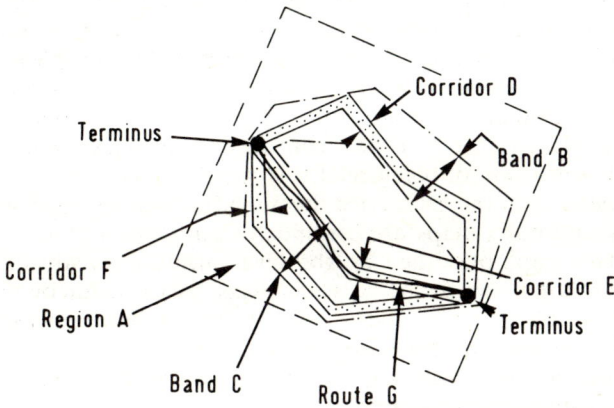

Fig. 1.1. A hypothetical route location problem showing the spacial relationships of various actions generated and evaluated[1]

The first step in the process requires fixing the end terminii, and then defining a *region* 'A' which will include these and all conceivably feasible routes between them. This region is then searched and a number of broad *bands* 'B' and 'C' are selected within which it is decided to concentrate further searches and selections. For a major rural road scheme, e.g. a rural motorway, a band might be 8–16 km wide. Within these bands, further search may result in the selection of corridors 'D', 'E' and 'F', each perhaps 3–8 km wide. A comparison of these corridors suggests that 'E' is the best and so *route* 'G' is generated within it; this route typically could be from 1–1·5 km wide. The next step then is to search this area and locate within it (not shown in Fig. 1.1) one or more different *alignments*, each perhaps 30 m wide and containing minor geometric differences. These alignments are compared, a final selection is made, and this is used for *design* purposes.

Note that the process involves continuous searching and selecting, using increasingly more detailed knowledge at each decision-making stage. Factors influencing the selection process at any instance include not only such 'tangibles' as topography, soil and geological details, land use, population distribution, travel demand, user costs, structure and maintenance costs, safety, etc., but also 'intangibles' such as political, social and environmental costs.

Physical surveys

To aid in the decision-making process, a classical approach has tended to be developed with regard to gathering physical information about the areas being evaluated. Generally, this has meant carrying out reconnaissance, preliminary line, and location surveys.

Reconnaissance survey. Reconnaissance is the process of evaluating the feasibility of one or more possible routes for a highway between specific points which may be many kilometres apart[2]. Good reconnaissance can be the greatest single money-saving phase in the construction of a new road. Hence the engineer should make ample provision both in time and finance for this stage of the location investigation. Existing maps and aerial photographs are normally the primary tools of the reconnaissance engineer. Britain is in the very fortunate position of being relatively well equipped in this respect.

Since about 1805 the Ordnance Survey have been engaged in mapping the country and these maps are continually being brought up-to-date and extended. The maps published by the Ordnance Survey are divided into three groups; these are the large scale plans, (with scales of 1:1250 and 1:2500) on national grid sheet lines, medium scale maps (1:10560 and 1:25000), and small scale maps (ranging from 1:63360 to 1:1250000). The determination of which map to be used at any stage depends on the detail of the information required.

It may be noted that the sites of many ancient monuments, generally protected by Acts of Parliament, are marked on these survey sheets, and the most up-to-date records of these should be obtained. In contrast, old Ordnance Survey maps can be very usefully perused in order to obtain information about the locations of concealed mineshafts, adits and wells; the sites of demolished buildings with possible concealed foundations and cellars; abandoned sewage farms; filled ponds, clay pits, quarries and sand and gravel pits; changes in landslip areas, etc.

The Institute of Geological Sciences (IGS), which incorporates the Geological Survey of Great Britain, is the principal source of geological data; information is readily available in the form of maps, geological handbooks, memoirs and papers. Very often also the Institute will be able to provide information regarding the groundwater conditions in a given area, as well as details regarding boreholes at particular points.

Land use, land classification, and soil survey data are also readily available from appropriate organizations which can provide a tremendous wealth of information of use to the highway engineer. The National Coal Board should be consulted if it is anticipated that the highway will penetrate areas where coal or oil-shale mining is/has been carried out.

The above are just a few of the sources of detailed information already available for the engineer to use in his reconnaissance survey. Details of these and other sources of information are readily available in an excellent publication in the literature[2].

As will be gathered from the above the first step in the reconnaissance survey is the location and acquisition of all maps and memoirs relating to the area, as well as the most suitable air photographs. These are then thoroughly studied; a visit to the area may be also desirable at this stage. Where appropriate, more relevant information may be obtained from IGS, other official bodies (e.g. local authorities and University geological departments), and the engineers of existing civil engineering projects in the area (e.g. site reports on roads, building foundations and pipelines). Typically, hills, waterways and land use are examined first. Low points or passes in the hills are potential fixed points in the location, as are river-crossing sites that afford suitable topography for approaches to bridges. Peat bogs and other marshy areas which may have to be avoided can also be detected at this stage. Sites for necessary fly-over structures can be found along intersecting roads and railway lines, and utility relocation problems may be anticipated.

Discovery of the foregoing controlling factors, together with the rapid appraisal of intervening grades and horizontal alignments, is possible with the modern topographic map. Photographs can be used, to a greater or lesser degree, to supplement the map, depending upon their quality. Stereoscopic techniques can afford quantitative data and, when applied by a skilled photo-interpreter, they can yield significant soil and subsurface information.

Next, armed with questions generated by the paper study, the reconnaissance engineer takes to the field, where he obtains an evaluation of the effect of unusual topographic features, hydrology, the nature of certain man-made works and subsurface conditions. Much can be gained from a simple inspection of geological exposures, and a study of the stability of slopes in local materials in old road or rail cuttings.

At this stage the engineer should have sufficient physical information which, when taken in combination with the traffic, economic, social and political data, will enable him to make one or more paper locations of feasible routes. If he is satisfied with the extent and quality of his data, it may be possible to make the necessary economic comparisons to aid in the selection of the best route (see Chapter 5, Vol. 1).

The results of these studies are presented in a *reconnaissance report*. In its barest essentials, this report should state the service and geometric criteria to be satisfied by the project, describe the chosen route(s), and present tentative estimates of cost. Also included should be provisional geotechnical maps for the locations under consideration, provisional longitudinal geological sections, and block diagrams showing the characteristics of the more important engineering features. Special situations which might lead to design or construction problems, or might require special attention in a later detailed site investigation, should also be pointed out.

Major highway projects such as motorways and trunk roads deserve a comprehensive report, particularly since public hearings may be required. In such cases the reconnaissance report may also be designed as a public-relations device which will inform interested persons about the advantages (and disadvantages) of a particular route. It should cover such matters as the exhaustive nature of the study, concern for landowner's interests, highway safety, and the expected benefits from the highway to both the road user and the communities served. The following has been recommended as a typical reconnaissance report outline[3]:

I SUMMARY (Conclusions and recommendations)
II REPORT
  A. Introduction
    1. Purpose of reconnaissance survey
    2. Traffic survey methods
    3. Study methods
    4. Design criteria
  B. General alignment details
    1. Choice of route
    2. Service to communities
    3. Interchange locations
    4. Land use and severance
    5. Safety considerations
    6. Drainage
    7. Soils and geology
  C. Project-cost estimate
    1. Roadway
    2. Structures
    3. Right-of-way
    4. Utility relocation.

Preliminary survey. The preliminary survey is a large-scale study of one or more feasible routes. It results in a paper location and alignment that defines the line for the subsequent location survey. This paper location and alignment should show enough ties to the existing topography to permit a location party to peg the centre-line. This much is a minimum; in many cases field detail for final design may also be obtained economically during the preliminary survey phase.

The preliminary survey is made for the purpose of collecting all the physical information which may affect the location of the highway. Thus within the established route area the shape of the ground, limits of catchment areas, positions and invert levels of streams and ditches and the positions of trees, banks and hedges, bridges, culverts, existing roads, power and pipe lines, houses and monuments are determined and noted. These are then translated into maps, profiles and (frequently) cross-sections which can assist the engineer to determine preliminary grades and alignments and prepare cost estimates.

Two approaches are available for preliminary-survey mapping; aerial surveys and ground surveys, either separately or in various combinations. The ground method is the best one to use in the situation where the alignment is well defined, narrow right-of-ways are contemplated, and the problems of man-made culture are few. Ground surveys, beginning with a traverse baseline, will probably furnish necessary data quite economically. Additional operations which can be quite easily included are the profile levels and cross-sections, and the ties to land lines and cultural objects.

Because the cost of an aerial survey does not increase in direct relationship to the area photographed, it is likely to be more adequate and economical than a ground survey in the following instances:

1. Where the reconnaissance was unable to approximate closely the final alignment. Illustrations of this are entirely new locations in rugged terrain or where land uses and values vary widely.

2. Where a wide right-of-way, such as that for a motorway, is necessitated.

3. Where it is desired to prevent the premature or erroneous disclosure of the details of probable location. The ground-control work required for aerial surveys reveals little of the highway engineer's intentions, thereby preventing any land speculation which might occur.

The choice of method should be an educated one, based on an advance cost analysis that takes into account the overall project schedule and the time requirements of the various techniques. The following discussion has to do primarily with the carrying out of the traditional type of *ground survey*. The use of the airphoto technique is described separately in the discussion on photogrammetry and aerial surveys.

The first step is the carrying out of a base-line traverse. This traverse may be simply a series of connecting straight lines if the curvature necessary to connect the tangents is such that the highway curves will not deviate greatly from the tangents. Where sweeping curves with large external distances are necessary, it may be necessary to include them as part of the original baseline so as to avoid unnecessary ground coverage. The base-line traverse should be stationed continuously from the beginning to the end of the survey, and the survey carried out to a degree of accuracy commensurate with the importance of the project and the nature of the topography being traversed. Angles between connecting lines should be measured in accordance with accepted highway-surveying procedures and every angle point should be carefully referenced to at least two points established well outside the area that might be occupied by the highway construction.

To furnish data for a profile of the base-line, levels should be taken at all marked stations, as well as at all important breaks in the ground. Elevations should also be noted at all cross-roads, streams and other critical points on the line. Levels should always be referred to the Standard Datum Plane of the country in which the works are being carried out. In Great Britain, the Ordnance Datum is used. This was originally referred to an approximate

mean sea level at Liverpool but was changed about thirty years ago to that
of the mean sea level at Newlyn in Cornwall, where it was fixed on the basis
of observations made over a number of years. Thus, old Ordnance Survey
maps refer to the Liverpool Datum while the present maps refer to the
Newlyn Datum. The difference between the two sets of levels varies over
the country from zero to about 600 mm as a result of corrections to
the original survey. When old maps are being used, care should be therefore
taken to determine on which datum it is based.

After the base-line has been pegged and levels run over it, the topography
elevations may be taken by one of several methods. The simplest method is
by cross-sections, and on fairly flat topography these are sometimes taken by
the level party at the same time as the profile levels. Observations are made
at right-angles from each station for as far as is considered necessary to
cover the expected construction area. Where wider deviations of the location
from the base-line are expected, the topography may be more rapidly taken
by tacheometry, or (less commonly) by plane table, or by a combination of
the two.

At the same time the locations of all trees, fences, buildings, etc. are noted
so that they can be shown on the preliminary map.

The preliminary map prepared from the data collected by the classical
ground-survey method is often termed a *strip* map because it is plotted on a
continuous roll of detail paper. The minimum information shown on the
map is the plot of the base-line and all planimetric detail; it may or may not
include the plotting of contours, depending on the complexity of the project.
In addition it may be desirable to have it supplemented with detailed maps
showing contours of those areas where complex structures or intersections
are to be located. The planimetric strip map should also show surface and
sub-surface information that might affect the location. This includes data
regarding fence lines, trees, limits of bogs and other undesirable soil and
geological features, and overhead and subsurface utility lines.

Paper location.   The selection of a horizontal alignment, and the extent to
which it may be selected to fit the ground economically, depends primarily
upon the geometric design standards adopted for the highway. These in turn
may range from the relatively low criteria for widths, curvature, grades and
sight distances sufficient to accommodate sparse rural traffic at moderate
speeds, to those criteria necessary for large volumes of heavy traffic travelling at high speeds. With the service classification established by the traffic
requirements, and the appropriate standards of alignment thereby fixed, the
combination of tangents and horizontal curvature is then sought that will
best fit the topography while yet remaining within the limits of these criteria.

At the same time consideration will have to be given to factors other than
the ground fit. These have already been listed in the outline of the principles
of location.

Factors affecting the choice of the horizontal alignment will also affect
the vertical alignment. One which deserves some further mention, however,

is that of the earthworks required, as this is perhaps one of the most important factors to be considered with reference to construction. When the needs of all the other more important factors have been satisfied, then the best line is the one that gives the minimum total cost for earthworks. The ideal situation leads to the minimum quantities of excavation so balanced in sections with the quantities of embankment as to require a minimum of haulage without overhaul. This is usually an extremely difficult and rare condition to achieve and compromises are frequently necessary. Many factors affect earthwork decisions. For instance, the unit cost of excavation may be cheaper in one classification of material as against another; or the length of haul and the amount of required overhaul may make wastage and borrow desirable as an economical alternative. The solutions of these problems lie in economic analyses of the road-user benefits and costs of the alternatives.

Final location survey. This survey serves the dual purpose of definitely fixing the centre-line of the road, while at the same time physical data are collected which are necessary for the preparation of plans for construction purposes. If the previous surveys have been properly carried out, it will be found that the final location survey can be reduced essentially to the pegging and referencing of the centre-line, and obtaining critical elevations not already available.

As with the reconnaissance and preliminary surveys, the trend in carrying out the final location survey is to make as much use as possible of aerial photography and photogrammetric maps in order to reduce tedious field study. Large-scale maps procured in this way for the preliminary survey will usually provide much, if not all, of the information needed for the final location, the preparation of construction drawings and the other operations necessary to advance the project to the construction stage.

Electronic computers are further aids to efficiency at this stage of a highway project. Very significant savings both in time and labour can be effected by placing survey data on computer cards or tapes and applying them as direct input on the electronic computer.

The following are the general features of this survey. However, it should be emphasized that part of the requirements specified here may already have been well satisfied in the earlier surveys.

Pegging the centre-line. The centre-line established on the ground at this stage should follow closely the paper location on the preliminary survey map, conforming as much as possible to the major and minor control points and the alignments prescribed. In practice, however, this is not always possible, and local deviations will be necessary to best fit the line to the topography and to allow for incomplete or innacurate preliminary survey information. The centre-line is pegged with reference to the preliminary traverse or base line if conventional ground-surveying methods were used in the preliminary survey, or with reference to the control traverse if aerial-survey methods were used.

Centre-line levelling.   Profile levels are taken along the centre-line at each station and at all intermediate points where there is any significant change in the slope of the ground, so that a truly representative profile is obtained. This profile should normally extend for at least 150 m beyond the beginning and end points of the scheme so as to allow for transitions to existing or new facilities. Where temporary bench marks come within the limits of the construction width, new ones should be set up clear of the works. These are usually located about every kilometre along the centre-line and, in addition, there should be at least one such bench mark within about 60 m of each highway structure.

Cross-sections.   Cross-sections should be taken at each station, points of significant change in ground slope, and for a reasonable distance beyond the beginning and end points of the project. (Where large-scale aerial photographs are available, this detail work can be considerably curtailed, since both profile and cross-section data can be developed in the stereoplotter to an accuracy comparable with that obtained by the ground-survey procedure.)

Property lines.   The positions of all property corners, lines, fences, buildings and other man-made improvements are accurately determined and noted during the final location survey. The exact extent to which all property owners are directly affected by the new location should also be determined.

Intersecting roads.   The directions with respect to the pegged centre-line of all intersecting roads should be measured. Profiles and cross-sections of the intersecting roads should be taken for some distance on both sides of the new centre-line. It should be remembered that it is always easier to design a new junction when there are too many cross-sections rather than too few.

Ditches and streams.   All ditches and streams within the area of construction should be carefully located with respect to the pegged centre-line. In addition stream-bed profiles should be taken for some distance up and down each stream; usually about 60 m will be adequate for all but the larger streams. Cross-sections may also have to be taken of larger streams to provide information required for the hydraulic design. Detailed information should be obtained on all existing culverts or bridges including the type, size, number of openings or spans, elevations of culvert flow lines and stream-beds under bridges and, where available, high-water elevations.

Special site surveys.   In order to obtain the detailed information needed for the design of large culverts, bridges, roundabouts, interchanges and other complex intersections or structures, special site surveys will have to be made. Information obtained should include such items as alignments of streams and intersecting roads, topography, profiles, cross-sections, elevations of grade-controlling points, foundation conditions and—for bridges and large culverts—data on adjacent structures upstream and downstream.

Site surveys for borrow pits and quarries should provide information on the quality and quantity of materials, as well as their availability and accessibility. If particular pits or quarries are specified, elevations along lines perpendicular to one or more arbitrary base-lines may be taken in each case for use in later computations of the quantities of materials removed.

BRIDGE SURVEYS

In the not too distant past, it was common practice, when a new highway had to cross a river or estuary, to determine first where the most favourable bridge site was located and then to 'bend' the highway so that the crossing could be made at this point. The term 'favourable' as used in this context usually meant the cheapest right-angle crossing. Today, however, the bridge location is determined upon a more desirable basis, it being realized that the crossing having the least initial cost is not necessarily the most economical one. It is accepted that the bridge is an integral and important part of the highway, but not the controlling one. Thus the approach now taken is first to secure the most favourable location and alignment for the highway, and then to require the bridge engineer to furnish all necessary bridges to fit this scheme. This, of course, usually results in more expensive structures, since skew bridges are more costly than right-angled ones and, all other things being equal, longer bridges are more expensive than short ones. In the long run, however, a better and more economical highway is obtained.

Surveys for large bridges, as with other transportation projects, can be carried out in the three classical phases, i.e. reconnaissance survey, preliminary survey and the final location survey. The following is but a brief description of what these entail; for further details the reader is referred to an excellent report which is readily available in the literature, and on which this brief discussion is based[3].

The *reconnaissance survey* for a major water-crossing structure is carried out to obtain information regarding the location of all possible sites, their comparison with respect to feasibility, and the preparation of preliminary estimates of the cost of construction. Field surveys in depth are rarely required in this investigation, most of the required information being already available from existing maps, boring records and aerial photographs. However, in certain instances it may be desirable to inspect the sites and make precursory subsurface explorations.

The reconnaissance survey having established the need for a water-crossing within a limited zone, the *preliminary survey* is carried out to define its exact alignment, and to obtain physical data needed for the design of the project and for the aquisition of necessary lands. The operations usually include a preliminary triangulation to establish the starting co-ordinates for traverses at the bridge-heads and along the approaches, topographic surveys along these traverses, and the design of a more elaborate triangulation system that can be used later during the construction phase of the project.

In addition to the foregoing surface operations, the preliminary survey will also normally include a hydrographic survey of the project site. This is carried out to determine the hydraulic or flow characteristics of the water at the proposed crossing as well as the nature of the channel or estuary bottom containing the water. Also included are engineering and geological surveys of the subsurface foundation conditions. The nature of the hydrographic data required will, of course, vary somewhat with the type of crossing and the location and exposure of the structures. In general, however, it will be necessary to obtain information on such items as the water datum, tide or river flood stages, tidal and river or other currents, water and—in cold climates—ice erosion characteristics, together with a complete estuary or river-bottom profile with cross-sections. Where embankments or causeways are being constructed using hydraulic fill procedures, some underwater studies for material borrow areas will be required.

The *final location survey* can be considered as two stages, the preconstruction survey and the construction survey. The preconstruction survey operations are performed in advance of the actual construction but are directly related to it. Examples are the establishment of the horizontal control stations and bench marks that will constitute the fundamental framework for defining lines and grades. Much of this work can be accomplished during the preliminary survey but generally some later refinement and strengthening is necessary to improve its accuracy. The construction survey operations provide the intermediate and final positioning, both horizontally and vertically, of the various components of the bridge structure. This is a most important stage since, for instance, the exact location and span distance between two main piers of a cantilever bridge is considerably more critical than the general positioning of these two piers and the remainder of the bridge within a given area. Also these operations must be closely coordinated with the construction schedule to avoid any unnecessary and costly delays to the contractor.

The importance of attaining high accuracy in this survey cannot be over-emphasized; it must be attained if wasteful delays and needless expenditure of money are to be avoided. For instance, to locate properly bridges of a thousand metres or more in length, surveys having an accuracy of 1 part in 100 000 may be required. This type of accuracy can usually only be attained by using special geodetic equipment[4, 5].

LOCATION SURVEYS IN URBAN AREAS

The determination of a suitable location for a major highway in an urban area can be one of the most complex problems imaginable. While, in theory, the traditional sequence of reconnaissance, preliminary and final location surveys are carried out, in practice there are usually only two phases; these are a combination reconnaissance-preliminary survey, and the final location survey.

The final location survey is similar to that described for a highway location survey in a rural area, except that it is usually much more difficult to carry out. Very rarely in an urban area will it be possible to directly peg a continuous centre-line. So many obstructions are met with that, until the ground is cleared for the actual road works to begin, it is always necessary to set out the centre-line by means of complicated off-setting and referencing. The taking of profile and cross-section levels is always a complicated and indeed, sometimes, a dangerous problem during which the operators are not only faced with considerable sighting problems but are subject to harrying by traffic on the existing roads.

Notwithstanding the difficulties associated with the final location survey, it is in the earlier reconnaissance-cum-preliminary survey that the main problems arise. This is normally an office-based study since, in an urban area, there is usually ample information available from previous surveys, street improvements, property locations, and utility installations to render superfluous any but the most necessary field investigations. It should be pointed out that in this instance aerial surveying procedures can be invaluable in indicating features which may be invisible to the completely ground-based investigator.

The reconnaissance-preliminary survey leading to the location and preliminary design of a major highway in an urban area can be divided into the following six inter-related steps: 1. Determine the approximate traffic load along a general route suggested by traffic desire lines. 2. Select the type of highway, the number of lanes needed to accommodate the approximate traffic load and the type of service to be provided. 3. Make plane and field sketches to establish one or more preliminary lines that approximate the desire-line location, and make sketch preliminary designs including interchange locations. 4. Assign traffic to one or more of the selected locations to determine design traffic volumes. 5. Adjust line and complete sketch preliminary plans for major alternate locations. 6. Analyse and compare alternative locations for selection of the preferred one by making cost estimates, analyzing road-user benefits and considering other controls and factors.

Most of these factors have already been discussed elsewhere in this text, particularly in the chapter on Highway Planning (see Vol. 1), so the following discussion is confined to the *controls* in step 6 which may be taken into account.

Urban location controls

There are really very limited possibilities for locating new major highways, or improving existing streets to carry out the same functions, in an urban area. The town is an established entity and it is usually only on the very outskirts that undeveloped land can be obtained. The streets necessary for the free flow of traffic are fixed in location and size by the natural topography and by the very buildings which they service. Thus the location of a new highway or the substantial improvement of an old one must inevitably result in the elimination of, or changes in, portions of the established city culture and this, just as inevitably, complicates the problem of finding a suitable location.

While the anticipated traffic is a major factor controlling the location of a highway, it must always be qualified by an evaluation of its effect upon, and relationship with, land use and other similar town-planning considerations. Town planning is concerned with the present and future needs of the business, industrial, residential and recreational elements of a town so that a pleasant and functional whole is assured now and in the future. Thus the highway engineer is actively participating in town planning when locating (and designing) a new facility and must be prepared to bow to policy and social needs where necessary and possible.

The parking problem is acute in most cities, as is the traffic congestion in the streets, and these can be alleviated or accentuated by the location of a new major route. Thus it is generally desirable to locate the new highway as close as possible to existing or potential parking areas. This is particularly important in or near central areas where, of necessity, vehicle travel and consequent traffic congestion has to be minimized. If the most suitable route location does not meet this criterion, then congestion can often be reduced by the judicious location of the on- and off-ramps which connect the major highway to appropriate town streets.

The existing transport systems are most important controls affecting the location and type of urban arterial highways. Pedestrians, cars, lorries, buses and trams use the streets and obviously any new facility must be integrated with the existing road system if optimum usage is to be obtained. Equally important, however, are the other transport media, e.g. railways and their terminal stations, waterways and docks, and airports; all of these facilities need servicing by major roadways if they are to fulfil their functions. With the present tendency towards cutting railway services, it may be that a railway line may offer a location possibility of value for highway purposes. Usually also the area along a railway is either rundown or sparsely developed so that it can be obtained at prices well below alternative locations.

As in rural areas, the topography and physical features of a town can be major controls influencing the location of a highway. Soil and ground-water conditions may also affect location. For instance, subsurface conditions, such as a high water table or rock close to the surface, will normally preclude a depressed highway, even though it may be desirable from other points of view. Poor subsoil conditions may also preclude the construction of elevated highways, or render more difficult and expensive the building of an at-grade roadway.

In an urban area, existing public utilities, such as storm and sewer pipes, water, gas and electric lines and their appurtenances, can present severe difficulties which may affect the location of a major highway. Whether they be for at-grade or depressed highways, some adjustments in utilities are always necessary; in the case of a depressed highway the utility relocation costs may be so great as to make another site more attractive.

PHOTOGRAMMETRY AND AIRPHOTO INTERPRETATION

Photogrammetry is the technique of obtaining reliable measurements by means of photography.[6] This definition does not include the 'interpretation of photographs' which is a function of equal importance; as will be discussed later, the ability to recognize and identify a photographic image is as valuable to the highway engineer as the ability to measure it accurately.

Photogrammetry is usually divided into categories according to the type of photographs used or the manner of their use. Thus the type of photogrammetry used when the photographs are taken from points on the ground surface is called ground or terrestrial photogrammetry; if the optical axis of the camera is horizontal they are called horizontal photographs. The term aerial photogrammetry is used to describe the use of photographs which have been taken from air-borne vehicles: these can be either vertical or oblique photographs.

High oblique photographs are aerial photographs taken with the axis of the camera up to about 20 deg from the horizontal, so that the horizon is included. In appearance these resemble a ground photograph taken from a commanding height; they provide true perspective views of the land surface and as such they may be analyzed in terms of the laws of perspective. In practice, high oblique photographs are used primarily for pictorial and illustrative purposes because of the facility with which they can provide a panoramic view in familiar perspective.

Low oblique aerial photographs are taken with the camera axis inclined at such an angle as not to show the horizon. If the angle as measured from the vertical axis is large, a perspective view is obtained similar to that which is obtained with a high oblique. If the angle is small, the result obtained very closely resembles a vertical photograph.

As is implied by the name, a vertical photograph is one taken with the optical axis of the camera coinciding with the direction of gravity. In fact, however, true vertical photographs are rarely obtained. The major factor which invariably causes non-verticality is the inability of the pilot to keep the plane exactly level, with the result that the camera axis deviates from the vertical. Nevertheless 'vertical' aerial photographs are the best known and most used in aerial surveying for highway purposes, and so the following discussion is concerned with them and the manner in which they are used in practice.

Vertical aerial photography

Aerial photography can be defined as the science of taking a photograph from a point in the air for the purpose of making some type of study of the surface of the earth. Before describing how and where these studies are carried out it is necessary to obtain an understanding of various features of air photography and air photographs. One way of doing this is to consider part of the terminology which is commonly used.

Photographic scale. A most important characteristic of a photograph is its scale. Unlike a map drawing, on which any measured distance bears a *fixed* and uniform relationship with the corresponding distance on the ground, the scale on a vertical photograph will rarely remain constant but will vary from photograph to photograph and from point to point on a photograph. A photograph is actually a perspective projection, not an orthographic one, and thus areas on the ground lying closer to the centre of the photograph will appear larger than corresponding areas lying farther away from the camera. Secondly, the scale of a vertical photograph will vary from location to location depending on the elevations of the ground area. This is illustrated in Fig. 1.2, which represents relationships established when a photograph is taken at a fixed height H above some datum plane, say sea level.

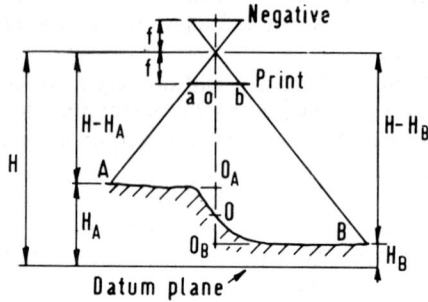

Fig. 1.2.   Determining the scale of a photograph

The focal length of the camera is f and the principal or nadir point of the photograph is at o. The elevation of the ground point A is H_A and it appears as an image on the photograph at a; point B has an elevation of H_B and appears on the photograph at b. By similar triangles the scales at a and b can be determined; thus

$$\frac{ao}{AO_A} = \frac{f}{H-H_A} \text{ and } \frac{ob}{O_B B} = \frac{f}{H-H_B}$$

Therefore $S_h = \dfrac{f}{H-h}$

where S_h = scale at any elevation h,
            f = focal length of camera,
and H = flying height above the datum plane.

   In planning photographic studies an average scale is used for rough calculation purposes. This average scale is given by the equation

$$S_{ave} = \frac{f}{H-h_{ave}}$$

where h_{ave} is the average elevation of the terrain being photographed.

Ground control. If accurate horizontal and vertical measurements are desired from the air photographs it is necessary to establish the positions and heights of a number of air-visible ground control points. The horizontal control points are used to determine the scale and true orientation of each photograph, to provide a basis for correcting cumulative errors which arise when strips of photographs are assembled, and to link the photographs to the standard geographic (National Grid) co-ordinates. The vertical control points are needed before contour lines or point-elevations can be obtained from the photographs.

Ground-control establishment is divided into two parts. The first part, called the *basic control*, establishes a basic reference network of pillars and beacons, which include triangulation and traverse stations and bench marks, over the survey area as a whole. The second phase, the *photo control*, establishes the horizontal locations and the elevations of particular points which can be identified on the photographs and related to the basic control points.

The geodetic survey for the purpose of establishing ground control can be a most expensive and time-consuming process; in many instances its cost may account for as much as 50 per cent of the total of an aerial survey. Hence great care must be taken in choosing the number of points required and their exact location. The most important consideration is that the points must be easily and clearly identifiable on both the ground and the photograph. As a result, the control points are usually chosen after the aerial survey has been carried out. Another main factor is their accessibility. Choosing an ideal point on a photograph is of little value if it cannot be reached and referenced in the field. A vertical photo control point should always be selected so that it is part of an area of no elevation change. Typical suitable points are the intersection of two existing road centre-lines, or the base of a telephone pole in a flat field.

The number of photo control points which have to be established depends entirely on the purpose for which the aerial survey is being carried out. For instance, if a controlled mosaic of an area is to be constructed, no vertical control points are required while only a limited number of horizontal photo control points may be necessary. On the other hand, if a topographic map is to be constructed from the air photographs, four vertical and two horizontal control points may be required on each overlap area of the initial and final pair of overlapping photographs. In these cases the vertical control points are placed at the corners of the overlap in order to establish the datum above which the elevation can be measured. The two horizontal control points are located towards the centre of overlap so as to fix the scale of the map. In between, ground control should be maintained at the rate of one horizontal point near the centre of every fifth photograph and two vertical points, one near each edge, of every third or fourth photograph. Additional control points may then be 'bridged-in' to intervening pictures with a first order stereo-plotting machine. If no vertical bridging is done, then two points will be needed for each photograph.

Mosaic. A mosaic is an assembly of individual aerial photographs fitted together systematically so that they provide a composite view of the entire area covered by the photographs. Thus the mosaic gives the appearance of a single photograph of the earth's surface that is both comprehensive and complete, and which can generally be used and appreciated without need for special photogrammetric training. It is particularly valuable both for obtaining a complete picture of the terrain—this cannot be duplicated by a map—and as a public relations instrument to explain highway schemes to the interested public.

In order to provide distinction between the many ways of compiling mosaics, they are generally classified as being either controlled or uncontrolled. An *uncontrolled mosaic* is a compilation of photographs so oriented as to match as closely as possible corresponding images on adjacent photographs, but with no attempt being made to tie-in particular horizontal control points. They are useful when a rapid composite picture of an area is required, and where relative rather than absolute positions of terrain features are required. A *controlled mosaic*, on the other hand, is composed of an assembled group of photographs, each of which has been corrected for major tilt distortions beforehand so that ground control points are more accurately positioned.

A mosaic that is of particular interest to the highway engineer is the strip mosaic. This is prepared from a single strip of photography taken along a single course or flight strip. It can be either controlled or uncontrolled, although for technical work the controlled strip mosaic is obviously of most value.

Planimetric and topographic maps. A planimetric map is one which illustrates the horizontal locations of particular features, whereas a topographic map also shows topographical relief, very often in the form of contour lines. A contour is defined as a line of equal elevation on the terrain. The chief characteristics of contours are as follows:

1. All points on any one contour have the same elevation.

2. Every contour closes on itself, either within or beyond the limits of the map. If a contour does not close within the limits of the map, it will run to the edge of the map.

3. A contour that closes within the limits of the map indicates either a summit or depression. In depressions there will usually be found a pond or a lake. Where there is no water the contours are usually marked in some special way to indicate a depression.

4. Contours can never cross each other except where there is an overhanging cliff, and here there must be two intersections, i.e. where the lower ground contour line enters and emerges from below the overhanging cliff-line.

5. On a uniform slope, contours are equally spaced. The steeper the slope, the smaller is the spacing; the flatter the slope, the larger is the spacing.

6. On a plane surface they are straight and parallel to each other.

7. In crossing a valley, the contours run up the valley on one side and, turning at the stream, run back on the other side. Since the contours are always at right-angles to the lines of steepest slope, they are always at right-angles to the thread of the stream at the point of crossing.

8. Contours also cross ridge lines (watersheds) at right-angles.

9. In general, the curve of the contour in a valley is convex toward the stream.

10. All contour lines are noted as multiples of their intervals.

Digital Ground Models.   It should be pointed out here that as a result of the very significant developments which have taken place in electronic computing over the past decade, there is an ever-increasing requirement today for photogrammetric output data to be provided in digital form in which the shape of the terrain is represented as a mathematical model instead of just being graphically represented by contours. Thus the tendency in recent years is toward producing survey information in the form of Digital Ground Models (DGM), as well as in graphical form. The DGM is simply a close net of spot heights determined at set intervals throughout the area under consideration; the basic data are stored on magnetic tape or punched cards so that, when fed into the computer, a whole range of output information can be provided (e.g. grade lines; cross sections which fit different parameters of straights, transitions curves, cuttings and embankments; mass haul diagrams; earthwork quantities, i.e. areas and volume of cut and fill; stake lines; seeding areas; and even perspective drawings of the constructed highway[7]), and not just that required for location purposes. The graphical production shows the physical detail features with perhaps a coarser contour interval used only to assist the engineer as a visual aid.

Indeed, it is not unlikely that future years will see this process extended—provided current real costs of equipment and program development can be reduced—whereby the need for any graphical information will be eliminated, and it will be common practice for the highway engineer to use the computer[8] in combination with automated draughting machines to produce limited graphical information only after all the alignment (and design) work is completed in digital form.

Flight line, flight height and overlap. A flight line is the path taken by the plane as the photographs are being taken. Along this flight line, photographs are taken at continuous intervals, so that successive ones overlap each other by about 60 per cent and thus alternate photographs will overlap each other by about 20 per cent. This forward overlap of consecutive photographs is necessary for stereoscopic examination purposes; the overlap between successive alternate photographs is required because of the need to have an area common to two consecutive photographs in order to extend horizontal and vertical control by photogrammetric means.

For mapping purposes, flight lines are generally parallel to each other, and adjacent flight strips normally overlap each other by about 25 per cent. This ensures complete coverage of the ground, and ties-in adjacent strips.

The flying height for survey photography is dependent on the following factors: 1. Desired scale of mapping or profiling for which photography will be used. 2. Vertical interval of contouring or height accuracy required. 3. Intended interpretation usage, e.g. geology, vegetation, traffic, soils, land use, etc. 4. Operational ceiling of aircraft above ground level (this may be

TABLE 1.1. *Criteria governing the usage of air photography*[9]

Photo scale recommended	Map scale	Approx width of coverage (single strip), km	*Contour interval, m	*Spot height accuracy, m	Geological interpretation for engineering purposes	Tree identification	Land-use	Soil boundaries	†Approx aircraft ceiling, m	Photographic cost factor per km² (excluding positioning)
1/3 000	1/1 000	0·5	0·5	0·15	Detailed site survey	Detailed species	Very detailed	Very detailed	450	25
1/6 000	1/2 000	1·25	1	0·25	Fairly detailed site survey	Detailed species	Very detailed	Very detailed	900	8
1/12 000	1/4 000	2·50	2	0·5	Site location	Some species	Detailed	Detailed	1 800	3
1/15 000	1/5 000	3·00	2·5	0·6	Site location	Selected species	Detailed	Detailed	2 300	2·5
1/25 000	1/10 000	5·5	5	1	Location of construction materials	General forest types	Regional	Detailed	3 800	1
1/50 000	1/20 000	11	10	2	Regional only	Regional	Regional	Semi-detailed	7 600	0·5

* Assumes the use of high-order stereo-plotting equipment.
† Assumes the camera employs a modern colour corrected lens on a 230 mm square format with focal length approximately 152 mm.

governed by prevailing air traffic regulations in the area). 5. Type of photo-grammetric plotting equipment available, and techniques for mapping. 6. Desired film material, i.e. colour, infra-red (or false) colour, or panchroma-tic. 7. Cost. Table 1.1 summarizes the inter-relationships between these criteria.

Stereoscopic vision. Stereoscopic perception is the facility which makes it possible for a person to see the image of a scene in three dimensions. Normally two-eyed vision is needed to realize and measure the depth dimen-sion. An understanding of how this occurs can be gained by considering Fig. 1.3. Cover up the two black dots b and b' in Fig. 1.3(a) and look hard at the dots a and a', i.e. focus the eyes as if beyond the page. It will be found that four images will initially be seen but that the two middle ones can be made to superimpose to form another image in between the two real dots a and a'. A little practice is required before the fusion is clear; holding a post card between them and at right-angles to the page may help. Next uncover b and b' and look at all four dots in the same way. It will be found that the additional four 'middle' dots which appear can be made to fuse into two. To be sure that all four dots are represented in the superimposed image, check that the letters a and a' are adjacent to the upper dot while b and b' should be beside the lower one. When the fusions have taken place, it will be seen quite dramatically that the upper dot seems to be floating in space relative to the lower one. Figure 1.3(b) illustrates why this is so. In this figure, a_H and a'_H represent the upper pair and b_H and b'_H the lower pair of dots as seen when viewing the plane of the paper as an edge. The rays from the eyes to a and a' intersect at A and those to b and b' intersect at B. These intersections, A and B, are the positions in space at which the stereoscopic images appear to be, with the B intersection located behind A.

If two air photographs are taken of the same land area, but from two different exposure stations, they can be oriented and viewed simultaneously so that, as illustrated in Fig. 1.3, the observer has the impression of viewing a three-dimensional model of the topography. In other words, the two posi-tions of the camera, perhaps a kilometre apart, substitute for the observer's eyes.

In order to facilitate three-dimensional examination of pairs of photo-graphs, a viewing instrument known as the *Stereoscope* is used. There are two main types of stereoscope; the direct lens or refraction type, and the reflection mirror type. The direct type is the simpler and more compact, consisting only of two matched magnifying lenses mounted on supports and separated by a distance corresponding to that between the eyes. These ster-eoscopes are quite suitable when relatively small-sized photographs are being studied. When larger photographs are being used, they have to be spread farther apart than the normal eye distance and this necessitates using the mirror type of stereoscope. This consists simply of four mirrors set at 45 degrees to the plane of the photographs and, as illustrated in Fig. 6.4(b), this enables a broader field of vision to be obtained.

To the highway engineer, the most important feature of stereoscopic

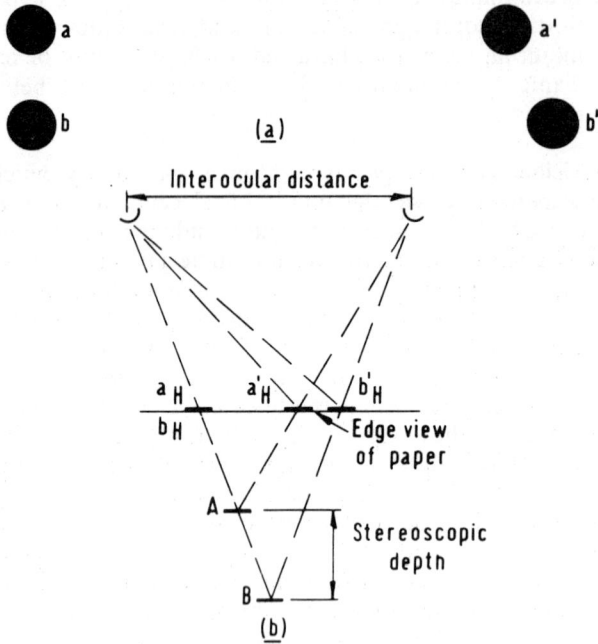

Fig. 1.3.   Illustration of stereoscopic fusion

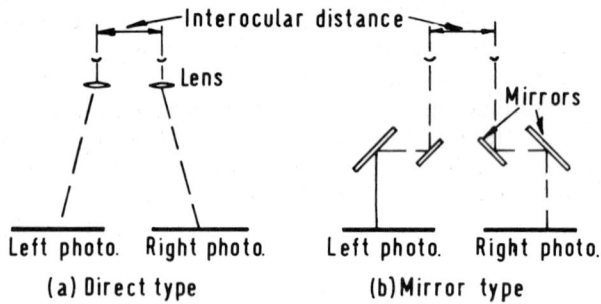

Fig. 1.4.   Basic types of stereoscope

vision is that it enables contour lines to be drawn directly from the vertical air photographs. A crude way of doing this is to sketch the contour lines under the stereoscope with the aid of numerous known spot elevations on each photograph. Much better results are obtained with the aid of mechanical contouring devices. Without going into detailed discussions of the functioning of these instruments, it is possible to describe briefly the basic manner in which the contours are drawn.

A pair of consecutive or parallel air photographs are viewed in a binocular system, with each photograph being held in a mechanical reconstruction of the original air camera system. When the two 'cameras' are correctly oriented, the observer is provided with a three-dimensional view of the terrain in which the depth dimension appears exaggerated. The reconstruction of the ray paths from objects on the ground through the air camera lens is simulated in the plotting instrument by mechanical rods, each of which pivots about a point corresponding to the perspective centre of the original taking camera. The monocular system for each eye contains a reference dot which is engraved in the optical train. Viewed stereoscopically, the two dots fuse together to provide a floating reference mark which can be moved horizontally or vertically. This is, of course, only a subjective impression since, in fact, the marks remain stationary and it is the upper ends of the mechanical space rods which move. Thus, when the space rods intersect at a corresponding image point in the model, the fused dot appears to the observer to be in contact with the ground at this point. By this means it is possible to trace a contour line, or any other such topographical detail, by keeping the reference dot always in contact with the ground surface as it appears in the model. The contour trace can be then passed to an instrument such as the 'D-Mac' digitiser in which the contour lines are 'traced' by a reference marker and the x, y and z coordinates of these lines are automatically recorded on tape or punched card ready for entry into the computer.

Air photo interpretation. Interpretation from aerial photographs differs from direct observation in scope, perspective and time relations. The large area imaged permits the stereoscopic observer to perceive and make deductions regarding relations between objects and their surroundings which to a ground observer might not be evident. The stereoscopic effect permits the interpreter to perceive the shape of objects; in this respect the exaggeration of vertical distance, which is present in stereoscopic pairs, is particularly helpful in emphasizing small but important differences in elevation and separating objects from their background. The permanence and fidelity of the photographic image permit the interpreter to conduct a close and careful study of an area under more favourable conditions than could be obtained by direct observation; in addition the photographs lend themselves well to comparative and before-and-after studies.

Airphoto interpretation resembles direct observation, however, in one most important respect. The amount and reliability of the information obtained depends entirely upon the training and aptitude of the observer and on the nature of the scene observed.[10] It is really only through practice in interpretation that an observer having a solid base of technical knowledge can develop a proper understanding of the meanings of the characteristics and patterns of photographic images.

The size of an object is one of the most useful clues to its identity. Simply by measuring an unknown object on an aerial photograph, an interpreter can eliminate from consideration whole groups of possible identifications. To an interpreter experienced with industrial facilities, the vertical view of a factory can tell considerable detail about its function. Shadows are particularly helpful in the identification of man-made features. For instance, tall

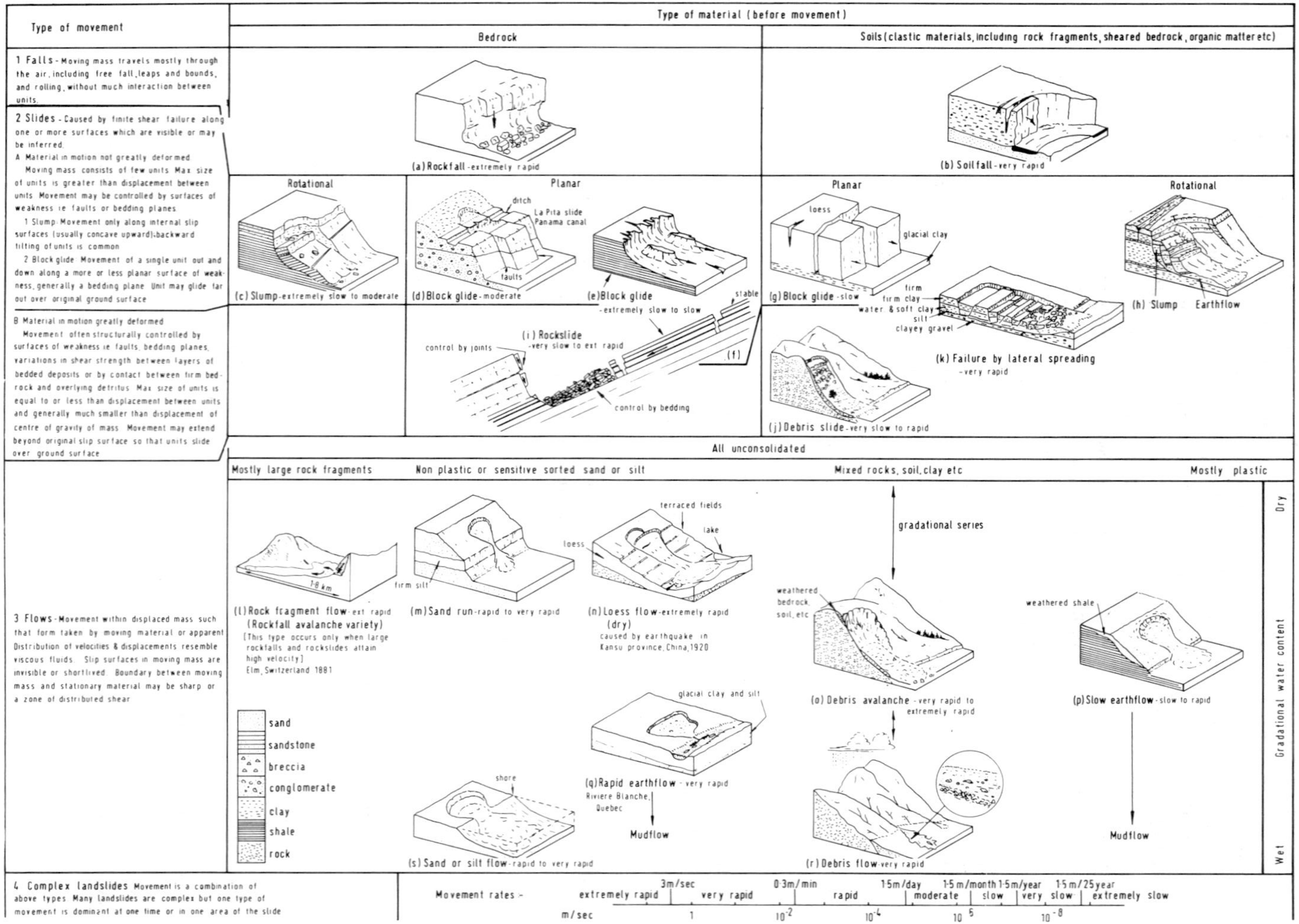

Type of movement	Type of material (before movement)		
	Bedrock	Soils (clastic materials, including rock fragments, sheared bedrock, organic matter etc)	

1 Falls - Moving mass travels mostly through the air, including free fall, leaps and bounds, and rolling, without much interaction between units.

2 Slides - Caused by finite shear failure along one or more surfaces which are visible or may be inferred.

A Moving mass not greatly deformed. Moving mass consists of few units. Mass size of units is greater than displacement between units. Movement may be controlled by weakness ie faults or bedding planes.

1 Slump. Movement usually along internal slip surfaces (usually concave upward) backward tilting of units is common.

2 Block glide. Movement of a single unit out and down along a more or less planar surface of weakness, generally at a bedding plane. Unit may glide far out over original ground surface.

B Material in motion greatly deformed. Movement often structurally controlled by surfaces of weakness ie faults, bedding planes, variations in shear strength between layers of bedded deposits or by contact between firm bedrock and overlying detritus. Mass size of units is equal to or less than displacement between units and generally much smaller than displacement of centre of gravity of mass. Movement may extend beyond original slip surface so that units glide over original ground surface.

3 Flows - Movement within displaced mass such that form taken by moving material or apparent distribution of velocities & displacements resemble viscous fluids. Slip surfaces in moving mass are invisible or shortlived. Boundary between moving mass and stationary material may be sharp or a zone of distributed shear.

4 Complex landslides Movement is a combination of above types. Many landslides are complex but one type of movement is dominant at one time or in one area of the slide.

Rotational
(a) Rockfall - extremely rapid
(c) Slump - extremely slow to moderate

Planar
La Pita slide, Panama canal
(e) Block glide - extremely slow to slow
(i) Rockslide - very slow to extremely rapid — control by bedding

Rotational
(b) Soilfall - very rapid
(h) Slump Earthflow

Planar
loess, glacial clay, firm clay, silt, clayey gravel, water & soft clay
(d) Block glide - moderate
(g) Block glide - slow
(j) Debris slide - very rapid slow to rapid

(k) Failure by lateral spreading - very rapid

All unconsolidated

Non plastic or sensitive sorted sand or silt

Mostly large rock fragments

(l) Rock fragment flow - very rapid (Rockfall avalanche variety) (This type occurs only where large rockfalls and rockslides attain high velocity) Elm, Switzerland 1881

firm silt, loess
terraced fields, lake
(m) Sand run - rapid to very rapid
(n) Loess flow - extremely rapid (dry) caused by earthquake in Kansu province, China, 1920

shore
(o) Sand or silt flow - rapid to very rapid

Mixed rocks, soil, clay etc — gradational series

weathered bedrock, soil etc
(o) Debris avalanche - very rapid to extremely rapid

glacial clay and silt
Mudflow
(q) Rapid earthflow - very rapid Rivere Blanche Quebec

Mostly plastic

weathered shale
(p) Slow earthflow - slow to rapid → Mudflow

(r) Debris flow - very rapid

Gradational water content — Dry / Wet

sand, sandstone, breccia, conglomerate, clay, shale, rock

Movement rates: - extremely rapid / very rapid / rapid / moderate / slow / very slow / extremely slow

m/sec — m/sec — 3m/sec — 0.3m/min — 15m/day — 15m/month 15m/year — 15m/25 year

10^2 — 1 — 10^{-2} — 10^4 — 10^6 — 10^{-8}

Fig 1.5(a)-(s). Characteristics of landslides (11)

slender objects such as church spires, water towers, tanks and smoke-stacks would frequently be almost indistinguishable but for their shadows.

Joints, faults, shear zones and brecciated zones all have features which readily reveal their presence to the skilled photogeologist. Thus, for instance, very recent faults can generally be recognized where graded surfaces, such as alluvial fans or pediments, have been displaced; rocks on both sides of a fault will fail to match in type or attitude, or both. Old landsides and areas susceptible to landslides are readily detected from aerial photographs by characteristic steep scarps, hummocky surfaces, ponded depressions and disturbed drainage conditions. Types of landslide, their characteristic shapes and some factors affecting their movements are illustrated in the diagrams in Fig. 1.5.

Soils can be interpreted in aerial photographs by studying the patterns created by the nature of the parent rock, the mode of deposition, and the climatic, biotic and physiographic environment. Drainage patterns, such as

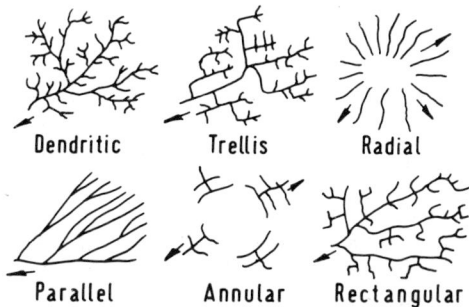

Fig. 1.6.   Some basic drainage patterns

those indicated in Fig. 1.6, suggest the origin and composition of the underlying geological formation. A dendritic drainage pattern is commonly found in glacial topography. Trellis drainage is to be found in areas of folded sedimentary rocks; the characteristic pattern is due to the ridges being cut by stream gaps. The radial pattern can reflect water draining into an enclosed basin but, more often, it indicates a high dome or hill, or a volcanic cone. A parallel drainage pattern usually indicates a regional sloping terrain condition or a system of faults or rock joints. The classical annular drainage system is usually indicative of tilted sedimentary rock about an igneous intrusion. Whenever a drainage pattern appears rectangular, it is nearly certain that rock is at or close to the ground surface, and that the drainage is controlled by rock joints, fissures or faults.

Colour perception can be also most important in airphoto interpretation. In black-and-white photography, distinctions between hues are lost and objects are observed in tones of grey; to the skilled soil engineer interpreter, these are major clues which enable him to classify soils or to identify particular geological features. On the positive (reverse) image of infra-red colour, water shows up as a strong cyan blue which, against a magenta background of living vegetation, will indicate pools of stagnant water or sections of a river visible between a forest canopy. Colour can also be used to identify particular types of tree-form which in turn act as indicators of swamp areas, narrow watercourses, etc.

Erosional features, particularly the shapes of gullies, provide clues to soil texture and structure, profile development, and other conditions. For instance, U-shaped gullies are found in wind-deposited loess soils or similar type, but stratified, sandy or silty soils; the characteristic and near vertical sides of the gullies reflect the fact that there is relatively little clay present and that these soils crumble fairly easily. A broad saucer-shaped gully is indicative of a non-plastic and cohesive clay or silty clay soil, while V-shaped gullies are found in semi-granular soils with some cohesiveness.

Vegetation can also be an important indicator of soil conditions, with areas having the same natural plant cover tending to contain the same soils. Similarly, land use is often directly related to soil type, profile development, porosity and other conditions of interest to the highway engineer.

Uses of aerial surveys

Aerial photographs have become most valuable tools in the hands of the highway engineer. They are very useful visual aids in technical conferences and at public hearings, as well as being very valuable in seeking the most suitable topographic location for the highway. Additional qualifying factors which can be determined by a skilled interpreter are as follows[3]: 1. Location of sand and gravel deposits, borrow pits, and rock areas suitable for quarrying. 2. Rate soil-bearing capacities as good, fair and poor. 3. Outline soils as to their texture, and estimate the depths of organic deposits. 4. Classify bedrock as to physical types and their relative depths below the surface. 5. Estimate depths of glacial drift and wind-blown material. 6. Estimate hydrological factors. 7. Delineate active and potential areas involving problems such as sink holes, slides, rock falls and frost heave. 8. Locate underground utilities placed by cut-and-cover methods.

New aerial photographs provide the most up-to-date information regarding current land use, and this is of particular importance at the reconnaissance phase of a highway location study, particularly if it is being carried out in an urban area. Similarly these photographs can be easily used to locate important intermediate control points, so that possible band-routes can be selected. At this stage of the reconnaissance a small-scale mosaic will provide sufficient detail to enable elimination of all but a couple of possible bands.

To enable the choice to be narrowed, an enlarged mosaic can be viewed in conjunction with a topographical map. A useful technique is to assemble one or more mosaic bands and to overlay them with topographic maps that are enlarged to the same scale and drawn on transparent material. Trial lines for each route can then be sketched on the mosaics and profiles plotted directly from the contours of the overlay. When route profiles are found which indicate alignments that might satisfy the highway's design standards, they can be compared on a benefit-cost basis to aid in the final selection.

Large scale photography is necessary at both the *preliminary* and *final location* phases of a location survey. The manner in which the data can be automatically processed has been already described, and so will not be further discussed here.

Air-photo interpretative methods can be used to cut down the amount of soil survey field work which is normally carried out during the final location survey. As part of the routine field studies, it is customary to carry out auger investigations at fixed intervals along the location line, and then to carry out other detailed studies as indicated by the results obtained. Much of this routine augering and subsequent laboratory analysis may be eliminated if the air photos are used to delineate ground areas with similar physical characteristics. Then it is only necessary to go to the field and test a limited number of points within each delineated area. Similarly, sources of suitable aggregates are more easily determined from the photographs and the amount of field investigation and laboratory testing minimized.

When large-scale aerial photography is used to prepare the necessary maps, it may be possible to complete the plans and specifications for a highway and to let the construction contract without actually placing a centre-line peg in the ground. When this occurs, the final location line need not be pegged until the contractor is about to begin construction, in which case the final location survey becomes a combined location and construction survey.

The most important *advantages* of aerial photogrammetric procedures in highway engineering can be summarized as follows:

1. On a large highway scheme the time required to locate the facility and prepare plans for contract is considerably accelerated.
2. Cost savings of from 30 to 40 per cent may be gained by comparison with conventional ground survey methods.[12]
3. Skilled technical personnel are released from dull routine work and can be used more profitably on more demanding problems.
4. Topographic maps prepared from air photographs can be more reliable than those produced from ground surveys.
5. The ability to survey large areas of land ensures that the most suitable location is not accidentally overlooked.
6. Highway profiles and cross-sections can be measured without physically encroaching on private lands. Thus there is the minimum of obstructive field work, land-owners are not upset, and land values are not affected.
7. A most complete inventory of all surface features is obtained. This is always available for use and may be referred to at any time. Furthermore, it can be analysed at any time, and this is not affected by poor weather or lighting conditions.
8. Public relations are considerably facilitated and enquiries aided by the use of aerial photographs intelligible to the layman.

In general it can be said that the use of aerial photographs greatly facilitates the location and design of highways. Nevertheless, there are certain *disadvantages* associated with their use.

1. First and most important, the taking of the photographs cannot be guaranteed at any particular time. Good photography requires clear atmospheric conditions and photographs taken on wet or misty days, or in haze or smoky conditions, may be of little value.
2. A topographic map cannot be accurately obtained if the ground is covered by snow or obscured by the leaves of trees. Because of the latter, air photographs of, for instance, heavily wooded tropical areas can be of considerably less value than might normally be expected. In temperate areas, where the ground has a deciduous cover, the taking of aerial photographs must await a suitable time of the year.
3. If the area to be studied is relatively small, it will usually be more economical to use ground survey methods rather than aerial procedures.
4. Skilled photo interpreters are still relatively uncommon, and this may result in uneconomical or improper usage.

SUBSURFACE EXPLORATIONS

Subsurface explorations are made for the purpose of obtaining information regarding the types, location and extent in plan and profile of soil and rock which will be encountered in connection with the proposed highway location. They are an essential part of a preliminary engineering survey for location and design purposes inasmuch as they provide pertinent information on the following subjects:[13]

1. The location of the road, both vertically and horizontally.
2. The location and selection of borrow materials for fills and subgrade treatment.
3. The design and location of ditches, culverts and drains.
4. The need for subgrade treatment and the type of treatment required.
5. The design of the roadway cross-section.
6. The location of local sources of construction materials for the pavement.
7. The selection of the type of surface and its design.

A subsurface exploration cannot and must not be considered as a separate detached function in the location and design of a highway, but rather as an integral part of the development of the scheme. Thus, from the moment that consideration is given to the approximate location of the proposed new highway improvement until the highway is actually constructed, information is continually required regarding subsurface conditions so that the proper analyses and decisions can be made.

Although subsurface investigations are, naturally enough, concerned primarily with soils and soil conditions, they should not be considered as soil surveys only. A comprehensive subsurface investigation for highway purposes may well consist of not only a study of the critical soil layers and the extent of adverse ground conditions such as swamps or peat bogs, but also of the depths to bedrock and the water table. In areas of proposed deep rock cuts, geological cross-sections may also be required. Where an old roadway is already in existence data may also be obtained which will relate the pavement condition with the characteristics of the underlying materials, provide possible reasons for failure, and suggest corrective measures for the project under examination.

Sequence of operations

The sequence of operations in carrying out a subsurface investigation can be divided into four main parts:[14] 1. Preliminary work. 2. Site examination. 3. Determination of the soil profile and collection of samples. 4. Location of water table.

The first two of these operations usually occur during the reconnaissance for location purposes, while the latter pair are carried out mainly during the preliminary survey and as necessary during the final location survey.

Preliminary work. Prior to starting field work, the literature should be reviewed to obtain general information regarding the conditions which may be met.[2] The Geological Survey Maps and the appropriate *Memoirs* can be most helpful in this respect. These maps are usually published in two editions, Drift and Solid; the former shows the superficial deposits of glacial origin and the solid rocks exposed at the surface, while the latter shows the solid rocks as they would appear if the cover of superficial deposits were removed. It is also desirable to consult the authorities at the Geological Survey and Museum, as they often have useful unpublished information. For instance, by the Mining Industry Act of 1926, it is a statutory obligation that the Geological Survey be notified of all sinkings and borings to a depth of more than 30·5 m for minerals and, by the Water Acts of 1945 and 1946, to more than 15·25 m for water; thus a very large amount of unpublished but valuable underground information is collected and available for preliminary study.

Of particular value at this stage are the pedological maps prepared by the Soil Survey of Great Britain. These maps, which are prepared primarily for agricultural purposes, provide most useful information regarding soil conditions to a depth of about 1·25 m below ground level.

Subsurface explorations made for other engineering projects in the vicinity may also provide helpful information. In the case of existing roadways, highway maintenance records should be scanned for information concerning areas which have given trouble because of frost heave, seepage water, pumping of pavement joints, excessive or uneven settlement of the roadbed, landslides, or any other difficulties associated with subsurface behaviour. Knowing the positions of such areas in advance will enable the highway engineer to make better decisions affecting the preliminary line, while at the same time being guided towards obtaining information of the kind and extent which will be needed at a later time by the highway designer.

Site examination. In areas where preliminary information is limited, essential terrain information can be obtained by on-the-site examination. The soils observed along roads in the vicinity of the proposed highway line should be studied and changes in the soil (or rock) profile noted as they occur in highways cuts. These notes should include a complete description of the soil (or rock) profile for each type observed. A correlation of this information with the type of parent material, the range in slope, the topographic position, the drainage conditions, and the land form or airphoto soil pattern[15] can be used to help identify different soil types in the area. It will be found that similar soil conditions are encountered where similar soil profile characteristics exist; that changes in soil conditions accompany similar changes in relief and parent material; and that similar subsurface conditions have similar airphoto soil patterns for regions where no appreciable changes occur in climate or vegetation.

Some indications of the depth of the water table below the surface can often be obtained from an inspection of streams or ponds, while lush growths of vegetation, such as rushes, cotton-grass or willows, indicate that the water table is near the surface for at least part of the year. Bracken and gorse growths, on the other hand, indicate that the soil is well drained.

Geological faults may be indicated by such topographical features as a step in the line of an escarpment or a depression where the shattered rock has been more rapidly eroded than the surrounding rock. These places should receive detailed inspection, as igneous intrusions may involve excavation while, in certain circumstances, a fault may also give rise to constructional difficulties. Potential rock-fall and soil-fall slides can often be foreseen simply by recognizing geological conditions that are likely to produce overhanging or oversteep cliffs. Some of the geological settings that fall in this category are illustrated in Fig. 1.5 and reference 24.

Careful note should also be made, especially on sloping clay formations, of any irregularities in the ground that may be due to landslides or lack of stability.

At this stage it may be desirable to prepare a preliminary strip map covering the route or routes in which it is considered that the new highway may be located. In complex terrain, especially if adverse ground conditions exist, this map will be most useful in establishing preliminary lines for location purposes. Usually it will be found that it is most convenient to make these maps on an airphoto base prepared from airphoto enlargements. Correlation of airphoto soil patterns with soil types and parent materials from geological and agricultural maps, supplemented with limited ground reconnaissance, should enable the engineer to prepare a subsurface map showing the distribution of the major conditions likely to be encountered during subsequent more detailed subsurface investigations. This strip map, if prepared at the proper scale, can later be converted into a detailed engineering subsurface map by accurately locating all auger borings and soil boundaries determined during the more detailed survey made for the final road location in the area.

Determination of soil profile. The field work for this phase of the investigation consists of making examinations of the subsurface by means of auger borings, test pits, road cuttings or, if necessary, by geophysical surveying methods. There is no hard and fast rule to follow as to where these tests should take place, except that the profile should be examined at close enough intervals to determine the boundaries of each significant soil type occurring on the scheme. This interval, which is normally uniform, may be varied:

1. If the soil profile is uniform, the interval may be increased.
2. When the character of the soil profile changes, intermediate borings should be made until it is clear that all variations have been mapped.
3. Where the topography is rolling, and the grade changes rapidly from cut to fill, detailed data are necessary only in cuts.
4. Where the original ground line or old highway grade is to be covered with fill material, no examination is normally necessary except to determine the character of the support.

One practical recommendation[2] is that tests for motorways should be carried out every 300 m in uniform soils, and every 16 m for quickly changing ground such as glacial deposits.

All subsurface investigations should normally be carried to a depth of 1·25 m to 1·5 m below existing ground level when little cut or fill is required. For cuttings, they should be carried out to the same depth below the formation level. The investigation depths may vary, however, according to the following stipulations:

(a) When the highway lies within uniform layers of the soil profile, the study should be extended down to the first layer below the ditch-line which would stop percolation, or through the pervious layer which would carry water.
(b) When embankment material is to be borrowed from ditches or pits alongside the road, the investigations should be extended to the estimated depth of borrow.
(c) If frost action is a problem, studies should extend to the mean depth of frost in those materials showing a high affinity for frost accumulation.
(d) Where high embankments ($>7·5$–9 m) are concerned, they will produce appreciable bearing pressures and so soil borings should be extended to a depth of at least $2\frac{1}{2}$ times the height of the embankment.
(e) With low embankments (<3 m) the ground pressures are not very great and exploration to a depth of 3–4·5 m will usually be sufficient.
(f) When the new location line follows an already existing road, the subsurface may be adequately mapped by examining exposed cuts, but supplemental investigations should be carried out if areas show adverse pavement performance.

If the proposed centre-line crosses a peat bog, soundings should be normally taken at 15 m intervals along the centre-line and 7·5 m to the right and left for a distance of 30 m. If there is an appreciable difference in the depth of swamp material between 15 m intervals, the boundaries of change should be established by additional soundings.

In the case of bedrock, sufficient data should be obtained to outline accurately the bedrock contact occurring in all road cuts. Where necessary, samples of the bedrock should be examined to determine the uniformity and nature of the underlying rock.

Sampling.   At least one sample should be taken from each stratum found at each test location and sent to the laboratory for further investigation. The soil tests usually carried out are those required for identification and classification purposes, i.e. mechanical analysis, liquid limit, plastic limit, compaction and, where necessary, strength tests.

Records.   Complete and systematic records of all subsurface investigations should be kept for every highway scheme. The description of the site should indicate each test location, the vertical location and identification of each sample taken for laboratory analysis, the nature of the topography and the origin of the parent material and landform.

Wherever feasible, the agricultural soil name of the profile should be given for possible future correlation purposes. The number, sequence and thickness of each soil layer should be noted. Each layer should be described according to texture, structure, organic content, relative moisture content, and degree of cementation.

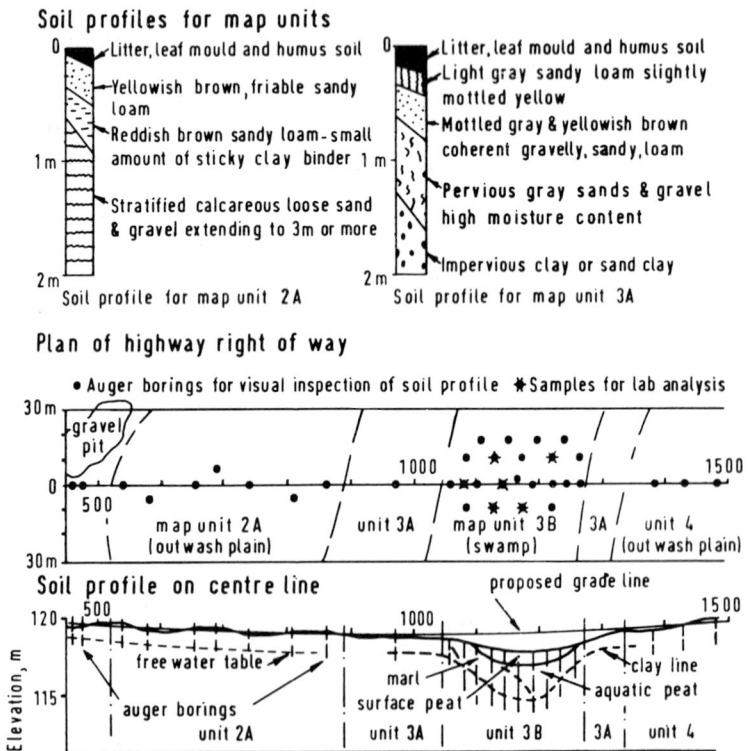

Soil profiles for map units

Soil profile for map unit 2A

Soil profile for map unit 3A

Plan of highway right of way

Soil profile on centre line

Fig. 1.7.   Example of subsurface profile mapping, illustrating the determination of the limits of soils identified by soil profiles (After reference 13)

Note should also be made of the location of seepage zones, the depth to bedrock, the nature and type of bedrock encountered, and the position of the free water table. If the soil is so stony that a reasonably representative sample cannot be obtained, a note should be made of the approximate proportion of stone in the soil and the range of sizes found.

Mapping the profile.   The data obtained from the subsurface investigation should be plotted on a profile sheet to assist in the making of location and design recommendations. One excellent method of so doing—this is indicated in Fig. 1.7—is to establish the range of subsurface profile characteristics for each type (map unit) identified during the survey, and then to indicate the boundaries of the different types on a plan or strip map of the proposed highway location. The location of borings, points where samples were taken and supplemental notes on drainage, landform, and soil or rock types are usually included. Normally a profile sketch is provided in addition to the plan view to show the subsurface conditions that are likely to influence the position of the grade line.

Location of water table. If no trace of soil moisture is found in a borehole at its full depth, it can be assumed that the water table is below this depth, and the investigation continued elsewhere. If, however, water is indicated, the hole should be left for 12 to 24 hours to allow the water to rise to its final level. The most convenient way of then measuring the depth of the water level is to lower a measure until its lower end just touches the water surface.

It occasionally happens that a boring will pierce a layer of impervious clay which is covering a permeable layer in which the water is under considerable pressure. This artesian condition is indicated by water rising *above* the level at which it was first found seeping into the hole. It very often requires special drainage facilities, and thus additional borings should be made nearby to check the magnitude and extent of the water condition. These checks should be carried out in all instances where free water is encountered.

The colour of the subsoil will often indicate whether the water table is fluctuating. While it is desirable to make the water table determination in winter time when the water level can be expected to be at its highest, obviously this is not always possible and a mottled subsoil will assist in the identification of a location with a fluctuating water table. In clay soils this mottling is very noticeable as grey and blue colorations interspersed with brown, yellow or rust colorations.

Exploration methods

Many methods have been devised for securing information regarding the subsurface, and each one of them has particular advantages. Thus it is not necessary to recommend one method as against another, but rather to state that the method, tools and equipment used in any given case will vary with the location and topographical features encountered. In fact, in many instances it will be found desirable to use more than one method on a given scheme.

Before discussing the various methods[16, 17] one further point should be emphasized in relation to obtaining and analyzing subsurface data. Subsurface exploration has always been a highly specialized art in which there is little room for the unsupervised or inexperienced. Although this type of exploration is based on sound engineering principles, it is still heavily dependent on considerable experience if valuable results are to be obtained. This experience and ability cannot just come from a book, which—however erudite—can only provide a very general background to a vast field of study.

Test Pits. If it is possible to say that one method of obtaining data is 'better' than others, then certainly it is that of digging test pits and examining the profile *in situ*. Test pits, dug by hand or machines, are open excavations large enough to permit an investigator to enter and examine formations in their natural condition. Although this method makes it possible to obtain samples in either a disturbed or undisturbed condition, it can be a relatively expensive procedure, which may make it impractical on many sites. Test pits can be excavated very rapidly with the aid of a backhoe, in non-flowing soils, to depths of up to about 5 m, while bulldozers can dig pits or trenches to depths of over 3 m. It is usually very expensive to excavate pits by hand labour, but this may be necessary in particular instances, e.g. when examining the profile of an existing roadway.

Soil auger borings. Two basic types of hand-operated tools are used for auger borings. One of these is the farm posthole auger, which consists of a spiral-shaped bucket, about 100 mm in diameter by 150 mm long, fastened to a shaft of narrow piping with a T-handle at the top for rotating the auger. The second type consists of a helical-shaped bit of about 25–37·5 mm in diameter fastened to a piping with a T-handle at the top. In either case, the tool is rotated into the ground until the bit has penetrated a distance equal to its length, after which it is withdrawn directly and the retained soil examined and sampled as necessary. The process is then repeated until the required depth is obtained. If the hole caves in, it may be necessary to use a section, or sections, of pipe of slightly larger diameter than the auger to form a casing to protect the walls.

On large projects it may be economically feasible to use augers operated by motor power. These are capable of obtaining soil from depths of up to 6 m.

Subsurface exploration by means of auger borings has many disadvantages. The principal one is that in stony or granular soils, or at sites with a high water table, the soil wall usually flows into the hole and the excavated material falls from the auger as it is being withdrawn. Furthermore, not only is the sample obtained in very disturbed condition, but it is often contaminated with scrapings from the side of the hole. Nevertheless, auger boring is probably the most effective 'value-for-money' method of obtaining subsurface information and, hence, is used extensively for location and design purposes. It is a particularly useful way of determining the depth to a relatively shallow (< 3 m) water table.

Wash borings. In wash borings, a 170–200 mm diameter casing is driven into the soil, one length at a time, by a drop-hammer action similar to that used in pile driving. After each length of casing has been driven, the contained soil is cleaned out by forcing water under pressure through hollow rods or a narrow pipe inside the casing. A chisel-shaped chopping bit is attached to the end of these thin pipes and the whole unit is then alternatively raised and dropped and turned so that the resulting churning-cum-jetting action loosens the soil within the casing and it is washed to the surface.

An experienced operator handling the wash pipes is able to 'feel' a significant change in material and, when this occurs, the churning is stopped and the casing washed clean of all loosened soil. After the wash rods or pipes are raised from the hole, a sampling tool is substituted for the churning bit and driven to obtain a sample. After the sample has been removed for further examination, the sample tool is replaced by the chopping bit and the drilling continued until another material change is noticed. When there are apparently no distinct changes in the soil profile, samples are taken at about every 1·5 m anyway.

Rotary drilling. This is very similar to wash boring except that no churning-cum-jetting process is used to loosen the soil. Instead the soil is loosened by rotation of a heavy string of rods, while continuous downward pressure is maintained through the rods on a bit at the bottom of the hole. A number of different types of bit are used in practice, many of which are able to reduce stone or the most compact layers of soil to small particles. Water under pressure is then forced down the rods to the bit, and the particles are washed to the surface.

Rotary drilling is most often used for deep investigations in relatively hard material, whereas wash borings are most useful for fairly shallow holes in unconsolidated soils. In addition, a fairly large and powerful rig is required for rotary drillings, whereas the wash boring equipment is fairly light and more easily transported over rugged topography.

Rock core drilling. When rock is met within a cut section, its nature and condition must be established in sufficient detail for a reasonable estimate of the cost of excavation to be made. A rotary drilling process, either the shot drilling method or the diamond/tungsten carbide core method, is usually employed to obtain these specimens. The process is essentially the same in both. An annular-shaped bit is screwed on to the bottom of a cylindrical core-barrel and the bit and core-barrel are then fastened to rod sections of hollow steel tubing. These are rotated by the drill while water is forced into the hole through the rods and barrel. The water serves a dual purpose, that of cooling the bit and of bringing the pulverized rock particles to the surface. As the bit cuts into the rock, the core feeds into the barrel, and this is raised at regular intervals to the surface for extraction.

The bit used in diamond core drilling is set with commercial diamonds, which makes the process a relatively expensive one. As against this, it is generally accepted that the best cores are obtained using diamond drill bits. Tungsten carbide insert and sawtooth bits are useful for drilling through, but not sampling, well-compacted overburden materials; the tungsten carbide insert bit also gives quite good samples in soft rock. In shot drilling, chilled steel shot is fed down through the hollow rods with the water and then beneath the rotating bit where the particles act as cutting agents. Unfortunately, however, shot drilling produces poor core recovery in many rock formations, especially if the core holes are of small diameter. This, taken in combination with the cost of commercial diamonds, results in shot drills being mostly used to obtain cores greater than about 75 mm in diameter.

Soundings. Also known as probings, these consist of simply pushing or driving sections of steel rod down through the soil as far as they can be made to penetrate. The simplest type of sounding equipment consists of a 25 mm diameter pointed steel rod which is driven into the ground by a man wielding a sledge hammer; mechanically operated hammers may also be used for this purpose. More elaborate sounding equipment is available which provides a pipe surrounding the rod to eliminate side friction, advances the pipe by jetting with water, and measures the resistance to penetration of the rod by recording the number of blows or force of a jack required to advance it a given distance. The penetration resistance readings obtained with these penetrometers are then compared with the results obtained in known profiles for interpretation purposes.

The results which can be interpreted from soundings are limited, especially with regard to subsurface boundaries. When the penetrometer is driven to refusal, the only real information obtained is that bedrock does not exist above the depth to refusal—what is not known is whether the refusal is due to bedrock or a boulder or perhaps a bed of gravel.

Geophysical methods. Geophysical exploration is a form of field investigation utilizing a blend of physics and geology in that physical measurements made at the surface by specially developed instruments are interpreted in terms of the subsurface geological conditions. Two methods of geophysical exploration, the electrical resistivity and seismic refraction methods, have found most application to the shallow depths met with in highway work. They can be very usefully employed in determining the amount of rock to be removed from road cuts, resolve doubts as to whether subsurface obstructions are bedrock or boulders, and worked or unworked coal seams,[18] facilitate deductions of subsurface irregularities which often are missed by borings, locate buried sources of roadfill material, i.e. buried sand pits, gravel pits, etc., and aid in determining the extent of known sources of road materials. They can be used to determine bedrock profiles under water as well as on land[19], but the strata of mud, silt, sand and gravel which may overlie the bedrock under water cannot be delineated with accuracy as there is insufficient contrast between these materials when water is present.

Geophysical survey methods offer a means of quick and relatively inexpensive subsurface exploration and as such they are very attractive to the engineer faced with a comprehensive investigation programme for a large highway scheme. However, he should consider carefully before attempting to apply these procedures, even though a skilled instrumentalist be at hand. Although the tests may be carried out flawlessly, they are of little value unless the results are interpreted correctly, and correct interpretation can only be expected from experienced persons with a basic understanding of geophysical theories and procedures, and of the branches of engineering dealing with soils and geology.

Where the scientific and technical 'know-how' is available, however, the procedures can be very gainfully employed in reducing the large number of test pits, borings and soundings which might otherwise have to be carried out for a large scheme. While it is necessary to correlate the results of the geophysical survey with those obtained from direct investigations over the site, the direct explorations need not be carried out prior to the geophysical survey. Instead a preliminary interpretation can be made on the basis of the geophysical results, and this enables a more judicious siting of necessary direct test holes, with a consequent large reduction in the total number required.

Electrical resistivity method

The resistivity of a substance is defined as the electrical resistance offered by a unit cube of the substance to the flow of an electrical current perpendicular to one of its faces. It is usually denoted by the symbol ρ and expressed in units of ohm-centimetres.

With the exception of certain metallic minerals which are very good conductors, the constituent minerals of the earth are more or less insulators, and the resistivities of the various subsurface formations are almost entirely dependent on the salinity of the moisture contained in them. In the case of a rock, the more porous or the more jointed and fissured the rock, the more moisture it will hold and the lower will be its resistivity. Thus, in general, igneous rocks have a much higher resistivity then sedimentary rocks and, of the latter, limestones and sandstones are more resistant than shales and compacted clays. Gravels, sands and other loose soils have high resistivity values as compared to ordinary moist soils with moderate clay or silt contents. Thus, it is by noting the differences in the resistivities of different materials at the site and comparing these with results obtained for known conditions that predictions can be made regarding the subsurface materials.

To carry out an electrical resistivity test, an electric current is applied to the ground through two metal rods called current electrodes; these are connected by insulated wires through a milliammeter to a battery or generator. As illustrated in Fig. 1.8(a), the current flows through the ground so that the lines of flow make a pattern similar to the lines of magnetic force about a bar magnet. In theory the lines of current flow extend to an infinite depth, but, since the intensity diminishes quite rapidly with depth, for practical purposes the current can be considered as confined within a depth equal to about one-third of the distance between the current electrodes. Two further electrodes, called potential electrodes, are next inserted in the ground so that the four electrodes are in a straight line and spaced at equal intervals. The potential electrodes are actually porous pots containing solutions of copper sulphate, some of which seeps into the ground so that good electrical contacts are made. The pots are connected by insulated wires to a potentiometer so that voltages can be measured.

To determine the resistivity of the ground, the current I flowing from the battery and through the ground between the current electrodes is measured on the milliammeter. At the same time the voltage drop V between the potential electrodes is measured on the potentiometer. The resistivity of the material is then given by an equation such as

$$\rho = \frac{2\pi A V}{I}$$

where ρ = resistivity, ohm-cm,
             A = distance between electrodes, cm,
             V = potential drop, volts,
and I = current, amps.

The manner in which the data are interpreted can be illustrated by the following simple example. A soil of relatively low resistivity is overlying a rock with high resistivity. When the electrode spacing is very small compared with the thickness of the upper layer, practically all the current will flow through the uppermost depth and the resistivity measured will approximate closely the actual resistivity of that layer. If the electrode spacing is increased, the depth of current penetration increases and the resistivity measured represents an average for all material existing within the depth A below the surface. As the electrode spacings are progressively expanded, the current flow lines eventually encounter the underlying rock formation. This

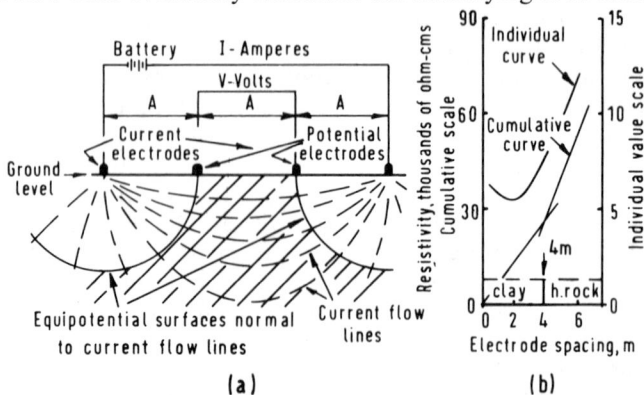

Fig. 1.8.   Diagrams illustrating the use of the electrical resistivity method[20]

material, which has an appreciably higher resistivity than the overlying soil, then begins to affect the average resistivity values; as the spacings are further increased the effect of the lower bed increases proportionally until, eventually, the resistivity measured approaches closely the resistivity of the rock.

When the resistivity ρ is plotted as an ordinate against the electrode spacing A as the abscissa, a curve is obtained similar to the individual-test curve in Fig. 1.8(b). When the values of apparent resistivity are plotted on a cumulative curve, a straight line (or a curved line of gentle curvature) is obtained as long as the 'effective' current flow remains within the surface layer. As the electrode spacing is expanded, the cumulative curve shows an increasing upward curvature, reflecting the influence of the higher resistivity of the rock. If straight lines are then drawn through as many points as possible on the cumulative curve, the intersection in the region of the increasing curvature gives a good approximation of the thickness of the surface material.

Seismic refraction method

The seismic method is based primarily on the measurement of the time required for a sudden elastic disturbance, set up in the ground by exploding a small charge of gelignite, to be picked up by a number of vibration detectors or seismometers located on the ground surface at known distances from the shot point. The seismometers transform the ground vibrations into electrical impulses which are transmitted to a recording apparatus where they are amplified and recorded on photographic paper moving at a high speed. The instant of explosion and time markings are also recorded and the interval of time between the instant of explosion and the arrival of the first vibration impulse at each seismometer can then be determined. Knowing these data, the velocity at which the seismic waves are moving through the various underlying strata can be graphically calculated, after which the thicknesses of the various layers can be determined from a simple mathematical relationship.

The basis of the method is most easily understood by considering an example.[21] Figure 1.9(a) represents the advance of a wave-front in a two-layer structure consisting of about 24·5 m of gravel on top of bedrock. The wave travels about 1525 m/sec in the gravel, and about 6100 m/sec in the bedrock. In such a condition the wavefront has approximately the shape of a sphere of 15·25 m radius ·01 sec after it leaves the shot point. About ·016 sec after the shot, it reaches the rock surface at point A. From this point of contact a wavelet starts into the rock with a velocity of 6100 m/sec, so that ·02 sec after the shot the wavelet has reached the distance noted by that circle in the diagram, not only vertically below A, but radiating in all directions within the rock; thus the ·02 sec wavefront reaches the point B along the gravel-rock interface. Along the interface, however, the wave in the rock starts a new series of wavelets into the gravel, directed towards the gravel surface. The ones at B and C are but two of many such wavelets, and their envelope is the new refracted wave DE as shown ·025 sec after the shot.

The bottom part of Fig. 1.9(b) shows the continued advance of the wavefront through the gravel until more than ·04 sec after the explosion. It is clear that the wavefront travels to the nearer seismometers at a speed of 1525 m/sec. Equally obvious is the fact that, at some distance along the line of seismometers, a refracted wave will travel the shorter distance to the surface and be picked up before the original wavefront reaches the same point through the gravel. Beyond this 'critical distance' the refracted wave is always picked up first by the seismometers.

The top part of Fig. 1.9(b) is a plot of the arrival times at the seismometers versus distance from the shot point. The earlier seismometers record the velocity of the gravel; note that the plotted points fall along a straight line, the slope of which is 1525 m/sec. The farther seismometers

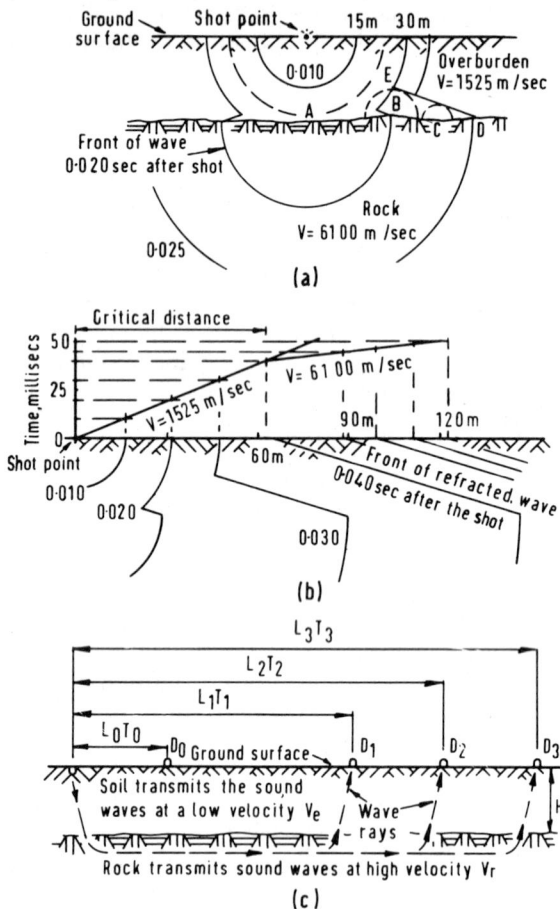

Fig. 1.9.   Diagrams illustrating the use of the seismic refraction method

record values which plot along a line whose slope equals 6100 m/sec; this is the velocity of the bedrock. The distance from the shot point to the point of intersection of these two lines is controlled by the thickness of the gravel, and the greater this is the greater will be the distance of the intersection from the shot point.

In seismic tests the velocity of the transmitted sound waves generally increases with an increase in the density of the transmitting medium. Wave velocities in loose unconsolidated soil layers range from 180–460 m/sec. Velocities in more compact subsurface layers range from 600–2750 m/sec, the lower ranges of 600–1050 usually being associated with clay materials and the higher ranges of 1225–2750 with compact gravels, badly broken or weathered rock, and soil-boulder mixtures. Predicting the material from its velocity requires considerable skill and judgement, and so calibration tests over known subsurface formations are essential for a successful interpretation of the data obtained.

In the interpretation of seismic data it is convenient to represent the propagation of the seismic waves by means of wave rays. By definition a wave ray is a wave path which is normal to the progressing wavefront at every instant and is, therefore, a path of minimum time. Fig. 1.9(c) shows the path of the wave rays between the shot point and four seismometers spaced along the ground for the simple level two-layer structure already discussed. The equation for the determination of the depth to solid rock may be developed as follows:

$$V_e = L_0/T_0$$

and
$$T_3 = H/V_e + L_3/V_r + H/V_e$$
$$= 2H/V_e + L_3/V_r$$

and
$$T_2 = H/V_e + L_2/V_r + H/V_e$$
$$= 2H/V_e + L_2/V_r$$

Therefore
$$V_r = \frac{L_3 - L_2}{T_3 - T_2}$$

and
$$H = \frac{V_e T_3}{2} - \frac{V_e L_3}{2V_r}$$

where H = depth to solid rock, m,
   V_e = velocity of the sound wave in the soil, m/sec,
   V_r = velocity of the sound wave in the rock, m/sec,
   T_0, T_1, T_2, T_3 = time for sound wave to reach seismometers
   D_0, D_1, D_2, D_3, sec,
and L_0, L_1, L_2, L_3 = distance to seismometers D_0, D_1, D_2, D_3, m.
In this equation V_e is calculated from the results obtained from the seismometers located within the critical distance. The velocity V_r is determined from results obtained at seismometers beyond the critical distance.

A simple graphical method of determining the depths to the underlying layers has also been developed and is discussed in the literature.[22]

Comparison of the geophysical methods. The seismic test is particularly useful for determining the presence or absence of dense, solid rock since the high velocities associated with such formations make the determination quite dependable (to within 10 per cent of the depth). Although the resistivity method will, in most instances, indicate the depth of overburden to a high-resistivity formation such as rock, it cannot always, in the absence of confirming geological data, furnish completely dependable evidence for predicting the presence of rock. Sand and gravel can show surface anomalies similar to those shown by solid rock. In areas where solid rock layers are interbedded with less dense materials such as shales, the resistivity test is the better tool, since it can detect the change from the hard rock to the softer and less resistant shales. The seismic test under such conditions is limited to an indication of the depth of overburden above the high-velocity sandstone or limestone, and the lower-velocity shales cannot be located.

Provided saturated soils do not exist above the water table, the depth to the water table in soil can be determined using either the seismic or resistivity method. Generally, however, only the resistivity method can be used to locate the water table in bedrock.

The resistivity test offers a practical means for the rapid investigation of large areas in search of localized deposits of gravel, sand or other granular materials useful in highway construction. The seismic method is not so well adapted to an area survey, but is best applied to the determination of conditions at a single designated spot or limited area. Even in the limited areas, if differentiating weathering has resulted in pinnacles and deep valleys in the subsurface rock—a condition sometimes encountered in limestone formations—the resistivity test may prove the more valuable method.

An important advantage of the seismic method is that the seismic compression velocity of a rock or soil has been correlated[23] with its rippability with a certain size of tractor. This permits the calculation of excavation costs and plant allocation prior to excavation. However, the use of explosives is not desirable in thickly populated areas, and this may place a handicap on the use of the seismic method. On the other hand, stray currents emanating from underground railway systems in large urban areas, and from buried utilities such as water and gas pipes, can be troublesome when making a resistivity survey in these locations.

The time required for conducting a seismic test can vary from 1 to 3 hours, depending on local conditions, while resistivity tests can be carried out at the rate of about 3 per hour to depths of up to 18 m in the most rugged terrain.

PREPARATION OF PLANS

The drawings for a highway scheme are, in effect, the graphical instructions as to how the roadway is to be constructed. As such they are an integral part of the contract drawn up between the highway authority and the contractor. Hence great care must be taken in their preparation so that the contractor knows exactly what he has to do and costly changes are not necessary after the contract is awarded.

Component parts

A complete set of plans for a highway improvement may contain the following:

1. A location map of the proposed improvement. This shows the location of the proposed line, major topographical features, possible detour routes, and other general features of introductory interest.

2. A detailed plan of the proposed highway. This should show the chainages and exact locations of all pegs; bearings of tangent lines, radii and other geometrical data affecting the layout of all horizontal curves, boundaries of the right-of-way, and streams, railways, buildings, fences, public utility lines, and other structures contained within it; existing and proposed drainage structures; locations of bench marks; and any other details to be considered in the course of construction. Contours showing the topographical nature of the terrain may also be included.

3. A profile section of the longitudinal line of the road. This should show the required surface or formation line of the highway and the natural ground or existing road line; all necessary existing and required elevations at marked pegs; all required vertical curve data; percentages of grade for each continuous grade; elevations of floors of culverts and bridges, and the beds of streams; and any other information necessary for the vertical design of the highway.

4. Earthwork cross-sections at all necessary locations. Each cross-section should be clearly located and indicate the elevations and line of the existing ground and the proposed roadway formation, and the areas of cut and fill for each section.

5. Typical roadway cross-sections at selected locations. These should provide all information required to construct the pavement, shoulders, footways, sideslopes, drainage ditches, and other such structural items.

6. A mass-haul diagram. This indicates the amount of cut or fill to particular locations along the line of the roadway, and shows where excavated material may be economically moved to create embankments.

7. Construction details of such items as culverts, bridges—these are very often separate contracts—guard rails and fences, traffic islands, kerbs, gutters and drainage inlets, pavement joints, superelevation, and other such items.

8. Special requirements of the scheme.

Steps in preparation

A basic approach to the preparation of highway plans is as follows:
1. Prepare the location map.
2. Draw the highway plan.
3. Plot the profile section and establish the vertical alignment of the road.
4. Develop preliminary concepts relative to cross-sections, drainage requirements and minor structures.
5. Visit field and test feasibility of concepts.

  6. Prepare cross-section details.
  7. Estimate earthwork quantities.
  8. Prepare construction details of all structural items.
  9. Complete and check plans.
 10. Visit field and verify that the plans are operational.

The first four stages involve what might be termed the preparation of the preliminary plans. After they have been prepared, it is good practice to visit the field and, by inspection, check the extent to which such features as proposed drainage structures are feasible, and how they and other such features affect, and are affected by, field conditions. At this stage also any further data required to complete the plans are obtained.

The remaining stages involve the preparation of the final plans. When they are considered complete, the designer again traverses the entire length of the scheme, and verifies that each item is as desired.

Establishing the vertical alignment. The first step in the establishment of the vertical alignment is the determination of the ruling grade for each section of the highway. As discussed in the chapter on Geometric Design (Vol. 1, Chapter 6), there are maximum ruling grades for highways; these depend on the type and location of the roadway. Below these maximum values, however, there are a number of other conditions which influence the choice of grade at particular locations.

In certain situations the grade chosen for a particular section will be controlled by adjacent topographic features or man-made culture. For instance, when the terrain is low-lying and swampy it is necessary to establish a line well above the existing ground level. Bridges, intersections with other roadways, etc. are fixtures which require that the highway be graded to meet them. In urban areas, the grade is controlled by the requirements of street intersections, footpaths, drainage needs, and the value of adjacent property.

Where the existing ground grades are within the maximum permitted for the highway, the establishment of the grade lines is primarily a matter of fitting the new grades to the existing ones so that a comfortable riding surface is obtained which produces a pleasing appearance with the minimum of earthworks. Good drainage is also obtained in this way. In rolling topography, particularly in rural areas, the grades may be chosen so as to balance the quantities of excavation against the amounts required in embankment. If the grades selected result in an alignment where the earthwork quantities just balance, a most economical roadway can be constructed.

A practical—non-computerized but nevertheless effective—method of establishing the vertical alignment is as follows. First, a trial grade line is established by visually attempting to balance the cut and fill volumes. This can be done by stretching fine thread over the profile and holding it in place with pins located at intersection points; the use of pins permits easy changes to study the effects of different grades without the need for tediously erasing pencil lines. In choosing suitable grades, the cross-section data should be frequently inspected since the centre-line elevation is very often different from the side elevations. When the grades have been decided on, vertical curves are calculated, located on the profile drawing and then checked to ensure that the sight distances are adequate; often the vertical curves have to be flattened in order to lengthen the available sight distance.

Estimating earthwork quantities. The determination of earthwork quantities is usually based on field cross-section data. The cross-section data show the extent of the excavation in cuttings and the filling for embankments at regular intervals—usually about every 15–30 m—and where major surface irregularities occur along the centre-line of the proposed highway.

Determining cross-section areas. When the ground surface is level or regular, the area of a cross-section is most easily determinable by dividing the enclosed space into triangles and trapesiums and using standard formulae in the calculations. If the ground surface is irregular, the planimeter provides a very easy means of measuring area. Alternatively, the area of an irregularly shaped cross-section can be obtained using the co-ordinate method. By

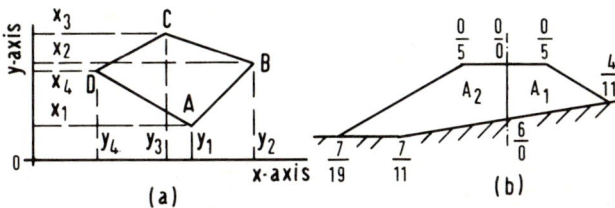

Fig. 1.10.   Determining area by the co-ordinate method

geometry it can be shown that the area enclosed in Fig. 1.10 is given by

$$ABCD = \tfrac{1}{2}[y_1(x_4-x_2)+y_2(x_1-x_3)+y_3(x_2-x_4)+y_4(x_3-x_1)]$$

From this the following simple rule can be postulated:

Moving about the closed figure in an anti-clockwise direction, multiply each ordinate (*y*-value) by the algebraic difference between the prior and following abcissas (*x*-values), find the algebraic sum of the products and divide the result by 2. If the enclosed figure is to the left of the *Y*-axis, movement should take place in a clockwise direction.

The use of the above relationship in calculating a cross-section area is illustrated by the following example:

PROBLEM.   Determine the cross-sectional area of the embankment shown in Fig. 1.10(b), which is to be constructed across sloping ground. The width of the embankment is to be 10 m.

Solution.   For convenience in calculation the embankment will be divided into two parts, A_1 and A_2. The co-ordinates of each corner are written in fraction form and are referenced with respect to the centre-line and the top of the embankment. Thus the 'fraction' $\tfrac{7}{11}$ means that the point referenced is 7 m below the top of the embankment and 11 m from the centre line.

Considering area A_1, begin at any corner and proceed anti-clockwise about the enclosed area so as to obtain

$$A_1 = \tfrac{1}{2}[4(0-5)+0(11-0)+0(5-0)+6(0-11)]$$

$$= -43 \, \text{m}^2.$$

Considering area A_2, begin at any corner and proceed clockwise about the enclosed area so as to obtain

$$A_2 = \tfrac{1}{2}[7(0-19)+7(11-5)+0(19-0)+0(5-0)+6(0-11)]$$

$$= -157 \, \text{m}^2.$$

$$\therefore \quad A_1 + A_2 = -200 \, \text{m}^2.$$

The minus value has no significance other than it indicates that an embankment is being considered. If the calculations are carried out for a cutting, i.e. with the centre-line and the bottom of the embankment as the reference lines, then a plus answer will be obtained. In the case of a side-hill section involving both cut and fill, the vertical reference axis should be taken through the point of intersection of the formation or carriageway and the natural ground level so that the areas in cut and fill can be determined separately.

Determining earthwork volumes. When the cross-section areas of excavation and embankment are known, the volumes can be calculated. The simplest and most common procedure for measuring volume is by means of the Trapezoidal or Average-End-Area method. This merely involves the estimation of each of the end cross-section areas of a length of roadway, the calculation of their mean and the multiplication of the mean area by the distance between the ends. Thus, in Fig. 1.11, if the area *abcd* is denoted by A_1 and *ijkl* by A_2, then

$$V = \frac{D}{2}(A_1 + A_2)$$

where V = volume, m^3,

D = distance between end areas, m

and A_1 and A_2 = end areas, m^2.

For a series of successive cross-sections spaced at uniform distances D apart, with areas A_1, A_2, A_3, ... A_n, the volume enclosed between the first and last sections is given by the expression

$$V = \frac{D}{2}(A_1 + A_2) + \frac{D}{2}(A_2 + A_3) + \frac{D}{2}(A_3 + A_4) + \ldots \frac{D}{2}(A_{n-1} + A_n)$$

$$= \tfrac{1}{2}(A_1 + A_n) + A_2 + A_3 + A_4 + \ldots + A_{n-1}$$

The above formula is exact only when the end areas are equal. When they are not, as is usually so, the results given by the equation are larger than the true values. In practice, however, it is found that the total error on a long line is rarely more than a few per cent; hence this very simple method is almost invariably used to calculate earthwork quantities.

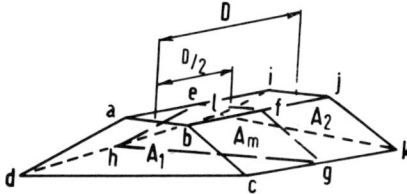

Fig. 1.11.   Determining volumes by the average-end-area and prismoidal formulas

When more precise results are required—and when the field data are sufficiently exact to warrant them—volumes may be determined by means of the Prismoidal formula. Referring again to Fig. 1.11, if A_1 and A_2 are as indicated before and the mid-section area *efgh* is denoted by A_m, then

$$V = \frac{D}{6}(A_1 + 4A_m + A_2)$$

Note that A_m is not the average of the end areas; it is the area of the cross-section at the point midway between the end areas. For a series of successive and equally spaced cross-sections with areas $A_1, A_2, A_3 \ldots A_n$, and when n is an *odd* number, the volume enclosed between the first and last sections is given by

$$V = \frac{D}{6}\left[A_1 + 4A_2 + 2A_3 + 4A_4 + 2A_5 + \ldots + 2A_{n-2} + 4A_{n-1} + A_n \right]$$

Where such precise results are required as to justify the use of the prismoidal formula, they will also warrant correcting the volumes on curves for errors involved in assuming the centre-line to be straight. On curves, the cross-sections are taken as near radially as possible, so that the volume being estimated is a curved solid between two non-parallel plane ends; this is shown in Fig. 1.12. This volume can be closely determined by means of Pappus' second theorem which states that 'if a plane area rotates about an axis in its own plane which does not divide it into two parts, the volume of the solid thereby formed is equal to the area multiplied by the length of the path of the centre of gravity of the area'. In Fig. 1.12, let A_1 and A_2 be two adjacent end areas that are at a distance D from each other. Let the centroids of A_1 and A_2 be distances of e_1 and e_2 respectively from the centre-line, which has a radius of curvature R. If the average area is $\frac{1}{2}(A_1 + A_2)$, then by Pappus,

$$V = \tfrac{1}{2}(A_1 + A_2) \times \text{arc distance between the centroids.}$$

But $D = R\theta$

where D = length of curve,

   R = radius of curve,

and θ = angle subtended at the centre of the curve.

Therefore $\theta = \dfrac{D}{R}$ radians.

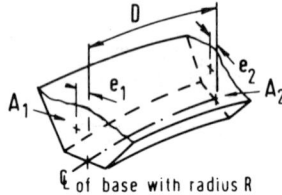

Fig.1.12.   Determining volumes on a curved centre-line

The average eccentricity is $\frac{1}{2}(e_1 + e_2)$, hence the arc distance between the centroids is given by

$$\text{Centroid arc distance} = \theta[R + \tfrac{1}{2}(e_1 + e_2)]$$

$$= D\left[1 + \frac{1}{2R}(e_1 + e_2)\right]$$

Thus the volume is given approximately by

$$V = \frac{D}{2}(A_1 + A_2)\left[1 + \frac{1}{2R}(e_1 + e_2)\right]$$

In the above equation the correction factor is actually $D/4R(e_1 + e_2)$. Thus it can be seen that, if D is quite small and R is large, the correction can be relatively unimportant. If the eccentricity of the centroid of the average section is within the inside of the curve, the correction factor is subtracted from, instead of being added to, the volume as determined by ignoring the curve.

Distributing earthwork quantities. As discussed earlier, the location of a highway and the selection of its vertical alignment should ideally be such that the volume of material excavated within the limits of the scheme is equal to that required in embankment. If all the excavated material can be hauled from the cuttings to the embankments, then wastage and the need to borrow materials are eliminated and a most economical design is obtained. In order to cost the scheme, it is necessary therefore to analyse the relationships between cut and fill materials along the proposed line of the road.

Shrinkage.   When calculating earthwork quantities, an allowance has to be made for the excess of excavation required to form a given embankment volume. The term shrinkage is almost universally used to explain the condition whereby a unit volume of excavation material will occupy less space when placed in a compacted embankment. The *Shrinkage Factor* is the term used to describe the relationship between the two volumes. This can, however, be a very confusing term as its value depends upon how it is determined. If the shrinkage factor is defined as the ratio of a volume of embankment to a fixed *volume of excavation* containing the same weight of dry material, then

$$\text{S.F.} = V_f/V_e$$

where   S.F. = shrinkage factor,
          V_f = volume of embankment,
and V_e = volume of excavation.
But $W = D_f V_f = D_e V_e$
where W = weight of dry material to fill a given volume of either excavation or embankment,
          D_f = dry unit weight of embankment material,
and D_e = dry unit weight of excavation material.

Therefore $\text{S.F.} = \dfrac{W/D_f}{W/D_e} = \dfrac{D_e}{D_f}$

This means that if the unit weight of the undisturbed excavation material is 1520 kg/m^3, and the unit weight of the embankment is expected to be 1680 kg/m^3, then 1 m^3 of excavated material will theoretically occupy 0.905 m^3 in the embankment; in other words S.F. = 0.905. The per cent shrinkage, expressed on the basis of a unit excavation volume, is given by

$$\% \text{ shrinkage} = (1.000 - \text{S.F.})\, 100$$
$$= (1.000 - 0.905)\, 100$$
$$= 9.5$$

Payment for earthworks is usually based on excavation quantities; hence, for design purposes, the earthwork calculations are often determined on the basis of the volume of excavated material required to occupy a *given volume of embankment* containing the same weight of material. Using the same method of analysis as before, it can be shown that on this basis

$$\text{S.F.} = \frac{D_f}{D_e}$$

Thus, using the same example as above,

$$\text{S.F.} = \frac{1680}{1520} = 1.111$$

In other words 1.111 m^3 of excavation is needed to obtain 1 m^3 of embankment. The per cent shrinkage expressed on this basis is

$$\% \text{ shrinkage} = (\text{S.F.} - 1.000)\, 100$$
$$= 11.1$$

Whatever the basis of comparison shrinkage is relatively small when volumes of sand and gravel are being considered, since their unit weights do not differ greatly from the natural to the compacted state. However, shrinkage can be very high if the soil is clayey and contains much organic matter; not only will the unit weight of the embankment material be much greater than the material in its natural condition, but in addition much of the excavation material will have to be deliberately wasted because of its humus content.

Swell.   When a given volume of excavation occupies a greater volume of embankment, the per cent change between cut and fill is called the per cent swell (or bulking). The ratio of embankment volume to excavation volume (or vice versa) is called the *Swell Factor.*

When excavated *soil* is placed in an embankment, the final volume is nearly always less than the 'natural' volume; here it is the shrinkage factor which is of importance. When, however, *rock* is excavated, it will always occupy a greater volume in embankment, and it is the swell factor and the per cent swell which must be estimated.

Haul, free-haul and overhaul.   In earthwork calculations the term Haul has a dual meaning. It is used to describe the *distance* over which material is moved and also, the *volume-distance* of material used.

In earthwork contracts, it is usually stated that the contractor will be paid a specified price for excavating, hauling and dumping material, provided that the haulage distance does not exceed a certain amount. This distance, which is called the Free-haul, can be as little as 150 m on small road schemes and 350 m or more on large motorway construction schemes. Within the free-haul distance the contractor is paid a fixed amount per cubic metre of material, irrespective of the actual distance through which it is moved. When the haulage distance is greater than the free-haul, however, the contractor is usually paid at a higher rate for the Overhaul. The unit overhaul price is based on the cost per station metre of moving material beyond the free-haul distance, i.e. the length measurement begins at the end of the free-haul distance.

Economic-haul.   When the haul distances are large it may be more economical to waste excavation material and borrow from a more convenient source than pay for overhauling. On any given scheme, the Economic-haul distance will vary considerably, as it depends both on the availability of suitable borrow materials and of nearby sites where excavated material can be wasted.

The economic-overhaul distance can be determined by equating the cost of roadway excavation plus overhaul and tipping in embankment with the cost of borrow pit material (includes original cost as well as cost of excavating, hauling and tipping borrow in embankment), plus excavation, haul and wasting of roadway material within the free-haul distance.

Thus, if a = cost of roadway excavation per m^3,

b = cost of overhaul and tipping per m^3 per station,

c = cost of borrow material per m^3,

and L = economic-overhaul distance in stations,

then
$$a + bL = c + a$$

Therefore
$$L = \frac{c}{b} \text{ stations.}$$

If the free-haul distance is denoted by F stations, then the economic-haul distance is given by

$$F + L = F + \frac{c}{b} \text{ stations.}$$

Mass-haul diagram. To enable the contractor to have a knowledge of the amount and extent of free-haul and overhaul on a project, so that he can submit a bid, it is common practice to include a mass-hauling diagram in the plans for the scheme. This diagram is a graphical representation of the amount of earthworks involved in the highway scheme, and the manner in which they may be most economically handled. It shows accumulated volume at any point along the proposed centre-line and from this the economical directions of haul and the positioning of borrow pits and spoil heaps can be estimated.

An example of a mass-haul diagram is shown in Fig. 1.13. It consists of a graph showing the algebraic summation of excavation and embankment as one proceeds along the centre-line of the roadway. The accumulated total at each station is the summation of the differences between cut and fill. Prior to summation, however, the embankment areas were adjusted so that all volumes are expressed in terms of excavated volumes.

Before discussing how it is used in practice, it is worthwhile outlining the characteristics of the diagram:

1. The ordinate at any station along the curve represents the earthwork accumulation to that point.

2. The maximum ordinate $(+)$ indicates a change from cut to fill as one proceeds along the centre-line from an arbitrarily assumed origin. The minimum ordinate $(-)$ represents a change from fill to cut. These maximum and minimum points may not necessarily coincide with the apparent points of transition as indicated by the profile section; this depends on whether or not there are side-hill transitions at these points.

3. A rising curve at any point indicates an excess of excavation over embankment material at this point. A falling curve indicates the reverse.

4. A steeply rising or falling curve indicates heavy cuts or fills. Flat curves show that the earthwork quantities are small.

5. The shapes of the loops indicate the direction of haul. A convex loop shows that the haul from cut to fill is from left to right, while a concave loop indicates that the haul is from right to left.

6. Since the ordinates of a curve are plotted from cut volumes and

adjusted fill volumes, then any line parallel to the base line which cuts off a loop intersects the curve at two points between which the amount of cut is equal to the fill. Such a line is called a 'balancing line' and the intersection points are called 'balancing points'.

7. The area between a balance line and the mass-haul curve is a measure of the haul (in station-cubic metres) between the balance points. If this area is divided by the maximum ordinate between the balance line and curve, the value obtained is the average distance that the cut material must be hauled in order to make the fill. This distance can also be estimated by drawing a horizontal line through the mid-point of this maximum ordinate until it intersects the loop at two points; the length of this line is very close to the average haul distance when the shape of the loop is 'smooth'.

8. Balance lines need not be continuous; the vertical break between any two balance lines merely indicates unbalanced earthwork between two adjacent points of termination of the lines. Adjacent balance lines should never overlap, as this means using the same part of the mass-diagram twice.

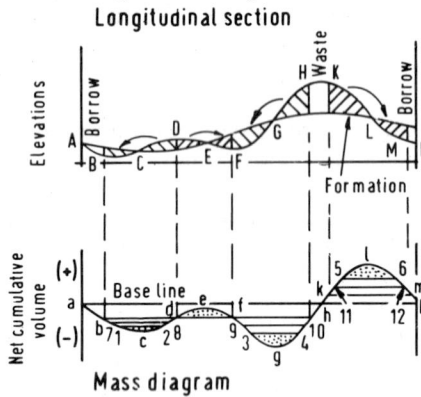

Fig. 1.13.   Example of a mass-haul diagram

One way of using a mass-haul diagram is illustrated in Fig. 1.13. This figure consists of two parts, the top being a profile section showing the ground contour and the proposed road formation, while the bottom is the mass-diagram illustrating the volumes of fill and excavation plotted as an additive curve beginning at *a*. First of all the limits of economic-haul are noted; these are drawn as the balance lines *bd, fh* and *km* between *B* and *D*, *F* and *H*, and *K* and *M*. Therefore the earthwork quantities are not only balanced in volume but economically so as well. The manner in which haulage should take place is indicated by the arrows; note that haulage takes place downhill to the embankments so that the empty vehicles can travel uphill to the excavation sites.

The limits of free-haul are indicated by the balance lines 12, 34 and 56. The free-haul station-meterage is indicated by the dotted areas 1*c*2, 3*g*4, and 5*l*6. In this case, by pure coincidence, the balance line of *df* is equal to the free-haul distance and as a result the area *def* is also free-haul station-meterage.

The overhaul volume for the section BCD is given by the ordinate difference between c to bd and c to 12. The average length of overhaul is estimated by drawing the balance line 78 through the median of the overhaul ordinate. Since the curve is smooth, the points 7 and 8 will lie directly below the centres of gravity of the overhaul volumes, and thus the average distance that this excavated material is moved is given by the distance 78. Since the free-haul is given by 12, the average overhaul is equal to the distance 78 less the distance 12.

It is rarely that the first attempt to choose an economical grade line results in the earthworks being completely balanced from beginning to end. In Fig. 1.13 the earthworks are not balanced and so material must be borrowed in order to build embankments between A and B, and M and P. The earthwork quantities involved are given by the ordinates at b and m. Between H and K the excavated material will have to be wasted on a spoil-heap since it is not an economical proposition to overhaul and use it in the preparation of embankments.

PREPARATION OF QUANTITIES

When a road improvement has been located and designed and the final plans drawn, the next phase is the drawing-up of a bill of quantities for the scheme. A bill of quantities consists of a tabulation of all items of work expected to be met with during the course of construction, the estimated quantities of each, and the unit and total cost of every item. On the basis of these quantity data, the engineer can estimate whether the cost of the scheme will be within the budget at his disposal.

When the work is to be advertised for tenders from interested contractors, a bill of quantities is normally included as a part of the plans for the scheme. In this, the unit prices and total costs are not, of course, included. It is up to the contractor to make his own estimation of these costs for tendering purposes. When preparing bills of quantities it is usual to organize the presentation according to a detailed procedure available as a guide in the literature.[25]

SPECIFICATIONS

Whereas the highway plans provide the graphical information necessary for the construction of the highway, the specifications are the written instructions which set the standards to which the work must be carried out. These standards are based on the result of experience and research knowledge acquired over many years as to the quality of materials and workmanship which can be demanded, and expected, on particular types of scheme. It is most important that the specifications should describe every construction item which enters into the contract, the materials to be used and the tests they must meet, methods of construction in particular situations, the method of measurement of each item and the basis on which payment should be calculated.

Because of the repetitive nature of substantial parts of highway schemes, most highway authorities have at their disposal standard specifications for highway construction.[26] These specifications are usually printed in permanent book form. On any given scheme, however, it will be found that particular problems arise which necessitate deviating from the standard specifications and thus the final specifications will include supplementary provisions that cover the conditions peculiar to the project.

Specifications are integral parts of the contract documents and their preparation should not be taken lightly. Because of their legal importance, the preparation of specifications requires not only considerable knowledge of highway practice, but also expertise in the law of contracts, so legal opinion should always be sought before specifications are placed for contract. To enable his legal advisers to prepare the proper documents so that later recriminations are avoided, the engineer engaged in writing a specification should know exactly what is required and what is economically attainable. Careless and loose wording can result in the use of poor quality materials and workmanship, and the specifications should be tightly written so that this can be avoided. On the other hand, the requirements specified should not be so exacting that they are either impossible to attain or the costs of so doing are not justified by the results.

SELECTED BIBLIOGRAPHY

1. TURNER, A. K. Route location and selection, Lecture 2 in the *Proceedings of the Conference on Computer Systems in Highway Design* held in Copenhagen on Sept. 2–9, 1972, and Organized by the Royal Technical University, Copenhagen, Planning and Transport Research and Computation, London, and Laboratory for Road Data Processing, Copenhagen.
2. DUMBLETON, M. J. and WEST, G. Preliminary Sources of Information for Site Investigations in Britain. *RRL Report* LR 403, Crowthorne, Berks., The Road Research Laboratory, 1971.
3. COMMITTEE ON ENGINEERING SURVEYING. *Report on Highway and Bridge Surveys.* New York, American Society of Civil Engineers, 1962.
4. ADOLFSSON, B. The use of electronic distance and direct reading tacheometers and survey systems based on them, Lecture 5A in the *Proceedings of the Conference on Computer Systems in Highway Design* held in Copenhagen on Sept. 2–9, 1972. (See also ref. 1.)
5. CANNING, A. S. Modern methods of surveying at large scales, *Journal of the Institution of Highway Engineers*, 1969, **16,** No. 11, 17–19.
6. DUMBLETON, M. J. and WEST, G. Air-Photograph Interpretation for Road Engineers in Britain. *RRL Report* LR 369, Crowthorne, Berks., The Road Research Laboratory, 1970.
7. McNOLDY, C. E. Highway design using photogrammetric terrain data, *Highways Design and Construction*, 1971, **39,** No. 1743, 14–18.
8. WITHEY, K. H. The Optimization of the Vertical Alignment of the M5 Motorway from Chelston to Blackbrook. *TRRL Report* LR 473, Crowthorne, Berks., The Transport and Road Research Laboratory, 1972.
9. MOTT, P. G. (Hunting Surveys and Consultants, Ltd., Boreham Wood, Hertfordshire.) Personal communication to the author. Aug. 1972.

10. AMERICAN SOCIETY OF PHOTOGRAMMETRY. *Manual of Photographic Interpretation.* Washington, D.C., The Society, 1960.
11. VARNES, D. J. Landslide types and processes, *Highway Research Board Special Report* No. 29, 1957, 20–47.
12. KASPER, H. and W. BLASCHKE. Aerial survey and road construction; translation by D. Kennedy of 'Luftbildmessung und Strassenbau', *Brucke und Strasse*, Vol. 1–3, 1960.
13. A.A.S.H.O. Standard methods of surveying and sampling soils for highway purposes; A.A.S.H.O. Designation T86–54. *Book of Standard Specifications for Highway Materials and Methods of Sampling and Testing*, Part II. Washington, D.C., American Association of State Highway Officials, 1961.
14. ROAD RESEARCH LABORATORY. Soil survey procedure, *Road Research Technical Paper* No. 15. London, H.M.S.O., 1954.
15. DOWLING, J. W. F. and F. H. P. WILLIAMS. The use of aerial photographs in materials' surveys and classification of land forms. *Proc. Conference on Civil Engineering Problems Overseas, 1964.* London, Institution of Civil Engineers, 1964.
16. HARDING, H. J. B. Site investigations including boring and other methods of sub-surface exploration, *Proc. Inst. Civ. Engrs*, 1949, **32**, No. 6, 111–157.
17. HVORSLEV, J. Subsurface exploration and sampling of soils for civil engineering purposes, *Report on a Research Project of the American Society of Civil Engineers.* New York, The Engineering Foundation, 1962.
18. COWAN, D. R. Geological and Geophysical Investigations on Part of the M.90 Kelty By-pass. *RRL Report* LR 255, Crowthorne, Berks., The Road Research Laboratory, 1969.
19. STEWART, M. The Use of Seismic Refraction in a Route Feasibility Study in St. Lucia. *RRL Report* LR 424, Crowthorne, Berks., The Road Research Laboratory, 1971.
20. MOORE, R. W. Applications of electrical resistivity measurements to subsurface investigations, *Public Roads*, 1957, **29**, No. 7, 163–169.
21. LINEHAN, Rev. D. Seismology as a geologic technique, *Highway Research Board Bulletin* 13, 1948, 77–85.
22. MOORE, R. W. Geophysical methods of subsurface exploration in highway construction, *Public Roads*, 1950, **26**, No. 3, 49–64.
23. CATERPILLAR TRACTOR CO. *Handbook of Ripping.* Peoria, Ill., The Caterpillar Tractor Co., 1966.
24. HOEK, E. and BOYD, J. M. Stability of slopes in jointed rocks, Journal of the Institution of Highway Engineers, 1971, **18**, No. 12, 16–18.
25. Department of the Environment. *Method of Measurement for Road and Bridge Works.* London, H.M.S.O., 1971.
26. MINISTRY OF TRANSPORT. *Specification for Road and Bridge Works.* London, H.M.S.O., 1969.

2 Surface and sub-surface moisture control

Modern road engineers recognize that the entire serviceability of a highway is greatly dependent upon the adequacy of its drainage system. Water standing on the carriageway is a danger to high-speed traffic which is accentuated when freezing temperatures occur. Water seeping into the pavement and subgrade leads to the development of soft spots which result in the break-up of the surfacing and the need for expensive maintenance work. The wash-out of a pipe culvert across a motorway is just as much a hindrance to traffic as the collapse of a large bridge. Thus, it can be said with total justification that proper drainage design is an essential and integral part of economical highway design; indeed it is a consideration which should not only be taken into account during design but should also, when necessary, influence the location of the highway itself.

Road drainage considerations can conveniently be divided into those relating to the flow of surface water and of subsurface water. *Subsurface moisture control* is concerned with the flow of water within soils. *Surface drainage* is concerned with the measures taken to control the movement of water over the ground on and adjacent to the highway, so that it is directed to suitable disposal points without any detrimental effects to the roadway.

SURFACE DRAINAGE

There are three aspects of surface drainage design in which the highway engineer is particularly interested. First of all, he is concerned that precipitation falling on the road surface be removed as quickly as possible so as to minimize the danger to moving vehicles. This is achieved by crowning the carriageway, inserting gulleys (in towns) and, on rural roads, sloping the side-shoulders. The second consideration has to do with the bridging of rivers and relatively large streams. Here the highway engineer is well advised to seek the advice and co-operation of the appropriate waterway authorities when determining the exact form that these bridging structures should take; in Great Britain this information is best obtained from the local River Authority. Thirdly, the engineer designing a long stretch of roadway or a road system is continually faced with the twin problems of intercepting and disposing of the water in the myriads of permanent minor streams and temporary water courses which are interfered with by the road construction.

It is convenient to divide a surface drainage study into two parts. First of all, there is the problem of deciding the amount of water which must be catered for, i.e. the amount of water arriving at the inlet, drainage ditch or culvert, and secondly the design of the facility needed to handle this water. The first part of such an investigation is termed the hydrological study, while the latter is called the hydraulic study.

Hydrological study

Hydrology can be described as the science which deals with the operations governing the circulation of moisture in its various forms, above, on, and beneath the earth's surface. As such it involves the study of the various phases of the hydrologic cycle, i.e. precipitation, run-off, infiltration, evaporation and condensation. The two main phases of the hydrologic cycle in which the highway engineer is most interested are precipitation and run-off.

Precipitation. Water can be precipitated in the form of rain, hail, snow or sleet. In practice, however, the engineer is usually concerned only with the moisture which falls in the form of rain; it is only in countries which are normally very cold and subject to rapid upward changes in temperature that special consideration has to be given to the provision of extra drainage for the amounts of snow which fall and accumulate on the ground. To enable estimates to be made of the amount of run-off water which will have to be catered for in any one instance, the engineer must have information regarding the rainfall intensities within the catchment area being studied.

A fundamental feature of rainfalls is that the intensity of rainfall throughout a given storm is inversely proportional to the length of the storm, i.e. as the duration of a rainfall increases, its average intensity decreases. This is perhaps to be expected since the meteorologic forces which cause a heavy rainfall are also continually causing it to move quickly from one area to another. A further point to be noted is that at a given site, and for a given storm length, the intensity can vary considerably, with the least heavy rainfalls occurring most often.

While an engineer designing a large bridge or a complete flood-control system is mainly interested in storms which cover large areas and last for hours and perhaps days, the highway engineer engaged in designing for a culvert or a drainage ditch is primarily interested in the high intensity, short duration storms, i.e. the storms which result in the peak rate of run-off from the catchment area. Internationally, the storm duration which is most often chosen for design purposes is based on the assumption that the maximum discharge at any point in a drainage system occurs when the entire catchment area tributary to that point is contributing to the flow, and that the rainfall intensity producing this flow is the average rate of rainfall which can be expected to fall in the time required for a raindrop that falls on the most remote point of the catchment area to flow to the point under investigation.

As used above, the 'most remote point' refers to the point from which the time of flow is the greatest; this may or may not be at the greatest linear distance, depending on the topography and other conditions. The time taken by the raindrop to make this trip is called the *Time of Concentration.* This may have two components, the entry time and the time of flow. If the drainage point being considered is at the entry to the drainage system, then the entry time is equal to the time of concentration. If, however, the design point is within the drainage system, then the time of concentration is equal to the entry time plus the time required by the raindrop to traverse the drainage system to the point under study. Figure 2.1 illustrates in graphical form some entry time values which are used for relatively flat land in the United States. Table 2.1 gives some entry time values used for typical agricultural catchment areas in rolling topography; they apply to catchment areas with about 5 m of fall per 100 m, and with lengths about twice their average widths. In urban areas in Great Britain, an entry time of 3 to 5 min is normally used. When using data such as those in Fig. 2.1 it should be kept in mind that, when the furthermost raindrop has to travel over different types of surface to reach the point of entry, the time of entry should be determined by summing the times determined for flow over the lengths of the various surfaces.

If complete weather data are available for the site in question, curves can

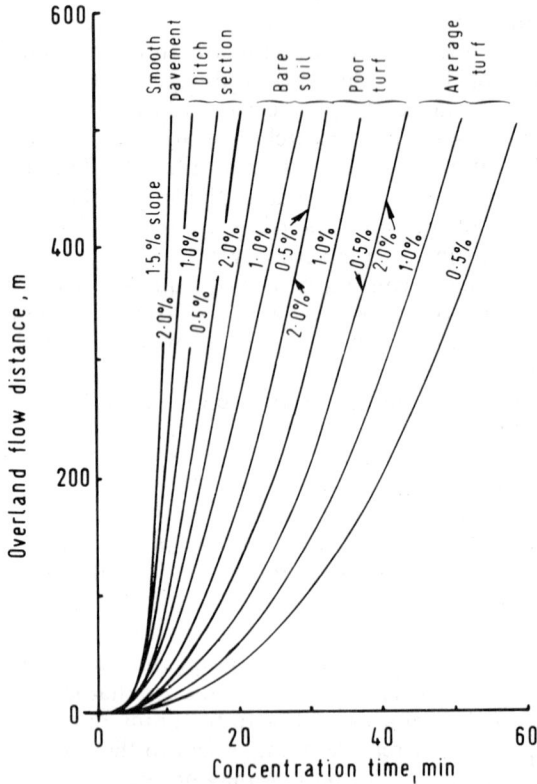

Fig. 2.1.   Concentration (entry time) values for relatively flat land

be drawn to relate intensity with duration of rainfall. Once the time of concentration has been determined, the next step involves determining the design intensity of rainfall for a storm length equal to the time of concentration. Before this can be done, however, a decision must be made as to the storm frequency which is to be catered for in the design.

There is considerable debate as to what storm frequency should be used for design purposes. In Great Britain the usual practice is to design for the most intense storm of duration equal to the time of concentration which is likely to occur once a year or once every two years. In the United States, design frequencies of 10 to 25 years are most commonly used for major rural highways. While the British approach generally results in less costly drainage structures than in the U.S., it also means that a British drainage system is more susceptible to flooding and, therefore, has a greater likelihood of incurring greater economics losses in the long run.

TABLE 2.1 *Concentration (entry time) values for typical agricultural catchment areas in rolling country*

Size of catchment area, ha	Minimum concentration time, min	Size of catchment area, ha	Minimum concentration time, min
0·4	1·4	40	17
0·8	3·0	80	23
2	3·5	120	29
4	4·0	160	35
8	4·8	240	47
12	8·0	320	60
20	12·0	400	75

Ministry of Health formulas.   While the best practice in selecting the design intensity is to prepare rainfall curves relating intensity and duration, there are occasions when this is not possible. In such instances, recourse is often made to the use of simple empirical formulas which attempt to relate intensity to duration of rainfall. These formulas usually take the form

$$I = \frac{a}{t+b}$$

where I = intensity of rainfall, mm/h,

t = duration of storm, min,

and a and b = constants.

Care should be taken in using formulas of this type, since there is no apparent reason why intensity and duration should be related by a simple mathematical law; at best they have only local application, i.e. the constants will probably change from location to location.

Such formulas commonly used in Britain are known as the Ministry of Health Formulas. These were devised as a result of an inquiry carried out by a Departmental Committee of the Ministry of Health in 1929–30. The formulas are:

$$I = \frac{762}{t + 10}, \text{ where } t \text{ is between 5 and 20 minutes}$$

$$\text{and} \quad I = \frac{1016}{t + 20}, \text{ where } t \text{ is between 20 and 100 minutes.}$$

These formulas approximate to those likely to occur once every two years for storms of short duration (<about 30 minutes) and once a year for storms of long duration (30–120 minutes). When recommending these equations, the Departmental Committee emphasized that they were based on average data, and that exceptional cases should be considered on their own merits where adequate records were available of the frequency and intensity of storms.

Modified Bilham formula. In 1935, E. G. Bilham published[1] the results of his analysis of rainfall data obtained from 10 stations in England and 2 stations in Wales for the period 1925–34 and established that, for a given amount of rainfall, the storm frequency was directly proportional to its duration. The relationship between intensity, duration and frequency derived by Bilham was subsequently modified by the Meteorological Office[2] following examination of data from a further 10 stations, and it is this modified relationship which is now officially recommended for use in the design of drainage systems in Great Britain. For convenience, the intensities predicted by the modified Bilham approach for various rainfall duration and return periods are given in Table 2.2.

As can be seen from Table 2.2, a fundamental feature of rainfall is that the intensity throughout a given storm is inversely proportional to the length of the storm, i.e. as storm duration increases, its average intensity decreases. This is perhaps as might be expected since the meteorological forces which cause a heavy rainfall are also continually causing it to move quickly from one area to another. A further point to be noted is that at a given site, and for a given storm length, the intensity can vary considerably with the least heavy rainfalls occurring most frequently.

It should be pointed out that a basic assumption underlying the validity of the data in Table 2.2 is that there is a fixed rainfall intensity associated with each storm duration and return period combination, e.g. if a 5-min storm occurs once a year, then the rainfall intensity is always 50·8 min/h irrespective of the location where the rainfall occurs. Recent work[3] has shown, however, that the modified Bilham approach tends to give rainfall intensities which are most closely related to 'average' conditions in the country. The most consistent significant discrepancies are at sea coast locations and at certain upland areas.

Run-off. Ultimately the highway engineer is concerned with the maximum rate of surface run-off from the catchment area. This again is a most difficult problem, since the maximum rate is dependent upon a great number of conditions, is never constant over a given catchment area, and varies considerably during a single storm. The principal factors governing the amount

TABLE 2.2. *Rainfall intensity (mm/h) in Great Britain as a function of rainfall duration and frequency*

Duration,	Return period, years						
min or h	1	2	5	10	20	50	100
minutes							
2	69·2	85·8	107·8	124·6	141·5	164·2	181·5
2½	65·2	80·9	101·8	117·8	134·1	156·0	172·9
3	61·6	76·5	96·5	112·0	127·8	149·1	165·5
3½	58·5	72·7	91·9	106·9	122·3	143·0	159·1
4	55·6	69·3	87·9	102·4	117·4	137·7	153·5
4½	53·1	66·2	84·3	98·4	113·1	133·0	148·5
5	50·8	63·5	81·0	94·9	109·1	128·7	143·9
5½	48·7	61·0	78·1	91·6	105·6	124·8	139·8
6	46·8	58·7	75·4	88·6	102·4	121·3	136·1
7	43·4	54·7	70·6	83·3	96·6	115·0	129·4
8	40·5	51·3	66·5	78·8	91·7	109·6	123·7
9	38·0	48·3	63·0	74·9	87·4	104·8	118·6
10	35·9	45·7	59·8	71·4	83·5	100·6	114·1
11	33·9	43·4	57·1	68·2	80·1	96·8	110·1
12	32·2	41·3	54·6	65·4	77·0	93·4	106·4
13	30·7	39·5	52·3	62·9	74·2	90·3	103·1
14	29·3	37·8	50·2	60·6	71·7	87·4	100·0
15	28·1	36·2	48·4	58·4	69·3	84·8	97·2
16	27·0	34·8	46·6	56·5	67·1	82·3	94·6
17	26·0	33·5	45·0	54·7	65·1	80·1	92·2
18	25·1	32·3	43·5	53·0	63·2	77·9	89·9
19	24·3	31·2	42·2	51·4	61·5	76·0	87·7
20	23·5	30·2	40·9	49·9	59·8	74·1	85·7
25	20·4	26·1	35·6	43·8	52·9	66·2	77·2
30	18·2	23·2	31·5	39·1	47·6	60·1	70·5
35	16·5	21·0	28·4	35·4	43·3	55·2	65·1
40	15·1	19·2	26·0	32·4	39·8	51·0	60·5
45	14·0	17·7	24·0	29·9	36·9	47·6	56·7
50	13·1	16·5	22·3	27·8	34·4	44·6	53·3
55	12·3	15·5	20·9	26·0	32·2	42·0	50·4
60	11·6	14·6	19·7	24·5	30·3	39·7	47·8
70	10·5	13·2	17·7	22·0	27·3	35·8	43·4
80	9·6	12·1	16·2	20·1	24·8	32·7	39·8
90	8·9	11·2	14·9	18·5	22·9	30·1	36·8
100	8·3	10·4	13·9	17·2	21·3	28·0	34·3
110	7·8	9·7	13·0	16·1	19·9	26·2	32·1

TABLE 2.2 (contd)

Duration, min or h	Return period, years						
	1	2	5	10	20	50	100
hours							
2	7·3	9·2	12·3	15·2	18·7	24·6	30·2
3	5·6	7·0	9·3	11·4	14·1	18·5	22·7
4	4·6	5·7	7·6	9·4	11·5	15·1	18·5
5	3·9	4·9	6·5	8·0	9·8	12·9	15·8
6	3·5	4·3	5·7	7·0	8·7	11·3	13·9
7	3·1	3·9	5·1	6·3	7·8	10·2	12·4
8	2·9	3·5	4·7	5·8	7·1	9·2	11·3
9	2·6	3·3	4·3	5·3	6·5	8·5	10·4
10	2·5	3·0	4·0	4·9	6·0	7·9	9·6
11	2·3	2·8	3·7	4·6	5·6	7·4	9·0
12	2·2	2·7	3·5	4·3	5·3	6·9	8·5
18	1·6	2·0	2·7	3·3	4·0	5·2	6·3
24	1·3	1·7	2·2	2·7	3·3	4·2	5·2

of run-off are as follows:

1. *Type and condition of the soil with respect to infiltration.* All the rainfall from a given storm does not reach the culvert or drainage ditch. For instance, if the soil is granular and/or very dry, little precipitation from the initial part of the storm will run off. On the other hand, once the ground is saturated, the greater part of the precipitation will flow easily over the ground surface and the run-off may be more than twice the initial flow. Similarly, if the ground is frozen, the run-off is high.

2. *Kind and extent of cultivation and/or vegetation.* These also change the rate of overland flow. This is indicated indirectly in Fig. 2.1 where it can be seen that the concentration time is considerably affected by the type of vegetation.

3. *Length and steepness of slopes.* The steeper and shorter the slopes, the less chance there is of the precipitation not being part of the run-off.

4. *Number, arrangement, slope and condition of the natural and man-made drainage channels in the catchment area.*

5. *Irregularity of ground surface.* Depressions in the ground surface will hold back run-off water until they are filled.

6. *Size and shape of catchment area.* If the catchment area is square in shape, all the run-off may reach the culvert within a relatively short period of time. If the shape is long and narrow, the run-off from the zone near the culvert may have drained through it before the water from the more distant parts arrives.

7. *Temperature of air and water.* If the air temperature is warm, quite a significant amount of precipitated water may return to the atmosphere before it reaches the culvert or ditch. If the temperature is relatively low, the water is more viscous and the rate of overland flow will be slowed.

8. *Changes in land use.* It must be kept in mind that changes in land use during the life of the drainage structure may increase the rate of run-off by as much as 100 per cent.

In view of the previous considerations, it can be seen why it is that there is great difficulty in arriving at any one design figure which will take all factors into proper account. The approach therefore usually taken is for the engineer to utilize a compromise factor which will adequately meet the most severe design requirements but still be within the economic scope of the scheme.

TABLE 2.3. *Suggested values of ultimate impermeability of various types of surfaces*

Type of drainage area	Impermeability factor
Concrete or bituminous surfacings	0·8 to 0·9
Gravel or macadam surfacings	0·4 to 0·7
*Bare impervious soils	0·4 to 0·7
*Impervious soils, with turf	0·3 to 0·6
*Bare, slightly pervious soils	0·2 to 0·4
*Slightly pervious soils, with turf	0·1 to 0·3
*Pervious soils	0·1 to 0·2
Wooded areas	0·1 to 0·2

* These values are applicable to relatively level ground. When the slopes are greater than 2%, the impermeability factor should be increased by 0·2 (to a maximum of 1·0) for every 2% increase in slope

Many authorities have examined the run-off problem and most have concentrated on obtaining values of relative imperviousness, i.e. ratio of run-off to rainfall, since this is the most important consideration. In these studies, most emphasis has been placed on run-off determinations for urban areas, and relatively few thorough investigations have been carried out for rural conditions. An examination of these data leads to the values given in Table 2.3 which relate run-off to rainfall; these data are only applicable to rural conditions. It is not possible to give such detailed recommendations for urban conditions, since many of the studies carried out have been in countries other than Britain, where the housing conditions are very different. Perhaps the most common practice in Britain with respect to built-up areas is to assume that all garden and unpaved land is completely permeable and that all paved and roofed surfaces have an ultimate permeability of 0·8.[4]

Lloyd-Davies formula. Many formulas have been developed which attempt to measure the run-off from a storm. The one which has gained most favour with engineers in Britain is that developed by D. E. Lloyd-Davies, first published in 1906.[5] Lloyd-Davies related run-off to rainfall by means of the following equation:

$$Q = \left[0.167 \times \frac{60}{t} \times r \right] \times A \times p$$

where　Q = discharge, m^3/min,

　　　　t　= time of concentration, min,

　　　　r　= total rainfall during time of concentration, mm,

　　　　A = catchment area, ha,

　　　　p　= proportion of total rainfall running off after allowing for soak-
　　　　　age and evaporation,

and 0.167 = run-off in m^3 per min per ha for a rainfall of 1 mm per hour on
　　　　a completely impervious surface.

Now if I is the rainfall intensity, in mm/h, for a storm of duration t min, then

$$r = \frac{t}{60} I$$

If this value for r is inserted in the above equation, then

$$Q = 0.167 A I p$$

where Q, A, I and p are as defined before.

　　　The Lloyd-Davies formula for run-off is essentially the same as the one
known as the Rational formula, which is used in the United States.

　　　It is as well at this stage to emphasize some of the assumptions underlying the use of both the Lloyd-Davies and Rational formulas, so that the
reader can realize why it is that in practice the calculated results are very
often qualified by engineering judgement.

　　　1. The rainfall intensity I is selected for a storm of duration equal to the
time of concentration. The entry time component of this time is usually
selected by measurements based on the assumption that the raindrop moves
perpendicularly to the ground contour-lines. If too low a time of concentration is chosen, this will result in the selection of too high a rainfall intensity
and the over-estimation of the run-off.

　　　2. The rainfall intensity value used is assumed to be the same throughout
the catchment area during the storm. If the area is very large and/or the
topography is very variable, this assumption may well not be valid.

　　　3. The impermeability factors which are used in the calculations, such as
those given in Table 2.3, only take into direct account the type of surface and
the slope, and assume that the other influencing factors are covered by these.

　　　4. Finally, both the Lloyd-Davies and Rational formulas assume that
the maximum run-off occurs with the same frequency as the design storm. In
fact, this assumption is only valid when the ground has been well wetted
beforehand by a preliminary rainfall, or where the precipitation falls on an
impervious surface. Thus, for instance, if in a rural area a design storm with a
frequency of 25 years is chosen and the ground is well wetted before such a
storm but once in every three occasions, then it is likely that the maximum
run-off may occur only once in 75 years.

　　　From these and other considerations it can be seen that the design
precision of the Lloyd-Davies and Rational methods is somewhat limited. In
general they can tend to over-estimate the maximum run-off, and the extent to

which this occurs increases with the size of the catchment area. As a result it is probable that use of the formulas should be confined to urban areas and relatively small catchment areas in rural locations. Suggested upper limits on the size of the catchment area have ranged from a low of about 5 to a high of about $50\,km^2$.

EXAMPLE. A culvert is to be designed to carry the run-off from a 40 ha catchment area across a new motorway which is to be constructed. Some 24 ha of the catchment area are on rolling glacial till topography and are used for farming purposes. The remaining 16 ha are on an old flood plain area and the slope is about 0·5 per cent. Determine the amount of water which the culvert must be capable of handling.

Solution. The first step is to determine the time of concentration, which in this instance is equal to the time of entry. Ideally this should be obtained by visiting the site during some previous heavy rainfalls and measuring the time for markers such as paper confetti to reach the entry point from the furthermost location. Here this has not been done so the data in Fig. 2.1 and Table 2.1 will be used to estimate the concentration time.

From Table 2.1 it is seen that the minimum time of concentration for 24 ha of agricultural land is about 13 minutes. Examination of the survey maps shows that, while the shape of the area meets criteria on which the table is based, the average slope is slightly less. An overland time of 15 minutes will therefore be used for this area.

For the remaining 16 ha, examination of the survey maps shows that the length of overland flow is approximately 183 m. Entering Fig. 2.1 and reading the 0·5 per cent curves for average turf conditions, the time of overland flow for this section is found to be 38 minutes.

Combining the two overland flow values, the time of concentration is therefore 53 minutes. Entering Table 2.2 with a length of storm equal to this value, and—since this is for a motorway—selecting a 10 year storm frequency, the design intensity of rainfall is found to be 33 mm/h.

The impermeability factor has next to be determined. After visiting the site, and then examining the data in Table 2.3, a value of 0·3 is chosen for the flat land and 0·4 for the farmland. The average factor for the total area is therefore

$$\frac{0·3 \times 16 + 0·4 \times 24}{40} = 0·36$$

Substituting these values in the Lloyd-Davies formula,

$$Q = 0·167 A I p$$
$$= 0·167 \times 40 \times 26·7 \times 0·36$$
$$= 64·3\,m^3/min.$$

Hydraulic design

Once the quantity of run-off has been determined for a particular point of concentration, the next step is the hydraulic design of the ditch, culvert (or pipe) required to carry it. It is convenient to discuss this by treating the ditch and culvert designs separately.

Ditches. Two types of ditch are of particular interest to the highway engineer. These are the side ditches which run approximately parallel to the roadway and the open-channel drainage ditches, in which road culverts lie, that are intercepted by the roadway.

Side ditches. These are normally placed alongside roadways in order to intercept surface water running off the carriageway and shoulders. In cut sections they also serve to prevent water running down the back slopes and invading the roadway. Side ditches are usually V-shaped or trapezoidal in cross-section. On low-cost roads the V-ditch is very often favoured because it can be very economically formed and maintained with a blade grader. On high-type roads, such as motorways and dual carriageways, the trapezoidal section is to be preferred, not only because of its greater drainage capacity, but also because the need for erosion control makes blading of the ditch sidewalls objectionable. A further major consideration influencing the selection of trapezoidal sections for high-type roads is that from a safety aspect they form less of a hazard to vehicles which intrude upon them.

Normally small roadside ditches are not hydraulically designed, due to lack of economical justification. Instead the ditch side walls are simply cut to the natural angle of repose of the soil and to a depth which is usually between 0·3 and 0·6 m; in this latter respect great care should always be taken to ensure that at the very least the depth is such that sustained flow in the bottom of the ditch never rises above the bottom of the pavement. On roads such as motorways, however, side ditches should always be checked for their hydraulic capabilities to ensure that they are able to handle the expected flows without danger either to traffic or the road structure itself. This is especially important if the ditches carry water from adjacent back-slopes as well as from the roadway. As discussed in the chapter on Geometric Design (Vol. 1, Chapter 6), vehicle-safety considerations usually govern the ditch side-slopes chosen on these roads, preference being given to the use of relatively flat slopes, especially on the side closest to the carriageway.

Motorist safety considerations do not normally enter into the design of ditches which are non-parallel to a roadway. In these instances, there is greater versatility in the choice of cross-section and the features which govern the design are primarily hydraulic ones.

Open channel design. By definition, an open channel is a conduit which is open to the atmosphere and, as a result, flow takes place not because of a pressure difference between two points—as in a horizontal closed pipe—but because of a difference in elevation between two points. The basic hydraulic formulas for determining the flow of water in open channels are written as follows:

$$V = C\sqrt{RS} \text{ and } Q = AV$$

where Q = discharge, m³/sec,

         C = a coefficient of roughness whose value depends upon the nature of the surface over which the water is flowing,

         R = average hydraulic radius, m, = A/P,

         S = slope of hydraulic gradient, m/m,

         A = cross-sectional area of flow, m²,

         P = wetted perimeter, m,

and V = average velocity of water, m/sec.

This formula, known as the Chézy formula after the French engineer and mathematician Antoine Léonard de Chézy who lived from 1718 to 1798, is the basis of practically all the open-channel hydraulic formulae which are in use today. The great array and complexity of these formulae can often be very confusing to the engineer who is not a hydraulic specialist. Fortunately or unfortunately—depending on how one looks at it—there is often relatively little agreement between the results obtained by very many of these formulae. Hence, under the many-variable conditions prevalent in the field, the engineer is wise to settle on some basic formulas for use in routine work and use them sufficiently often to become well acquainted with the effects of the various coefficient values used in their equations.

Perhaps the most widely investigated formula used in highway drainage design is the Manning formula.[6] This formula gives the value for C in the Chézy formula as:

$$C = \frac{R^{1/6}}{n}$$

where n = Manning's roughness coefficient,

and R = hydraulic radius, m.

The simplified Manning formula is

$$V = \frac{R^{2/3}S^{1/2}}{n} \text{ and } Q = AV = \frac{AR^{2/3}S^{1/2}}{n}$$

Values for n for various channel surfaces which are recommended by the U.S. Corps of Engineers are given in Table 2.4. In the reference from which this table is abstracted, the bare soil values were given for soils classified by the Unified classification system; here, for convenience, the value given for each soil is its nearest equivalent (and very similar) Extended Casagrande classification. Also shown in Table 2.4 are the maximum permissible velocity values for various types of ditch lining. Velocity values in excess of these will cause erosion in the ditches, which will not only increase the maintenance costs but also, in the case of side ditches, may weaken the road structurally and be a possible cause of hazard to motorists.

Before illustrating the manner in which Manning's formula is used in ditch design problems, there is one further item which must be discussed; this has to do with the term *critical velocity* and its relationship to the *critical discharge* and *critical slope* of an open channel. In a uniform open channel the depth of flow at any point is dependent on the shape of the cross-section, the roughness of the channel, the slope of the channel and the discharge. This depth of flow, called the normal depth, is relatively great when the channel slope is gentle and the water is in tranquil or subcritical flow. If the slope is increased, a normal depth of flow — called the critical depth — will be reached; at steeper slopes the flow becomes supercritical. At this stage, for instance, a discontinuity of flow in the form of a hydraulic jump may be caused downstream which will result in the generation of intense local turbulence at and just downstream of the jump; this may be very erosive and cause severe damage to the walls and bed of the ditch. The gradient at which the change occurs is called the critical slope, the velocity at this slope is called the critical velocity, and the discharge is termed the critical discharge.

Technically, critical flow at a point in an open channel can be said to occur when the discharge is a maximum for a given energy head or, alternatively, when for a constant discharge the energy head is a minimum. The energy head at any point in the channel is the depth of water plus the velocity head, $V^2/2g$, where V is the velocity of the water at that point and g is the acceleration due to gravity. Increasing the slope of the channel beyond the critical slope does not increase the discharge but simply makes the water flow faster at a depth less than the critical depth. In any channel in which (in theory) the water is flowing at the critical depth, the velocity head is equal to one-half the mean depth of flow where the mean depth is defined as the water cross-section area divided by the width of the free-water surface. In other words, super-critical flow does not occur when the velocity head is less than one half of the mean depth of flow.

Once the discharge volume has been estimated by means of hydrologic study, the next stage is the hydraulic design of the ditch which has to carry this volume. A suitable design procedure which utilizes the Manning formula may be outlined as follows:

Step 1. For the soil in which the ditch will be running select the maximum permissible velocity to avoid erosion, the Manning roughness coefficient n and the side slopes of the channel. The first two of these values can be obtained from Table 2.4. The side slopes are normally controlled by the angle of natural repose of the soil; a commonly used value which suits most soils is a slope of 1 vertical to 2 horizontal. If it is a side ditch which is being evaluated, flatter side slopes should be used because of traffic safety considerations.

Step 2. Using the Manning formula, calculate the maximum permissible hydraulic radius.

Step 3. Using the equation of continuity, calculate the minimum permissible cross-section area required by the given discharge and permissible velocity.

Step 4. Calculate the wetted perimeter for this area.

Step 5. Using the expressions obtained in steps 3 and 4 solve simultaneously for the bottom width of the ditch and the depth of flow.

Step 6. Check that the depth of flow is greater than the critical depth.

Step 7a. If the depth of flow is greater than the critical depth, add a suitable freeboard and modify the section for practicality. In the case of a

TABLE 2.4. *Manning's ' n ' and maximum permissible velocity of flow in open channels*[7]

Ditch lining	Manning's ' n '	$V_{max.}$ m/sec
1. Natural earth		
A. Without vegetation		
(i) Rock		
(a) Smooth and uniform	0·035–0·040	6·1
(b) Jagged and irregular	0·040–0·045	4·5–5·5
(ii) Soils:		
GW	0·022–0·024	1·8–2·1
GP	0·023–0·026	2·1–2·4
GC	0·020–0·025	0·6–1·5
GF	0·024–0·026	1·5–2·1
SW	0·020–0·024	0·3–0·6
SP	0·022–0·024	0·3–0·6
SC	0·020–0·023	0·6–0·9
SF	0·023–0·025	0·9–1·2
CL and CI	0·022–0·024	0·6–0·9
MI and ML	0·023–0·024	0·9–1·2
OL and OI	0·022–0·024	0·6–0·9
CH	0·022–0·023	0·6–0·9
MH	0·023–0·024	0·9–1·5
OH	0·022–0·024	0·6–0·9
Pt	0·022–0·025	0·6–0·9
B. With vegetation		
(i) Average turf		
(a) Erosion resistant soil	0·050–0·070	1·2–1·5
(b) Easily eroded soil	0·030–0·050	0·9–1·2
(ii) Dense turf		
(a) Erosion resistant soil	0·070–0·090	1·8–2·4
(b) Easily eroded soil	0·040–0·050	1·5–1·8
(c) Clean bottom with bushes on sides	0·050–0·080	1·2–1·5
(d) Channel with tree stumps		
d1. No sprouts	0·040–0·050	1·5–2·1
d2. With sprouts	0·060–0·080	1·8–2·4
(e) Dense weeds	0·080–0·120	1·5–1·8
(f) Dense brush (flood plains)	0·100–0·140	1·2–1·5
(g) Dense willows (flood plains)	0·150–0·200	2·4–2·7

TABLE 2.4 (contd)

Ditch lining	Manning's 'n'		V_{max}. m/sec
2. Paved			
A. Concrete, w/all surfaces	Good	Poor	
(i) Trowel finish	0·012–0·014		6·1
(ii) Float finish	0·013–0·015		6·1
(iii) Formed, no finish	0·014–0·016		6·1
B. Concrete bottom, float finished, w/sides of:			
(i) Dressed stone in mortar	0·015–0·017		5·5–6·1
(ii) Random stone in mortar	0·017–0·020		5·2–5·8
(iii) Dressed stone or smooth concrete rubble (Rip-rap)	0·020–0·025		4·6
(iv) Rubble or random stone (Rip-rap)	0·025–0·030		4·6
C. Gravel bottom, sides of			
(i) Form	0·017–0·020		3·0
(ii) Random stone in mortar	0·020–0·023		2·4–3·0
(iii) Random stone or rubble (Rip-rap)	0·023–0·033		2·4–3·0
D. Brick	0·014–0·017		3·0
E. Asphalt	0·013–0·016		5·5–6·1

side ditch, the freeboard should at least equal the height above the bottom of the pavement. For other channels, an arbitrary value of about 0·5 m is often chosen for ditches other than side ditches.

Step 7b. If the depth of flow is less than the critical depth, then surface discontinuities in the form of a hydraulic jump, slug or roll-wave flow, or air entrainment are likely to occur downstream provided that the channel characteristics do not change. In such instances consideration should automatically be given either to reducing the slope of the channel or to determining how it might be protected by the installation of linings. If a large section of the channel is to be lined, it will be advisable to steepen the side slopes to angles of 60 degrees and obtain a new design on the basis of the most 'economical section', i.e. a design based on the channel cross-section which gives the greatest flow for the minimum of excavation. This design treatment can be found in detail in any good textbook on hydraulics and so is not further discussed here.

EXAMPLE 1.   Determine the ditch cross-section necessary to carry a design flow of 3 m³/sec. The ditch will be cut in material which is classified as a CL-soil. The natural slope of the ground is approximately 0·04 per cent. Assume a trapezoidal channel with 2:1 side slopes.

Solution.   For this ditch the following coefficients apply:

$$S = 0·04 \text{ m/m}, \ V_{max} = 0·66 \text{ m/sec}, \ n = 0·022$$

Thus $0·66 = \dfrac{1}{0·022} R^{2/3}(0·0004)^{1/2}$

Therefore $\qquad R \leq 0.6275\,\text{m}$

But $\qquad A \geq Q/V = 3/0.66 = 4.5\,\text{m}^2$

and $\qquad P = A/R \geq 4.5/0.6275 = 7.17\,\text{m}$

However $\qquad A = h(b+2h) = 4.5$

and $\qquad P = b+2h(1+2^2)^{1/2} = b+4.48\,h = 7.17\,\text{m}$

where $\qquad h$ = depth of flow, m

and $\qquad b$ = bottom width, m

Solving these two simultaneous equations gives $b = 3.09\,\text{m}$, and $h = 0.915\,\text{m}$.
  The critical depth $= V^2/g = (0.66)^2/9.81 = 0.045\,\text{m}$. Obviously in this case the flow is supercritical (i.e. depth $> 0.045\,\text{m}$), and so no secondary design (e.g. lining of the channel) is necessary. Adding $0.5\,\text{m}$ for freeboard, the desired depth of channel is $1.4\,\text{m}$.

EXAMPLE 2. An existing highway in cutting in glacial till topography is to have side ditches to drain both the roadway and back slopes. The maximum discharge in each side channel is expected to be $1\,\text{m}^3/\text{min}$. Each ditch is to have a trapezoidal shape, with side slopes of $3:1$ and bottom width of $1\,\text{m}$. The bottom of the ditch will be $10\,\text{cm}$ below the pavement and on a gradient of 4 per cent. Determine whether the side channel is capable of handling this discharge without the water eroding the sides or entering the pavement.

Solution. The soil is classified as GF and the ditch will be grassed after construction. Therefore use a value of $n = 0.070$. Then

$$V = \frac{R^{2/3}S^{1/2}}{n} = \frac{1}{n}\left[\frac{(b+sh)h}{b+2h(1+s^2)^{1/2}}\right]^{2/3}S^{1/2}$$

$$= \frac{1}{0.070}\left[\frac{(1+3\times0.1)0.1}{1+2\times0.1(1+3^2)^{1/2}}\right]^{2/3}(0.04)^{1/2}$$

$$= 0.529\,\text{m/sec}.$$

This is well below the maximum permissible velocity so erosion should not be a problem here.

$$Q = AV = (b+sh)hV$$

$$= (1+3\times0.1)0.1\times0.529\times60$$

$$= 4.12\,\text{m}^3/\text{min}.$$

Therefore the ditch is well capable of discharging the design volume without water penetrating the pavement and gaining ready access to the subgrade.

Culverts. A culvert can be defined as a conduit which conveys water through an embankment. Since a bridge performs the same function, it might be said that a culvert is also a bridge, although it is not normally considered as such.

Perhaps the easiest distinction between the two is that a bridge surface forms part of the carriageway whereas the top of a culvert is always beneath the carriageway. One further difference is that culverts may very often be designed to flow full, whereas bridges are normally designed to pass floating debris and, possibly, boats.

As with road pavements, there are two types of culvert; flexible culverts and rigid culverts. Flexible culverts are either thin-walled steel pipes or galvanized corrugated metal pipes; they rely only partly on the strength of the pipe walls to resist the external loads, and instead they are designed to deflect under the loads. When deflection takes place, the horizontal diameter of the culvert increases and compresses the soil at the sides and in this way employs the passive resistance of the soil to help support the applied loads. When failure of a flexible culvert does occur it is primarily because of excessive deflection. In contrast to the flexible pipes, rigid culverts are composed of reinforced concrete, cast iron or vitrified clay and their load-carrying ability is primarily a function of the stiffness of the walls of the culverts. When failure does occur it is usually due to rupture of the walls of the culvert.

Culverts are utilized in pipe, arch or box form, although pipe culverts are by far the most common. Pipe and arch culverts are formed from both rigid and flexible materials, whereas box culverts —they have rectangular cross-sections —are always made rigid. Most of the larger culverts are provided with concrete headwalls, irrespective of the material from which the culverts themselves are made. The function of a headwall is to aid in the smooth movement of water from the ditch to the culvert. In so doing it not only increases the design capacity of the culvert, but, by reducing the vortices caused by secondary flow at the entrance, it also protects the surrounding soil from excessive erosion. Endwalls and aprons are very often provided at the culvert exits so that the emerging water is prevented from scouring the ditch into which it empties.

Location.   Proper location is a prime prerequisite to the efficient and economic operation of a culvert within a drainage system. Improper location can lead to the softening and, possibly, failure of a roadway, as well as undesirable ponding and damage to adjacent property on the upstream side of the embankment. Great care should always be taken therefore to ensure that a culvert is properly located, due consideration being given to both its horizontal and vertical alignment.

A culvert is simply an enclosed channel which serves to carry an open stream under a highway. If it is to be an efficient substitute for the open-ditch section it must be placed so that the water has both a direct entrance and a direct exit. Thus, as is illustrated in Fig. 2.2(a), the culvert is usually placed in the natural ditch bed so that its alignment conforms closely to that of the original situation. Often this means that the culvert must be located at a skew angle rather than at a right angle to the centre line of the road. If the stream has a tendency to meander, every effort should be made to get it across the roadway as quickly as possible. As illustrated in Figs. 2.2(b), (c)

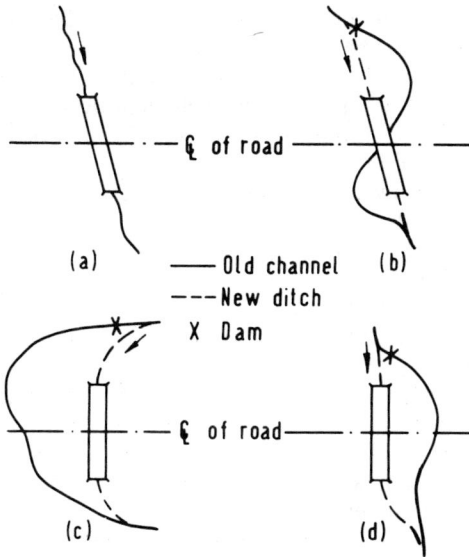

Fig. 2.2.   Various methods of locating culverts in roadways

and (d), this will usually involve cutting a new ditch into which the water can be diverted prior to the culvert. Once the water has passed through the culvert it should normally be returned to its original channel as quickly as possible.

The slope of a culvert should normally conform as closely as possible to the natural grade of the stream; this is usually the one which produces least silting or scouring in conjunction with an economic length of culvert. The silt-carrying capacity of a stream varies as the square of its velocity and hence considerable sedimentation can occur if the velocity of the water is reduced by changing the slope. If sedimentation is expected, as for instance when draining freshly-graded bare soil areas, it may be advisable to set the invert of the culvert several centimetres or more above the stream bed but at the same slope, so that the carried material is prevented from being deposited *within* the conduit. Indeed if the embankment is high and ponding of the water is allowable, the culvert can very often be placed at a level above the stream bed, so that its length is reduced significantly and its replacement or relocation, should this ever become necessary, is simplified. It is generally considered that culverts should be placed at a minimum slope of about 0·5 per cent if sedimentation is to be avoided.

If the slope of the culvert is greater than the natural slope of the water-course, then the increased water velocity may cause scouring of the sides and base of the ditch at the outlet to the culvert. In this latter respect special care has to be taken when an embankment is constructed on the side of a steeply sloping hill to ensure that erosion of the downstream side slope does not take place as water emerges from the culvert.

Hydraulic design. If a culvert were the same size and shape as the drainage ditch which it services, then its hydraulic design would be relatively easy. Unfortunately, however, this is not so and a culvert normally has the effect of a constriction to the stream flow. The efficient design of this constriction may be very difficult depending on the governing factors. These factors and some of their effects are as follows:

1. *The depth of the headwater pond.* For a given design flood, the depth of the pond formed at the entrance is a function of the size and shape of the culvert. Conversely, the manner in which flow takes place in the culvert is affected by the head of water available at the inlet. Laboratory experiments indicate that the entrance to a culvert can be considered as hydraulically submerged when the depth above the invert of the inlet is greater than about 1·2 times its diameter.

2. *The depth of the tailwater pond.* If both the inlet and outlet are submerged during the design flood, the culvert will usually run full, although there are certain instances when it will not do so. On the other hand, if the outlet is not submerged, it is still possible for the culvert to run full if the inlet is submerged and the culvert is long.

3. *The type of entrance.* If the culvert has a poorly designed entrance, then considerable turbulence will occur at the inlet and energy will be dissipated which would otherwise be available for moving the water through the culvert. Studies have shown that in this respect culverts with flared or wing-type entrances are much more efficient than ones with straight square-edge headwalls or where the culverts project into the headwater pond. For instance, culverts with submerged inlets having flared entrances may flow full whereas those with poor inlets will not. Good entrance design is a most important feature of the design of short culverts and long culverts on steep slopes, but is of less importance in the design of long culverts on gentle slopes.

4. *The roughness of the interior walls.* Other things being equal, a rough-textured culvert which runs full will discharge less water than a smooth one also running full, since much of the energy head is used up in overcoming the resistance to flow. The longer the culvert, the more important is the roughness factor.

5. *The length of the culvert.* As discussed above, the length of the culvert dictates whether or not the type of entrance and the roughness of the interior walls are major or minor features of the hydraulic design.

6. *The slope of the culvert.* Other considerations being equal, it is the culvert slope which dictates whether or not the culvert operates as a free surface channel. If the slope is too gentle, the culvert will tend to flow full and the depth of the headwater pond will be increased.

As can be gathered from the few examples given above, the manner in which the hydraulic flow occurs in a culvert is dependent on a number of inter-related factors. Some of the more important of these inter-relationships are illustrated in Fig. 2.3.

Figure 2.3(a) illustrates the case, commonly experienced in flat or slightly rolling country, where both the inlet and outlet of the culvert are submerged.

The culvert then normally operates as a pipe. It will certainly do so if the level of the tailwater pond plus the head loss due to friction in the culvert is greater than the elevation of the crown of the culvert at the inlet. It will not operate as a pipe if the culvert has a poorly designed inlet and is on a steep slope. In this instance the inlet acts as a sluice and the culvert will not run full except beyond the hydraulic jump which will form near the outlet; in time, however, the air space created between the entrance to the culvert and the jump will gradually disappear due to the air-entraining action of the hydraulic jump. If the velocity of the water is sufficiently high the hydraulic jump may be formed outside the outlet with the result that it is not submerged.

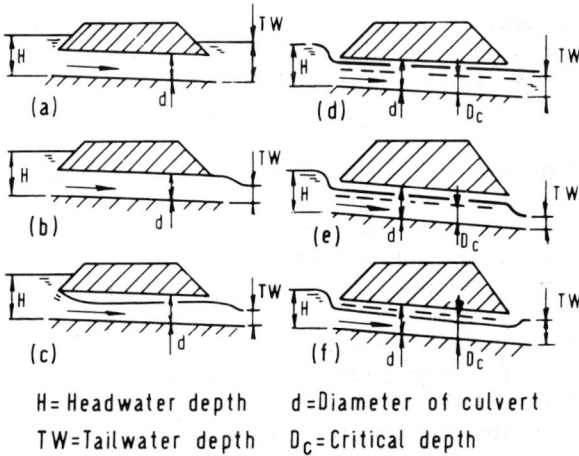

H = Headwater depth   d = Diameter of culvert
TW = Tailwater depth D_c = Critical depth

Fig. 2.3.   Some conditions under which culverts operate

Figures 2.3(b) and (c) illustrate the situations which are likely to occur in hilly topography where the slope of the downstream ditch is sufficiently steep to carry the water away from the outlet and prevent the formation of a tailwater pond. Normally a culvert with a well-shaped inlet will flow full in such a location provided that the inlet is hydraulically submerged. If, however, the culvert has a sharp-edged projecting or square-edged inlet, the flow entering the culvert may be contracted to a depth less than the height of the culvert. The culvert is said to be hydraulically 'short' if it is sufficiently steep and/or short for the water to be carried through so quickly that part-full flow will occur through its entire length; this is illustrated in Fig. 2.3(c). If the inlet is poorly designed but the slope of the culvert is very gentle while its length is considerable, then the velocity of flow may be sufficiently low to allow the depth of flow to increase so that the free water surface reaches the top of the culvert and full flow occurs; in this case the culvert is said to be a 'long culvert'.

Figures 2.3(d) and (e) illustrate conditions where both the inlet and the outlet are not hydraulically submerged. The condition shown in Fig. 2.3(d) can be found when the depth of the tailwater pond is between the critical depth and the crown of the culvert at the outlet. Since the slope is less than the critical slope and the flow is tranquil, this condition is only likely to occur at locations where the culvert empties into a deep narrow outlet ditch on a flat slope. When the culvert discharges into a channel which has a relatively flat slope and a wide flat floodplain, the flow condition shown in Fig. 2.3(e) may be experienced. Here, although the depth of the tailwater pond is less than the critical depth, the flow is still tranquil.

Figure 2.3(f) illustrates a situation, commonly found in hilly topography, where neither the inlet nor the outlet are submerged and where the slope of the culvert is greater than the critical slope.

Once the importance of the basic features influencing flow through a culvert is understood, the actual hydraulic design becomes relatively straight-forward. It consists essentially of the following steps:

1. Decide upon the allowable headwater depth above the inlet.
2. Determine the slope of the culvert from an examination of the ditch profile.
3. Determine the length of the culvert by examining the embankment cross-section.
4. Determine the elevations of the inverts of the inlet and outlet of the culvert. This can also be obtained from the embankment cross-section.
5. Select a size and type of culvert which it is considered will be suitable at the particular location.
6. Identify the hydraulic method of operation.
7. By means of an appropriate hydraulic equation, calculate the head-water depth for the design flood and culvert discharge.
8. If the calculated depth is greater or significantly lower than the allow-able depth, a new culvert is selected and the procedure repeated as before.

It is not possible to describe here all the equations and nomographs which have been developed for use in step 7. For these the reader is referred to an excellent and detailed publication by the U.S. Bureau of Public Roads.[8] For practical purposes, however, it is very often sufficient to obtain an approximate solution by using the charts shown in Fig. 2.4. These charts are applicable to problems involving culverts flowing partly full and having square-edged entrances. When the culverts have rounded entrances under average conditions, the value of H/d may be roughly estimated by the follow-ing expressions, in which H/d refers to the ratio of headwater to barrel height for a culvert with a square-edged entrance:

Type	$H/d < 1.0$	$1.0 < H/d < 1.5$	$H/d > 1.5$
Circular	$0.87\ H/d$	$0.87\ H/d$	$1.09 + 0.10\ H/d$
Box	$1.00\ H/d$	$0.36 + 0.64\ H/d$	$0.62 + 0.46\ H/d$

These data give headwater heads which are as high as is likely to occur under adverse conditions; as such they are to a certain extent conservative.

Gulleys on urban roads. A road gulley is a waterway designed to drain water

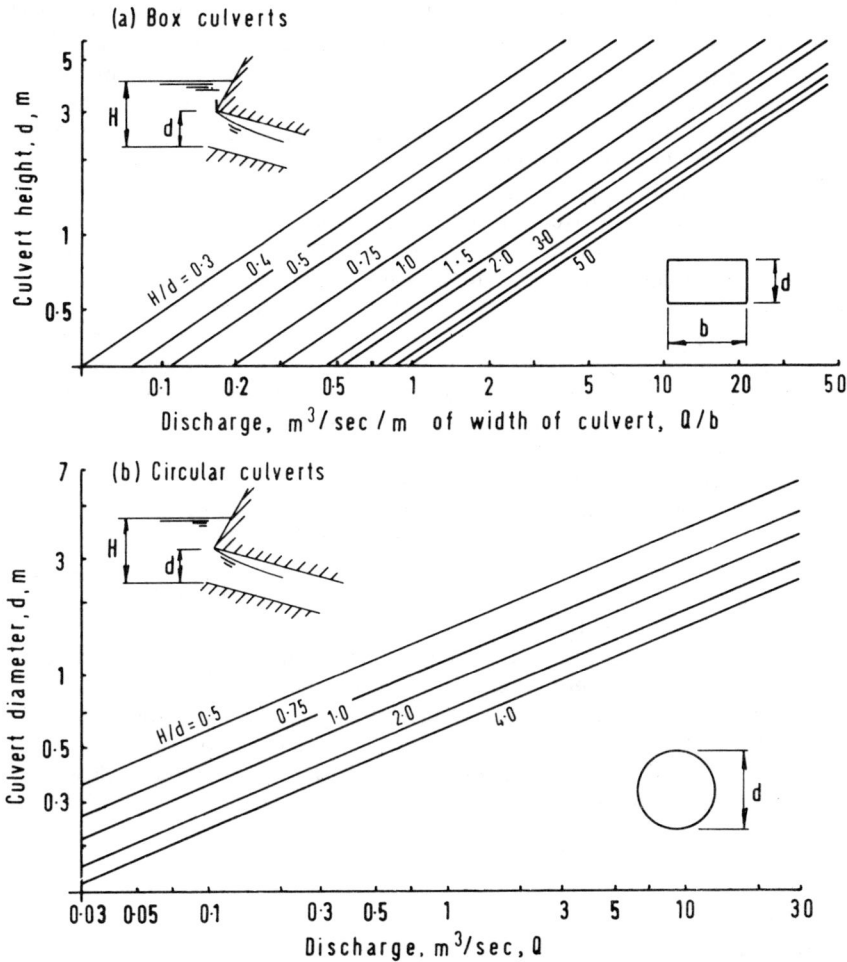

Fig. 2.4. Charts for estimating headwater depth on culverts with square-edged entrances, flowing partly full (After reference 9)

from the carriageway surface. It comprises a gulley pot (which acts as a trap for silt and small debris) connected by a pipe — usually about 150 mm in diameter — to an underground 'storm' drain or channel, and a steel frame fitted with a cover or grating which bridges the gulley pot. In urban areas, the gulleys are normally fed by surface channels, formed by the kerb and gutter, which are laid most commonly to the longitudinal slope of the road.

The gratings/covers used on gulleys in Britain[10] are as follows:

A. Heavy duty gratings able to carry heavy wheel loads of up to 11·4 Mg. They take the form of double triangular gratings of either cast iron or steel with waterway areas of 916 cm³ or 1393 cm³, respectively.

B. Medium duty gratings and covers for carriageways carrying normal commercial vehicles of up to 14·2 Mg. There are three types: 1. A straight bar grating (Type E-12) with a minimum waterway area of either 668 cm^2 or 780 cm^2. (Either grating can be used on carriageways with gradients of less than 1 in 50); 2. A curved bar grating (Type E-13) with a minimum waterway area of either 471 cm^2 or 619 cm^2: used with gradients of 1 in 50 or steeper; 3. A kerb inlet (Type E-14) normally made to fit in line with the road kerb so that water enters from the side.

In relation to the above, it may be noted that the kerb inlet is normally the most inefficient of all the gulley types, particularly on steep longitudinal slopes. The reason for this is that the small hydraulic head acting at right angles to the initial direction of flow is unable to move much water over the side weir, and so much of the flow by-passes the gulley. The action of this gulley can, however, be much improved by shaping the approach channel so that the inlet is more directly in the path of the water flow. The other types of gulley can collect up to 95 per cent of the flowing water provided that the width of flow does not exceed 1·5 times the width of the grating.[11]

There is no simple overall recommendation about where gulleys should be located. Generally, however, they are placed at low points or 'sags' on the roadway where water will naturally accumulate. They should be also inserted just prior to road junctions so as to intercept the water flowing in the kerb-gutter channels before it reaches the pedestrian crossings.

A rough guide as to the locations of intermediate drains can be obtained from the recommendation[12] that the impermeable area per gulley on housing estate roads should not exceed 200 m^2. A more rational approach to the design of gulley spacings has been put forward by the Transport and Road Research Laboratory;[11] this makes use of tables incorporating data *re* the efficiencies of different inlets combined with data on the water flow over various widths of the road.

SUBSURFACE MOISTURE CONTROL

Methods of control

As is discussed later in the text, all scientific pavement design procedures base their designs on measurements of the subgrade soil. The thickness design suggested by any procedure can only be validly utilized when the soil conditions assumed at the time of testing are similar to those that will actually pertain in the field. For instance, if the moisture content used in the test is exceeded in the field, then the design conditions no longer apply and the pavement may well fail. Similarly if, as is the practice in the United States, the test specimens are saturated prior to testing, and if saturation is not likely to occur under natural conditions, then the pavement may be over-designed.

It should be clear therefore that the subgrade conditions on which the thickness design is based must remain substantially unchanged throughout the design life of the pavement. From a drainage aspect this means that substantial moisture control installations may have to be incorporated into the roadway design. This is not always easy to do; in fact it can be a major expense item for, as is illustrated in Fig. 2.5, there are very many ways in which moisture can enter and leave the subgrade of a highway.

Seepage from high ground. This condition is likely to occur in hilly topography and in highway cuttings where a layer of permeable soil overlies an impermeable stratum. Moisture can find easy access to the subgrade via the permeable layer, thus causing it to soften. In fact, it is not unusual to have springs suddenly appear in the subgrade when this occurs in a highway cutting.

The best solution to this problem is to intercept the seepage water on the uphill side of the roadway. If the seepage zone is close to the surface, interception can sometimes be carried out by means of open ditches. Normally, however, surface drainage cannot be utilized since the seepage water must be kept at least 1–1·25 m below the formation level; this may require a deep drainage ditch which, if it is close to the edge of the pavement, will have to be relatively wide and have gentle slopes which can be traversed by vehicles leaving the carriageway. In hilly country or in roadway cuttings this can result in construction costs that may often not be justified by the importance of the highway.

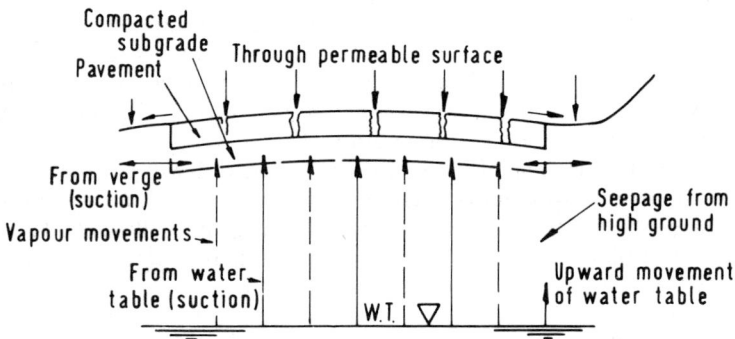

Fig. 2.5.   Ways in which water can enter and leave road subgrades

The more usual way of tackling the problem of seepage is to utilize a subsurface drain to intercept the water. If the seepage zone is not very deep, it is usual to install the subdrain to the depth of the underlying impervious stratum and thus completely cut off the entering water. This solution is illustrated in Fig. 2.6(a). If, on the other hand, the seepage zone extends relatively deeply, it is not practical to attempt to intercept all of the seepage water. In this case the subdrain should be sunk sufficiently deep to keep the free water level at least 1·25 m below the bottom of the pavement.

High water table. The water table must always be kept at least 1–1·25 m below the formation also. Not only is it undesirable to have free water anywhere near the pavement, but the compacted subgrade must not lie within the zone through which the capillarity (suction) moisture is capable of rising. This is particularly important when the subgrade is composed of a fine-grained soil.

A high water table can often be lowered by the installation of longitu-dinal drains on either side of the carriageway (see Fig. 2.6(b)). The depth to which the drains should be laid and the need for any particular type of system are functions of the width of the carriageway and the soil type. The most practical way of determining the effect of any particular system is to carry out simple field trials. For instance, for the installation illustrated in Fig. 2.6(b), two parallel trenches, each about 15 m long, should be dug to a depth of about 0·6 m below the level to which it is desired to lower the water table, along the line of the proposed drains. A transverse line of boreholes at intervals of 1·5–3 m are then sunk between the two ditches and extended on either side for a distance of about 4·5 to 6 m. Observations are then made of the levels of the water table in the boreholes before and after pumping the water out of the trenches, allowing a sufficient period of time for equilibrium conditions to be established. By plotting the observations an estimate can be made of the effect of the drawdown effects of the ditches, thereby enabling decisions to be made regarding the correct depth and spacing of the drains. The size of the drain pipes can be estimated on the basis of the rate of pumping necessary to keep the trenches free of water.

It must be pointed out that the level of the water table will not normally be lowered to the level of the drains themselves, except immediately adjacent to the drains. As illustrated in Fig. 2.6(b), there is a natural tendency for the water table to rise as the distance from the point of moisture release in-creases. The steepness of the water table adjacent to the drain is dependent on the soil type; if the soil is coarse-grained the shape of the water table will be relatively flat, whereas if the soil is fine-grained it will be steep.

A further point to mention here is that a single longitudinal drain ben-eath the centre of the roadway may sometimes be sufficient. The most obvious examples of this are when the water table is already relatively deep and/or when the subgrade soil is coarse-grained.

Fine-grained soils cannot, however, be drained by means of gravity drains.[13] While sophisticated methods utilizing the theory of electro-osmosis are available to drain moisture that cannot be removed by ordinary means, they are expensive and require such specialized technical expertise that they are impracticable for use on most highway projects. Therefore, instead of attempting to lower the water table in such soils, an embankment may be constructed so that the bottom of the pavement is raised the desired distance above the water level.

The construction of an embankment is the usual remedy when the road-way crosses low-lying saturated soil. In this case the solution may be to partially lower the water table by means of surface drainage ditches and then to construct a small embankment to obtain the remaining necessary clearance.

Permeable surface. If the wearing course of the pavement is composed of soil aggregate or an open-graded type of bituminous macadam, it can be ex-pected that moisture will enter the pavement through the road surface and eventually find its way to the subgrade. In concrete-covered roads the rain water may enter through joints and cracks that are not adequately sealed. If this moisture is not removed immediately it may cause the subgrade to soften and, eventually, the pavement to fail.

When it is expected that moisture will enter through the surface, a 100–150 mm thick sub-base composed of granular material should be interposed between the subgrade and the roadbase. The purpose of the sub-base here is to act as a drainage layer to enable the water to move freely and quickly to the sub-drains after it has penetrated through the roadbase. This concept is illustrated in Fig. 2.6(c).

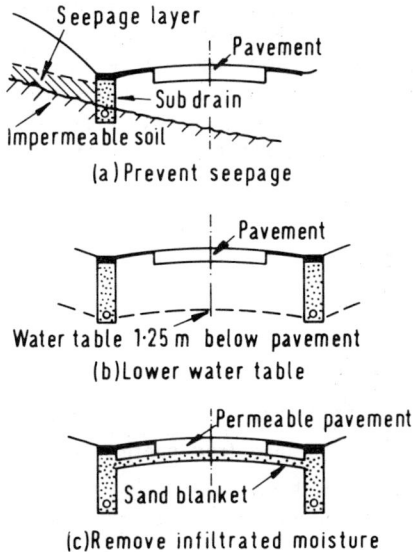

(a) Prevent seepage

(b) Lower water table

(c) Remove infiltrated moisture

Fig. 2.6.   Diagrammatic illustrations of the use of sub-drains to prevent moisture intruding into the subgrade

If the moisture is to move freely and swiftly to the sub-drains the granular sub-base material will have to meet design permeability requirements and the top of the subgrade must be carefully shaped and compacted.

Verge moisture-movements. Two problems are associated with the effect of roadside verges on subgrade moisture. The first and more obvious one is that which occurs in winter when the verges are much wetter than the subgrade. This sets up a suction potential which causes moisture to transfer to the subgrade, causing detrimental softening to take place. With a very expansive clayey subgrade this may cause the edge of the carriageway to be lifted sufficiently to initiate longitudinal cracking of the pavement.

The second effect, which is just as important, occurs in very dry weather when the subgrade soil is significantly wetter than the verge soil. Moisture conditions such as are illustrated in Fig. 2.7 will cause soil shrinkage beneath the edges of the carriageway. If the subgrade soil is a fine-grained material, this in turn may lead to longitudinal cracking of the pavement.

Drying-out of the soil near the edge of the carriageway is accelerated if fast-growing trees and shrubs are allowed to grow adjacent to the roadway. This is accentuated in urban areas where much of the ground area is covered anyway, with the result that, especially in times of drought, the vegetation can have great difficulty in getting adequate moisture.

It must be pointed out, however, that the removal of large trees from the sides of roads already constructed has also been known to cause pavement trouble. Here the problems arose as a result of upsetting moisture conditions which had become stable over the course of time.

The movement of moisture to and from the verges can be prevented by interposing a waterproof membrane between the verges and the roadbase, sub-base and compacted subgrade. Another approach, which from a stability aspect is most desirable, is to extend the roadbase, sub-base and subgrade into the verge to form a continuous hard shoulder. This shoulder will then act as a buffer zone to absorb the most detrimental differential effects.

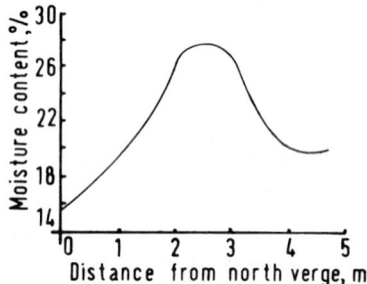

Fig. 2.7.   Variation across the width of a road of the average moisture content of the top 450 mm of the clay subgrade during a severe drought in Britain[14]

Vapour movement. The movement of moisture in vapour form is associated with differences in vapour pressure arising from differences in temperature and/or moisture contents in various vertical position in the subgrade soil. Vapour movement resulting from temperature differences can assume considerable importance in climatic areas where there are substantial fluctuations in the daily temperature; because of the temperate climate prevalent in Britain, however, this form of vapour movement is not significant. Vapour movement because of moisture differences is not normally a problem in this country either. For it to occur to any large extent the soil must be relatively dry and again this is a moisture condition that is not very common in this temperate climate.

When moisture movement in the vapour phase does take place it will usually be in coarse rather than fine-grained soils. For this reason the often advocated introduction of a coarse-grained cut-off layer between the subgrade and the roadbase will do little or nothing to protect either the roadbase or the subgrade. Perhaps the only effective method of combating vapour movement when it is expected to occur is to interpose a horizontal impermeable membrane between the top 0·5 m of the subgrade soil and the underlying uncompacted soil.

Sub-drain design

As noted previously, the installation of a sub-drainage system is economically justified only when there is 'free' water to be collected and if the sub-drains are properly designed and inserted.

Sub-drain installations now used in highways contain pipes to remove the drainage moisture quickly. These pipes are placed in trenches containing carefully selected coarse-grained backfill material. Normally the drainage

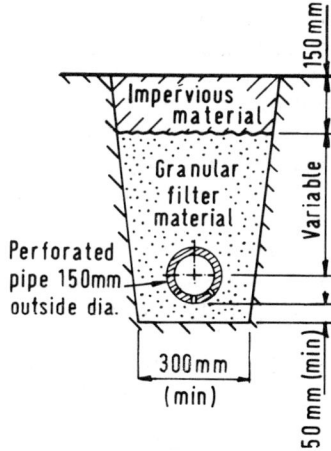

Fig. 2.8. Typical sub-drain cross-section

pipes are of perforated metal, vitrified clay or concrete, of porous concrete, or of unperforated concrete or vitrified clay laid with open joints to allow the water to enter. Perforated pipe is less liable to silting and so is to be preferred. The perforations should be laid downward in order to minimize clogging by soil-fines. If the underlying soil is relatively impervious it will be necessary to bed the pipe in at least 50 mm of granular material to avoid plugging of the perforations; this bedding layer is usually not necessary if the soil is coarse-grained.

In the past it was common practice to extend the filter material up to the surface, the theory being that the sub-drain could serve both as a drain for the subgrade and as a means of removing the storm-water from the surface. While this may seem at first sight to be economic practice, it is in fact to be deplored. Very often all it does is provide means by which surface moisture can gain easy access to both the subgrade and the pavement. In addition, substantial amounts of fine material are washed from the surface down into the filter soil and eventually it cloggs the drain. To prevent this happening the top 150 mm of the sub-drain should consist of a layer of compacted clay or some such impervious covering (see Fig. 2.8).

Sub-drains are usually nearly parallel to the centre-line of the roadway. If the intention is to intercept seepage water, then one drain will usually be adequate for this purpose. If it is the water table that has to be lowered, then a sub-drain on either side of the carriageway may be required. If the soil is relatively pervious and if the amount that the water level has to be reduced is not too great, then a single drain beneath the centre-line may be sufficient; in this case the sub-drain should be beneath the compacted layer at the top of the subgrade if differential movement of the pavement is to be eliminated.

The longitudinal sub-drains should be provided with outlets at regular intervals so that the collected moisture is able to escape. Normally these outlets should not be more than about 90 m apart. The slope of the sub-drain between the outlets should be as steep as possible, but certainly should never be less than 1 in 200 or 0·5 per cent.

Backfill criteria. The purpose of the backfill is to improve the interception ability of the drain, and to provide an effective water-collecting space adjacent to the pipe. To fulfil this function, the backfill material must be sufficiently coarse to allow water to have easy access to the pipe, yet be fine enough to act as a filter to prevent base soil from intruding into the pipe.

Theoretically, the *permeability* required of a filter can be determined from considerations of ground water flow; practicably, however, most recommendations are based on grain-size limits determined following comprehensive filter-design investigations. A recent excellent review of the literature[15] on this subject concluded that a filter material interposed between a pipe and any type of granular base soil should meet the following requirements to ensure that it will be sufficiently permeable at all times:

$$\frac{D_{15} \ (\text{Filter})}{D_{15} \ (\text{Protected soil})} \geqslant 5$$

where D_{15} represents the 15 per cent size (mm) of either the filter material or the surrounding protected base soil.

Many recommendations have been made with regard to *filtration* requirements for sub-drains. The most reliable for granular soils is divided into four parts:

(i) $$\frac{D_{15} \ (\text{Filter})}{D_{85} \ (\text{Protected soil})} \leqslant 5$$

and $$\frac{D_{15} \ (\text{Filter})}{D_{15} \ (\text{Protected soil})} \leqslant 20$$

and $$\frac{D_{50} \ (\text{Filter})}{D_{50} \ (\text{Protected soil})} \leqslant 25$$

(ii)   If, however, the soil is uniform, i.e.

$$\frac{D_{60} \ (\text{Protected soil})}{D_{10} \ (\text{Protected soil})} \leqslant 1·5$$

then use $$\frac{D_{15} \ (\text{filter})}{D_{85} \ (\text{Protected soil})} \leqslant 6 \ (\text{instead of 5 as in (i) above})$$

(iii)   If the soil is well-graded, i.e.

$$\frac{D_{60} \; (\text{Protected soil})}{D_{10} \; (\text{Protected soil})} \geqslant 4$$

then use $\dfrac{D_{15} \; (\text{Filter})}{D_{15} \; (\text{Protected soil})} \leqslant 40$ (instead of 20 as in (i) above.)

(iv)   At no time should a filter material be used if it is gap-graded (as it will tend to segregate during placement), nor should more than 5 per cent of its weight pass the 75 micron BS sieve (as otherwise the filter fines will migrate into the pipe). If the protected soil is gap-graded, then the filter design (and, consequently, D_{10}, D_{15}, D_{50} and D_{85}) should be based on the soil portion finer than the gap in the grading. If the protected soil contains layers of fine material, the filter design criteria should be selected to protect against the intrusion of the finest layer.

Many clay soils contain non-cohesive particles which are liable to migrate to the sub-drain; however, the low water velocity and the natural cohesion of these soils will allow only groups or aggregates of clay particles about the drain to be washed away. For these reasons clay soils, and also silt soils, are best protected by a filter composed of a clean sand, (e.g. in Great Britain, concrete sand to BS 882, Zone 2). Such a sand backfill will also normally be sufficiently permeable to easily transmit all the water flow from the soils.

EXAMPLE: Design a sub-surface filter material to protect a coarse-grained soil with a gradation as shown in Figure 2.9.

Solution.   Check first for filtration. In this case, the uniformity coefficient of the protected soil is given by:

$$U_c = \frac{D_{60} \; (\text{Protected soil})}{D_{10} \; (\text{Protected soil})} = \frac{0 \cdot 15}{0 \cdot 075} = 2$$

The exceptions noted in (ii) and (iii) above do not therefore apply, so use the criteria in (i), thus:

$$\frac{D_{15} \; (\text{Filter})}{D_{85} \; (\text{Protected soil})} \leqslant 5 = \frac{D_{15} \; (\text{Filter})}{0 \cdot 21}$$

Therefore, D_{15} (Filter) $\leqq 1 \cdot 05$ mm.

Alternatively,

$$\frac{D_{15} \; (\text{Filter})}{D_{15} \; (\text{Protected soil})} \leqslant 20 = \frac{D_{15} \; (\text{Filter})}{0 \cdot 085}$$

Therefore, D_{15} (Filter) $\leqslant 1 \cdot 7$ mm.

Decision: Use the lower D_{15} (Filter) value, i.e. $1 \cdot 05$ mm.

$$\frac{D_{50} \text{ (Filter)}}{D_{50} \text{ (Protected soil)}} \leqslant 25 = \frac{D_{50} \text{ (Filter)}}{0.14}$$

Therefore, D_{50} (Filter) $\leqq 3.5$ mm.

A filter material should be chosen which will meet the above D_{15} and D_{50} criteria. In this instance a suitable backfill could well have a D_{85} size of about 9.5 mm (see Figure 2.9).

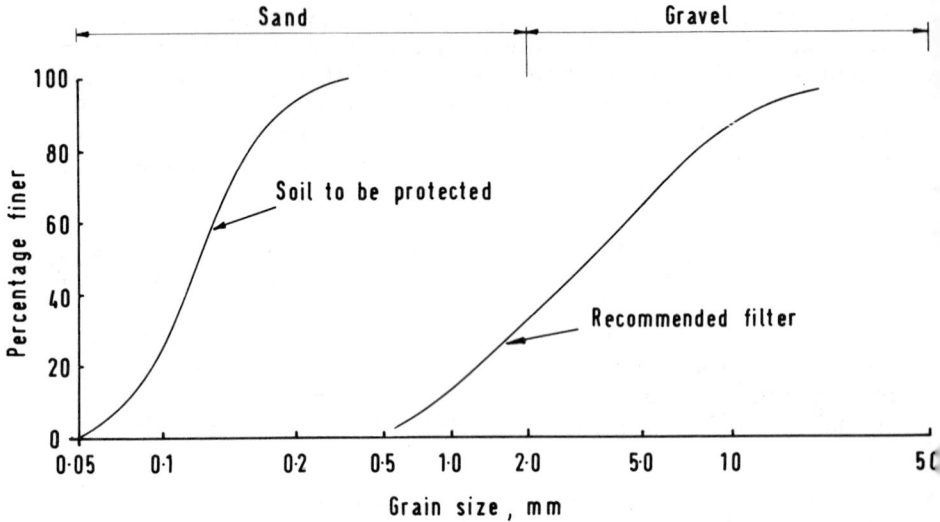

Fig. 2.9. Example of filter design to protect a coarse-grained soil

Now check that this filter material will meet the permeability requirements, thus

$$\frac{D_{15} \text{ (Filter)}}{D_{15} \text{ (Protected soil)}} = \frac{1.05}{0.085} = 12.35 \text{ (which is} > 5)$$

This material will be sufficiently permeable.

Pipe criteria. The length and diameter of pipe should be selected so that (a) the pipe will not run full near its outlet and flood the surrounding filter material, and (b) it must be able to intercept all of the water entering the drain without causing a high head in the filter material, as this will reduce both the depth to which the water table can be lowered and its rate of lowering. In the case of a perforated pipe, the holes must be also sufficiently small to prevent the filter material from being washed into the pipe and clogging it.

At this time no authorative single recommendation can be made with regard to pipe diameters and length (although, in practice, 152 mm diameter pipes and outlet spacings of about 90 m are commonly used). The following criteria have been recommended[15] with regard to the size of the pipe perforations:

$$\text{Max. size of circular holes} \quad = D_{85} \text{ (Filter)}$$

$$\text{Max. width of slotted holes} \ = 0.83 \, D_{85} \text{ (Filter)}$$

Special compaction of the backfill will not normally be required if these criteria are followed—although it is possible that the hole size limits could be increased if the filter material is compacted.

Although approaches based on various assumptions are suggested there is, as yet, no definitive method of estimating the numbers of perforations per unit length of pipe.

EXAMPLE HOLE-SIZE CALCULATION.   Determine the size of pipe perforations which can be used with the filter material shown in Figure 2.9.

Solution.   From this figure, D_{85} (Filter) = 9.5 mm. Therefore,

$$\text{Max. size of circular hole} \ = 9.5 \text{ mm.}$$

$$\text{Max. width of slot} \qquad = 0.83 \, (9.5) = 7.9 \text{ mm.}$$

If the only available pipes have perforations which are larger than the above, then an appropriate coarser material must be placed next to the pipe. In such a case, however, the gradation calculations for the new intervening filter material would have to be carried out with the old filter material (see Figure 2.9) considered as the soil to be protected.

EXAMPLE CALCULATION FOR THE NUMBER OF HOLES.   As noted above there is no definitive procedure for determining the number of perforations per unit length of pipe. The following calculation[15] is simply one example of an approach towards determining the number of pipe perforations.

Consider the cross-section shown in Figure 2.10, and assume that $D = 1$ m, $W = 12$ m, and $d = 0.06$ m at the time that drainage is started. Assume the subgrade soil has a permeability of 1×10^{-4} m/sec (such as might occur in a saturated fine sand). Determine the order of magnitude of the water flow involved, and a likely number of 5 mm diameter circular perforations per metre length of pipe.

Solution.   The flow rate into the drains will be greatest just as drainage begins. Thus using a d/D ratio of 0.06, the flow into each pipe can be obtained from

$$0.8 = \frac{q}{KD} = \frac{q}{10^{-4}(1.0)}$$

Therefore $q = 8 \times 10^{-5}$ m³/sec per metre length of pipe, where $q = $ flow in the filter toward each metre length of pipe. However, the flow intercepted per metre length can also be approximated (from Bernouilli's equation) by

$$q = n.a.C_d \, (2gh)^{1/2}$$

where q = flow through perforations (assuming all water enters), m^3/sec

            n = no. of perforations per metre length

            a = area of each perforation, m^2

            C_d = coefficient of discharge of each perforation (assume 0·8 in this instance)

            g = acceleration due to gravity = $9·81\,\text{m/sec}^2$

and h = hydraulic head on each perforation (assume 5 mm in this instance)

Therefore $q = 8 \times 10^{-5} = n.a.(0·8)(2 \times 9·81 \times 0·005)^{1/2}$

Hence $n.a. = 32 \times 10^{-5}\,\text{m}^2$ per metre length of pipe

            $= 320\,\text{mm}^2$ per metre length

Water table before drainage

Water table

during drainage W

q = Discharge per unit length per unit time

t = Time since beginning of drainage

k = Coefficient of permeability

y = Volume of drainable water per unit volume of soil

Dimensionless ratios		
$\dfrac{tkD}{yW^2}$	$\dfrac{d}{D}$	$\dfrac{q}{kD}$
0·001	0·06	0·80
0·01	0·37	0·47
0·1	0·79	0·25

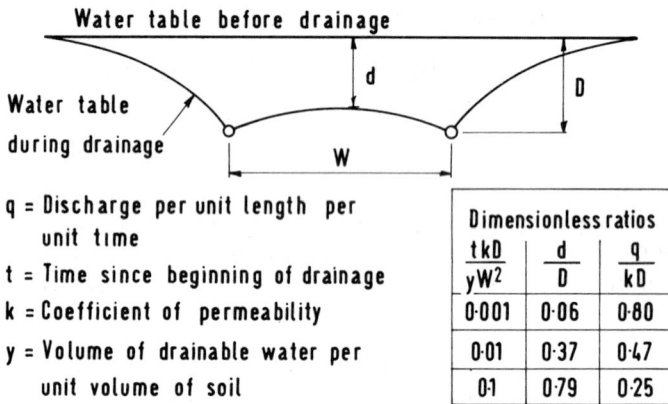

Fig. 2.10. Example pipe perforation problem involving the use of McClelland's dimensionless ratios for drainage by two parallel pipes (as noted in ref. 15).

If the perforations are 5 mm diameter, then $a = \pi(2·5)^2\,\text{mm}^2$.

Therefore $n = \dfrac{320}{\pi(2·5)^2} = 16·3$

Sixteen 5 mm dia holes per metre length of pipe should be adequate for drainage purposes, since the permeability value used is significantly higher than that which would be normally encountered.

SELECTED BIBLIOGRAPHY

1. BILHAM, E. G. Classification of heavy falls in short periods, *British Rainfall*, 1935, p. 262.
2. *Appendix to Hydrological Memoranda No. 33* Bracknell, Berks. The Meteorological Office. Aug. 1968.

3. ASHWORTH, R. and O'FLAHERTY, C. A. Validity of the modified Bilham equation in predicting rainfall intensities in England, paper submitted for publication in the *Journal of the Institution of Municipal Engineers*, 1973.
4. ESCRITT, L. B. *Sewerage and Sewage Disposal.* London, C. R. Brooks, 1964.
5. LLOYD-DAVIES, D. E. The elimination of storm-water from sewerage systems, *Proc. Inst. Civ. Engrs*, 1906, **164,** 41–67.
6. MANNING, R. On the flow of water in open channels and pipes, *Trans. Inst. Civ. Engrs of Ireland*, 1891, **20,** 161–207.
7. U.S. CORPS OF ENGINEERS. Planning, site selection, and design of roads, airfields and heliports in the theatre of operations, *U.S. Army Technical Manual* TM5–330. Washington, D.C., Department of the Army, 1963.
8. BUREAU OF PUBLIC ROADS. *Highway Drainage Manual.* Washington, D.C., U.S. Government Printing Office, 1953.
9. CHOW, V. T. *Open-Channel Hydraulics.* New York and Maidenhead, McGraw-Hill, 1959.
10. B.S. 497:1967. *Specification for Cast Manhole Covers, Road Gulley Gratings and Frames for Drainage Purposes.* London, British Standards Institution, 1967.
11. RUSSAM, K. The Hydraulic Efficiency and Spacing of B.S. Road Gulleys. *RRL Report* LR 277, Crowthorne, Berks., The Road Research Laboratory, 1969.
12. PARKER, B. C. and BROWN, K. R. Housing estate road design and construction. *Journal of the Institution of Highway Engineers*, 1968, **15,** No. 12: 23–28.
13. RUSSAM, K. Sub-soil Drainage and the Structural Design of Roads, *RRL Report* LR 110, Crowthorne, Berks., 1967.
14. MACLEAN, D. J. The problem of clay in road construction, *Final Report of Public Works and Municipal Services Congress and Exhibition*, 1950, 209–248.
15. SPALDING, R. Selection of Materials for Sub-surface Drains, *RRL Report* LR 346, Crowthorne, Berks., 1967.

3 Pavement materials

The materials used in the construction of a highway are of intense interest to the highway engineer, in contrast to many other branches of Civil Engineering where the engineer need not be very deeply concerned with the properties of the materials being used. All highways have to be founded on the soil, and all require the efficient usage of locally available materials if economically-constructed facilities are to be obtained. This requires not only a thorough understanding of the soil and aggregate properties which affect pavement stability and durability, but also of the properties of the binding materials which may be added to improve these pavement features.

The most important pavement materials are soil, rock and slag aggregates, bituminous binders and cement, so these materials—with the exception of soil—are discussed in some detail in this chapter; because of the all-important role which soil plays in highway construction, it is discussed separately in another chapter. The properties of the binders and modifiers other than cement and the bituminous materials, which are used in the stabilization of soil for highway purposes, are dealt with in the chapter on Soil Stabilization.

BITUMINOUS ROAD BINDERS

Terminology

The reader of the scientific literature relating to bituminous road binders may come to the conclusion that there is much confusion with regard to what is meant by the word 'bitumen', and to a certain extent he will be right. Chemically it is accepted that all bituminous road binders are primarily hydrocarbons, i.e. combinations of hydrogen and carbon. Practically, however, there is debate as to whether the term bitumen should be confined to naturally occurring materials, or whether it should include both the distillates of petroleum and the tars.

In what may perhaps be described as the most authoritative treatise on bituminous binders,[1] an analysis of the views concerning the scope of the term bitumen shows that the interpretations applied to it in different countries and by different bodies may be grouped into the four following classes:

1. Bitumens—naturally occurring hydrocarbons. This represents the layman's interpretation of the word, and is based on the dictionary definition which for generations confined the term to the hydrocarbon substances

which occur in nature. This interpretation distinctly *excludes* hydrocarbon substances which are produced artificially.

2. Bitumens—naturally occurring hydrocarbons, and the residues obtained from the distillation of petroleum. This is the view held in Great Britain.

3. Bitumens—naturally occurring hydrocarbons (which may be gaseous, liquid, semi-solid or solid), residues obtained from the distillation of petroleum, and artificial hydrocarbon substances, e.g. tars and pitches. This definition of bitumen is one which is now generally applied to a 'bituminous material', i.e. a material which contains bitumens, or resembles bitumens, or constitutes the source of bitumens. Its particular difference from the second view is that it accepts tars and pitches as being bitumens.

4. Bitumens—only those components of the third view which are soluble in carbon disulphide (CS_2). This interpretation is the one which is held in the appropriate standards and literature in the United States, and by the Association Internationale Permanente des Congrès de la Route (Paris).

To avoid any misinterpretation, the definition of the word bitumen used in this text will be that which is accepted in Britain. Thus tar is discussed as a substance which is different from bitumen. When the term 'bituminous binder' is used, it will however refer to *both* bitumen and tar binders.

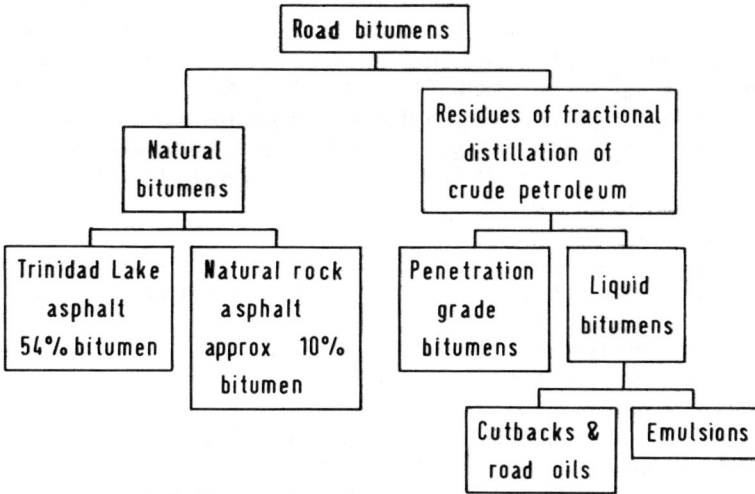

Fig. 3.1. Road bitumens used in Great Britain

Bitumen

Bitumen is a viscous liquid or solid material, black or dark brown in colour, having adhesive properties, consisting essentially of hydrocarbons, derived from petroleum or occurring in natural asphalt, and soluble in carbon disulphide.[2] There are two main categories of bitumens, namely, those which occur naturally and those which are by-products of the fractional distillation of petroleum at a refinery. Of the, literally, hundreds of possible types falling into these categories, the ones which have tended to be used for highway paving purposes in Britain are illustrated in Fig. 3.1.

Lake asphalt. The largest natural deposit of bitumen in the world occurs on the Island of Trinidad off the north coast of South America. The bitumen is found in conjunction with a substantial amount of solid mineral matter and the *combination* is known as asphalt. (In the United States literature, the term asphalt is used to refer to what in Britain is simply called bitumen; in this country, the term asphalt is reserved for mixtures of bitumen and mineral particles).

The asphalt lake in Trinidad covers an approximately circular area of about 44·5 ha, and is located in a shallow crater at the crest of a hill 42 m above sea level. The lake has a diameter of about 600 m, and is at least 41 m deep at the centre, becoming shallower towards the edges. In recent years 10 million tons of asphalt have been removed from the lake, during which time the level of the lake has sunk about 1·5 m.

The origin of the Trinidad lake asphalt is still not exactly known, but it is believed that it is derived from the remnants of an oil deposit originally present in an anticline of Cretaceous rocks. Lowering of the earth's surface led to an incursion of the sea, and a considerable deposit of sedimentary material covered the bitumen. At a later date, part of the silt and clay penetrated the bitumen and saturated it, forming the now familiar asphalt mixture of bitumen, silt, clay, gas and water. Upon subsequent elevation of the land surface, the covering layer of soil was removed by erosion and the asphalt mixture exposed to its present state.

As excavated from the lake, Trinidad asphalt is remarkably uniform in composition. Specimens taken from the various portions of the lake's surface, after pulverizing and drying to constant weight in air at room temperature, average:[1]

Soluble in carbon disulphide, 53·0% to 55·0%
Free mineral matter, 35·5% to 37·0%
Water of hydration, etc., 9·7%

After excavation from the lake, the asphalt is heated to 160°C to drive out the moisture, after which it is run through strainers to remove vegetable debris and then poured into light wooden barrels for export under the name of Trinidad Epure or Refined Trinidad Lake Asphalt. Some of the asphalt is blended in the stills with a heavy petroleum oil to obtain a penetration-grade asphalt suitable for direct road use; this is usually exported in steel drums.

In Great Britain it is considered more convenient to import the Epure and to flux it at the paving plant as required. This is usually done by melting the Epure in tanks of up to 9000 l capacity and adding to it an appropriate proportion of refinery bitumen of flux oil. For road purposes, the refinery bitumen added is usually of nominal 200 or 300 penetration-grade. Where it is desired to soften the Trinidad asphalt with a liquid flux, the oil used is normally a petroleum oil having the characteristics set out in Table 3.1.

The actual amount of fluxing material used in a particular situation depends primarily on the fluxing value of the oil employed and the penetration desired. When liquid fluxes are used, the mixture may be composed of 85 per cent Epure and 15 per cent liquid flux. When it is desired to blend Epure with refinery bitumens—the commonest procedure—the mixture may include between 20 and 50 per cent Epure. For rolled asphalt (hot process) road surfacings a 50–50 mixture of Epure and refinery bitumen gives optimum results in most normal cases. Details regarding the use of Trinidad Lake Asphalt in road surfacings are included in five British Standards.[4, 5, 6, 7, 8]

TABLE 3.1. *Typical petroleum oil used to flux Trinidad Epure*[3]

Specific gravity at 25°C	0·920
Viscosity (Redwood 1) at 60°C	210 sec
Soluble in carbon disulphide	99·9%
Loss after 5 hours at 163°C	1·5%
Closed flash point	185°C
Open flash point	210°C
Fire point	225°C

TABLE 3.2. *Composition of crude rock asphalt obtained from the Val de Travers region*[1]

	%
Soluble in carbon disulphide	10·50
Calcium oxide	54·98
Magnesium oxide	0·15
Carbon dioxide	42·85
Iron and aluminium oxides	0·23
Silica	0·32
Alkali, sulphates and loss	1·47

Natural rock asphalt. This consists of a granular material, usually limestone or sandstone, which in its natural state contains bitumen intimately dispersed throughout its mass. The bitumen content of natural rock asphalt varies from 4 to 18 per cent, the remainder of the material being solid mineral matter.

The natural rock asphalt used in Great Britain is imported from France and Switzerland. The Swiss deposit, which is located in the Val de Travers region, is believed to have been produced by the decomposition of marine animal and vegetable matter. On the other hand, the bitumen content of the material obtained from France, mined in the Department of Gard, was probably introduced into the rock as a bituminous petroleum, the lighter fractions of which evaporated with time and left the bitumen behind. Both the French and Swiss deposits contain about 10 per cent bitumen. A typical road material from the Val de Travers region is shown in Table 3.2.

Natural rock asphalts are now used only to a very limited extent in Britain. Details regarding their usage are given in two British Standards.[5, 9]

Refinery bitumens. Bitumens artificially produced by the industrial refining of crude petroleum oils are known under a number of names such as residual bitumens, straight-run bitumens, steam-refined bitumens and—as is now most commonly accepted—refinery bitumens. Not all petroleum crudes contain a sufficient quantity of bitumen to enable straight reduction to specification road bitumen. Those which do are known as asphaltic-base crudes. Crudes which contain high proportions of simpler paraffinic compounds, with little or no bituminous bodies present, are known as paraffinic-base crudes. Some petroleum crudes exhibit characteristics of both the previous categories, and these are known as mixed-base crudes. Most of the

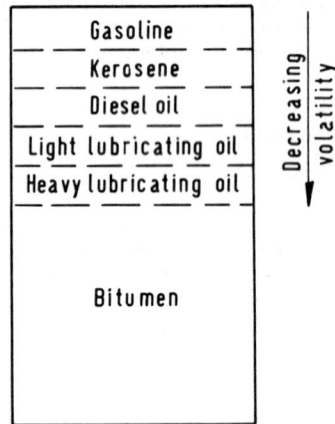

Fig. 3.2.   Schematic illustration of the components of an asphaltic-base petroleum crude

petroleum crudes refined in Great Britain are obtained from the oilfields of the Middle East and Central America. The Central American crude oils are particularly noted for being asphaltic-based, whereas the Middle East crudes tend to be paraffinic.

The composition of an asphaltic-base petroleum crude is shown in Fig. 3.2. It should be clear that this is a schematic diagram, and that the proportions in any particular instance will vary with the crude. Bitumen is obtained from the crude by a refinery distillation process which involves condensation in a fractionating column. For bitumens utilized in Great Britain the first distillation is normally carried out in refineries at the oil-fields; there the crude is heated to temperatures not greater than 350°C at atmospheric pressure in order to drive off the light gasoline and kerosene fractions. The 'topped' oil is then shipped to Britain for further refining to remove additional fractions as required.

Penetration-grade bitumens.   On arrival at the refinery, the topped oil is heated with the aid of reduced pressures and steam injection in the fractionating columns. Heating under atmospheric conditions is not possible as the temperatures required to drive off the heavier oils would be so high that 'cracking' or chemical change of the vapours would take place; this imparts undesirable paving properties to the bitumen residue.

Bitumens obtained by fractionating are mostly what are called penetration-grade bitumens. These are also known by the names 'asphaltic bitumens' and 'asphalt cements'. They vary in consistency from semi-solid at room temperature to semi-liquid under the same conditions. Asphaltic bitumens are classified according to hardness as indicated by the depth that a specified needle is able to enter the samples when standard penetration tests are carried out. Details of the grades and properties of the asphaltic bitumens manufactured in Britain are given in Table 3.3. Note that the penetration number reflects the hardness of the bitumen; thus the 15 pen bitumen is the hardest and the 450 pen bitumen the softest of those shown. The penetration-grade bitumen most commonly used in rolled asphalt surfacings on high-quality roads in this country is the 50 pen one.

In a particular refinery, a bitumen of a desired penetration may be obtained either by controlled refining such as has been described, or by fluxing a harder grade with an oil of high boiling range such as is produced at the lower end of the fractionating column during distillation of the topped crude oil. Quite often a refinery will not manufacture all the penetration grades shown in Table 3.3 because of the large tankage which this would require. Instead it is common practice to prepare and stock large quantities of two asphaltic bitumens, one of a low penetration and the other a high one. All intermediate penetration-grades are then prepared by blending appropriate quantities of the two stock grades.

Several standards for highway surfacing materials include specifications regarding the properties of the asphaltic bitumen binders to be used in specific situations.[4, 5, 6, 7, 8, 11]

Cut-back bitumens.   The penetration-grade bitumens range from being very viscous to semi-solid in consistency at normal temperatures, so it is normal for them to be heated before use in road construction. There are many instances, however, where it is neither desirable nor necessary to use a hard bitumen and preference is given to the use of liquid binders such as the cut-back bitumens.

Cut-backs differ from penetration-grade bitumens in that the asphaltic bitumen is dissolved in a liquid solvent which makes it suitable for direct application and manipulation in road construction. While the solvent is primarily a substitute for heat, in many instances it is more useful than heat since its liquifying effect lasts over a longer period of time.

After a cut-back bitumen has been spread on the particles it is intended to bind, the solvent will dissipate itself by evaporation and/or photo-oxidation and leave behind the cementitious bitumen to tie the particles together. Thus the character and behaviour of a cut-back bitumen in any particular situation is largely dependent on the character and amount of solvent present. The more volatile the solvent, the shorter the curing period necessary after using the cut-back before the cohesive properties of the binder are utilized. The less volatile the solvent, the more of it is needed to bring the bitumen to a given degree of fluidity.

As indicated by Fig. 3.3, cut-back bitumens can be divided into three main types, depending on the type of solvent used to dilute the bitumen. These are commonly designated as Slow Curing (S.C.), Medium Curing (M.C.) and Rapid Curing (R.C.) cut-back bitumens.

Slow curing cut-back bitumens may be manufactured in either of two ways. In the first, the S.C. cut-back may be obtained entirely by direct distillation of the petroleum crude, in a manner similar to that by which asphaltic bitumens are obtained. In the case of cut-backs, however, many of the volatile oils are left in the mixture and so the liquid bitumens are usually much more fluid than the penetration-grade ones. In the second method, asphaltic bitumen is directly fluxed with slowly volatile oils to obtain particular cut-backs of the desired fluidity.

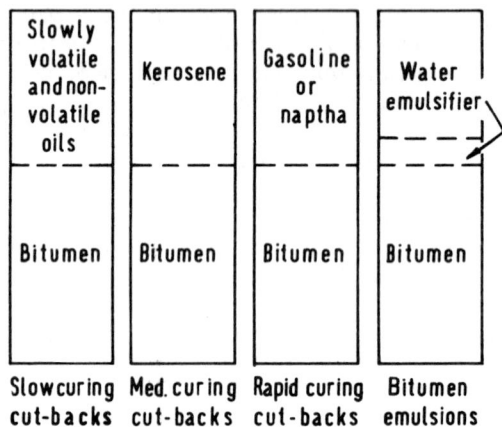

Slowly volatile and non-volatile oils	Kerosene	Gasoline or naptha	Water emulsifier
Bitumen	Bitumen	Bitumen	Bitumen
Slow curing cut-backs	Med. curing cut-backs	Rapid curing cut-backs	Bitumen emulsions

Fig. 3.3.   Schematic illustration of the components of liquid bitumens
Note: These diagrams are not proportional to composition

Whichever method is used, the S.C. cut-backs are liable to remain liquid in or on the roadways for a relatively long time and so the binding strength is developed correspondingly slowly. For this reason these cut-backs are best used with dense-graded aggregates which provide a strong interlocking framework and do not require immediate, strong, cementing action from the binder. For the same reason, an S.C. cut-back should never be used as a binder for open-graded road surfacings. Because of their slow curing qualities, S.C. cut-backs are also used on soil-aggregate roads in warm climates in order to keep the dry soil particles from creating a dust nuisance.

TABLE 3.3.   *Penetration-grade bitumens used for road purposes in Great Britain*[10]

Property	Grade								
	15 pen	25 pen	35 pen	50 pen	70 pen	100 pen	200 pen	300 pen	450 pen
Penetration at 25°C	15±5	25±5	35±7	50±10	70±10	100±20	200±30	300±45	450±65
Softening point °C, min.	60	55	51	47	43	40	33	29	25
Softening point °C, max.	76	67	62	56	52	49	42	38	34
Loss on heating for 5 h at 163°C (a) Loss by wt (%) max.	0·1	0·2	0·2	0·2	0·2	0·5	0·5	1·0	1·0
(b) Drop in penetration (%), max.	20	20	20	20	20	20	20	25	25
Solubility in carbon disulphide or trichloro-ethylene (% by wt), min.	99·5	99·5	99·5	99·5	99·5	99·5	99·5	99·5	99·5
Ash content (% by wt), max.	0·5	0·5	0·5	0·5	0·5	0·5	0·5	0·5	0·5

Prior to the 1960's six grades of S.C. cut-backs were available in the U.S. and in other countries where American road practice was followed. These grades were assigned numbers from 0 to 5, depending upon their consistency. The S.C.–0 grade, which contained the greatest amount of solvent (about 50 per cent), was somewhat similar to the 'topped' oil brought to this country for further refining. The S.C.–5 grade, containing about 18 per cent solvent, was the most viscous (and least fluid) grade. Following the revision of the system for describing cut-back bitumens in the United States four, rather than six, grades are now produced. The new designations are based

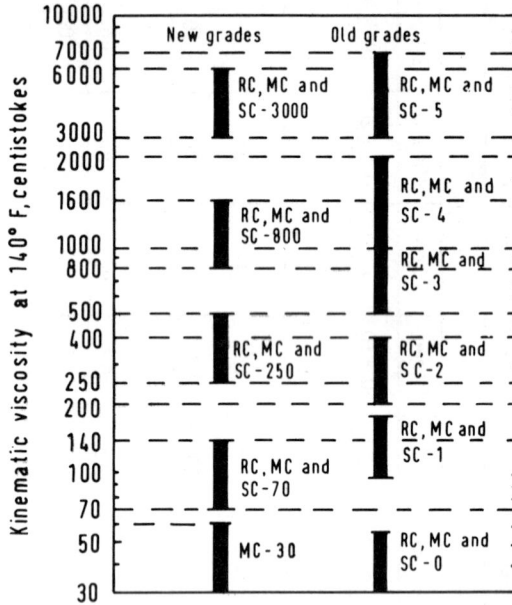

Fig. 3.4. Comparative viscosities of old, and new grades of cut-back bitumens discussed in the American literature[12]

on the kinematic viscosities of the mixtures, with the new designation numbers set at each grade's lower viscosity limit, while the upper limit is exactly double the lower limit. Because of the many references in the technical literature to these binders, their new and old designations are shown in Fig. 3.4. Note that the new designation of a cut-back is based on its kinematic viscosity, which is a fundamental scientific unit; the old designations were purely arbitrary in choice.

Some *slow curing cut-back bitumens* are manufactured in Great Britain for use in fine cold asphalt.[7] These cut-backs are frequently referred to as 'fluxed' bitumens. They are roughly equivalent to the American S.C.–800 and 3000 grades.

Medium curing cut-back bitumens are the variety mostly used in Britain. They are manufactured by combining 100/300 penetration-grade bitumens with a petroleum distillate, such as kerosene, or with a coal-tar creosote oil or anthracene oil, or with a mixture of these. Even though these solvents evaporate more rapidly than the oils used in the S.C. cut-backs, it may be a considerable time before the consistency of the base asphaltic bitumen is reached; in the well-diluted American M.C.–70 types, used in dense-graded road surfacings, this has been known to take as long as several years. Medium curing cut-backs have good aggregate coating properties, so they are very useful when fine-graded and dusty materials are incorporated in a road surface and it is desired to use a binder which is less viscous at the time of processing than after the mixture has cured for some time. They are also generally considered the most practical for use in the bituminous stabilization of soil.

The grades and properties of the medium curing cut-back bitumens commonly supplied in Britain are shown in Table 3.4. These are designated in terms of their viscosity limits expressed in seconds of flow in the Standard Tar Viscometer (S.T.V.) at 40°C;[13] they are roughly equivalent to the M.C.–800 and 3000 cut-backs described in the American road-surfacing specifications. The least viscous (more fluid) of the British grades is most often used for surface-dressing purposes, while the more viscous ones are used in bitumen macadam and fine cold asphalt surfacings.

Rapid curing cut-back bitumens are prepared by diluting a suitable penetration-grade bitumen with a very volatile petroleum distillate such as gasoline or naphtha. Since the volatile constituents of these cut-backs quickly evaporate, they are used when a quick change-back to the residual semi-solid binding agent is desired. Rapid curing cut-backs are little used in Britain, preference being given to the medium curing types. The volatility of the distillate, i.e. it has a relatively low flash point, can render their use hazardous, particularly when the very viscous grades—which may have to be warmed before admixing—are used in road construction.

Bitumen emulsions. An emulsion is a relatively stable suspension of one liquid in a state of minute subdivision, dispersed throughout another liquid in which it is not soluble. The liquid which is dispersed is often called the internal phase, while the surrounding liquid is known as the continuous or external phase.

Bitumen emulsions can be of two types. In the first and most common type minute globules of bitumen are dispersed in water. This is to all intents and purposes the only type manufactured in Great Britain. Another type is known as an *Inverted Emulsion* and consists of a bitumen–water mixture in which up to about 10 per cent water forms the dispersed phase and cut-back bitumen is the continuous phase. This latter type of emulsion has been used successfully in the United States for cold-mix road surfacings and mix-in-place soil stabilization; it is not used in Britain and so will not be further discussed here.

Emulsifiers. Most of the important properties of an emulsion are dependent on the amount and type of emulsifying agent used to promote the dispersal and stability of the bitumen–water mixture. Were it not for the emulsifier the dispersion of the bitumen, which may be brought about by

TABLE 3.4.   *Cut-back bitumens used for road purposes in Great Britain*[10]

Property \ Grade	50 sec	100 sec	200 sec
Viscosity (S.T.V.) at 40°C	50 ± 10	100 ± 20	200 ± 40
Distillation:			
(a) Distillate to			
225°C	1 max.	1 max.	1 max.
360°C	8–14	6–12	4–10
% by volume			
(b) Penetration of residue			
at 25°C	100–350	100–350	100–350
Solubility in CS_2 or trichloroethylene,			
% by weight	99·5 min.	99·5 min.	99·5 min.
Ash content, % by weight	0·5 max.	0·5 max.	0·5 max.

rapid stirring, would quickly re-form into a separate layer once agitation is stopped. When a suitable emulsifier is added, it forms an adsorbed film about each dispersed droplet of bitumen and gives to each a protective coating which provides resistance to coalescence.

Emulsifiers used in the preparation of bitumen emulsions belong to two main categories, known as anionic and cationic emulsifiers. The following are brief descriptions of some of their chief characteristics.

Anionic emulsifiers. By far the most common types of emulsifier are the anionic ones; the emulsions formed with them are called anionic or alkaline emulsions. The structure of an anionic emulsifying agent may be represented by the formula for sodium stearate, $CH_3(CH_2)_{16}COONa$. When this is dissolved, it dissociates into the long-chain stearate anion $CH_3(CH_2)_{16}COO^-$ and the sodium cation Na^+. The long-chain anions are soluble in bitumen and so they attach themselves to the droplets in such a way that each bitumen globule is coated with a negatively-charged stearate film which causes the individual globules to repel each other when they come into contact.

Cationic emulsifiers. These are emulsifiers which have only relatively recently been developed. They generally consist of amine salts made by reacting hydrochloric acid or acetic acid with an organic amine or diamine. When added to water, the emulsifying compounds dissociate in such a way that the long-chained amines—which in this case are the cations—attach themselves to the bitumen droplets so that each individual globule is coated with a protective positively-charged layer of amines. As with the anionic coatings, the result is that the positively-charged droplets tend to repel each other when they come into contact, which prevents coalescence.

Preparation and uses of emulsions.   The emulsification process consists of mixing water, bitumen and the emulsifying agent in either a colloid mill disintegrator or a high-speed mixer. A colloid mill is composed of a conical disc which very rapidly revolves within a stationary part called the stator, the clearance between the two usually being within the range 0·125–6·35 mm. When the separate streams of hot bitumen and hot water and emulsifier are forced through this minute clearance, dispersion readily takes place, the droplets are coated, and the emulsion is formed. With the high-speed mixer manufacturing process, the emulsion is formed by adding the hot bitumen to the hot mixture of water and emulsifier and then using a high-speed propellor-type agitator to achieve dispersion as the bitumen is admixed; emulsion batches of the order of 900 l per time can be prepared by this process. Whereas colloid mill emulsions can be produced as a continuous process with usual outputs of about 11 350 l/h, the high-speed mixer emulsions are output at the rate of 6000–9100 l/h.

All grades of bitumens can be emulsified, but the 100 and 350-penetration bitumens are most commonly used in highway engineering practice. When they are to be used with pre-mixed materials which may have to be stockpiled, emulsions are often made from cut-back bitumens.

Anionic road emulsions produced in Britain are classified as labile, semi-stable and stable. This classification,[14] is based primarily on the protection against coalescing of the globules given by the small amount ($<10\%$) of emulsifying agent. The following is a brief description of some of the properties and uses[15] of these emulsions.

Labile emulsions. These are characterized by a rapid breakdown on application as they contain only a minimum amount of emulsifier. They are divided into three sub-classes according to bitumen content, with class 1A containing not less than 60 per cent bitumen, class 1B containing not less than 53 per cent, and class 1C containing 30 to 50 per cent. Because of the ease with which labile emulsions break, they are not suitable for mixing with aggregates, but instead are used for surface dressing, tack-coating, grouting or penetration, and concrete curing purposes. Labile emulsions cannot be stored out of doors in cold weather since they will usually break on freezing and will not redisperse on thawing.

Semi-stable emulsions. These have sufficient emulsifier present to permit mixing with certain grades of aggregate before breakdown occurs. They are divided into two sub-classes, class 2A containing not less than 55 per cent bitumen and class 2B not less than 45 per cent. With semi-stable emulsions, certain aggregates can be pre-mixed before the emulsion breaks down and coagulation takes place. Premature breakdown occurs when the aggregate contains too many fine particles; in any particular instance, however, this can only be determined by trial and error. Semi-stable emulsions are also used in Britain for the 'retreading' of old road surfaces which were previously sealed with some form of bituminous binder. (The retread process consists essentially of breaking the top 76 mm of road, mixing the aggregate so obtained with the emulsion, re-shaping, compacting and then sealing the surface with a suitable surface dressing.)

Fully-stable emulsions. These have sufficient mechanical and chemical stability for all purposes involving mixing with aggregates, including those containing large proportions of fines or chemically active materials such as cement or hydrated lime. There are no sub-classes of this emulsion, only that it must contain at least 50 per cent bitumen. Fully-stable emulsions have a variety of uses. For instance, they can be used for pre-mixing of any type of aggregate, as well as for tack-coating purposes. Open-textured road surfacings can be sealed with a slurry of emulsion and fine sand or quarry fines.[16] Normally, fully-stable emulsions are the only ones which can be used to stabilize in-place fine-grained soils. In contrast with labile emulsions, fully-stable emulsions can be frozen and then thawed without any appreciable breakdown occurring; for this reason they are particularly useful for winter roadworks where the binder has to be stored out of doors.

At the time of writing, there is no British Standard for *Cationic* emulsions, although they are used to a limited extent in practice. American experience indicates that cationic bitumen emulsions can be used in any application where anionic ones are employed.[17] British experience suggests their usage in the following instances:

1. For tack-coating purposes. Here cationic emulsions are particularly advantageous since they are made to break immediately on contact with the road, and so are not susceptible to run-off from rain.

2. To seal gravel sub-bases and prime gravel surfaces before applying a surface dressing.

3. For coating wet aggregates, particularly open-graded gravel mixtures. For this a cut-back bitumen of low viscosity must be used in the preparation of the emulsion, since if a bitumen of higher viscosity is used the intermediate coating of the aggregate will prevent further mixing. Properly coated aggregates may either be used immediately or stock-piled for later use.

Mechanism of action.   Proper usage of bitumen emulsions is heavily dependent on a proper understanding of how 'breaking' of the emulsion takes place in the roadway. Breakdown of anionic emulsions is primarily dependent on the following four factors:

1. *Rate of evaporation of water.* The evaporation of water commences immediately on application of the emulsion. The rate at which it happens is dependent on the atmospheric temperature, relative humidity, wind velocity, and the rate and method of emulsion application. The initial breakdown is reversible; this means that an early fall of rainwater may result in dilution of the emulsion so that it is washed from the roadway. The rate of evaporation is by far the most important factor influencing breakdown.

2. *The physico-chemical reaction between the aggregate and the emulsion.* This is dependent on the charges carried by the coated droplets and the aggregate particles. When an aggregate is very siliceous, the negatively-charged anionic droplets may not attach themselves to the negatively-charged aggregate surfaces and use must then be made of a positively-charged cationic emulsion.

3. *The absorption qualities of the surface being covered.* Obviously the more porous and dry the road surface and/or the aggregate surface being coated, the more quickly water from the emulsion is absorbed by capillary action.

4. *The amount of mechanical disturbance applied during compaction.* As evaporation and moisture absorption take place, the emulsified bitumen particles come closer and closer together and eventually coalesce. This usually happens when the proportion of bitumen in the emulsion is about four-fifths; this change is identifiable by a switch from the normal brownish colour of the emulsion to the black of the bitumen. At this stage, the breakdown may be accelerated by compacting the aggregate–emulsion mixture, thereby causing further coalescence of the bitumen.

Unlike an anionic emulsion, the breaking of a cationic emulsion does not primarily depend upon the evaporation of water. Rather, breaking is mostly of an electrochemical nature in that the positively charged droplets are attracted to the negative surfaces of siliceous aggregates where the emulsifying agent is adsorbed and chemically retained; this then causes the emulsion to break.

Tar

Tar is as a viscous liquid, black in colour, with adhesive properties, obtained by the destructive distillation of coal, wood, shale, etc.[2] By destructive distillation is meant the subjecting of the raw material to heat alone, without access to air. This definition means that there are many forms of tar. In practice, however, the highway engineer is concerned only with tar produced from the destructive distillation of bituminous coal. Other tars have been used for road surfacing purposes, but generally they have been unsatisfactory in comparison with bituminous coal tars.

Preparation of crude tars. The first step in the manufacture of a bituminous road tar is the production of crude tar by high-temperature carbonization of bituminous coal. Crude tar is actually a by-product of this destructive distillation process, carbonization being primarily carried out in order to produce coke for the steel industry or to supply gas for municipal and industrial heating purposes.

Crude tar produced by the coke industry is called coke-oven tar. The crude is normally obtained by heating between 12 and 14 tonnes of coal at a time in narrow carton-shaped brick-lined ovens at temperatures above 1300°C, so that the volatile constituents of the coal are removed and only coke is left. The volatile products, which include tar, are collected and the tar removed with ammonia liquor; the tar is later separated from the liquor by decantation.

Crude tar manufactured in conjunction with the production of gas is called gas-works tar. In this case a highly volatile bituminous coal is heated in comparatively small cigarette-shaped fire-clay retorts which may be of the horizontal, inclined or vertical variety. Heat is provided by water-gas obtained by passing air and steam through incandescent coke derived from a previous charge of bituminous coal. The water-gas is burnt in flues surrounding the retorts, the process of combustion being controlled by the introduction of air in order to obtain a higher and more uniform temperature. Gases which are given off in the retort are collected, and the tar is removed in its crude form in a similar manner to that for the coke-oven crude.

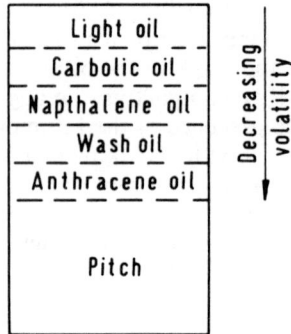

Fig. 3.5. Schematic illustration of the components of a dehydrated crude tar

Table 3.5 indicates the effects of temperature and type of production on the yield and characteristics of tar produced from an average grade of British coal. Although the highest tar yield is produced at relatively low temperatures, it is unsuitable for refining to road tar and cannot be used for surfacing purposes.

Road tars. While the crude tars obtained from the destructive distillation of bituminous coal are unsuitable for direct use on the roadway, they do contain many valuable ingredients. Figure 3.5 is a schematic illustration of the composition of a tar crude after it has been dehydrated and transported to a tar refinery. At the refinery, the distillation process by which the crude tar is refined is not only analogous to, but also similar to, the process of obtaining asphaltic bitumen from petroleum crude. In the distillation process the light oils such as benzene, toluene and xylene are first collected, and then the middle and heavy oils containing creosote oils and naphthalene are extracted. The residual material, called 'base tar' or 'pitch', is then fluxed back with tar oils to obtain road tars of the desired consistency and properties. Tar distillates of the creosote or anthracene oil types are most often used for this purpose. With both types, more viscous fluxing materials are used in Winter than in Summer.

Table 3.6 summarizes the main features of the road tars manufactured in Britain. As can be seen, there are two main types of tar, Type A and Type B, each of which has ten subdivisions based on viscosity characteristics. Type A tars contain the highest proportion of lower boiling oils and so are used extensively for surface dressing, as well as in the manufacture of some tarmacadam surfacings. Type B tars contain a smaller proportion of lower boiling oils and are used primarily in the preparation of tarmacadam surfacings. The manner in which these road tars should be used in particular surfacings is described in three British Standards.[20, 21, 22]

TABLE 3.5. *Crude tar production by various means from an average grade British coal*[1]

	Temperature °C	Tar l/tonne	Sp. gr. of tar at 77°F	Free carbon in tar %	Distillate to 315°C		Pitch % by wt
					Naphthalene %	Paraffin %	
Coke-ovens (narrow)	1300	37·9	1·210	20	30–35	0	72
Coke-ovens (ordinary)	1100–1250	43·9	1·200	15	20–30	0	66
Horizontal retort	900–1200	50·8	1·190	14	15–20	Trace	65
Inclined retort	1000–1100	55·4	1·155	12	5–12	5	58
Vertical retort	1000–1200	71·6	1·100	5·5	<5	13	48
Low-temperature carbonization	400–700	84·5	1·035	1	0	25	40

TABLE 3.6. *Types and properties of road tars used in Great Britain*[19]

Type A

Property \ Grade	A18	A22	A26	A30	A34	A38	A42	A46	A50	A54
Viscosity, equi-viscous temperature, °C	18±1·5	22±1·5	26±1·5	30±1·5	34±1·5	38±1·5	42±1·5	46±1·5	50±1·5	54±1·5
Water, % by weight (max.)	0·5	0·5	0·5	0·5	0·5	0·5	0·5	0·5	0·5	0·5
Distillation:										
(a) Oils below 200°C, % by weight (max.)	1·0	1·0	1·0	1·0	1·0	0·5	0·5	0·5	0·5	0·5
(b) 200–270°C, % by weight	12–17	9–16	8–15	6–14	4–13	3–10	2–8	2–7	1–6	1–5
(c) 270–300°C, % by weight (max.)	5–10	4–10	4–10	4–9	4–9	4–9	3–8	2–7	2–7	2–7
(b)+(c), % by weight (max.)	27	24	22	20	18	16	14	12	11	10
Softening point of residue:										
Ring and ball, °C (max.)	52	52	52	52	52	53	54	55	56	56
Ring and ball, °C (min.)	35	35	35	35	35	35	35	35	35	—
Phenols, % by volume (max.)	5·0	4·5	4·0	3·5	3·5	3·0	2·5	2·5	2·0	—
Naphthalene, % by weight (max.)	6·0	5·5	5·0	4·5	4·5	4·0	3·5	3·0	2·5	—
Matter insoluble in toluene, % by weight (max.)	21	22	23	23	24	24	25	25	26	26
Specific gravity at 15·5°C/15·5°C: (min.)	1·090	1·100	1·105	1·110	1·115	1·120	1·125	1·130	1·135	1·135
(max.)	1·230	1·240	1·245	1·250	1·255	1·260	1·265	1·270	1·275	1·275

Type B

Property \ Grade	B30	B34	B38	B42	B46	B50	B54	B58	B62	B66
Viscosity, equi-viscous temperature, °C	30±1·5	34±1·5	38±1·5	42±1·5	46±1·5	50±1·5	54±1·5	58±1·5	62±1·5	66±1·5
Water, % by weight (max.)	0·5	0·5	0·5	0·5	0·5	0·5	0·5	0·5	0·5	0·5
Distillation:										
(a) Oils below 200°C, % by weight (max.)	0·5	0·5	0·5	0·5	0·5	0·5	0·5	0·5	0·5	0·5
(b) 200–270°C, % by weight	4–11	3–10	1–8	1–6	1–5	0–4	0–3	0–3	0–2	0–2
(c) 270–300°C, % by weight (max.)	4–9	4–9	4–9	3–8	2–7	2–7	2–7	1–6	1–5	1–5
(b)+(c), % by weight (max.)	16	15	13	12	11	10	10	8	7	6
Softening point of residue:										
Ring and ball, °C (max.)	46	46	47	48	49	50	52	56	56	56
Ring and ball, °C (min.)	35	35	35	35	35	35	—	—	—	—
Phenols, % by volume (max.)	3·0	3·0	2·5	2·0	2·0	1·5	—	—	—	—
Naphthalene, % by weight (max.)	4·0	4·0	2·5	2·5	2·5	2·5	—	—	—	—
Matter insoluble in toluene, % by weight (max.)	23	23	24	25	25	26	26	28	28	28
Specific gravity at 15·5°C/15·5°C: (min.)	1·110	1·115	1·120	1·125	1·130	1·135	1·135	1·140	1·140	1·140
(max.)	1·250	1·255	1·260	1·265	1·270	1·275	1·280	1·280	1·285	1·290

Tar cut-backs and emulsions. As with the bitumen cut-backs and emulsions, tar cut-backs and emulsions may also be prepared. In practice, however, it has been found that these materials are comparatively hard to handle, and as a result they are now rarely used for highway work in Britain.

Comparison of bitumen and tar binders

There is considerable debate between engineers as to whether bitumen is 'better' than tar and vice versa. It is not intended to take sides in this controversy as, in fact, it can really be said with validity that each has particular advantages in certain situations.

Some properties of equivalent tar and bitumen materials are:

1. Both binders appear blackish in colour when viewed in large masses, but make a brownish stain and appear brown in colour when viewed in thin films.

2. Bitumens respond less readily than tars to small changes in temperature. Tar is liquid at lower temperatures and solidifies at comparatively higher ones.

3. Tars may be overheated and spoiled more easily than bitumens, but is much easier to get tar out of a road tanker.

4. Tar tends to penetrate more freely into open road surfaces.

5. Tar is not susceptible to the dissolving action of petroleum solvents or distillates. In parking areas, where petrol and oil are likely to drip or spill from vehicles, a tar surfacing may have a longer life than a bitumen one.

6. Bitumen is less brittle at low temperatures; this is because tar contains a higher percentage of free carbon.

Binder mixtures

In recent years, considerable research has been initiated into the effect of adding particular agents to the 'normal' road binders in order to improve their characteristics. Many additives have been considered, of which rubber and pitch are probably the most successful.

Rubber–bitumen/tar. Experiments involving mixtures of rubber and bitumen or tar have been carried out for nearly 50 years, but it is only since the development in 1938 of rubber powders made from latex that real progress has been shown. The following forms of rubber latex are now commercially available for use in bituminous road surfacings:

(a) *Latex.* This, the natural product of the rubber tree, is a suspension of rubber droplets in a watery serum concentrated and stabilized in such a way that the dry rubber content is between 60 and 70 per cent. Natural latex, concentrated by centrifuging to about 60 per cent of rubber and preserved with ammonia and other agents, is called *Centrifugal Latex. Evaporated Latex* is that which is concentrated by evaporation to more than 60 per cent of rubber and preserved with potassium hydroxide. Latex combines very well with bitumen emulsions but due to the presence of water, foaming may occur if it is added to a binder which is heated above 100°C. Foaming can be minimized by the addition of an anti-foaming agent in the binder or by efficient stirring during heating to provide a large exposed surface area of binder.

(b) *Unvulcanized powder.* This is a rubber powder containing 60 per cent latex rubber and 40 per cent by weight (19% by vol) of inert earth to prevent caking. The powder is made by spray-drying the latex and then adding the inert earth.

(c) *Vulcanized powder.* This is a powder containing about 95 per cent rubber which is derived from latex by crumbling lightly vulcanized coagulum.

(d) *Sheet rubber.* Many forms of sheet rubber are made from coagulated latex. For use on the road, the sheet rubber must first be milled and then dissolved in a fluxing oil. As a result sheet rubber is used only in the preparation of rubberized cut-back bitumen.

(e) *Graft rubber.* Used in the latex or powder form, this consists of a rubber to which is chemically grafted a proportion of a polymeric material (usually between 10 and 15 per cent of the natural rubber); this enables a stable mixture to be obtained with tar. Although considerable research work has been carried out, it has not yet been found possible to disperse an unmodified natural rubber directly into road tar and obtain a mixture that is sufficiently stable for road use.

Specifications have been published[18] which allow rubber to be used in rolled asphalt, bitumen macadam, and dense tar surfacings.

The field and laboratory investigations which have been carried out abroad[23] as well as in Great Britain,[24] indicate the following with regard to the use of rubber in bituminous materials:

1. The addition of certain percentages and types of rubber to certain types of bitumen in plant mix operations markedly increases cohesion and stability. When used with cut-backs in surface treatment work, tackiness is increased.

2. When the rubber is added with the mineral aggregates, the loss of the lighter fractions in the bitumen through weathering is retarded.

3. There is some evidence that aggregate–bituminous mixtures containing rubber are more resilient than ones without, and serve better to cushion vibrations and traffic shock.

4. The addition of rubber to bitumen increases its softening point, viscosity, elasticity and cohesion.

5. In cold areas, the addition of 5·5 to 7 per cent rubber by weight of bitumen reduces surface cracking due to temperature.

6. The addition of vulcanized rubber powder to mastic or rolled asphalt can very substantially reduce reflection cracking when bituminous surfacings are laid over joints or cracks in a concrete pavement.

7. A 50 per cent increase in life of bitumen macadam was obtained in one experiment when 4 per cent vulcanized powder was added to the binder. (A second study, with 2·5 per cent latex, was inconclusive.) Another limited full-scale experiment has suggested that the life of dense tar surfacing may be also increased by the addition of rubber powder.

8. No significant advantage is gained from using methcrylate-graft-rubber in tar surface dressing.

One further point which might be made here is that the addition of 4 per cent rubber can increase the cost of a wearing course by as much as 50 per cent. Thus, the use of rubber-binders is most likely to be justified on very heavily trafficked roads where it is particularly desirable to avoid the delays to traffic associated with the repair of cracks in ordinary surfacings.

Tar–bitumen. Recent years have seen the development of mixtures containing both tar and bitumen, the aim being to harness the favourable properties of both in one material, while minimizing their undesirable characteristics. Experiments to date indicate the following[26] with regard to tar-bitumen mixtures:

1. Tar–bitumen blends containing predominantly bitumen, but with a substantial proportion of tar, make highly satisfactory surface dressing binders. (The blends are now widely used commercially and are likely to be eventually accepted in national specifications.)

2. Tar–bitumen blends (65–70 per cent bitumen) can make satisfactory coated materials over a wide range of mixing temperatures. As with the blends used in surface dressing, a low brittle point similar to bitumen is a characteristic feature. The degree of hardening experienced by the blends during mixing and delivery of the coated materials depends upon the aggregate size and mixing temperature, and tends to be somewhat greater than bitumen. They will, however, readily withstand the high mixing temperatures often employed in cold weather. Fuming is at a low level also.

3. The good compaction achieved at lower temperatures with tar–bitumen coated materials indicates that considerable tolerance is available when laying during unfavourable weather conditions. A high resistance to deformation is exhibited on heavily trafficked roads.

4. Bitumen–tar blends (65% tar) have had only limited use in this country, and detailed conclusions regarding performance cannot yet be made.

Pitch–bitumen. The use of pitch–bitumen mixtures in road surfacings dates from 1950 when a patent was taken out which proposed the addition of coal-tar pitch to refinery bitumen in order to increase its susceptibility to oxidation; the intention was to produce a surfacing having a rougher surface texture and improved resistance to skidding. Research carried out at the Road Research Laboratory showed that weathering due to oxidation of bitumen is the principal factor determining the skid resistance of asphalt containing different bitumens. These experiments showed that the part played by the binder is primarily one of promoting a gradual wear of the sand-filler-bitumen mortar, and this is brought about by weather and traffic. The gradual erosion of the mortar—which still keeps a good micro-texture—exposes fresh unpolished aggregate, thereby maintaining a coarse, anti-skidding, texture. Prior to the advent of pitch–bitumen, only rolled asphalt wearing courses containing a substantial Trinidad Lake Asphalt content were permitted on motorways in Britain, i.e. the Trinidad material had the ability to constantly oxidise, where exposed to the atmosphere, so that the resultant thin film was removed progressively by vehicle wheels. It was then found that refinery bitumen could be made to oxidize at a rate equal to that of the Trinidad bitumen when modified by the addition of a certain amount of vertical retort pitch, with the result that the skidding resistance of the bitumen considerably improved. Furthermore, the adhesion between the aggregate and the binder was improved by the addition of this pitch, while resistance to stripping by water was also increased.

At the present time there is no British Standard for pitch–bitumen although it is now approved for use on all categories of roads previously approved for use of mixtures of petroleum bitumen and Lake asphalt. The pitch–bitumen specification for use with rolled asphalt states that it shall be composed of a mixture of 75–80 per cent petroleum bitumen (≤ 150 pen) and 20–25 per cent pitch derived as a result of the carbonization of coal at or above 600°C.

Because of their excellent qualities, the usage of pitch–bitumen surfacings in heavily-trafficked roads has grown considerably.[25] While it is initially more expensive than a straight-run bitumen surfacing, a pitch–bitumen surfacing costs less than one which uses a fluxed Trinidad Lake asphalt.

Binder tests and their significance

The most careful specifications with regard to the design and construction of a bituminous road surfacing are of little value if the properties of the bituminous binder used in the design are not adequately controlled. To aid the engineer in ensuring that the material obtained has the desired qualities, a number of tests have been devised which attempt to measure various binder properties for particular reasons. As is unfortunately the case with so many highway engineering test specifications, there are variations from organization to organization with regard to how exactly these binder tests should be carried out, although there is general agreement as to their significance. Since exact information on British practice in carrying out these tests is readily available in the literature (see refs 19, 27, 28, 29, 30, 31, 32, 33, 34, 35), they will be only briefly described, while more emphasis will be placed on factors relating to *why* the tests are carried out. Particular properties of bituminous binders will also be discussed in relation to the significance of the tests.

It is most convenient to discuss tests on binders by dividing them into the following four categories: 1. Consistency tests. 2. Composition tests. 3. Specific gravity test. 4. Flash and fire point tests.

Consistency tests. By consistency is meant the resistance of a material to flow. Since this property varies as the temperature changes from that required for processing at the construction site, i.e. up to about 177°C, to the much lower temperatures to which the road surfacing may be subjected in service, while the binders themselves may range from very thin liquids to semi-solids, it can be appreciated why there is no single method of test which can readily evaluate all bituminous binders for consistency over such a wide range. Instead there are a number of tests, each of which has certain advantages under specific conditions. The ones of importance are the penetration, viscosity and softening point tests.

Penetration test. This test consists of determining how far a standard steel needle will penetrate vertically into the binder under standard conditions of temperature, load and time. The standard test conditions are 25°C, 100 g and 5 sec. The results obtained are expressed in units of penetration, where one unit is equal to $\frac{1}{10}$ mm. The test is carried out on bitumens, rarely on tars.

Significance of test. The penetration test measures the consistency of semi-solid asphaltic bitumens so that they can be classified into standard grades. Since grade does not imply quality, the penetration test on its own has no relation to quality. However, penetration-grade refinery bitumens are known to reduce in penetration with age and to develop cracking tendencies. Penetration values below 20 have been associated with bad cracking of road surfacings, while cracking rarely occurs when the penetration exceeds 30. It is not known whether cracking is directly due to loss of penetration; it is more probable that both cracking and loss in penetration are due to a more basic change in the bitumen.

Surfacings containing penetration-grade bitumens must be premixed and laid hot. Generally the higher penetrations are preferred for use in colder climates, but good road surfacings have been produced with relatively low penetration materials in Canada and with relatively high penetration bitumens in the southern United States.[36] For normal roadworks in Britain, 35–450 pen bitumens are in common use. Mastic asphalts containing 15 pen bitumens are used at such locations as bus-stops, where traffic stresses are very high. Softer grades, up to about 60 pen, are used for most rolled-asphalt surfacings, while even softer ones are used for bitumen macadams and cold asphalts. Although it is more difficult to get lower penetration bitumens to adhere to aggregates, once adhesion has been established the cohesive bond is much stronger than if softer ones are used.

The penetration test, when carried out at different temperatures, can also determine the temperature susceptibility of a bitumen. Where resistance to flow is important, e.g. when bitumen is used to fill cracks in concrete road-slabs, a small change in consistency with wide change in temperature is desirable.

Viscosity tests. In basic terms it can be said that the viscosity of a liquid is the property that retards flow so that when a force is applied to a liquid, the slower the movement of the liquid, the higher the viscosity; in this sense viscosity is a 'pure' measure of consistency. The scientific definition of viscosity, or more correctly, the 'coefficient of viscosity' of a liquid may be explained as follows:

If the space between two parallel surfaces is filled with a liquid, and one of the surfaces is moved parallel to the other, a force which resists movement

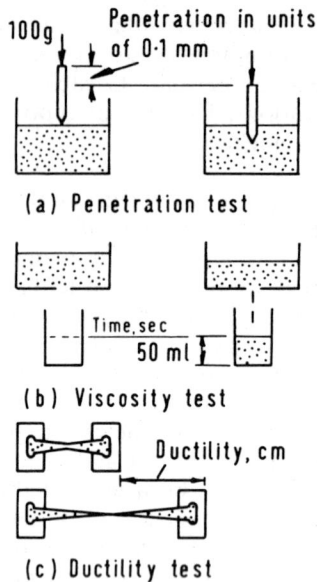

(a) Penetration test

(b) Viscosity test

(c) Ductility test

Fig. 3.6.　Schematic illustrations of some bituminous binder tests

is set up as a result of the presence of the liquid. If the force of resistance is denoted by F, the area of the surface by A, the velocity of movement of one surface relative to the other by v, and the distance between the surfaces by d, then F is proportional to Av/d. If a factor η, called the coefficient of viscosity, is introduced into the relationship, then the following equation can be written:

$$F = \eta \cdot \frac{Av}{d}$$

At a given temperature, η has a constant value for any one liquid, but generally it is different for different liquids. If the variables are measured in centimetre-gramme-second (c.g.s.) units, then the viscosity η is equal to the resistance in dynes offered by the liquid to the movement of a surface 1 sq. cm. in area moving with a velocity of 1 cm/sec at a distance of 1 cm from another fixed surface. In this system the unit of viscosity is known as the poise; thus if $\eta = 200$, the liquid is said to have an absolute viscosity of 200 poises. The absolute viscosity of water at 30°C is 0·01 poises.

At this time there are at least 58 instruments in use throughout the world for measuring 'viscosity'. They may be divided into three main groups, based on the following principles of operation:

1. The flow of a body through a liquid. 2. The flow of a liquid through a tube. 3. The rotation of one of two co-axial cylinders when the space between them is filled with a liquid.

While the truest measurements of viscosity are obtained with viscometers of the third group, they are rarely used in highway engineering practice. Instead, most bitumen binder specifications for roadworks are based on the results obtained with industrial viscometers which utilize the second principle. The one in most common use in Britain is the Standard Tar Viscometer (S.T.V.). Notwithstanding its name, this viscometer is used to evaluate both tar and bitumen viscosities.

The S.T.V. test measures the time, in seconds, for a fixed quantity of the binder liquid (50 ml) to flow from a cup through a standard orifice under an initial standard head and at a known test temperature. While two sizes of orifices are standardized at 10 and 4 mm, in practice the 10 mm one is most used. Because of the great variation of time required for a given amount of the different binders to flow through the orifice, it is not practical with this viscometer to determine the viscosity of all bituminous binders at the same temperature, so different temperatures are used for particular materials. Tar specifications require that the S.T.V. flow times lie between 10 and 140 sec, so the test temperatures for tars are chosen to ensure these conditions. With cutback bitumens the 10 mm orifice cup is used at 25°C for materials whose viscosities at that temperature and in that cup exceed 15 sec, and at 40°C for materials whose viscosities at 25°C exceed 500 sec. The 4 mm cup at a temperature of 25°C is used for cut-backs whose viscosities are less than 15 sec in the 10 mm cup at 25°C.

Since confusion can and does arise over employing several test temperatures, what is known as the Equi-viscous Temperature (e.v.t.) system has been devised to overcome this source of error. In this system the temperature in °C is specified at which the time of flow of the binder is 50 sec as measured on the standard tar viscometer. A particular advantage of the e.v.t. system with respect to tars is that it provides a single scale on which can be accommodated the viscosities of all tar products ranging from the very fluid ones right up to the hard pitches. As a result the viscosities of tars—but not bitumens—are now commonly reported as equi-viscous temperatures. This is possible with tars since the temperature susceptibilities of all road tars are similar, irrespective of the source, whereas the temperature susceptibilities of all bitumens are not.

If the viscosity of a tar as measured with the standard tar viscometer is known, and provided it lies between 33 and 75 sec, the e.v.t. of the tar can be determined by applying the appropriate correction factor given in Table 3.7 to the temperature at which the S.T.V. test was carried out. A more approximate translation may be carried out by using the relationships shown in Fig. 3.7.

Significance of viscosity tests. The viscosity of a bituminous binder is its most important physical characteristic. Hence viscosity measurements are useful not only in ensuring that the material with the desired properties has been obtained, but also as a means of selecting binders for specific uses. If a binder with too low a viscosity is pre-mixed with an aggregate, it may flow off the stone while en route from the mixing plant. Conversely, if the viscosity is too high, the mixture may be unworkable by the time it reaches the site. If too low a viscosity is used for surface dressing purposes, the result may be 'bleeding' or a loss of chippings under traffic. With low-viscosity

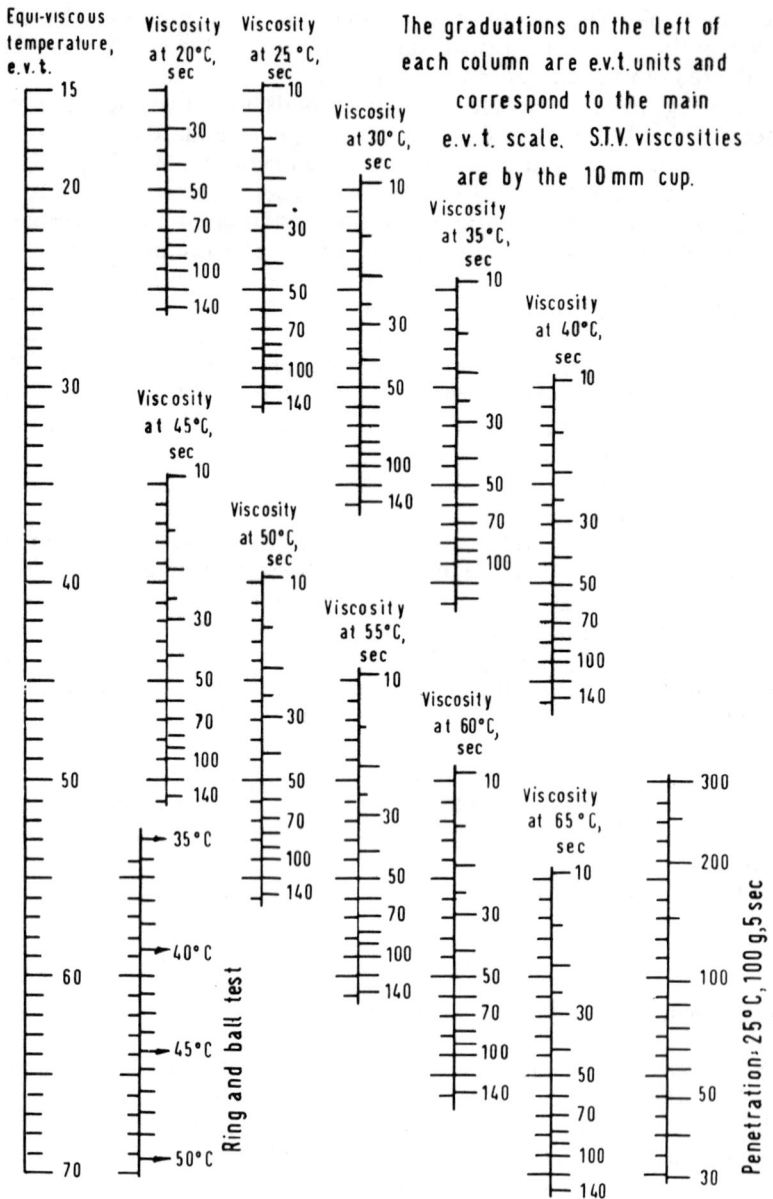

Fig. 3.7.   Chart showing approximate relationships between e.v.t., S.T.V. viscosity, ring and ball test and penetration[19]

TABLE 3.7. *Corrections, in °C, to be used in determining the equi-viscous temperatures of road tars whose standard tar viscosities at known temperatures are already available*[19]

S.T.V. vis-cosity sec	°C									
	0	1	2	3	4	5	6	7	8	9
30				−2·5	−2·3	−2·2	−2·0	−1·9	−1·7	−1·5
40	−1·4	−1·2	−1·1	−0·9	−0·8	−0·6	−0·5	−0·4	−0·3	−0·1
50	0	+0·1	+0·2	+0·3	+0·5	+0·6	+0·7	+0·8	+0·9	+1·0
60	+1·1	+1·2	+1·3	+1·4	+1·5	+1·6	+1·7	+1·7	+1·8	+1·9
70	+2·0	+2·1	+2·2	+2·2	+2·3	+2·4				

binders, there is generally less chance of pumping pipes becoming blocked, mixing and application temperatures can be kept lower, and aggregates are more easily coated.

Softening point test.   The ring and ball (R and B) softening point test is also extensively used to evaluate the consistency of bituminous binders. The test consists of placing a 9·53 mm diameter steel ball on a binder sample placed in a steel ring and immersed in a water bath. Heat is applied to the water and its temperature is raised until a value is reached when the test sample is sufficiently soft to allow the ball, enveloped in binder, to fall through a

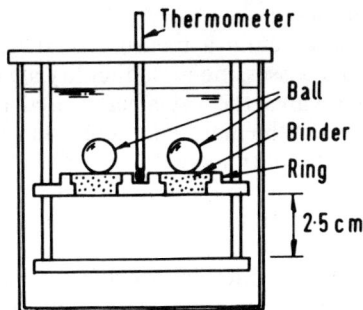

Fig. 3.8.   Apparatus for ring-and-ball test

height of 2·5 cm. The water temperature at which this occurs is read to the nearest 0·5°C and called the ring and ball softening point.

Significance of test. The softening point is not a melting point; bituminous binders do not melt but instead gradually change from semi-solids to liquids on the application of heat. The method of test is entirely arbitrary and must be exactly carried out if the results are to be of value. As the practical significance of the test is limited, the specifications of many binders for particular purposes are now often written without softening point requirements.

As indicated by Table 3.6, the R and B test is included in the standard specifications for road tar; the softening point of the pitch residue is a means of characterizing its rate of setting. It is useful also for determining the temperature susceptibilities of bitumens which are to be used in thick films, such as in crack fillers. When two bitumens have the same penetration value, the one with the higher softening point is normally less susceptible to temperature changes.

Ductility test. A ductile material is one which elongates when in tension. The ductility of a bituminous binder is expressed as the distance in centimetres that a standard semi-solid briquette will elongate before breaking. The minimum cross section of the briquette before testing is 1 sq. cm. Normally the test is carried out on semi-solid asphaltic bitumens and the residues from cut-back bitumens; they are tested at a temperature of 25°C and at an extension rate of 5 cm/min.

Significance of test. The ductility test is probably the most controversial of the many empirical tests found in the bituminous literature. Some authorities are of the opinion that the test is quite valuable, while others state it is useless. While most highway engineers agree that the property variously described as plasticity, elasticity, ductility and malleability is desirable in road bitumens in order to overcome surface movements induced by traffic and temperature stresses, there is considerable difference of opinion as to whether or not the elongation of a thin thread of bitumen is a measure of this desirable quality. Certainly correlation is negligible between the conditions existing in the standard test and the conditions to which the bitumen is subjected in a road surfacing.

The ductility test is actually a measure of the internal cohesion of a bitumen. Since bitumens possessing high ductility are normally cementitious and adhere well to aggregates, the test is better used as a measure of whether or not ductility is present in the material, rather than as a means of determining the exact degree of ductility available.[36] Thus a bitumen with 100-cm ductility might well be considered a better road surfacing constituent than one with 10-cm ductility, but a binder with 80-cm ductility is not necessarily better than a 60-cm one.

Bitumens possessing high ductility are also usually highly susceptible to temperature changes, while low ones are not. The lack of ductility does not necessarily indicate poor quality; indeed, bitumens of low susceptibility and low ductility are highly desirable as crack fillers in roadways. Tars and pitches are generally more ductile than bitumens of the same viscosity, while the harder grades of bitumen are less ductile than the softer ones. The extent to which ductility is dependent on the test temperature is demonstrated by the fact that changes of 5·5–11°C in the temperature at which a ductility test is carried out may reverse the relative ductility order of two different bitumens.

Composition tests. As is clear from Tables 3.3, 3.4, and 3.6, most binder specifications include criteria regarding composition. To ensure that these composition specifications are met, a number of tests have been developed to determine the proportions of the specific fractions and components of the bituminous binders.

Distillation tests. Distillation tests are used to determine the quantity and quality of the volatile constituents and the amount of non-volatile residues present in road tars, cut-back bitumens and binder emulsions. In emulsions, the volatile constituents are, of course, primarily water.

The distillation tests for cut-backs, tars and emulsions differ somewhat in procedure. Essentially, however, all involve heating a specified quantity of the binder in a standard flask or still at a specified rate, and then determining the amount of distillate removed at prescribed standard temperatures.

Significance of tests. The distillation tests are amongst the most valuable of the highway tests for bituminous binders. Most importantly, they enable a close check to be kept on the quality of the binders used on road schemes. This is simply illustrated by considering the specifications for road tar with regard to water, phenols and naphthalene. The moisture content of road tars must not exceed 0·5 per cent; water contents in excess of this will lead to undesirable foaming if the tar is heated beyond 100°C. Tar acids, or phenols, are to a limited extent desirable in road tars as they ensure good adhesive properties, but excess amounts can cause premature ageing through oxidation. An upper limit on the amount of naphthalene present in road tars is also desirable since it may crystallize out from the tar, thereby increasing the viscosity to an undesirable extent.

Distillation tests carried out on road tars and cut-back bitumens can provide useful information both on the type of volatiles in the binder, and on the rate at which these volatiles will be lost under field conditions. For instance, if a given cut-back is known to lose its volatiles too slowly under certain conditions, then under similar conditions a cut-back with higher boiling range volatiles—as determined by the distillation test—can be expeted to cure even more slowly.

The residue left from the distillation test can also provide useful information. While there is no guarantee that the residue is the same as either the base bitumen in the cut-back or the material left in the road surfacing after curing, it is sufficiently similarly to justify examination for characteristics measured by means of the standard consistency tests.

Water content test. While, of course, the moisture content of a binder is automatically determined when a distillation test is carried out, there are occasions when it is desirable to determine the moisture content alone, without carrying out a complete distillation procedure. In such instances the moisture content may be determined ' directly ' by mixing a specified amount of binder with a pre-determined amount of petroleum spirit (for bitumens) or coal-tar solvent (for road tars) with which it is immiscible, and distilling in a flask or still which is attached to a glass water-cooled reflux condenser and a graduated receiver. Distillation is continued until the volume of water in the receiver is onstant. This volume is then expressed as a percentage by weight of the original material.

Significance of test. Bituminous binders should only contain extremely low moisture contents if they are to be heated beyond 100°C, i.e. if significant quantities of water are present, foaming of the binder will occur.

Loss-on-heating test.   In this test, a 50 g sample of bitumen is placed in a small container and left for 5 hours in a revolving-shelf oven, the temperature of which is maintained at 165°C. At the end of the heating period, the sample is cooled to room temperature and weighed. The loss in weight of the sample is then expressed as a percentage of the original weight.

The penetration test is often carried out on the residue of the loss-on-heating test. The result obtained is expressed as a percentage of the penetration of the bitumen before heating.

Significance of test. Specifications for penetration-grade asphaltic bitumens may require that the maximum loss of weight on heating and the maximum drop in penetration should not exceed particular values. In practice, the losses in weight and penetration are almost always within the requirements.

This loss on heating is essentially an accelerated volatilization test. Although it is perhaps somewhat representative of heating conditions in field storage tanks, it is certainly not at all representative of conditions during plant mixing or distributor operation, because of differences in film thickness as well as temperature. Thus the test is of use only as a general indication of volatile content under the specified conditions of test. If the consistency of the residue is compared with that of the original, the amount of hardening resulting from this amount and manner of volatilization is all that is indicated.

Ash content test.   The ash contentof a bitumen is the percentage by weight of inorganic residue left after ignition of the sample. In the course of the test, a weighed amount of the sample is gently heated until it begins to burn, and then it is fired until the ash is free from carbon.

Significance of test. This test is carried out on both penetration-grade and cut-back bitumens. With refinery bitumens, the test is used to ensure that undesirable amounts of mineral matter are not present; this is particularly important with surface dressing materials. The presence of ash in a bitumen is not necessarily harmful in itself, as is evidenced by the use of lake asphalt in road surfacings either on its own or in blended mixtures. In such instances, the ash test can be useful in determining the composition of the binder.

Solubility tests.   In determining the percentage of binder present in either bitumen or tar, differed solvents are normally used. In the case of bitumens, the accepted solvent is normally carbon disulphide (CS_2) while in tars it is toluene. In either case a specified quantity of binder—usually about 2 g—is dissolved in a given quantity of solvent. After filtering the solution through a fine-porosity filter, the residue retained is determined and the percentage of soluble material is calculated by difference.

Significance of tests. A solubility requirement of 99·5 per cent in CS_2 is found in all British specifications for refinery bitumens. Carbon disulphide is highly inflammable and the much safer carbon tetrachloride or methylene chloride can usually be used for normal solubility tests without significant loss in accuracy. Tars are not completely soluble in these or, indeed, in any other solvent, and so it is very important to be aware of how much of the binder is soluble in the solvent used. For instance, tar refined from a vertical-retort crude is between 90 and 95 per cent soluble in toluene, whereas road tars manufactured from coke-oven or horizontal-retort crudes may only be 80 to 90 per cent soluble in toluene.

Insoluble material in refinery bitumens is normally dirt and like material picked up in the course of storage, or else salt which was not removed during the refining process; those in natural bitumens are finely-divided mineral materials. Insolubles in tar are measured in order to guard against an over-heated or 'burnt' tar being used, i.e. excessive heat can coke a tar and this is reflected in an abnormally high toluene insoluble content. Whichever binder is studied, solubility tests are most useful in establishing the amount of actual binder available for use in the road surfacing. Some authorities believe, however, that the ash content test is more reliable for this purpose as the solvents fail to remove dust in colloidal suspension.

Differences in the solubility of bitumens in carbon disulphide and carbon tetrachloride are due to the quantities of carbenes present. Carbenes are chemical compounds which are produced in refinery bitumens by overheating and cracking, and a content of 0·5 per cent may be indicative of detrimental overheating. However, because of lack of sensitivity, this carbene content determination is of doubtful value in detecting overheating.

Specific gravity test. The specific gravity of a bituminous binder is the ratio of the weight of a given volume of material at a given temperature to that of an equal volume of water at the same temperature. In the case of road tars, the test temperature is normally 15·5°C, but other temperatures may also be used as necessary. In all cases the temperatures of the binder and the water must be reported if the results are to have any comparative value.

Significance of test. The principal use of specific gravity determinations is in establishing the relation between binder weight and volume for transporting and billing purposes. Specifications for binders in road surfacings are normally expressed as percentages by weight whereas they are usually shipped and measured by volume.

Knowledge of the specific gravity is also very useful in differentiating between the different types of binders. The specific gravities of refinery penetration-grade bitumens normally lie between 1·00 and 1·05, while road tars, as indicated in Table 3.6, vary according to the manufacturing process by which the crude tar is produced. Vertical-retort tars have the lowest specific gravities—usually between 1·10 and 1·15—while horizontal-retort and coke-oven tars have the highest—between 1·18 and 1·25.

As discussed in the chapter on Bituminous Surfacings, it is also necessary to know the specific gravity of the binder in order to determine the percentage of voids in mechanically-designed mixtures of bitumens and mineral aggregates.

Flash and fire point tests.   The flash point test is carried out by heating a sample of the binder at a uniform rate, while periodically passing a small flame across the surface of the material. The temperature at which the vapours given off from the binder first burn with a brief flash of blue flame is called the flash point of the binder. If heating is continued until the vapours continue to burn for a period of at least 5 seconds, the temperature at which this occurs is called the fire point.

Significance of tests.   The flash and fire point tests are primarily safety tests although they may also be considered as indirect reflections of binder volatility. The flash point is the more important of the two, since it indicates the maximum temperature to which the binder can be safely heated. Safe practice requires special precautions when temperatures in excess of the flash point are being used. The flash points of most penetration-grade bitumens lie in the range 245–335°C, while rapid-curing cut-backs may flash at temperatures as low as 27°C. Medium-curing cut-backs usually flash between 52°–99°C, while slow-curing ones have flash points above 110°C.

The fire point is of little significance, and its use in specifications is negligible.

CEMENT

A cement is a material which, if added in a suitable form to a non-coherent assemblage of particles, will subsequently harden by physical or chemical means and bind the particles into a coherent mass. This definition, which is very broad in scope, allows such diverse materials as bitumen, tar and lime to be grouped together under the umbrella of ' cement '. In general practice, however, the term is used to refer only to the Portland and high-alumina cement binders.

Portland cement

This, the most commonly used cementing agent in the concrete building industry, was first made in England about 1825. The raw materials which constitute the make-up of the various types of Portland cement now in use, are calcium carbonate—found in the form of limestone or chalk—and alumina, silica and iron oxide—found combined in clay or shale. The first stage of the manufacturing process consists of mixing the clay or shale to a slurry with water, and then admixing the chalk or finely-ground limestone. The prepared slurry is then fed continuously through a long gently-sloping cylindrical kiln which is fired by burning pulverized coal blown into its lower end. As the slurry moves through the kiln it is gradually heated and successive changes take place. First the water is evaporated, and then the calcium carbonate decomposes into calcium oxide (quicklime) and carbon dioxide.

About three-fourths of the way down the kiln, when the temperature of the material is at about 1450°C, incipient fusion takes place and the components of the lime and the clay combine to form calcium silicates, calcium aluminates and the other compounds which constitute the burnt clinkers which leave the kiln. These clinkers, normally about 12·5 mm in diameter at this stage, are allowed to cool and then transported to ball-and-tube mills where they are ground to a fine powder. During grinding, a small amount—usually between 4 and 7 per cent—of calcium sulphate (gypsum) is added, to prevent the cement from setting too rapidly when hydrated and used.

Types of Portland cement. In essence all types of Portland cement are manufactured as described above. The various types which are available on the market differ only in that they are obtained by varying the proportions of the raw materials, the temperature of burning and the fineness of grinding. The following is a general description of the more important types of cement.

Ordinary Portland cement[37] is what might be termed the workhorse of the cement trade. It is this cement which is normally supplied by the manufacturer unless another type is specifically called for. Having a medium rate of hardening, it is suitable for most kinds of concrete construction, including road construction.

Rapid-hardening Portland cement[37] or, as it is also called, high-early strength cement, was first developed about 1920 for use in circumstances where concrete of high-early strength is required, e.g. when it is desired to allow traffic on a roadway as soon as possible. The principal difference between this and the ordinary cement is that the final cement clinker is much more finely ground. As a result, a much greater surface area is available for reaction with water, and so it is able to harden more rapidly. It is not, however, a quick-setting cement; the times for the initial and final sets of rapid-hardening cement are about the same as for ordinary cement. Rapid-hardening cement is more expensive than ordinary cement, but this disadvantage is offset by the fact that the 7-day strength of mixtures containing material are approximately equal to the 28-day strengths of normal cement. The final strength obtained with both cements are, however, about the same.

Portland-blast-furnace cement[38] is a mixture of finely-ground cement clinker and blast-furnace slag to which a little gypsum is added to retard the setting time. Granulated slag by itself is a relatively inert material; it reacts so slowly with water that it certainly cannot be regarded as a cementing agent. It has the important characteristic, however, that it is pozzolanic, i.e. it will react with lime in the presence of water and form a cementitious product. The hardening of blastfurnace slag cement is characterized by the superposition of two processes. The first of these is the hydration of the ground Portland cement clinkers. Secondly, as this takes place, there is a release of free calcium hydroxide which reacts with the slag as it hydrates.

Hence the properties of this type of cement are mainly dependent on the ratio of the two constituent materials. A Portland blastfurnace cement that is low in slag behaves very much like an ordinary Portland cement, while one with a high slag content—this is rarely more than 65 per cent—reflects the influence of the slag ingredients.

In general, slag cement hardens more slowly than ordinary cement, but in the long run there is little difference between the final strengths obtained with both types. The use of slag cement for road construction is most easily justified where economy is of primary importance and high initial strength is not. Usage of blast-furnace cement is most popular in Scotland, where there are a considerable number of blast-furnaces producing slags suitable for the manufacturing process.

Sulphate-resisting cement[51] in general, complies with the requirements for ordinary Portland cement. It is known that sulphates present in sea-water, some ground-waters, and in certain clay soils, cause the disintegration of concrete or soil cement pavements made with ordinary Portland cement. The advantage of sulphate-resisting cement is that this disintegration process, which is caused by the reaction between the sulphates and one of the compounds present in all Portland cements, is controlled to the extent that concrete or coarse-grained soil cement can be protected against particular sulphate (sulphur trioxide, SO_3) concentrations.

High alumina cement

High alumina cement[39] is very different, both in composition and properties, from Portland cement. It is manufactured by melting a mixture of limestone or chalk and bauxite in a reverberatory or electric furnace, allowing the molten mass to cool, and then grinding it to a fine powder. Bauxite is a naturally occurring lateritic material formed when an igneous rock is so heavily weathered that the silica, lime, magnesia and alkalis have been gradually removed in solution, leaving behind the hydrated oxides of aluminium, iron and titanium contaminated with residual silica. The bauxite used in high-alumina cement production in Britain is imported from France and Greece; the cement manufactured is often known as 'Ciment Fondu'.

The principal feature of a high alumina cement is its very rapid hardening property which enables it to attain a strength after 24 hours that is very close to its maximum, and which would take ordinary Portland cement about 28 days to achieve. In the course of this very rapid hardening a considerable amount of heat is generated and this enables the cement to be used at much lower air temperatures than ordinary Portland cement. High alumina cement has another advantage in that it is much more resistant to the attacking action of sulphates.

Although high alumina cement hardens very rapidly, it is also relatively slow-setting. By setting is meant the change in the cement paste which occurs when its liquid or plastic nature begins to disappear. The start of this change is called the initial set, and its completion is termed the final set.

Hardening, which is the development of strength, does not begin until setting is complete. Gypsum has the effect of retarding setting and it is for this reason that a small amount is added to cement in the course of its production. The setting time and rate of setting is of vital importance in the construction of roads since for good results to be obtained, concrete may have to be mixed, transported substantial distances, placed and compacted before substantial setting can occur. In the case of high alumina cement, the length of set is between 2 and 6 hours, so that there is normally more than adequate time to place the concrete in the roadway.

Chemical composition of cements

The four main chemicals present in cement are quicklime (CaO), silica (SiO_2), alumina (Al_2O_3) and ferric oxide (Fe_2O_3). Some typical cement compositions are shown in Table 3.8.

TABLE 3.8. *Main chemical constituents of cement*

Chemical	Type of cement		
	Ordinary and rapid-hardening Portland cement	Portland-blastfurnace cement	High-alumina cement
CaO	59–66	45–55	35–40
SiO_2	18–23	25–30	10–12
Al_2O_3	4·5–8	10–20	40–45
Fe_2O_3	2·5–4·5	3–5	1–20

From a practical point of view, however, the constituents of most importance in a cement are the mineralogical compounds formed during the fusion process in the kiln; it is these which, when hydrated, form the actual binding paste. The most important of these are:

Tricalcium silicate (C_3S)	$3\ CaO.SiO_2$
Dicalcium silicate (C_2S)	$2\ CaO.SiO_2$
Tricalcium aluminate (C_3A)	$3\ CaO.Al_2O_3$
Calcium aluminoferrite (C_4AF)	$4\ CaO.Al_2O_3.Fe_2O_3$

It is customary in cement chemistry to denote the individual clinker minerals by short symbols such as $CaO = C$, $SiO_2 = S$, $Al_2O_3 = A$ and $Fe_2O_3 = F$ and this enables the above formulas to be abbreviated to C_3S, C_2S, C_3A and C_4AF.

Little is really known about what happens when water is added to cement. As indicated by Table 3.9, the hydration of the C_3A component takes place very quickly; in fact whatever strength there is at the end of the first day is a function of this hydrated product. It is suggested that this hydration takes place in the following manner, without any lime being picked up or released:

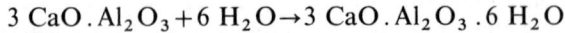

$$3\ CaO.Al_2O_3 + 6\ H_2O \rightarrow 3\ CaO.Al_2O_3.6\ H_2O$$

After the first day, the effect on strength of the hydrated C_3A compound is negligible. In contrast to this, the principal contributor to the strength gain at this time is the C_3S compound. The greatest C_3S strength gain is observed within 28 days, after which hydration continues but at a much slower rate. The reaction that takes place is believed to take the following form:

$$2[3\ CaO.SiO_2] + 6\ H_2O \rightarrow 3\ CaO.2\ SiO_2.3\ H_2O + 3\ Ca(OH)_2$$

As can be seen, this reaction involves the evolution of lime. The tricalcium disilicate hydrate which is formed is the final product of cement hydration and is believed to be the principal binding material. A similar product is believed to be formed when the C_2S compound is hydrated. Thus:

$$2[2\ CaO.SiO_2] + 4\ H_2O \rightarrow 3\ CaO.2\ SiO_2.3\ H_2O + Ca(OH)_2$$

This compound's final contribution to the long-term strength may ultimately be equal to that of the C_3S compound.

TABLE 3.9.　*Comparison of effects of main Portland cement compounds*

Principal constituent compounds	Rate of chemical reaction and heat generation	Most active period	Contribution towards final strength
C_3A	Fast	1st day	Small
C_3S	Moderate	2nd to 7th day	Large
C_2S	Slow	7th day onwards	Moderate

AGGREGATES

Mineral aggregates are the basic materials of highway pavement construction. Not only do they support the main stresses occurring within the pavement, but in addition the aggregates in the road surface must resist wear due to abrasion by traffic as well as the direct weathering effects of the natural elements. The manner in which they do so depends on the inherent properties and qualities of the individual particles and on the means by which they are held together, i.e. by interlocking, by cementitious binders, or by both.

Road aggregates may be used by themselves in road construction. In low-cost roads, where it is desired to make the most use of locally available materials such as gravels, they frequently form the entire pavement structure. In such instances, the gradation of the soil-aggregate materials is most important and particular attention must be paid to ensuring that maximum plasticity criteria are not exceeded. Aggregates are also used by themselves to form the sub-base beneath concrete pavements and as roadbase and sub-base materials in high-quality flexible ones. In the former, the load-bearing capacity of the aggregate sub-base is not of primary importance and emphasis is placed instead on achieving a gradation which will prevent pumping of the subgrade or intrusion of frost-susceptible materials while at the same time improving the subsurface drainage characteristics of the roadway. In a flexible pavement roadbase, stress-carrying capacity is the factor of primary importance, so the aggregate must be selected to serve this purpose.

Aggregates may also be used with binders, of which the two most important are cement and the bituminous binders. In high-quality bituminous road surfacings, aggregates may comprise up to about 95 per cent of the weight of the surfacing. It is the aggregate which is primarily responsible for any load-carrying capacity which the surfacing may have, while at the same time it provides the resistance to abrasion under traffic which is so important in preventing vehicles from skidding. Although there are very many types of bituminous surfacing, it has been stated that, in general, the ideal aggregates for their construction should have the following characteristics:[40] 1. Strength and toughness. 2. Ability to crush into chunky particles, free from flakes, slivers and pieces that are unduly thin and elongated. 3. Low porosity. 4. Hydrophobic characteristics. 5. Particle size and gradation appropriate to the type of construction.

These criteria are also important for concrete, particularly those relating to particle shape and size distribution, since they affect both water requirements and workability of concrete mixes as well as other important concrete properties. Also of importance are the textures and structures of the aggregates.

Types of aggregate

One definition of an aggregate is that it is a material such as broken stone, slag, gravel, sand or the like which, when held together by a binding agent, forms a substantial part of such materials as concrete, asphalt, coated macadam or the like. This definition means that any hard material, whether it be natural or artificial, may be classified as an aggregate. In practice, however, the materials *suitable* for use as road aggregates may be limited in a particular area.

Natural aggregates. By far the majority of road aggregates are formed from natural rock. Geologists have classified rocks into three main groups, based on their method of origin; these are known as igneous, sedimentary and metamorphic rocks. It is of interest to briefly examine these classification groups as it is the means by which they were formed which primarily determine the compositions and textures of the rock aggregates.

Igneous rocks were formed at or below the earth's surface by the cooling of molten material, called magma, which erupted from, or was trapped beneath, the earth's crust. Igneous rocks formed at the earth's surface when the magma came into contact with the atmosphere are called extrusive rocks, while those formed below the earth's surface are intrusive rocks. Extrusive magma cooled rapidly at the earth's surface and as a result the rocks formed are very often glassy or vitreous (without crystals) or partly crystalline and partly vitreous. This latter occurrence is due to the outer layers of the magma flow or lava cooling very rapidly and becoming vitreous, while the inner material cooled more slowly and became crystalline with the grain-sizes being very small indeed. As the magma escaped at the surface, the contained gases expanded, and hence an extrusive rock may contain cavities which give it a vesicular texture. In contrast with the extrusive rocks, the intrusive rocks are entirely crystalline, having been formed as a result of the magma cooling slowly under the protective cover of the earth's crust. In many instances the rock crystals may be sufficiently large to be easily distinguished by the naked eye.

The grain-size is most important in a road aggregate. If the grains are very coarse, i.e. > 1.25 mm, the particles are liable to be brittle and may break down beneath the crushing action of a compacting roller. On the other hand, if the grains are too fine, i.e. < 0.125 mm, and particularly if the rock is vesicular, they are also liable to be brittle and splintery. The best igneous roadstones normally contain medium grain-sizes between 0.125 and 1.25 mm.

Igneous rocks can also be separated on the basis of their being acidic or basic. Acidic rocks can be either extrusive or intrusive, depending on their silica content. As used here, the term 'silica' is meant to refer to SiO_2, which may be either in combination with alumina, etc., to form feldspars and other minerals, or free as in the mineral quartz. When a rock contains more than about 66 per cent silica it is described as acidic, while it is a basic rock if it contains less than 55 per cent silica; those containing between 55 and 66 per cent total SiO_2 are called intermediate rocks. Acidic aggregates are considered by many authorities to be undesirable road-making materials as they are hydrophilic or 'water-loving'. In preparing bituminous mixtures, these aggregates can be difficult to coat with binder in contrast with the hydrophobic or 'water-hating' aggregates formed from a rock such as basalt.

Sedimentary rocks were formed when the products of disintegration and/or decomposition of any type of rock were transported and redeposited, and then consolidated or cemented into a new rock type. Sedimentary rocks also include those formed as a result of chemical precipitation or the deposition of organic remains in water. They were, as the name implies, laid down in stratified layers and so are easily distinguishable from the massive structure of igneous rocks. The rock strata, at one time horizontal, may now be deformed and displaced through angles of up to 90 deg as a result of upheavals which continually occurred throughout geologic history.

Sedimentary rocks may be sub-divided by various means, but from the highway engineer's point of view the most convenient one is that based on the predominant rock mineral. This allows three main rock sub-classifications, the calcareous, siliceous and argillaceous groups.

Calcareous rocks were formed as the result of great thicknesses of the remains of small marine animals being deposited on the ocean floors. The predominant mineral is calcite, $CaCO_3$, and this renders the rocks basic and, in general, most suitable as an aggregate for bituminous surfacings.

Some types of calcareous rocks, are however too porous to be used as road aggregates. Generally the softer the rock the more porous it is, and porosity can be expressed in terms of the rock's saturation moisture content, e.g. for soft chalks this may be as high as 25 to 30 per cent while for hard limestone it will be less than 1 per cent. Tests[41] have shown that all crushed chalks are frost-susceptible with the magnitude of frost heave increasing linearly with the saturation moisture content of the chalk aggregate. Similarly, all oolitic and magnesian (dolomitic) limestones having an average saturation moisture content within the aggregate in excess of 3 per cent must be regarded as frost-susceptible and should not be used within a highway pavement.

In contrast with the calcareous rocks, siliceous rocks—also known as arenaceous rocks—were formed from deposits of sand and silt which became lithified as a result of pressure by overlying strata, or by the deposition of cementing material between the grains. The predominant mineral in these rocks is either quartz or chalcedony, both SiO_2, and this may tend to make adhesion between these aggregates and bituminous binders relatively difficult. These rocks are also much harder than calcareous rocks, but some are so brittle as to be unusable in road pavements, particularly in bituminous surfacings.

Argillaceous rocks are those in which the clay minerals predominate. They were formed when fine-grained particles of soil were first deposited as clays or muds and then consolidated by pressure from overlying deposits. As a result the argillaceous rocks are very fine-grained, highly laminated and very often are easily crushed into splinters. For these reasons, they are rarely used as road aggregates, and never in bituminous surfacings.

Also included under the title of sedimentary 'rocks' are the natural gravels and sands. These are described in the chapter on Soil Engineering for Highways and will not be further discussed here.

Metamorphic rocks are igneous or sedimentary rocks which, as a result of being subjected to tremendous heat (thermal metamorphism), or heat and great pressure combined (regional metamorphism), were transformed into new types of rock by the recrystallization of their constituents. Metamorphism in any particular instance may have been aided by the presence of permeating solvents which worked through the rock and promoted new mineral growth.

Thermal metamorphic rocks are almost always harder than the rocks from which they were originally transformed. In addition, they often show an interlocking of mineral constituents that renders them more useful as aggregates. As a result metamorphic rocks which were altered by heat alone are in considerable demand as road aggregates. In contrast to the thermally-formed rocks, the regional metamorphic rocks are relatively coarse-grained and some are highly laminated, with the result that they may be quite fissile and liable to be crushed when compacted with rollers. Hence they are generally less desirable as road aggregates.

Trade groups. Within the three broad geological classes discussed above, geologists have identified literally hundreds of different types of rock. Many of these rocks differ little from each other with regard to their road-making qualities, and so it is convenient to gather them together into groupings with common characteristics. The British Standard on mineral aggregates[42] recognizes ten 'trade groups' of natural rock aggregates, the members of each group being similar to each other in composition, grain-size and texture. These groupings and their general relationship with the geological groupings are shown in Fig. 3.9. The following is a brief description of each group:[43]

1. *Basalt group.* The most important rocks in this group are basalt, dolerite, basic porphyrite and andesite. Members of the group are primarily

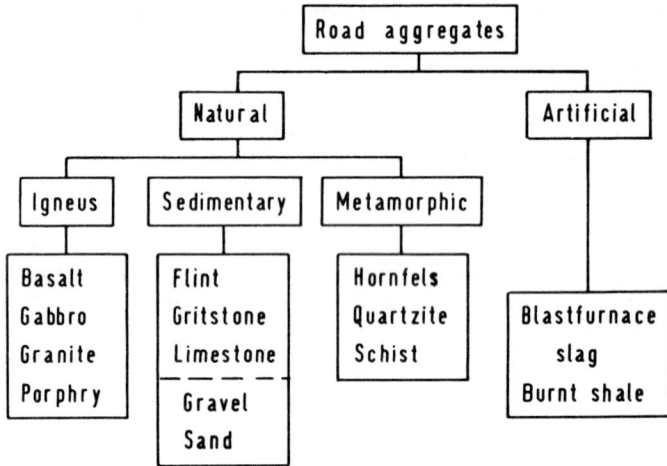

Fig. 3.9.   Principal groups of road aggregates

basic and intermediate igneous rocks of medium and fine grain-size, but some of their metamorphic equivalents are also included. They are composed principally of feldspars and ferro-magnesian minerals, the latter giving the rocks their characteristic dark colouring and a relatively high specific gravity. They are widely distributed in Britain and are extensively used as road aggregates.

2. *Flint group.* This group has only two members, flint and chert, which are very fine-grained sedimentary rocks consisting mainly of crypto-crystalline silica (chalcedony). These rocks vary from white to black in colour and have low specific gravities. Flint is widely distributed as gravel material and, in eastern and southern England, this is the principal local source of road aggregate.

3. *Gabbro group.* There are eight members of this group of which the most important are gabbro, basic diorite and basic gneiss. The gabbro group is composed primarily of basic igneous rocks, but some intermediate rocks of coarse grain size—as well as their metamorphic equivalents—are

included. The most common constituents are feldspars and ferro-magnesian minerals, and it is the latter which often predominate and give a dark colour and high specific gravity to the rocks. The rocks of this group are not widely distributed in Britain, but they are extensively used as roadstones where they occur.

4. *Granite group.* The members of this group are mostly acidic and intermediate igneous rocks of coarse grain size, but it also includes their metamorphic equivalents. The most important members of the group—there are seven altogether—are granite, quartz-diorite, gneiss and syenite. The predominant minerals are feldspars and quartz, but micas, pyroxenes or amphiboles may also occur. These rocks, which are light in colour and have specific gravities below 2·80, are widely distributed and extensively used for road aggregate purposes.

5. *Gritstone group.* This group is composed mainly of siliceous sedimentary rocks which are usually of medium or coarse grain size; they are sometimes cemented with fine-grained material which may be siliceous, calcareous, argillaceous or ferruginous. These rocks, typical examples of which are sandstone, tuff and greywacke, are usually medium or light in colour and have specific gravities below 2·80. They also are extensively used as road aggregates.

6. *Hornfels group.* This group includes all thermally metamorphosed rocks except marble and quartzite. They are medium to dark in colour and usually have fine to medium grain sizes. Hornfels rocks are dense and hard and have high specific gravities; they are not widely distributed in Britain but where they occur they are extensively used for roadmaking purposes.

7. *Limestone group.* This group includes rocks which are primarily composed of calcium carbonate. These are the sedimentary rocks, limestone and dolomite, and the metamorphic rock, marble. The limestone rocks have a medium to fine grain size, are medium to light in colour, and are intermediate with regard to specific gravity. Widely distributed, they are most extensively used as road aggregates.

8. *Porphyry group.* The ten members of this group are acid or intermediate igneous rocks of fine grain size. Typical examples are porphyry, granophyre, microgranite and felsite. Broadly similar to the rocks of the granite group, they are also widely distributed and much used as road aggregate.

9. *Quartzite group.* These are siliceous sedimentary or metamorphic rocks composed almost entirely of quartz. Members of this group, i.e. quartzite, quartzitic sandstone and ganister, are light in colour, have a grain size which is medium to fine, and are intermediate in specific gravity. Quartzite rocks occur in Britain both as solid rock and as gravel, and are very much used as roadstones.

10. *Schist group.* This group is composed of laminated rocks such as schist, phyllite and slate. They are rarely used as road aggregates because of their general instability.

Figure 3.10 is a map showing the locations of all major quarries and gravel pits in Britain. It clearly shows that it is only in particular parts of the

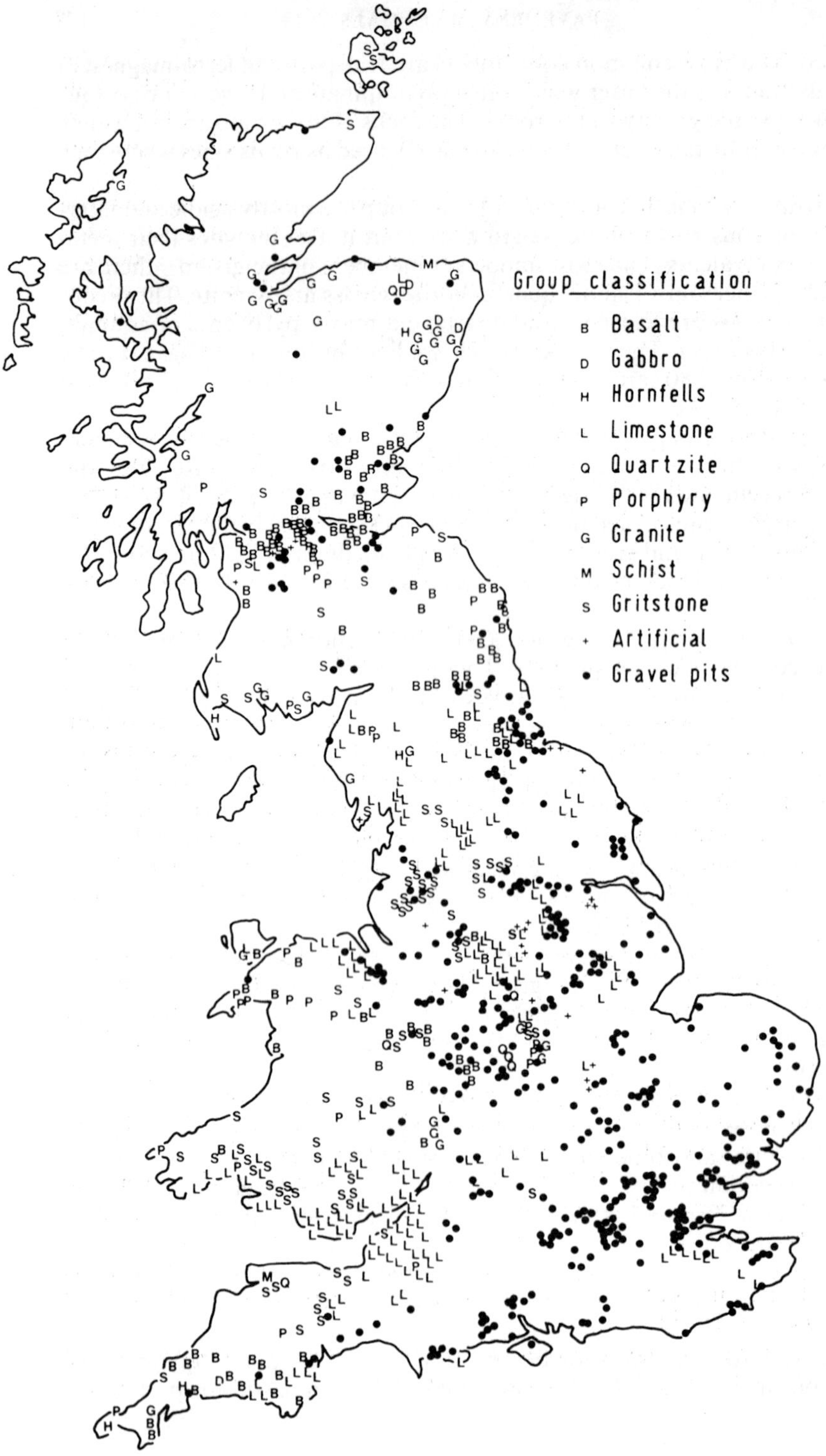

Group classification

B	Basalt
D	Gabbro
H	Hornfells
L	Limestone
Q	Quartzite
P	Porphyry
G	Granite
M	Schist
S	Gritstone
+	Artificial
•	Gravel pits

Fig. 3.10. Distribution of road aggregate quarries and pits in Great Britain[44]

country that there is any significant shortage of suitable aggregates for road construction. The area most lacking in natural rock aggregate is south-east England.

In particular localities it will be found that certain rock aggregates are known by traditional names which are very different from those recommended by the British Standard. This can be very confusing to those not familiar with local terminology, and it is strongly recommended that local names should not be used in official documents or scientific reports. Some of the more common traditional names and their British Standard equivalents are shown in Table 3.10.

TABLE 3.10. *Some common traditional rock names*[43]

Traditional name*	Appropriate group in B.S. classification for roadstone
Clinkstone	Porphyry (rarely Basalt)
Cornstone	Limestone
Elvan (Blue Elvan)	Porphyry (Basalt)
Flagstone	Gritstone
Freestone	Gritstone or Limestone
Greenstone	Basalt
Hassock	Gritstone
Hornstone	Flint
Pennant	Gritstone
Rag (stone)	Limestone (rarely Gritstone)
Toadstone	Basalt
Trap (rock)	Basalt
Whin (stone)	Basalt

* As the traditional names are often applied loosely, the information in this table cannot be precise.

Artificial aggregates. The principal artificially-formed aggregate used in road construction is air-cooled blast-furnace slag; this is a by-product of the smelting of iron ore in blast furnaces. Use of this material, however, is relatively limited at this time, as the accumulations of slag created during the past century have been used. Slag aggregate now utilized is taken almost entirely from current production.

When produced under controlled conditions, slag makes an excellent non-frost susceptible road aggregate. Desirably it should be reasonably uniform in density and quality, and contain few glassy fragments. Unfortunately this does not always occur as it cools very rapidly when removed from the furnace; this can often result in great differences in grain-size and porosity between the exterior—which may be glassy—and the more coarse-grained interior of the slag aggregate particle. A slag suitable for bituminous surfacing purposes will normally have a bulk unit weight greater than 1250 kg/m^3, sulphur content of less than 2·75 per cent and porosity, as reflected by water absorption, of not more than 4 per cent.[11, 20]

Aggregate tests and their significance

As aggregates obtained from different sources differ considerably in their constitution and properties, inevitably they differ also with regard to their engineering properties. It is necessary, therefore, to carry out various tests on aggregates to ensure not only that undesirable materials are excluded from highway pavements, but also that the best available aggregates are included.

Aggregate tests may be arbitrarily divided into four main groups; descriptive 'tests', non-destructive quality tests, durability tests and specific gravity tests. In the following discussion emphasis is placed on the reasons why these tests are carried out; detailed descriptions of the exact manner in which the tests are conducted are readily available in the literature.[42]

Descriptive tests. This title is intended to define the visual examination of an aggregate that enables it to be described in terms of both the shape and the surface texture of the particles. This results in subjective descriptions of these mineral aggregate characteristics. These descriptions may be catalogued under the headings described in Table 3.11 (a) and 3.11 (b).

Significance of tests. The descriptive 'tests' are most useful in classifying aggregates. The descriptive classifications are, in turn, very valuable guides relative to the internal friction properties of an aggregate. By internal friction is meant the properties which resist the movement of aggregates past each other. Thus, for instance, crushed basalt is generally considered an excellent road aggregate, since it has high internal friction as a result of having good interlocking qualities—because of the angular shapes of the particles—and a rough surface texture. In contrast with basalt, a rounded smooth aggregate such as gravel is relatively low in internal friction since particle interlock is not possible and surface friction is low. It is for this reason that most gravel-aggregate specifications require that the gravel be artificially crushed to produce jagged edges and surfaces before being used in a highway pavement.

Non-destructive quality tests. These non-destructive tests are carried out on the aggregate to determine its suitability for a specific use. The results obtained are normally compared with aggregate specifications to see whether they comply with the desired properties and characteristics. The tests of particular interest are the gradation, water absorption and shape tests.

Gradation test. Gradation, sieve analysis, screen analysis and mechanical analysis are synonymous terms which refer to the quantity expressed in percentages by weight of the various particle sizes of which a sample of aggregate is composed. This is determined by separating the aggregates into portions which are retained on a number of sieves or screens having specified openings[45] which are suitably graded from coarse to fine. The results obtained may be expressed either as total percentage passing or retained on each sieve or as the percentages retained between successive

TABLE 3.11. *Descriptive evaluations of mineral aggregates*[42]
(*a*) Particle shape

Classification	Description	Examples
Rounded	Fully water-worn or completely shaped by attrition	River or seashore gravel; desert, seashore and wind-blown sand
Irregular	Naturally irregular, or partly shaped by attrition and having rounded edges	Other gravel; land or dug flint
Flaky	Material of which the thickness is small relative to the other two dimensions	Laminated rock
Angular	Possessing well-defined edges formed at the intersection of roughly planar faces	Crushed rocks of all types; talus; crushed slag
Elongated	Material, usually angular, in which the length is considerably larger than the other two dimensions	
Flaky and elongated	Material having the length considerably larger than the width, and the width considerably larger than the thickness	

(*b*) Surface texture

Group	Surface texture	Characteristics	Examples
1.	Glassy	Conchoidal fracture	Black flint, vitreous slag
2.	Smooth	Water-worn, or smooth due to fracture of laminated or fine-grained rock	Gravels, chert, slate, marble, some rhyolites
3.	Granular	Fracture showing more or less uniform rounded grains	Sandstone, oolite
4.	Rough	Rough fracture of fine- or medium-grained rock containing no easily visible crystalline constituents	Basalt, felsite, porphry, limestone
5.	Crystalline	Containing easily visible crystalline constituents	Granite, gabbro, gneiss
6.	Honeycombed and porous	With visible pores and cavities	Brick, pumice, foamed slag, clinker, expanded clay

sieves. The total-percentage-passing method is very convenient for the graphical representation of a grading and is most widely used in graded-aggregate specifications.[46] The individual-percentages-retained-on-particular-sieves procedure is preferred in specifications for single-sized aggregates.[47]

Significance of test. Gradation is the characteristic of a road aggregate on which perhaps the greatest stress is placed in specifications; it is also probably the cause of the majority of controversies which arise between the supplier and road builder relative to the suitability of aggregates for specific uses. The proper grading of an aggregate is important because of its direct influence on both the quality and the cost of a completed pavement.

The limits placed on a particular gradation depend on the nature of the work in which the aggregate is to be employed. Thus, for instance, the grading of a material to be used in a dense bituminous surfacing—which depends considerably on gradation for its denseness and consequent stability—is more critical than the grading of an aggregate for use in macadam, in which stability is heavily dependent on the interlock of the aggregate particles. When well-graded aggregates have to undergo extensive handling or transporting, segregation of sizes may occur which can be relatively costly to remedy.

When aggregate particles are to be bound together by cement or bitumen or tar, a variation in the grading of an aggregate will result in a change in the amount of binder required to produce a material of given stability and quality. Proper aggregate grading contributes to the uniformity, workability and plasticity of the material as it is mixed. For example, in the case of concrete, improved workability generally permits a decrease in the amount of mixing water, which in turn either results in an increase in strength or allows the cement content to be reduced.

Shape tests. There are three mechanical measures of particle shape which may be included in the specifications for aggregates for road construction. These are the flakiness index, elongation index and angularity number.

The *flakiness index* of an aggregate is the percentage by weight of particles whose least thickness is less than three-fifths of their mean dimension; the mean dimension as used in each instance is the average of two adjacent sieve aperture sizes between which the particle being measured is retained by sieving. The flakiness test, which is not applicable to particles smaller than 6·35 mm in size, is carried out by first separating the aggregate into individual percentages retained on specified sieve sizes, and then passing at least 200 particles from the individual percentages through sieves or patterns having elongated slots whose widths are 0·6 times the individual mean dimensions. The flakiness index is then reported as the total weight of the material passing the various thickness gauges or sieves, expressed as a percentage of the total weight of the sample gauged.

The *elongation index* of an aggregate is the percentage by weight of particles whose greatest length is greater than 1·8 times their mean dimension. As with the flakiness test, the elongation test is not applicable to aggregate sizes smaller than 6·35 m; it is also similar in that first the aggregate sample is fractionated and then the individual particles from the fractions are passed through openings on a metal length-gauge. The elongation index is taken as the total weight of the material retained on the length-gauge, expressed as a percentage of the total weight of the sample gauged.

The *angularity number* of an aggregate is the amount, to the nearest whole number, by which the percentage of voids exceeds 33 when an aggregate is compacted in a specified manner in a standardized metal cylinder.

Significance of tests. The internal friction of an aggregate is the property which, by means of the interlocking of particles and the surface friction between adjacent surfaces, resists particle movement under the action of an imposed load. Use of the shape tests in specifications is based on the view that internal friction is influenced by the shapes of the particles. Thus, for instance, the standards for single-sized aggregates require that these materials should not have flakiness indices greater than 40 for 38·1 mm aggregate and larger, and 35 for 25·4 mm materials and smaller. Since, other factors being equal, an aggregate composed of smooth rounded particles of a certain gradation will contain less voids than one of the same grading but composed of angular particles, the angularity of an aggregate can be reflected in terms of the volume of contained voids when the aggregate is compacted. Measurements show that the angularity number may range from zero for a material of highly-rounded beach-gravel particles to 10 or more for newly-crushed rock aggregate.

Water absorption test. This test is normally carried out in conjunction with the specific gravity test. The procedure consists of soaking the aggregate sample in distilled water for 24 hours, surface-drying and weighing in air, and then oven-drying and weighing in air again. The water absorption is obtained by expressing the difference between the weights of the saturated and the oven-dried sample in air as a percentage of the latter.

Significance of test. Knowledge of the absorption properties of an aggregate is particularly important in bituminous surfacing design, since the porosity of the aggregate affects the amount of binder required and additional binder material may have to be incorporated in the mixture to satisfy the absorption by the aggregate after the ingredients have been mixed. On the beneficial side, porous aggregates usually show better adhesion to the binder due to the mechanical interlock caused by the binder penetrating the particles.

The water absorption values allowed for road aggregates normally range from less than 0·1 per cent to about 2 per cent for materials used in road surfacings, while values of up to 4 per cent may be accepted in roadbases. Water absorption values representative of the more important aggregates are listed in Table 3.12.

Durability tests. Two types of resistance test are carried out on road aggregates. These are the abrasion and the toughness tests.

TABLE 3.12.   *Summary of means and ranges of values for roadstone tests in each rock-group*[48]

Group classification (BS 812)		Aggregate crushing value	Aggregate impact value	Aggregate abrasion value	Water absorption (per cent)	Specific gravity	Polished stone coefficient
Artificial	Mean	28	27	8·3	0·7	2·71	0·59
	Range	(15–39)	(17–33)	(3–15)	(0·2–1·8)	(2·6–3·4)	(0·35–0·74)
	No. of samples	55	21	18	19	19	33
Basalt	Mean	14	15	6·1	1·1	2·80	0·62
	Range	(7–25)	(7–25)	(2–12)	(0·0–2·3)	(2·6–3·0)	(0·45–0·81)
	No. of samples	123	79	65	68	68	70
Flint	Mean	18	23	1·1	1·0	2·54	0·39
	Range	(7–25)	(19–27)	(1–2)	(0·3–2·4)	(2·4–2·6)	(0·30–0·53)
	No. of samples	63	32	45	24	24	7
Granite	Mean	20	19	4·8	0·4	2·69	0·59
	Range	(9–35)	(9–35)	(3–9)	(0·2–0·9)	(2·6–3·0)	(0·45–0·70)
	No. of samples	41	32	28	16	16	23

Gritstone	Mean	17	19	7·0	0·6	2·69	0·72
	Range	(7–29)	(9–35)	(2–16)	(0·1–1·6)	(2·6–2·9)	(0·60–0·82)
	No. of samples	81	45	31	33	33	32
Hornfels	Mean	13	12	2·2	0·4	2·82	0·45
	Range	(5–15)	(9–17)	(1–4)	(0·2–0·8)	(2·7–3·0)	(0·40–0·50)
	No. of samples	28	24	13	15	15	4
Limestone	Mean	24	23	13·7	1·0	2·66	0·43
	Range	(11–37)	(17–33)	(7–26)	(0·2–2·9)	(2·5–2·8)	(0·30–0·75)
	No. of samples	164	61	34	42	42	51
Porphyry	Mean	14	14	3·7	0·6	2·73	0·56
	Range	(9–29)	(9–23)	(2–9)	(0·4–1·1)	(2·6–2·9)	(0·43–0·71)
	No. of samples	62	29	23	30	30	23
Quartzite	Mean	16	21	3·0	0·7	2·62	0·58
	Range	(9–25)	(11–33)	(2–6)	(0·3–1·3)	(2·6–2·7)	(0·45–0·67)
	No. of samples	57	37	29	21	21	20
All groups	Mean	19	19	5·7	0·7	2·68	0·58
	Range	(5–39)	(7–35)	(1–26)	(0·0–3·7)	(2·3–3·4)	(0·30–0·83)
	Nos. of samples	724	370	311	313	313	292

Abrasion tests.   Many abrasion tests have been developed in order to evaluate the ease (or difficulty) with which aggregate particles will deteriorate under attrition. During pre-war years, the Deval attrition test and the Dorry abrasion test were the most widely used methods of evaluating the abrasiveness of road-stones. Since the war, however, the aggregate abrasion test and the accelerated polishing test have come to be the more widely used tests in Britain.

The *aggregate abrasion test* is carried out on 33 cm³ of aggregate passing the 12·7 mm and retained on the 9·52 mm sieve. The particles are mounted in, but project above, the surface of a setting compound contained in a small shallow tray. The tray is up-ended and the aggregate pressed in contact with a horizontal rotating steel disc; contact is maintained by a load of 2 kg for 500 revolutions at a speed of 28–30 rev/min. As the disc rotates, a standard abrasive sand is fed continuously to its surface just in front of the inverted tray. On completion of the 500 revolutions the test sample is removed and the percentage loss in weight of the aggregate is calculated; this is called the abrasion value of the aggregate. Table 3.12 includes aggregate abrasion values for representative materials from the various trade groups of roadstones.

The *accelerated polishing test* is essentially a skid-resistance test which is carried out on aggregate particles subjected to accelerated attrition. In the course of the test, 40–60 particles passing the 9·52 mm and retained on the 7·94 mm BS sieves are set in a sand–cement mortar and clamped on to the flat periphery of a 406 mm diameter wheel which is then rotated at a speed of 325 rev/min. A 200 mm diameter pneumatic-tyred wheel is brought to bear on the aggregate surfaces with a load of 40 kg, and water and sand (3 hours) and emery powder (3 hours) are fed continuously to the interacting surfaces. At the end of the six hours attrition, the state of polish reached by the aggregate specimen is measured in terms of the coefficient of friction between its surface and a rubber slider in a standard pendulum-type, portable skid-resistance tester. The measurement on this tester is called the polished-stone coefficient of the aggregate (see Table 3.12).

Significance of tests.   For an aggregate to perform satisfactorily in a highway pavement, it must be sufficiently hard to resist the abrasive effects of traffic over a long period of time. The abrasive tests are basically accelerated tests which attempt to eliminate aggregates which may be unsuitable from this point of view.

Whether the aggregate abrasion value test is a simulation of abrasion under actual traffic conditions is, of course, very debatable. Numerous testings have shown, however, that aggregates with abrasive values greater than 15 are too soft for the wearing course of a bituminous surfacing. The accelerated polishing test might be considered more representative of tyre-surfacing interaction on the roadway. In any case, it has been shown that the state of polish of the aggregate at the end of the 6-hour test approximates the state of polish reached after several months on a heavily-trafficked road or several years on a road carrying light traffic. Results indicate that an aggregate with a polished-stone coefficient of greater than 0·8 is likely to

TABLE 3.13. *Recommended polished-stone and sideways force coefficient values*

Type of site	Example locations	Polished-stone value	Skidding resistance	
			Test speed, km/h	Sideways force coefficient
A. Difficult	1. Roundabouts and their approaches 2. Bends of less than 150 m radius on unrestricted roads 3. Gradients 1:20 or steeper and longer than 91 m 4. Approaches to traffic signals on unrestricted roads	0·62	50	0·55
B. Average	1. Motorways 2. All other trunk and principal roads 3. All other roads in urban and rural areas, which carry more than 2000 veh/day	0·59	50	0·50*
C. Easy	1. Roads not included in types A or B but carrying more than 1000 veh/day	0·45	50	0·40

* For motorways the sideways force coefficient may be 0·45 if the test speed is 80 km/h

remain rough under any traffic conditions, while one with a value of 0·3 or lower will become so highly polished as to give rise to a dangerously slippery road surface when either wet or dry. As can be seen from the data quoted in Table 3.12, most British road aggregates have values well in excess of the minimum value of 0·30. (It might be noted, however, that, since these data were published, the polished-stone coefficient test has been slightly modified, so that the results shown here are now all about 10 per cent too high.) Table 3.13 shows the minimum polished-stone values recommended for aggregates used at particular highway locations with (a) surface dressings, or (b) embedded in the following bituminous surfacings—hot rolled asphalt, dense tar surfacing, fine cold asphalt, mastic asphalt.

Toughness tests. If toughness is defined as the power possessed by an aggregate to resist fracture under an applied load, then the tests in common usage which are reflective of this quality are the aggregate crushing test, the ten per cent fines test and the aggregate impact test.

The *aggregate crushing value* is a measure of the resistance of an aggregate to crushing under a gradually applied compressive load. The test is normally carried out on material passing the 12·7 mm and retained on the 9·52 mm BS sieve. The aggregate is placed in a standard mould using a specified procedure, and then a load of 40·64 tonne is gradually applied to the material over a period of ten minutes. The load is then released, and the amount of material passing the 2·36 mm sieve is determined. This weight, expressed as a percentage of the total weight of the sample, is termed the aggregate crushing value. Crushing values representative of the road aggregates in Britain are given in Table 3.12.

The *ten per cent fines test* is similar to the aggregate crushing test except that the load, in tonnes, is determined which causes 10 per cent fines (passing the 2·36 mm BS sieve) to be formed over a period of ten minutes. Preliminary estimates of this load are first obtained and then a simple formula is used to calculate that which gives the 10 per cent fines content.

The *aggregate impact value* gives a relative measure of the resistance of an aggregate to sudden shock or impact. The test is carried out by subjecting aggregate which has passed the 12·7 mm and is retained on the 9·52 mm BS sieve to 15 blows of a 13·6–14·1 kg hammer falling through a height of 381 mm. After impact, the material passing the 2·36 mm sieve is expressed as a percentage of the total weight of the original sample and termed the aggregate impact value.

Significance of tests. In all forms of flexible pavement, the aggregate must be tough enough to support the weight of the rollers during construction and the repeated impact and crushing actions of traffic. Thus the aggregate must have a durable resistance to both crushing and impact. The toughness tests just described are empirical attempts to measure this resistance; however, since they are empirical, the results obtained have to be correlated with field experience for them to have any significance.

The results obtained with the aggregate crushing test suggest that materials with values greater than about 35 are too weak to be utilized in a pavement. When aggregates with values greater than about 30 are evaluated, the crushing test begins to become insensitive due to the restricting cushioning effect which the fines formed in the early part of the test have on later aggregate break-down. The 10 per cent fines test was developed to evaluate these weak aggregates, and it is quite likely that it will eventually supplant the aggregate crushing test entirely. Values obtained with this test range from as low as 1 tonne for chalk to 41 tonne for the hardest aggregates; normally, however, road aggregates should not show a fines value of less than 8 tonnes. Values obtained with the aggregate impact test are, in general, numerically very similar to the aggregate crushing values. Exceptions to this are found with the fine-grained highly siliceous aggregates such as flint and quartzite; these are more sensitive to shock loads and have impact values which are significantly different from their crushing values.

Specific gravity test. This test is normally carried out in conjunction with the water absorption test. It consists of soaking a sample of the aggregate in distilled water for 24 hours, weighing it in water at the end of this period, surface-drying and weighing in air, and then weighing in air again after oven-drying for 24 hours. The bulk specific gravity of the aggregate is then obtained by dividing the weight of the oven-dry sample in air by the difference between the saturated sample weights in air and in water. Representative specific gravity values for British road aggregates are listed in Table 3.12.

Significance of test. As road aggregates are normally proportioned by weight, specific gravity is of vital importance in determining the proper

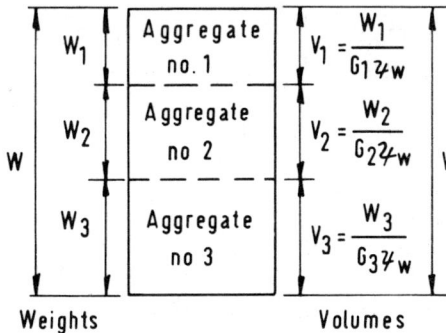

Fig. 3.11. A graphical representation of the components of an 'average' specific gravity calculation

mixture. Gradation specifications are valid only if the coarse and fine fractions have approximately the same specific gravities. If the value for the fine fraction is much greater than that for the coarse, the result is a mixture which, because of a lack of fines, may be too harsh. On the other hand, if the specific gravity of the coarse fraction is the greater, a mixture which is too rich in fines may be obtained. When these conditions are encountered in practice, arbitrary gradations should not be used, but instead various gradation mixtures should be analyzed carefully and evaluated on their merits.

The 'average' specific gravity of an aggregate composed of fractions of different specific gravities can be calculated from the individual values. Figure 3.11 is a schematic illustration of such an aggregate; it is composed of three fractions of known weights, volumes and specific gravities. From this diagram

$$G_{ave}\psi_w = \frac{W}{V}$$

However, the individual weight of each aggregate is always expressed as a percentage of the total weight W. Thus, taking W as equal to 100,

$$G_{ave}\psi_w = \frac{100}{\dfrac{W_1}{G_1\psi_w} + \dfrac{W_2}{G_2\psi_w} + \dfrac{W_3}{G_3\psi_w}}$$

Therefore

$$G_{ave} = \frac{100}{\dfrac{W_1}{G_1} + \dfrac{W_2}{G_2} + \dfrac{W_3}{G_3}}$$

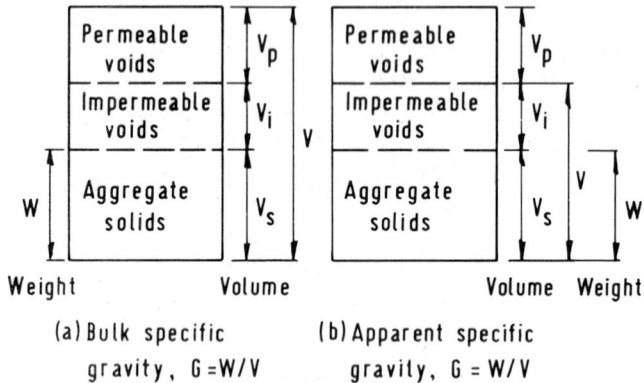

Fig. 3.12.   Diagrammatic illustration of the composition of an aggregate particle showing the weight and volumes used in specific gravity calculations

where G_{ave} = average specific gravity of the final aggregate, and W_1, W_2 and W_3 = weight percentages of the individual fractions.

The above has assumed that the specific gravity of an aggregate is a definite single physical property. In reality, this is not so as there are at least two specific gravities, depending on how the measurement is made. As illustrated in Fig. 3.12, these are the bulk and apparent specific gravities. The *bulk* specific gravity is the ratio of the weight in air of a volume of aggregate (this includes voids which are permeable and impermeable) to the weight in air of an equal volume of distilled water; it is this specific gravity test which is described at the beginning of this discussion. The *apparent* specific gravity is the ratio of the weight in air of a volume of aggregate (including impermeable, but not permeable, voids) to the weight in air of an equal volume of distilled water. In some laboratory design procedures, the choice of specific gravity can mean markedly different results if the aggregates are very porous.

Aggregate gradation calculations

Most aggregate gradation problems are straightforward and require no special mathematical knowledge to solve them. A typical gradation determination is shown in Table 3.14. This table shows that there are three ways of representing the grading of an aggregate; these are by an individual percentage which is retained on each sieve, by a cumulative percentage retained on each sieve, and by a total percentage passing each sieve.

TABLE 3.14. *Laboratory determination of the gradation of an aggregate (sample size = 2000 g)*

BS sieve size, mm	Individual weight retained on each sieve		Cumulative weight retained on each sieve		Total weight passing each sieve
	g	%	g	%	%
25·4	0	0	0	0	100
12·7	160	8	160	8	92
3·8	240	12	400	20	80
2·36	200	10	600	30	70
1·18	300	15	900	45	55
0·600	180	9	1080	54	46
0·300	180	9	1260	63	37
0·150	340	17	1600	80	20
0·075	220	11	1820	91	9

As illustrated in the chapter on Bituminous Surfacings, most standard specifications are expressed as the total percentage passing each sieve. In certain instances, however, an aggregate gradation may be submitted in which the limits are expressed in terms of the individual maximum and minimum amounts retained on each sieve, and it is then necessary to compare the gradation limits with the per cent passing specification; this can be troublesome for the inexperienced engineer and so one way in which it can be made is briefly described here. Another troublesome problem also discussed is that which arises when two or more aggregates have to be blended in order to meet a specification.

Comparing gradation specifications. The simplest way of comparing two specifications is to plot them on an aggregate chart and then see whether they are related to each other. Unfortunately, when one specification is given in terms of the individual amounts retained on each sieve, and the other is laid down in terms of the total percentage passing given sieve sizes, this direct comparison is not possible—except perhaps to the very experienced engineer who is able to determine instinctively by reading the two specifications just what exactly is the relationship between the two. There is, however, a very useful and simple method of translating specifications for individual contents retained into those based on total percentages passing[49] which enables the comparison to be made by even the most inexperienced; this is described here.

To demonstrate the translation procedure, the hypothetical individual

TABLE 3.15. *Translating an aggregate specification written in terms of per cent passing-and-retained into one of per cent passing*

(a) Assumed aggregate gradation requirements

BS sieve size, mm		Percentage of material
Passing	Retained on	
	25·4	0
25·4	12·7	25–45
12·7	4·76	10–25
4·76	2·36	6–15
2·36	1·18	6–9
1·18	0·425	8–13
0·425	0·180	7–13
0·180	0·075	7–12
0·075		2–8

(b) Calculations

BS sieve size	Cumulative per cent passing, by simple addition		Cumulative percentage retained		Cumulative per cent passing, by subtraction		Derived specification on a per cent passing basis
	Min. 1	Max. 2	Min. 1R	Max. 2R	Max. 3	Min. 4	5
25·4	71	140	0	0	100	100	100
12·7	46	95	25	45	75	55	55–75
4·76	36	70	35	70	65	30	36–65
2·36	30	55	41	85	59	15	30–55
1·18	24	46	47	94	53	6	24–46
0·425	16	33	55	107	45	−7	16–33
0·180	9	20	62	120	38	−20	9–20
0·075	2	8	69	132	31	−32	2–8

passing-and-retained specification shown in Table 3.15(a) will be used. The derivation of the per cent passing specification, which is shown analytically in Table 3.15(b), and graphically in Fig. 3.13, may be explained as follows:

Step 1. Calculate from Table 3.15(a) the cumulative percentages of material passing each sieve. First use the minimum limits throughout and then the maximum limits, starting with the smallest sieve size in each case. Enter the results in columns 1 and 2 of Table 3.15(b).

Step 2. Plot the minimum cumulative per cent passing as curve 1 in Fig. 3.13(a), and the maximum values as curve 2. Some of the calculated percentages are found to be greater than 100 per cent; they are disregarded as they are fictional values. Thus curves 1 and 2 define the limits of the intended grading in the fine sizes but not in the coarse sizes.

Step 3. Calculate from Table 3.15(a) the cumulative percentages of aggregate retained on each sieve. First use the minimum limits throughout, and then the maximum limits, starting with the maximum sieve size in each case. These results are entered in columns 1R and 2R of Table 3.15(b).

Step 4. Using the data in columns 1R and 2R, calculate the per cent passing each sieve by subtracting the percentage retained from 100. The results obtained are entered in columns 3 and 4 of Table 3.15(b).

Step 5. Plot the maximum cumulative per cent passing values from column 3 as curve 3 in Fig. 3.13(a) and the minimum values from column 4 as curve 4. Some of the calculated minimum values are found to be negative and so they are disregarded since they are fictional. These two curves define the limits of grading in the coarse sizes, but not in the fine sizes.

Step 6. Select the lower values of curves 2 and 3 and connect these with a solid line as illustrated in Fig. 3.13(b). Similarly, the higher values of curves 1 and 4 are selected and connected by a solid line. These curves define the limits of the intended grading band. The per cent passing values—also shown in column 5 of Table 3.15(b)—enclose an area identical in shape and size with that representing the passing-and-retained specification from which they are derived.

The two specifications are not yet completely identical however. There is a requirement in the passing-and-retained specification in Table 3.15(a) that for each fraction defined there is a definite control over the amount which passes any particular sieve and which must be retained on the next smaller sieve in the named series. This has the effect of placing maximum and minimum limits on the slope of each segment of any particular gradation curve that might be drawn within the outlines of the grading band. These maximum and minimum slopes vary as the grading limits pass from zone to zone, but they may be fixed in the manner described in the following steps.

Step 7. Minimum slopes are fixed by curves 1 and 3 in Fig. 3.13(b); these are parallel to each other throughout. Therefore each of the solid lines in the envelope curve that represents a segment of either curve 1 or curve 3 establishes the minimum slope for its zone.

Step 8. Maximum slopes are fixed by curves 2 and 4 in Fig. 3.13(b); they are also parallel to each other throughout (including the fictitious segments not shown).

Step 9. In the transition zone *A*, no segment of curves 1 and 4 was used as a part of the envelope curve; similarly in zone *B*, no segment of curves 2 and 3 was used as a part of the envelope curve. In these zones, proper control can be obtained by using the dashed segment of curve 1 to set the minimum permissible slope in zone *A*, and the dashed segment of curve 4 to set the

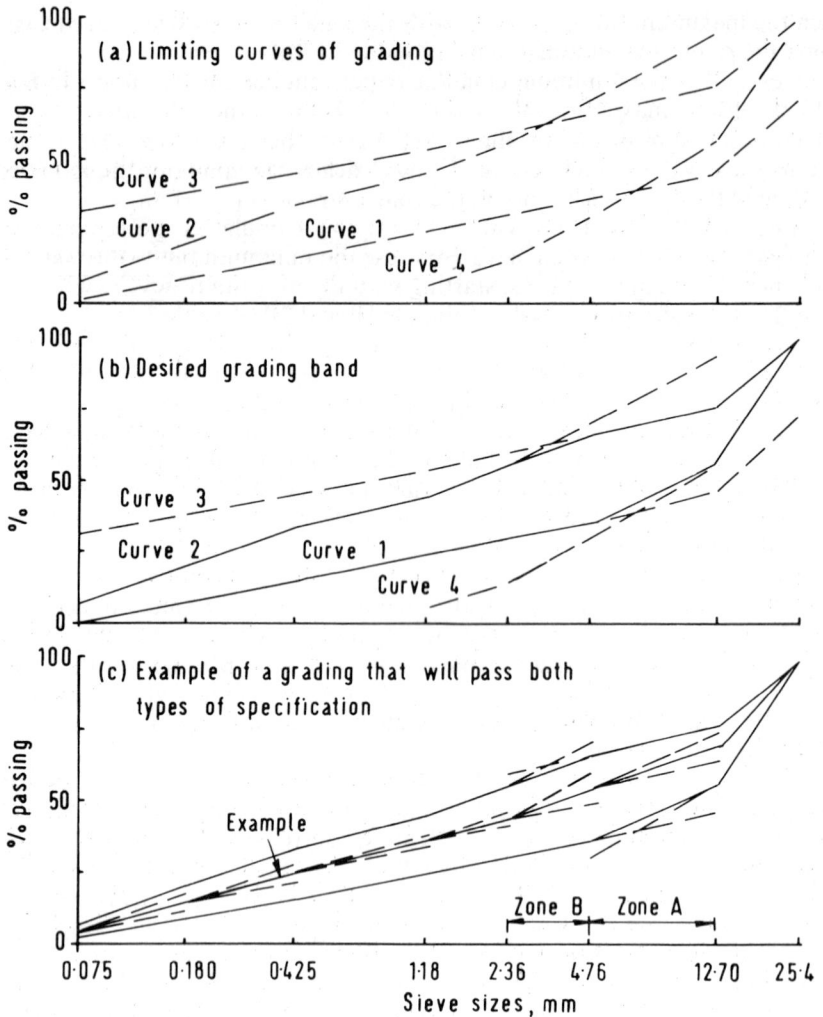

Fig. 3.13.   Graphical translation of an aggregate specification written in terms of per cent passing-and-retained into one of per cent passing

maximum permissible slope in that zone. Likewise, the dashed segments of curves 3 and 2 establish the minimum and maximum slopes respectively in zone *B*.

It may be noted that the number of different gradation curves which may be drawn within the boundaries of the grading band illustrated in Fig. 3.13(b) is greatly reduced by the slope limitations that are imposed in each zone by the passing-and-retained type of specification. There is, of course, no such restriction in the case of the ordinary type of per cent passing specification. As a result, unless some slope limit criterion is included in the grading requirements of the per cent passing type, a great many more grading curves may be included in the grading band under discussion than will meet the requirements of the passing-and-retained specification on which the grading band is based.

The solid line contained within the grading band limits of Fig. 3.13(c) represents a gradation curve that meets both the fractional and slope limits of the passing-and-retained grading specification of Table 3.15(a).

TABLE 3.16.   *Hypothetical gradations to be combined to meet specification limits*

| BS sieve size mm | Per cent passing given sieve size | | | | |
| | Fine aggregate A | Intermediate aggregate B | Coarse aggregate C | Specifications | |
				Limits	Mid-point
25·4	100·0	100·0	100·0	100	100·0
12·7	100·0	100·0	94·0	90–100	95·0
4·76	100·0	100·0	54·0	60–75	67·5
1·18	100·0	66·4	31·3	40–55	47·5
0·300	100·0	26·0	22·8	20–35	27·5
0·150	73·6	17·6	9·0	12–22	17·0
0·075	40·1	5·0	3·1	5–10	7·5

Combining aggregates. The grading required in a road aggregate depends very much on the form of construction which is going to be used. It is common practice therefore for the aggregate to be sorted beforehand into a number of closely-graded 'single-sizes' which may be subsequently remixed in desired proportions in order to meet the gradation specified for use. The blending of aggregates to obtain a desired gradation is a relatively simple exercise to the experienced engineer. To the young engineer, however, this can also be a most complicated problem, particularly when more than two aggregates are to be combined and their gradations overlap. While in the long-run the blending of aggregates to meet a particular specification re-solves itself into a trial-and-error procedure, the initial decision of 'where to start' can be made by using either of the following two methods, one math-ematical and the other graphical. To illustrate these methods it will be assumed that three aggregates are to be combined and that their gradations, and the required specification, are as listed in Table 3.16. It should be understood that neither the gradations nor the specification shown in Table 3.16 are in any sense typical or recommended values; they are simply hypothetical working data.

Mathematical method.   Before proceeding to the solution, a consideration of the following general mathematical relationships will help to clarify the method:

(a) An equation of the general form:

$$aA + bB + cC = T$$

can be obtained for each sieve size. In this equation, the lower-case letters are decimal values representing the proportions of the blend to be taken from each aggregate. When their values have been determined they may be expressed as percentages by multiplying each by 100. The capital letters represent the percentages either passing or retained on a particular sieve. For example, using the percentages passing the 4·76 mm sieve, the following is a valid equation:

$$100a + 100b + 54c = 67\cdot5$$

Or, using the percentages retained on the same sieve, the equation becomes

$$0a + 0b + 46c = 32\cdot5$$

(b) Equations obtained from any particular sieve can be combined either by addition or subtraction with equations obtained from one or more other sieves to produce other equations which are equally valid. Thus, combining equations obtained from the 1·18 mm and 300 μm sieves,

$$1\cdot18 \text{ mm B.S. } 100a + 66\cdot4b + 31\cdot3c = 47\cdot5$$

$$300 \text{ μm B.S. } 100a + 26b + 22\cdot8c = 27\cdot5$$

By addition, sum $\qquad 200a + 92\cdot4b + 54\cdot1c = 75$

By subtraction, difference $\quad 0a + 40\cdot4b + 8\cdot5c \quad = 20$

(c) A valid equation can also be obtained by summing the proportions of the individual aggregates. Thus,

$$a + b + c = 1$$

Proceeding now to the solution of the problem, it is seen that all the material larger than the 4·76 mm sieve is contained in the coarse aggregate. Using the percentages retained on that sieve, the following equation is obtained:

$$0a + 0b + 46c = 32\cdot5$$

Therefore, $\qquad c = 0\cdot71$

By subtracting the equation for the 300 μm sieve from that for the 1·18 mm sieve, the following equation is obtained:

$$0a + 40\cdot4b + 8\cdot5c = 20$$

Therefore, $\qquad 40\cdot4b + 8\cdot5 \ (0\cdot71) = 20$

and $\qquad b = 0\cdot35$

But $\qquad a + b + c = 1$

Therefore, $\qquad a + 0\cdot35 + 0\cdot71 = 1$

and $\qquad a = -0\cdot06$

Obviously a negative value cannot be accepted; it simply indicates that the mid-point specification values used in the calculation are exactly unobtainable. The computation is repeated using specifications of 70 per cent passing for the 4·76 mm sieve, and 45 per cent passing for the 1·18 mm sieve.

Then $\qquad 0a + 0b + 46c = 30$

and $\qquad c = 0·65$

Now, $\qquad 0a + 40·4b + 8·5c = 17·5$

Therefore, $\qquad 40·4b + 8·5(0·65) = 17·5$

and $\qquad b = 0·30$

But $\qquad a + b + c = 1$

Therefore $\qquad a + 0·30 + 0·65 = 1$

and $\qquad a = 0·05$

TABLE 3.17. *Example of results of mathematical method of combining aggregates to obtain a desired gradation*

BS sieve size, mm	Aggregate A × 5%	Aggregate B × 30%	Aggregate C × 65%	Combined
25·4	100 × 0·05 = 5·0	100 × 0·30 = 30·0	100 × 0·65 = 65·0	100·0
12·7	100 × 0·05 = 5·0	100 × 0·30 = 30·0	94 × 0·65 = 61·0	96·0
4·76	100 × 0·05 = 5·0	100 × 0·30 = 30·0	54 × 0·65 = 35·1	70·1
1·18	100 × 0·05 = 5·0	66·4 × 0·30 = 19·8	31·3 × 0·65 = 20·4	45·2
0·300	100 × 0·05 = 5·0	26·0 × 0·30 = 7·8	22·8 × 0·65 = 14·8	27·6
0·150	73·6 × 0·05 = 3·7	17·6 × 0·30 = 5·3	9 × 0·65 = 5·9	14·9
0·075	40·1 × 0·05 = 2·0	5·0 × 0·30 = 1·5	3·1 × 0·65 = 2·0	5·5

It is now necessary to check to see whether the proportioning values obtained are applicable to all the sieves. If they are, the specification will be met. This check is illustrated in Table 3.17 and, as can be seen, the assumed proportions are satisfactory since the combination in all respects lies within the specification limits. It is to be noted, however, that the amounts passing both the 150 μm and 75 μm B.S. sieves tend to be on the low side of the specification, so that it is advisable to adjust the proportion to increase these values. In this example it can be best done by increasing the proportion of aggregate A at the expense of aggregate B.

Graphical method. The use of simultaneous equations can be a very suitable method when it is necessary to combine two or perhaps three aggregates. Beyond this number, the method can be quite cumbersome due to the number of adjustment checks which may have to be made. In such instances the following graphical procedure can be most useful. This is best explained by referring to Fig. 3.14 and outlining the graphical procedure in a series of steps.

Step 1. Plot the per cent passing of fine material on scale *A* and the per cent passing of intermediate aggregate on scale *B*.

Step 2. Connect by a straight line the per cent passing for each respective sieve size. Note that the intersections of each sieve line with any vertical gradation line will define the composite gradation of the fine and intermediate aggregates for the blend proportions shown on the top and bottom horizontal scales.

Step 3. Lay off the specification limits as intercepts on each sieve line; these are the solid portions of the sieve lines shown in Fig. 3.14. The intercepts for any particular sieve line represent the range of proportions that will meet the specification limits.

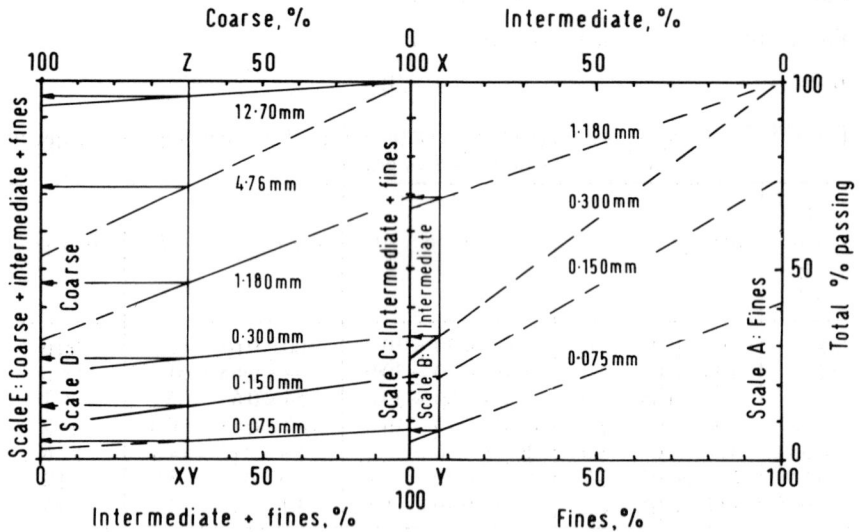

Fig. 3.14.   Graphical method of combining aggregates to obtain a desired gradation. (The data used are from Table 3.16)

Step 4. Select a vertical line that will strike the 'best average' through all of the specification intercepts shown. In this example, the vertical line selected represents 8 per cent fine material and 92 per cent intermediate aggregate. This is the line *XY* indicated in the figure.

Step 5. Project horizontally on to the vertical scale *C* the intersections of each sieve line with the gradation line *XY*. The numerical values obtained on scale *C* represent the per cent passing the respective sieve sizes for the combined gradation of the fine and intermediate aggregates.

Step 6. Plot the per cent passing of coarse aggregate on scale *D*, and repeat the process as outlined in steps 1 to 4 in order to determine the blend proportions of coarse with the mixture of intermediate plus fines. In this example the vertical line *XYZ* is selected; it represents 60 per cent coarse aggregate and 40 per cent intermediate plus fine aggregate, or

Coarse aggregate $= 60\%$
Intermediate aggregate $= 40 \times 0.92 = 36.8\%$
Fine aggregate $= 40 \times 0.08 = 3.2\%$

Step 7. Determine the gradation of the combined materials. This can be done either by projecting horizontally to scale E from the intersection of each sieve line with the vertical line XYZ and reading the values obtained, or by calculating mathematically the combined gradation.

The results obtained from the mathematical calculation are shown in Table 3.18 and, as can be seen, the blend selected meets the specification criteria in all instances. Again it is to be noted, however, that the amounts passing the $150\,\mu m$ and $75\,\mu m$ B.S. sieves are extremely close to the lower limits of the specification, so it will be necessary to adjust the selected blending percentages in order to increase the per cent passing for these sieve sizes. In this particular instance, it is best done by increasing the proportion of fine aggregate at the expense of the intermediate material.

TABLE 3.18. *Example of results of graphical method of combining aggregates to obtain a desired gradation*

BS sieve size, mm	Aggregate A $= 3.2\%$	Aggregate B $= 36.8\%$	Aggregate C $= 60.0\%$	Combined
25.4	$100 \times 0.032 = 3.2$	$100 \times 0.368 = 36.8$	$100 \times 0.60 = 60.0$	100.0
12.7	$100 \times 0.032 = 3.2$	$100 \times 0.368 = 36.8$	$94 \times 0.60 = 56.4$	96.4
4.76	$100 \times 0.032 = 3.2$	$100 \times 0.368 = 36.8$	$54 \times 0.60 = 32.4$	72.4
1.18	$100 \times 0.032 = 3.2$	$66.4 \times 0.368 = 24.4$	$31.3 \times 0.60 = 18.8$	46.4
0.300	$100 \times 0.032 = 3.2$	$26.0 \times 0.368 = 9.6$	$22.8 \times 0.60 = 13.7$	26.5
0.150	$73.6 \times 0.032 = 2.4$	$17.6 \times 0.368 = 6.5$	$9 \times 0.60 = 5.4$	14.3
0.075	$40.1 \times 0.032 = 1.3$	$5.0 \times 0.368 = 1.8$	$3.1 \times 0.60 = 1.9$	5.0

SELECTED BIBLIOGRAPHY

1. ABRAHAM, H. *Asphalts and Allied Substances*, Vols. 1–5. New York and London, Van Nostrand, 1960.
2. BS 892:1967. *Glossary of Highway Engineering Terms*. London, British Standards Institution, 1967.
3. BROOME, D. C. Trinidad Lake Asphalt and its use in road construction, *Roads and Road Construction*, 1964, **42**, No. 494, 41–45.
4. BS 594:1961. *Rolled Asphalt (Hot Process)*. London, British Standards Institution, 1968.
5. BS 1446:1962. *Mastic Asphalt (Natural Rock Asphalt Aggregate) for Roads and Footways*. London, British Standards Institution, 1966.
6. BS 1447:1962. *Mastic Asphalt (Limestone Aggregate) for Roads and Footways*. London, British Standards Institution, 1966.
7. BS 1690:1962. *Cold Asphalt*. London, British Standards Institution, 1968.
8. BS 2040:1953. *Bitumen Macadam with Gravel Aggregate*. London, British Standards Institution, 1953.
9. BS 348:1948. *Compressed Natural Rock Asphalt*. London, British Standards Institution, 1948.

10. BS 3690:1970. *Bitumens for Road Purposes*. London, British Standards Institution, 1970.
11. BS 1621:1961. *Bitumen Macadam with Crushed Rock or Slag Aggregate*. London, British Standards Institution, 1968.
12. ASPHALT INSTITUTE. *The Asphalt Handbook*. Manual Series No. 4. Maryland, The Institute, 1960.
13. BS 4693:1971. *Viscosity of Cutback Bitumen and Road Oil*. London, British Standards Institution, 1971.
14. BS 434:1960. *Bitumen Road Emulsion (Anionic)*. London, British Standards Institution, 1960.
15. BS 2542:1960. *Recommendations for the Use of Bitumen Emulsion for Roads*. London, British Standards Institution, 1960.
16. GRIFFITH, J. M. Emulsified asphalt slurry seal coats, *Roads and Streets*, 1957, **100**, No. 3, 162–164.
17. MERTENS, E. W. and M. J. BORGEFELDT. Cationic asphalt emulsions. In A. J. HOIBERG. *Bituminous Materials: Asphalts, Tars and Pitches*, Vol. II. New York and London, Interscience Publishers, 1965.
18. ROAD RESEARCH LABORATORY. Specification for the manufacture and use of rubberized bituminous road materials and binders, *Road Note* No. 36. London, H.M.S.O., 1967.
19. BS 76:1964. *Tars for Road Purposes*. London, British Standards Institution, 1964.
20. BS 802:1967. *Tarmacadam with Crushed Rock or Slag Aggregate*. London, British Standards Institution, 1967.
21. BS 1241:1959. *Tarmacadam and Tar Carpets (Gravel Aggregates)*. London, British Standards Institution, 1959.
22. BS 1242:1960. *Tarmacadam (Tar Paving)*. London, British Standards Institution, 1967.
23. LEGAULT, A. *Highway and Airport Engineering*. Englefield Cliffs, N.J., Prentice-Hall, 1960.
24. THOMPSON, P. D. and SZATKOWSKI. Full-Scale Road Experiments Using Rubberized Surfacing Materials. *RRL Report* LR 370, Crowthorne, Berks., The Road Research Laboratory, 1971.
25. SHOOP, S. Pitch-bitumen —Dotting the ' I's ', *Road Tar*, 1969, **23**, No. 2, 14–16.
26. LIVERSEDGE, F. Tar-bitumen blends in road construction and maintenance, *Road Tar*, 1972, **26**, No. 3, 4–7.
27. BS 4707:1971. *Loss on Heating of Bitumen and Flux Oil*. London, British Standards Institution, 1971.
28. BS 4710:1971. *Ductility of Bitumen*. London, British Standards Institution, 1971.
29. BS 4691:1971. *Penetration of Bitumen*. London, British Standards Institution, 1971.
30. BS 4692:1971. *Softening Point of Bitumen (Ring and Ball)*. London, British Standards Institution, 1971.
31. BS 598:1958. *Sampling and Examination of Bituminous Mixtures for Roads and Buildings*. London, British Standards Institution, 1966.
32. BS 4707:1971. *Loss on Heating of Bitumen and Flux Oil*. London, British Standards Institution, 1971.
33. BS 4690:1971. *Solubility of Bitumen*. London, British Standards Institution, 1971.
34. BS 4703:1971. *Flash Point (Closed) of Cutback Bitumen*. London, British Standards Institution, 1971.

35. BS 4688:1971. *Flash Point (Open) and Fire Point of Petroleum Products by the Pensky-Martens Apparatus.* London, British Standards Institution, 1971.

36. COMMITTEE ON SIGNIFICANCE OF TESTS FOR HIGHWAY MATERIALS. Significance of tests for highway materials — basic tests, *Proc. Amer. Soc. Civil Engrs, Journal of the Highway Division,* Sept. 1957, **83,** No. HW4.

37. BS 12:1971. *Portland Cement (Ordinary and Rapid Hardening).* Pt 2. London, British Standards Institution, 1971.

38. BS 146:1958. *Portland-Blastfurnace Cement.* London, British Standards Institution, 1967.

39. BS 915:1947. *High Alumina Cement.* London, British Standards Institution, 1962.

40. PAULS, J. T. and CARPENTER, C. A. Mineral aggregates for bituminous construction, *Symposium on Mineral Aggregates: Special Technical Publication* No. 83. Philadelphia, American Society for Testing Materials, 1948.

41. CRONEY, D. and JACOBS, J. C. The Frost Susceptibility of Soils and Road Materials. *RRL Report* LR90. Crowthorne, Berks., The Road Research Laboratory, 1967.

42. BS 812:1967. *Methods for Sampling and Testing of Mineral Aggregates (Sands and Fillers).* London, British Standards Institution, 1970.

43. SHERGOLD, F. A. The classification, production, and testing of road-making aggregates, *Quarry Managers Journal,* 1960, **44,** No. 2, 47–54.

44. ROAD RESEARCH LABORATORY and GEOLOGICAL SURVEY AND MUSEUM. *Sources of Road Aggregate in Great Britain.* London, H.M.S.O., 1960.

45. BS 410:1969. *Test Sieves.* London, British Standards Institution, 1971.

46. BS 882,1201:1954. *Concrete Aggregates from Natural Sources.* London, British Standards Institution, 1957.

47. BS 63:1971. *Single-sized Roadstone and Chippings.* London, British Standards Institution, 1971.

48. ROAD RESEARCH LABORATORY. Roadstone test data presented in tabular form, *Road Note* No. 24. London, H.M.S.O., 1959.

49. DALHOUSE, J. B. Plotting aggregate gradation specifications for bituminous concrete, *Public Roads,* 1953, **27,** No. 7, 155–158.

50. LIVERSEDGE, F. Tar-bitumen blends in road construction and maintenance. *Road Tar,* 1972, **26,** No. 3, 4–7.

51. BS 4027:1972. *Sulphate Resisting Portland Cement.* London, British Standards Institution, 1972.

4 Soil engineering for highways

Nearly all road improvement schemes, especially those in new locations, involve a considerable amount of earthworks. Soil is the foundation material for all highways, whether it be in the form of undisturbed *in situ* subgrade material or transported and reworked embankment material. In addition, the highway pavement itself, as well as the flanking shoulders, is very often composed of the cheapest raw material available, soil. Thus it can be seen that soil plays a most important role in highway construction and it is axiomatic that the highway engineer should have a thorough understanding of soils and how they behave.

THE ORIGIN OF SOILS

Definition of soil

The term soil stems from the Latin word 'solum' and this latter designation is still very much used by the agricultural scientist. Whereas in older times, only the agriculturalist's view of soil as a material was accepted, developments in other professional fields have resulted in the term now having various meanings and connotations. Thus, as illustrated in Fig. 4.1, the agricultural soil scientist (pedologist), the geologist and the engineer have three quite different interpretations of the word soil.

According to the pedologists, soil is 'a natural body, differentiated into horizons of mineral and organic constituents, usually unconsolidated, of variable depths, which differs from the parent material below in morphology, physical properties and composition, and biological characteristics'.[1] The parent material as used in this context is rarely considered as extending to more than 1–1·5 m below ground level. This definition is a natural one from the agricultural viewpoint, since applications of soil science in this area are mainly concerned with understanding and improving the material in which crops are grown.

To the geologist, the earth's crust is composed of rocks and the unconsolidated sediments which make up the regolith. As these sediments are exposed to the weather and biological activity, the uppermost part becomes considerably modified. This plant-growing stratum is given the designation soil by the geologist to distinguish it from unmodified sediments beneath.

To the engineer, the word soil means 'any naturally occurring loose or soft deposit resulting from weathering or breakdown of rock formations or from the decay of vegetation'.[2] This definition includes many materials of varied origin viz. detrital sediments such as gravels, sands, silts and clays, organic deposits such as peats, calcareous deposits like shell and coral sands,

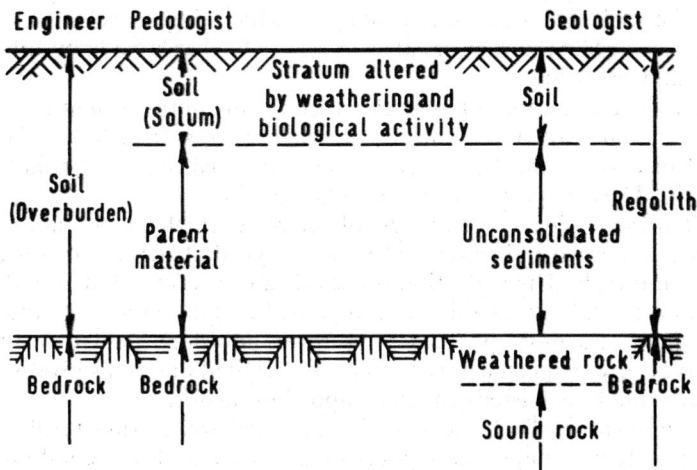

Fig. 4.1.   Pedological, geological and engineering understandings of the term soil

pyroclastics such as uncemented volcanic dust, and residual soils like the laterites. In other words, to the engineer the term soil means all unconsolidated materials above bedrock with which and upon which he builds his structures.

Formation of soil

The rocks which form the earth's surface form the basic materials for all engineering soils. Soil is, as it were, the incidental by-product of the physical and chemical weathering of these rocks which has been carried on relentlessly throughout multi-millions of years of geological time. *Physical* weathering is associated with the following disintegrating causes and effects:

1. *Temperature changes.* If temperature variations are sufficiently large and powerful, the temperature of the outside of a rock mass will change more rapidly than the inside, with the result that outside layers of the rock spall off and fractures appear in the remainder. Stresses are also set up within the rocks as a result of the unequal expansion and contraction of adjacent minerals caused by repeated daily and seasonal temperature fluctuations. Eventually the mineral particles are pried loose and the ingredients of soil are formed.

2. *Ice and vegetation growth.* Roots of trees and bushes grow within the cracks and crevices of rock masses and as time progresses they grow larger and act as wedges forcing the rock to disintegrate. Similarly, water resting in the cracks expands on freezing and causes the rock to break apart.

3. *Abrasion.* Although the winds, glaciers and rivers are of primary importance because of their transportation facilities, the materials which they carry convert them also into soil-forming agents. Thus sand-laden winds and rivers and moving glaciers actually create soil particles as they scour and scrape the rock strata and fragments over which they travel.

4. *Erosion.* The rains, rivers and seas cause further erosion by continuous impact and over-flow.

The *chemical* weathering of rock which results in the formation of soil is carried out simultaneously with the physical process. It includes the chemical and physio-chemical processes of solution, hydration, carbonation and oxidation. These may be briefly explained as follows:

1. *Hydration.* This is the process of decay by which water combines with the rock minerals and enters into their crystal structures to form other hydrous minerals. The hydration process is more common in humid than in arid climates and is very often accompanied by an increase in volume and the creation of a more porous structure. In addition, the hydrous mineral may have a lower specific gravity than the original one. This process, which is a reversible one, hastens the decomposition action.

2. *Carbonation.* This refers to the chemical weathering resulting from oxides of calcium, magnesium, potassium, sodium and iron combining with the carbon dioxide of the soil, air and water to form carbonates, which are readily soluble, particularly under the action of carbonic acid (H_2CO_3) present in the rain water. This process is most noticeable in soils found in regions of high rainfall.

3. *Oxidation.* This is the process of soil formation resulting from oxygen being added to the rock. Of the various chemical changes due to weathering, this is usually one of the first to be manifested. It is particularly noticeable in rocks carrying iron, which is an element very easily oxidized in the presence of moisture. For instance, the oxidation of pyrites, which are composed of iron and sulphur, results in the formation of sulphuric acid which attacks rocks so that the oxidized iron compounds, which are almost insoluble, are generally deposited.

4. *Solution.* The solvent action of water, and of the other ions it carries, furthers the weathering process as it moves through and around rocks and minerals.

As will be discussed in more detail later, chemical weathering is primarily responsible for the clay-size particles in a soil, whereas physical weathering results in the formation of the larger particles. The physical forces result in disintegration of the rock without materially affecting the composition of the particle constituents. Chemical weathering, on the other hand, causes decomposition of the rock so that definite chemical changes take place, soluble materials are released and new minerals are synthesized or left as resistant end-products.

Types of soil

While most soils originated from bedrock, it is obvious that the mechanical forces of nature have been instrumental in moving many of them from place to place. It is possible to group the inorganic soils according to the means by which they were naturally transported and deposited in their present locations.

The following are brief descriptions of the various soils falling into these groups.

Residual soils. These are sedentary soils which have been formed from the underlying bedrock and which, as the name implies, have never been moved. Typically these soils are relatively fine-grained near the surface but become more coarse-grained, since they contain angular fragments of rock, with increasing depth. Residual soils are rarely found in glaciated areas as they have either been removed or are buried beneath other overlying glacial soils.

Colluvial soils. These are simply accumulations of rock debris or talus which become detached from the heights above and are carried down the slopes by the forces of gravity. Soils formed from these accumulations are usually coarse and stony; this is to be expected since they result from physical rather than chemical weathering.

Cumulose soils. These are *organic* soils such as peats or mucks which have been formed in place as a result of the accumulation of chemically decomposed plant residues in shallow ponded areas. Although easily noted when on the surface, they are often encountered in glacial areas at considerable depths below the surface. Both peat and muck are highly organic, but muck is considerably older. While the fibres of unoxidized plant remains are obviously visible in peat, they are largely oxidized in muck soil.

Glacial drift. Soils of the glacial drift are those which have been transported and deposited by the great glaciers of the Ice Ages. Glaciers were several thousand feet thick and covered vast areas of the northern hemisphere. When they retreated deposits composed of boulders, rock fragments, gravel, sand, silt and clay were left behind. Particular landforms associated with glacial topography which are of particular interest to the highway engineer are moraines, kames, eskers and drumlins.

Moraines. These are irregular hummocky hills which usually contain assorted glacial materials. Four types of moraines are generally recognized; these are known as terminal, lateral, recessional and ground moraines.

Terminal moraines are those long low hills, perhaps 1·5–3 km wide and 30 m high, which mark the southernmost limit of the glacial advance. Roughly perpendicular to the direction of glacial movement, they are believed to have occurred as the ice sheet started to melt as it reached a warm climate. As the front of the ice sheet meltedted, the material carried was dropped. At the same time the ice front was continually forced forward by the pressures from the colder regions, so that more drift material became added to that which had already accumulated. These accumulations are known as terminal or end-moraines.

Lateral moraines are similar in nearly every way to terminal moraines. They were formed along the lateral edges of the ice sheet as glacial material was dropped through the melting ice, and as a result are roughly parallel to the direction of ice movement.

As the temperature became warmer, the ice front melted faster than its rate of advance, with the result that the glaciers began to 'recede'. This recession was constantly interrupted of course, with the result that minor 'terminal' moraines were formed at irregular intervals. These hills are known as recessional moraines. When the topography is characterized by a series of recessional moraines interspersed with many depressions containing small lakes, it is sometimes called a 'knob-and-kettle' topography.

When the ice sheet melted rapidly, material continually dropped at the edges of the receding ice sheet; this material is referred to as ground moraine.

Morainal deposits are also known as boulder clays or glacial tills. These can be described as unstratified sedimental deposits which were directly deposited by and underneath the ice, were neither transported nor sorted by water, and contain materials ranging from boulders to gravel to finely pulverized clays.

Kames.   Of particular interest to the highway engineer are the conical-shaped hills or mounds known as kames. These hills mainly contain poorly stratified accumulations of gravel and sand, but some clay is also usually present. The materials found in kames are generally quite suitable for road-building purposes.

Eskers.   These are long sinusoidal ridges, usually with flat tops and steep sides, which contain highly stratified gravel and sand eminently suitable for highway construction. Eskers are relatively low, usually not more than about 18 m in height, but may be many kilometres in length. They are the end-products of accumulations of glacial material which were deposited in crevices or tunnels within the thin wasting ice. The lack of fine-grained materials in eskers is explained by the fact that they were washed away by streams flowing towards the edge of the ice front.

Drumlins.   These are elongated hills of glacial drift with ellipse-type contour lines, oriented with their major axes parallel to the direction of ice movement. Drumlins may be as much as 2 km long, over 400 m wide and 100 m or more high. It is believed that their smoothly rounded shapes may have been formed as a result of irregular accumulations of drift material being overridden by the moving ice. Although many drumlins contain some sand, they are usually composed of drift materials very rich in clay.

Glacial outwash deposits. At the time of the Ice Age, the weather was obviously warmer at particular times of the year than at others, just as it is today. When this occurred, huge volumes of ice melted and rushed out from the wasting ice field. These streams were charged with large quantities of glacial materials which spread out over the uncovered land outside the glaciers. In narrow valleys, through which these waters were able to flow rapidly and freely, the materials deposited tended to be coarse-grained, whereas on plains they were extremely assorted.

In many instances, the outwash waters fed into enclosed valleys within which lakes were formed. Great quantities of debris were deposited in these lakes, ranging from coarse to fine in content. Because of the different rates of settlement of the various materials carried by the waters, extreme stratification took place and, particularly towards the middle of the lakes, uniform horizontal and alternating layers of silt and clay were laid down. Over the centuries, these layers accumulated and formed lacustrine deposits now given the name of 'varved' deposits. The thickness of each varve or layer can vary from being infinitesimally small to perhaps 10 mm thick, each pair corresponding to one year's deposit. Varved soils can give rise to troublesome foundation problems and should be avoided when locating a highway.

Alluvial stream deposits. Alluvial soils are ones which have been deposited by moving river waters. They are generally regarded as falling into the following three categories; flood-plain soils, alluvial fans and deltas.

Flood-plain soils.   These are found to a certain extent alongside every river, but are most evident along mature streams. Older rivers are usually on gently inclined beds and as a result they tend to use up their energies in horizontal meandering rather than in cutting deeper channels. In the course of a stream's meanders, coarse-grained material is deposited at the inside of the meander curves while the moving waters cut into the opposite banks. These deposits at the insides of bends can be very valuable as construction materials and may influence the location of a highway as well as its pavement design.

With time, the mature river continually changed its path, with the result that old meanders were cut-off and the now-familiar half-moon shaped 'ox bows' were formed. In other words, these ox bows are simply old meander beds which were left as depressions after the river changed its course. Over the years, these depressions became filled with water during wet weather and dried out again during fine weather. As moisture ran into the ox bow lakes, fine-grained material was carried in also, so that with the passage of time the depressions became full of silty and clayey materials. Since they are not continuous but rather occur as pockets of clay-like materials, ox bows should be avoided when locating a highway as they are liable to cause differential heaving and/or settlement of the pavement.

When a stream is carrying an excess of water, due to heavy rainfalls or perhaps sudden snow melting, it will very often—this is particularly true of the more mature rivers—burst its banks and the water will spread out over the adjacent land. If the land is wide and flat it becomes a 'flood plain' over which the velocity of the flooding water decreases the farther it gets from the river bed. This decrease in velocity results in sand and gravel being deposited adjacent to the river banks, silt farther out, and the finest materials still farther away. Thus, natural levels or ridges of coarse-grained materials may be found adjacent to rivers in flood plains, while poor draining swampy lands are to be found farther back. The coarse-grained materials can be used for construction purposes on highway projects, but usually they must be washed beforehand to remove excess fines.

In many instances it has happened that, for some geological reason, an old meandering river suddenly becomes 'rejuvenated' and begins to incise a new channel in the old formed alluvium, leaving what are called 'terraces' on either side. A number of these raised horizontal plains can be seen along the sides of river-bearing valleys, indicating that there were periods when the streams were at these heights. Again, these terraces can be sources of good construction materials. In addition, they very often provide good locations for highways through valleys, as they reduce the need for expensive excavation into the sides of the valley, while at the same time keeping the highway well above the possible flood waters of the river.

Alluvial fans.   Other good sources of highway construction materials are alluvial fans. This is a landform which is to be found at the foot of a stream which emerges suddenly from high land on to a plain. As the stream rushes on to the flat ground, the rapid change in gradient causes it to spread out in the shape of a fan. Sediment carried in the water is deposited in this characteristic shape also. Because of the higher velocity of the water as it enters the plain, fan material is generally coarse-grained and well drained, and contains little or no fines.

Delta deposits.   Although much of the material carried by a river is deposited over its flood plains, a considerable amount obviously still reaches the mouth of the stream, where it is discharged into the body of water such as an ocean or lake to which it is a tributary. By this time, the sediment carried may be mostly fine-grained material which has not had the opportunity to settle previously. If the stream flows into a standing body of water, where currents or wave action are minor, the suspended sediment eventually drops to the bottom to form a fan-shaped promontory known as a delta. The most highly developed deltas are generally to be found where rivers enter the ocean. When the body of water into which the stream discharges is a lake, the deposits are known as *lacustrine* soils.

Marine soils. As in the case of deltaic soils, much of the material carried by rivers is deposited in the ocean. In addition to this, however, the ocean itself is continually eroding the shoreline at one location and depositing the eroded material at another. These processes have been going on for millions of years with the result that, as oceans have receded from land areas, great areas of the land surface of the world have become exposed to the weathering elements. These exposed deposits range from sands to silts to clays, depending on whether the exposed land was originally close to a shoreline or farther out to sea. It is difficult to make more definite comments about these marine deposits, varying as much as they do.

Aeolian soils. Aeolian soils are those which have been transported and laid down by the wind. The wind-blown soils are usually subdivided into two groups, dune sands and loess soils.

Dune sands. These soils are to be found both in arid regions and in temperate zones adjacent to large bodies of water where large quantities of beach sand are deposited. When the wind blows consistently from one direction in these areas, the sand particles 'skip' along the ground until eventually they lodge together and dunes begin to form. These dunes have a relatively gentle slope on the windward side, whereas the leeward slopes are at the very much steeper angle of repose of the sand.

It is because of their tendency to shift that sand dunes become a problem for the highway engineer. When a sand dune is sufficiently formed for the amount of sand deposited on it by the wind to balance the amount taken away from the leeward side by the same wind, the dune is caused to migrate as sand particles are blown up and over the top. Highways which pass through migrating dune sand areas can be continually covered and uncovered by these sands and in some instances—particularly when the roads are low-cost and unsurfaced—rendered impassable.

Loess soils. These soils are to be found farther away from the beaches on which they draw for their resources. They are composed primarily of silt-sized particles which have been picked up from the beaches by the wind and transported through the air to their ultimate destinations. As found at particular locations in the field, loess soils are composed of very uniform particles which get smaller and smaller the farther the distance from the beach source. They are very porous, and their dry unit weights *in situ* rarely exceed $1440 \, \text{kg/m}^3$. Loess soils are both free-draining and rich in calcium carbonate, and this combination makes it possible to excavate highway cuttings with near vertical sides. The cementing action of the calcium carbonate is lost, however, with manipulation of the loess, and hence embankments formed from loessial soils must have similar side slopes to those formed from other soils.

Soil phases

As found in nature, soil can be considered as an assemblage of solid particles interspersed with void or pore spaces which may or may not contain water. This conception of soil is a most important one since the highway engineer is most often concerned with obtaining a soil condition where the void spaces are kept to a minimum and the volume occupied by the mineral particles is at the maximum possible.

It is easiest to visualize the various soil phases, and the relationships which exist between them, by representing in graphical form a soil sample in which the solid, liquid and gaseous phases are segregated. Figure 4.2 is a schematic diagram in which it is assumed that a sample of soil is placed in an equi-volume cylinder of unit cross-section. Such an arrangement makes it possible to consider the component volumes as represented by their heights, so that it is a simple matter to develop some important relationships.

Void ratio. The void ratio of a soil is the ratio of the volume of voids to the volume of solids present in the soil. This relationship pays no regard to the proportions of water, air or other gases which may constitute the pore spaces. Thus,

$$e = \frac{V_v}{V_s}$$

where　e = voids ratio,
　　　　V_v = volume of voids,
and　　V_s = volume of solids.

In a saturated soil, the voids ratio is of course directly proportional to the moisture content, m. In this case the volume of voids V_v is equal to the volume of water V_m. Then,

$$e_s = \frac{V_m}{V_s}$$

$$= \frac{W_m}{W_s} \times \frac{\psi_s}{\psi_w}$$

$$= m \times \frac{\psi_s}{\psi_w}$$

where　W_m = weight of water,
　　　　W_s = weight of solids,
　　　　ψ_s = weight per unit volume of solids,
　　　　ψ_w = weight per unit volume of water,
　　　　m = moisture content of the soil,
and　　e_s = voids ratio when soil is saturated.

Porosity. The porosity of a soil is defined as the ratio of the volume of the voids to the total volume of the soil. As with the voids ratio, the porosity pays no regard to the constituent volumes of the pore spaces. In contrast to the voids ratio, porosity is usually expressed as a percentage value. Thus,

$$n = \frac{V_v}{V} \times 100$$

where n = porosity, %,

V_v = volume of voids,

and V = total volume of soil mass.

The porosity of a soil can also be expressed by the formula

$$n = \frac{e}{1+e} \times 100$$

when n and e are as defined before.

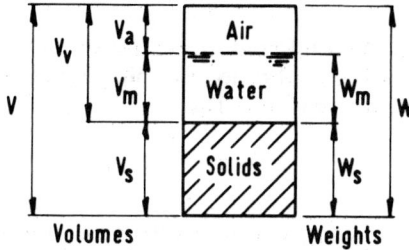

Fig. 4.2. Diagram illustrating the three phases of a soil

Per-cent air voids. The percentage of the total volume of the soil which is occupied by the air in the voids is referred to as the per-cent air voids. Thus,

$$n_a = \frac{V_a}{V} \times 100$$

where n_a = per-cent air voids,

V_a = volume of air voids,

and V = total volume of the soil mass.

Per-cent water voids. The percentage of the total volume of the soil which is occupied by the water in the voids is referred to as the per-cent water voids. Thus,

$$n_m = \frac{V_m}{V} \times 100$$

where n_m = per-cent water voids,

V_m = volume of water voids,

and V = total volume of the soil mass.

Obviously the sum of the per-cent air voids and per-cent water voids is equal to the porosity. Thus,

$$n = n_a + n_m$$

Degree of saturation. The extent to which the voids present in a soil are filled with water is termed the degree of saturation. It is the ratio of the volume of water to the volume of voids, expressed as a percentage. Thus,

$$S = \frac{V_m}{V_v} \times 100$$

where S = degree of saturation, %,

V_m = volume of water voids,

and V_v = volume of air plus water voids.

Degree of compaction. Also referred to as relative unit weight, this refers to the extent that the existing void ratio of a soil is representative of the range of void ratios possible for that soil. Thus,

$$D_d = \frac{e_{max} - e}{e_{max} - e_{min}} \times 100$$

where D_d = degree of compaction or relative unit weight, %,

e_{max} = void ratio of soil in its loosest state,

e_{min} = void ratio of soil in its densest state,

and e = existing void ratio.

Thus a soil which is in its most compact state will have a relative unit weight of 100 per cent, while one in its loosest state will have a relative value of zero.

A clear distinction should be made between the term *degree of compaction* and *relative compaction*, which has an entirely different meaning. While this is brought out more clearly in the discussion on the moisture-unit weight relationship, it is appropriate to note here that relative compaction is the ratio, expressed in per cent, of the dry unit weight of the soil *in situ* to the maximum dry unit weight of the soil as determined by particular laboratory compaction tests.

Solid constituents of soil

Soil is a porous mixture of inorganic particles, decaying organic matter, air and water. The solid phase consists of particles of varying degrees of subdivision, ranging in size from boulders to colloids. The coarseness or fineness of a soil is reflected in terms of the relative fractions of the different separates present; those most commonly used are sand, silt and clay.

The terms sand, silt and clay have different meanings to different scientific organizations, and it is necessary to be aware of these differences when reading and interpreting the technical literature on soils. All organizations accept that the most convenient way to define the separates is on the basis of particle size; the differences arise over what exactly these sizes should be. Figure 4.3 shows the limiting sizes adopted by some of the agencies which are of particular interest to the highway engineer.

Atterberg's size-fractions. To the Swedish soil scientist Atterberg can be given the credit for what is perhaps the most important early attempt to put on a scientific basis the choice of the limiting sizes of the various soil fractions. It is his 1908 classification system which is still essentially used by the International Society of Soil Science today.

Atterberg classified gravel particles as those which were between 20 and 2 mm in size. He suggested that these were the limits within which no water is held in the pore spaces between the particles and where water is weakly held in the pores. Sand was stated to be between 2 and 0·2 mm in size, the

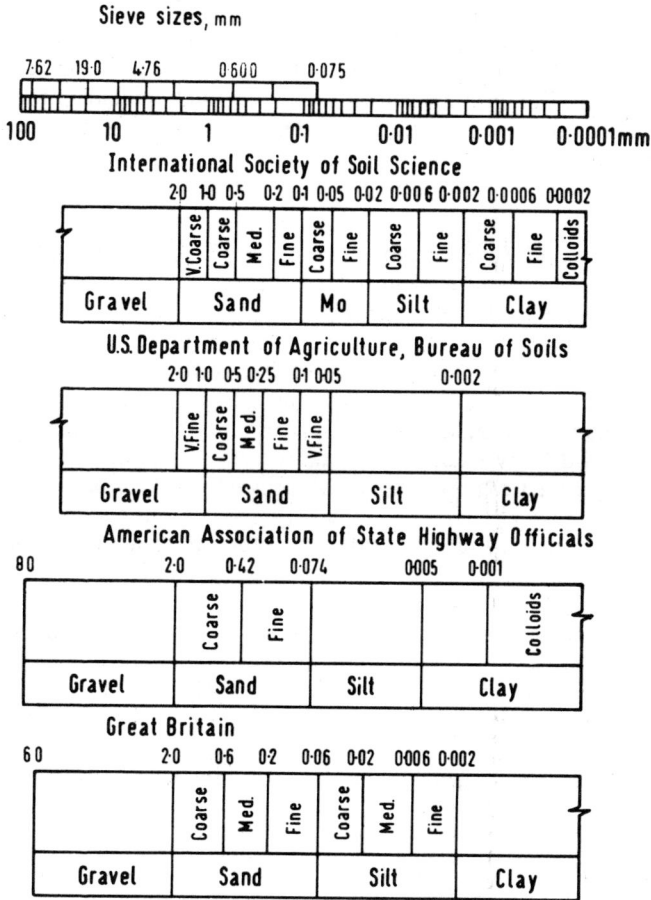

Fig. 4.3.   Systems of particle-size separations

lower limit being set at the point where water is held in the pores by capillary action. The lower limit of Atterberg's silt group was given the theoretical significance of being the size beyond which smaller particles could not be seen with the naked eye, did not have the usual properties of sand and could be coagulated to form 'crumbs'. Thus silt was visualized as being the material which ranges from where sand begins to assume clay-like features to the upper limit of clay itself. The choice of 0·002 mm as the upper limit of the clay fraction was based on the premise that particles smaller than this exhibited Brownian movement when in aqueous suspension.

Physical and mineralogical characteristics. The inorganic soil solids are composed of primary mineral fragments in conjunction with the secondary mineral products of weathering. (The term 'secondary mineral', as used here, refers to any mineral created after the original rock was formed from which the soil was derived.) The primary mineral quartz dominates the soil fraction derived by physical weathering, i.e. the sand and coarse silt fraction. Quantities of other primary silicate minerals such as feldspars, hornblende and the micas are also present at this stage. The primary minerals become of lesser significance in the fine silt particles until, with the clay fraction, it is found that the secondary minerals are the dominant ones.

Sand. Since sand is primarily composed of quartz particles, it is quite inactive chemically and it is its physical characteristics which are of most interest. In contrast to clay particles which are flat or flake-shaped, sand particles are generally bulky in shape. The individual grains may be further classified as angular, subangular, rounded or subrounded, depending on the degree of abrasion received prior to final deposition. For instance, residual sands are usually angular, whereas river and beach sands are generally rounded. Wind-blown sands are usually very fine and well-rounded, while ice-worn sand particles may have flat faces which have been scoured by the ice.

Table 4.1 shows that the stability potential of a sandy soil is significantly influenced by its state of compaction, gradation, and particle shape, i.e. note that the angle of internal friction increases with the degree of compaction, and with the angularity and size of particle. Not shown here is the effect of moisture content upon stability.

Clean sand particles do not exhibit any cohesive properties and so are little influenced by changes in moisture content. Because of the general bulkiness of the particles, the pores between them are relatively large; thus sandy soils are very permeable and well-drained and consolidation effects are small.

Silt. Silt particles can be considered as transitional between sand and clays. Physically they are similar to sands in that they derive their stability from the mechanical interaction between the particles. Coarse silt particles are essentially miniature sand particles and so tend to have similar bulky shapes as well as the same dominant mineral. Unlike sands, silts possess a certain limited amount of cohesion due to inter-particle water films. Although they are generally classed as permeable, moisture can only move through the (small) pore spaces relatively slowly. Where the smaller-sized particles pred-

TABLE 4.1. *Some approximate shearing resistance results for coarse-grained soils*[3]

Grain-size	Degree of compaction	Angle of internal friction, deg	
		Rounded grains, uniform gradation	Angular grains well graded
Medium sand	Very loose	28–30	32–34
	Moderately dense	32–34	36–40
	Very dense	35–38	44–46
Sand and gravel			
65%G + 35%S	Loose	—	39
	Moderately dense	37	41
80%G + 20%S	Dense	—	45
	Loose	34	—
Blasted rock fragments	—	40–55	

ominate, silts exhibit strong clay-like tendencies and may undergo considerable shrinkage and expansion when exposed to changes in moisture content.

Clay. The particles in the clay fraction differ from the sand and silt particles both in their chemical make-up and physical properties. Physically they are different in that they are flat and elongated, or lamellar, and thus have a much larger surface area per unit weight than the bulky-shaped silts

TABLE 4.2. *Some physical characteristics of soil separates*[4]

Name	Diameter mm	No. of particles per gramme	Surface area of 1 gramme of each separate, sq. cm
Fine gravel	2·00–1·00	90	11·3
Coarse sand	1·00–0·50	722	22·7
Medium sand	0·50–0·25	5,777	45·2
Fine sand	0·25–0·10	46,213	90·7
Very fine sand	0·10–0·05	722,074	226·9
Silt	0·05–0·002	5,776,674	453·7
Clay	<0·002	90,260,853,860	11,343·5

Note: Each particle is assumed to be a sphere having the maximum diameter of each group.

and sands. A measure of the difference in external surface areas exposed by the various fractions can be obtained by assuming that the individual separates are spherical in shape and then calculating the surface areas involved. Table 4.2 presents a summary of such calculations. Since the intensity of the physico-chemical phenomena associated with soils is a function of the exposed surface area, this table illustrates why it is that the clay fraction exerts an influence on a soil's behaviour which can appear to be very much out of proportion to its weight or volume in the soil.

Any study of the clay fraction of a soil is to a large extent a study of its colloidal ingredient. It is the soil colloids that are primarily responsible for the cohesiveness of a plastic soil, the facility with which it can undergo considerable changes in volume with changes in the moisture content, and its ability to solidify into a hard mass upon drying. In addition, the movement of air and moisture through a soil is considerably influenced by the amount and form of its colloidal content.

There is no uniform designation as to what constitutes a soil colloid. In theory it can be said that a colloid is any particle which exhibits Brownian movement when in an aqueous suspension. In practice, however, the term is usually applied to particles smaller than 1 micron (0·001 mm) even though it is known that the Brownian phenomenon is applicable to particles up to 4 microns in size.

The importance attached to the colloidal fraction is associated with the electrical charges which they carry on their surfaces. The reasons for these charges can perhaps be briefly explained by considering the manner in which ions, atoms or molecules are combined to form a crystalline solid. These chemical units tend to combine to a definite pattern so as to accomplish the greatest possible degree of electrical neutralization. The simplest grouping of these atoms is called a unit cell, while the term 'crystal' is applied to a three-dimensional repetition of unit cells. Within the structural arrangement of a crystal—this is called a space lattice—the electrostatic bonds or valencies of the atoms are completely neutralized, whereas those at the edges, corners and surfaces are not. The net result is the unbalanced negative charge associated with the colloidal particle.

A second major cause of the electrical charge carried by a soil colloid is the substitution of one ion for another within the crystal lattice. This phenomenon, which is known as isomorphous substitution, is the main source of the electrical charges associated with montmorillonitic and illitic soil colloids; this is in contrast with kaolinitic colloids, where the unsatisfied valency phenomenon is the main cause. In montmorillonite, for instance, the substitution of magnesium (Mg^{++}) for aluminium (Al^{+++}) on the basis of one Mg^{++} for one Al^{+++} results in an unsatisfied valency.

Clay minerals. Not so many years ago, clays were believed to be amorphous materials having no definite recurring pattern of atoms, ions or molecules. Research on clay minerals started about 1925 and since then it has been clearly established that clay particles have overwhelmingly crystalline structures. Thus clays are now considered as essentially composed of minute flakes and in these flakes, as with all crystalline substances, the atoms are arranged in a series of units which are either tubular or in the form of sheets.

There are two fundamental building blocks for the clay mineral structures. One is a silica tetrahedral unit in which four oxygens or hydroxyls having the configuration of a tetrahedron enclose a silicon atom. As illustrated in Fig. 4.4(a) the tetrahedra are bonded horizontally in a sheet structure so that the oxygens of the bases of all the tetrahedra are in a

TABLE 4.3.   *Classification of the crystalline clay minerals*[5]

A.   Two-layer type (sheet structures composed of units of one layer of silica tetrahedrons and one layer of alumina octahedrons).
    1.   Equidimensional
        (*a*)   Kaolinite group, i.e. kaolinite, nacrite. etc.
    2.   Elongate
        (*a*)   Halloysite group.

B.   Three-layer types (sheet structures composed of two layers of silica tetrahedrons and one central dioctrahedral or trioctrahedral layer).
    1.   Expanding lattice — Equidimensional
        (*a*)   Montmorillonite group, i.e. montmorillonite, sauconite, etc.
        (*b*)   Vermiculite.
    2.   Expanding lattice — Elongate
        (*a*)   Montmorillonite group, i.e. nontronite, saponite, hectorite.
    3.   Non-expanding lattice
        (*a*)   Hydrous mica (illite) group.

C.   Regular mixed-layer types (ordered stacking of alternate layers of different types).
    1.   Chlorite group.

D.   Chain-structure types (horneblende-like chains of silica tetrahedrons linked together by octahedral groups of oxygens and hydroxyls containing Al and Mg atoms).
    1.   Attapulgite.
    2.   Sepeolite.
    3.   Palygorskite.

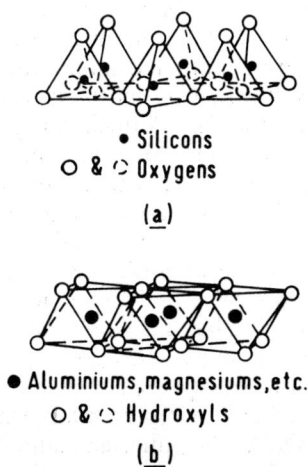

• Silicons
O & ◌ Oxygens

(**a**)

• Aluminiums, magnesiums, etc.
O & ◌ Hydroxyls

(**b**)

Fig. 4.4.   Basic building blocks of the clay minerals

common plane, and each oxygen is shared between two tetrahedra. In fact the silica tetrahedra sheet may be viewed as a layer of silicon atoms located between a layer of oxygens in the base and a layer of hydroxyls at the tips of the tetrahedra.

The second fundamental building block is an octahedral unit in which an aluminium, iron or magnesium atom is enclosed by six hydroxyls having the configuration of an octahedron. Again, as illustrated in Fig. 4.4(b), these octahedral units are bonded into a sheet structure in which each hydroxyl ion is common to three octahedral units. In this case the sheet can be viewed as two layers of densely packed hydroxyls with the aluminium cations between the sheets in octahedral co-ordination.

As indicated by the classification data in Table 4.3, the different clay minerals are formed through bonding together of two or more of these molecular sheets. From an engineering point of view, the ones which are of most importance are the kaolinite, montmorillonite, and the hydrous mica (illite) groups. Kaolinitic clays which generally tend to be associated with well-drained soils, were derived under humid climatic conditions from severely weathered crystalline rocks. Montmorillonites, on the other hand, are associated with poorly-drained and organic soils, were derived from rocks high in ferromagnesian minerals, dark igneous rock, or volcanic ashes. Illites are associated with soils derived from shales and marine deposits, micaceous rocks, and glacial debris from crystalline rocks.

Kaolinite. This group includes a number of clay minerals of which the most common, as well as important, is kaolinite. As indicated schematically in Fig. 4.5, the kaolinite structural unit is composed of an aluminium octahedral layer with a parallel superimposed silica tetrahedral layer inter-grown in such a way that the tips of the superimposed silica sheet and one of the layers of the alumina unit form a common sheet. Thus the kaolin unit can be viewed as a succession of layers of oxygens, silicons, oxygens and hydroxyls, aluminiums and hydroxyls, which is about 7 ångstroms thick (1 ångström, abbreviated Å, is 1×10^{-7} millimetres), and—in theory—extends infinitely in the other two directions. The kaolinite mineral is composed of a stacking of these 7 Å thick sheets such that its structure might be considered akin to a book in which each leaf is 7 Å thick. Successive 7 Å layers are held together by hydrogen bonds which allow the mineral to cleave fairly easily into very thin platelets. In clays these platelets occur in thicknesses of from 100 Å to 1000 Å and in widths from 1000 Å to 20 000 Å.

The kaolinites are considered to be very stable clays from an engineering aspect. The hydrogen bonds between the elemental sheets are sufficiently strong to prevent water molecules and other ions from penetrating, and hence the lattice is considered to be restricted or non-expanding. As with the other clay minerals, the kaolinite platelets carry negative electrical charges on their surfaces which attract thick layers of adsorbed water. Because of the non-expanding lattice the effective surface area to which the water molecules can be attracted is restricted to the outer faces and it is because of this that the plasticity, cohesion, and shrinkage and swelling properties of kaolinite are very low when compared with other silicate clays.

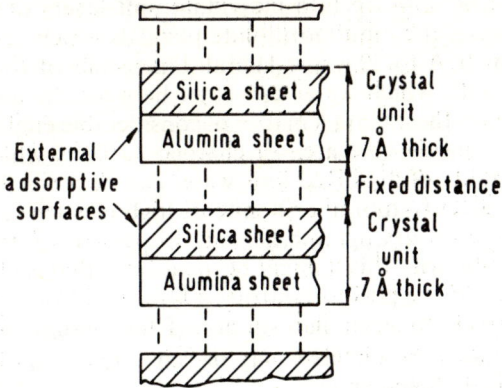

Fig. 4.5.   Schematic diagram of the structure of typical kaolin crystals

Montmorillonite. This group also contains several clays of which montmorillonite is by far the most important. The schematic diagram in Fig. 4.6 presents an edge view of two of the many units which may make up a crystal of the clay. Each crystal unit is composed of two silica tetrahedral sheets and one octahedral sheet, giving a 2:1 type of lattice structure. The octahedral

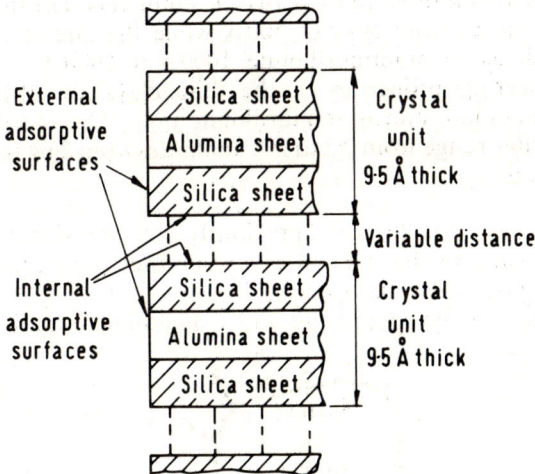

Fig. 4.6.   Schematic diagram of the lattice structure of typical montmorillonite crystals

sheet is between the silica sheets, with the tips of each tetrahedral sheet and an hydroxyl layer of the octahedral sheet intergrown to form a single layer. The thickness of each crystal unit is about 9·5 Å and the dimensions in the other two directions are indefinite. These 9·5 Å sheets are stacked one above the other, like the leaves of a book. There is very little bonding between successive crystal units and as a result water molecules and other cations can readily enter between the sheets. In the presence of an abundance of water, the mineral can be split up into individual unit layers of 9·5 Å thickness. Normally, however, the montmorillonite platelets which occur in clay have thicknesses from 10 Å to 50 Å and lateral dimensions of 1000 Å to 5000 Å.

The facility with which water can enter between the montmorillonite's crystal units makes these clays a matter of considerable engineering concern. Because of the considerable area of charged surfaces which are exposed, substantial amounts of dissolved ions/water can be attracted to these soils and this facility gives to montmorillonite its high plasticity and cohesion, its marked shrinkage on drying, and a ready dispersion of its fine flaky particles. In connection with this it might be mentioned that each thin platelet of montmorillonite has the power to attract a layer of adsorbed water approximately 200 Å thick to each flat surface. Thus, assuming zero pressure between the surfaces, the platelets may be separated by a distance of 400 Å and still be 'joined' together.

Illite. Like montmorillonite, illite has a 2:1 lattice structure. It differs in that there is always a substantial (about 20 per cent) replacement of silicons by aluminiums in the tetrahedral layers. The valencies vacated by this substitution are satisfied by positively charged potassium ions which lie between the structural units and tie them together. The strength of this potassium bond is intermediate between the hydrogen bond of kaolinite and the water linkage of montmorillonite, and the net result is that illites have properties intermediate between those of these two clay minerals. The thickness of illite clay particles varies from 50 Å to 500 Å while the lateral dimensions are about the same as for montmorillonite, 1000 Å to 5000 Å.

The engineering properties of the illite clays are also intermediate between the kaolinite and montmorillonite ones. The specific gravities of montmorillonites range from 2·2 to 2·7, kaolinites from 2·64 to 3·0, and illites from 2·60 to 2·68.

Exchange capacity. As discussed previously, colloidal silicate-clay particles usually carry negative charges and as a result they attract to, or adsorb on, their surfaces positively charged cations such as hydrogen and the base metals sodium, calcium, potassium and magnesium. Figure 4.7 illustrates

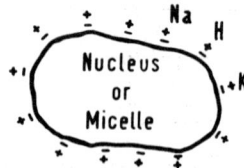

Fig. 4.7. Schematic representation of a colloidal particle

what is known as the Helmholtz double-layer concept of the make-up of a colloidal particle. As can be seen, the inner part consists of an insoluble nucleus or micelle—the inner sheath of negative charges is part of the wall of the nucleus—surrounded by a swarm of positively-charged cations. These cations are in equilibrium at different but infinitesimally small distances from the surfaces of the colloid.

A clay in which the principal adsorbed ion is hydrogen is called a hydrogen clay, while it is called a calcium clay if the adsorbed ions are mostly calcium. If a clay having adsorbed ions of one particular type is brought into contact with ions of another kind, some of the first type of ions may be released and some of the second type adsorbed in their place. This exchange of positively-charged ions, termed *Cation Exchange*, is at the basis of the stabilization of soils with certain chemicals and by electro-osmotic phenomena.

The efficiency with which ions can replace each other is dependent upon the following factors:

1. *The relative concentration or numbers of ions.* This is an application of the chemical law of mass action, whereby the greater the number of ions in the solution, the better the statistical chances of exchanges taking place.

2. *The number of charges on the ions.* The higher the valency, the greater the number of charges on an ion and—other things being equal—the greater its replacing power and the more difficult it is to displace when already attached to the colloidal particle. The exception to the rule is hydrogen which behaves similarly to a divalent or trivalent ion.

3. *The speed of movement or activity of the different ions.* With ions of the same valency, the replacing power tends to increase as the effective diameter of the ion increases. In other words, the smaller ions are less tightly held than the larger ones. This may seem at first sight to be at variance with Coulomb's Law, which states that the force of attraction between two ions varies inversely as the square of the distance between them; this means that the smaller ions should be held more tightly. In a soil solution, however, the ions are all hydrated, with the result that a water hull surrounds each ion. Because of Coulomb's Law, the smaller ions attract the greatest number of water molecules and as a result they end up by having the thickest water hulls. Therefore, the highly hydrated smaller ions have got larger effective sizes than the more weakly hydrated but initially larger ions. The greater effective size reduces the ease of movement (migration velocity) of the hydrated ion, as well as the degree of tightness with which it is held to the colloidal particle.

4. *The type of clay mineral.* As explained previously, clay minerals differ considerably in the amount of surface area exposed. Thus montmorillonites, which have expandable lattice structures, have high exchange capacities, whereas kaolinites have relatively low ones.

The actual manner in which the cations are exchanged can be explained by considering colloidal particles in a solution. Owing to heat movement and the Brownian motion, the adsorbed ions are never at rest but are in fact continually moving back and forth within a limited range from the surface of the particle. If electrolytes are added to the solution, cations are set in random motion because of the Brownian effect, and numbers of them may slip between the negative wall of the nucleus and the adsorbed but oscillating ions. These electrolytic cations then become preferentially adsorbed, while some surface ions are released and remain in the solution as exchanged ions. Obviously the more loosely the surface ions are held the greater is the average distance of oscillation and hence the greater the possibility of ion adsorption and/or replacement.

The ease with which cations may be exchanged and adsorbed is expressed in terms of what is known as the *Cation Exchange Capacity* of a soil. This is the number of milli-equivalents (m.e.) of ions that 100 grammes of the soil can adsorb. By definition, a milli-equivalent is one milligramme of hydrogen, or the amount of any other ion that will combine with or displace it. Thus, if the cation exchange capacity of a soil is 1 m.e., it means that every 100 g of dry material is capable of adsorbing and holding 1 mg of hydrogen or its equivalent of other ions.

Montmorillonite normally has a cation exchange capacity of between 80 and 100, illites are usually between 15 and 40, while kaolinites have values between 5 and 15 m.e. per 100 g.

Soil structure

The term structure in connection with soil usually refers to the manner in which the individual particles in the soil mass are, or can be, arranged with respect to each other. In the technical literature, however, two types of structure are differentiated; the above definition meets the criterion for *primary structure*, while the term *secondary structure* is used for the natural systems of cracks, fissures and other discontinuities which sometimes develop subsequent to a soil's formation and deposition.

Primary structure. Karl Terzaghi, the founder of modern soil engineering, has classified soils with regard to primary structure as single-grained, honeycomb and flocculent. This classification recognizes that structure is dependent on particle size and shape, as well as on the minerals composing the individual grains.

Single-grained structure. In a single-grained structure, each individual soil particle can be considered as being in direct 'point-to-point' contact with several adjacent particles, or, by analogy, as having come to rest as might a mixture of marbles poured into a container. In other words, the arrangement of particles is entirely accidental and has no particular pattern. In addition there is virtually no tendency for the particles to form aggregates, but each particle functions as an individual (see Fig. 4.8(a)).

Although found to a lesser extent with the fine-grained soils, a single-grained structure is usually associated with the coarse-grained sands and gravels. The individual mineral particles may be deposited in a loose state having a high void ratio, or in a dense state having a low void ratio; whichever it is, the structural designation is the same. From an engineering viewpoint, an important characteristic associated with single-grained soils is their ability to deform. Dense soils increase in volume when caused to deform, while loose soils decrease in volume when deformed by, e.g., vibration which causes the particles to slide or roll into a more dense state. The finer-grained cohesionless particles deposited in a loose state, say in an embankment, and then saturated form a particularly unstable construction. When the material is deformed by a sudden vibration, the void ratio decreases. Because of the lower permeability of the denser material, the water cannot escape from the soil quickly, with the result that the water has to carry the load and liquefaction takes place.

Honeycomb structure. As illustrated in Fig. 4.8 (b), a honeycomb structure is composed of particles which touch each other at only a few points. The configuration would obviously be unstable were it not for the binding action of the molecular forces which occur at the points of contact of the particles.

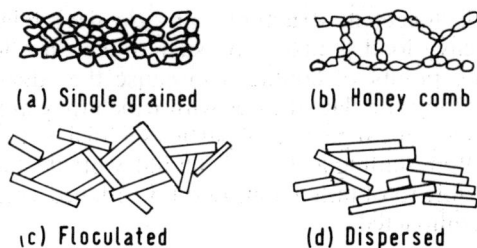

(a) Single grained (b) Honey comb

(c) Floculated (d) Dispersed

Fig. 4.8. Types of structure found in natural soils

Honeycomb structure is most often associated with water-deposited silt-sized particles. The structure was created by the silt particles settling out of suspension as individual grains. When the grains came together however, the molecular forces at the contact points were sufficiently strong to prevent the particles from rolling down over each other under the gravitational influence of the submerged weight. In this way several adjacent particles formed chains or strings which became miniature beams or arches capable of bridging relatively large void spaces.

Honeycomb soils have high void ratios and so have to be treated with respect when placing engineering loads upon them. In general they tend to have a critical loading below which excessive volume changes will not occur. When, however, the applied load is sufficient to break down the honeycomb structure, considerable settlement can occur, since the material then acts as if it were single-grained and deposited in a loose state.

Flocculent structure. This type of structure is generally associated with clay soils formed in large bodies of water. It is easiest to describe the structure in relation to a soil formed in water.

When the clay is formed as a result of individual particles coming into contact whilst still in suspension and prior to deposition, e.g. in salty water, then the soil is said to have a *flocculated-type* structure in which particles are arranged in an edge-to-edge or 'card-house' manner (see Fig. 4.8 (c)). If, on the other hand, the individual particles settle independently from a dispersed suspension, the sediment will have a *dispersed-type* or *oriented* structure in which the platelets tend toward face-to-face contact (see Fig. 4.8 (d)). In reality, clay soils tend to be composed of a more random mixture of the above two idealized forms of flocculent structure, i.e. some particles are face-to-face, some are edge-to-edge, and others have intermediate orientations.

The individual particles can also combine into aggregates or 'clusters' within which they may be, for example, randomly in contact with each other. If these clusters then form cementitious bonds with each other—these are much weaker than the bonds within each cluster—a flocculent honeycomb type of soil structure will be formed. This is termed a cluster-type of structure. Because of the relatively large volume of inter-aggregate pores in cluster-type clays, they tend to be more permeable than the other flocculent clays.

Clay soils with flocculent structures can also have relatively high void ratios. When a heavy load is placed on such a deposit, the high pressures experienced at the points of contact can cause the edges of the platelet particles to slide along the flat surfaces with which they are in contact; this results in a more dense arrangement of particles with a consequent reduction in the void ratio. Remoulding a clay soil similarly breaks down its structure and makes it possible to obtain a considerably reduced void ratio when the reworked soil is compacted.

Secondary structure. The main systems of secondary structure found in soils are the arrangements of fissures, cracks, slickensides and concretions. Clay soils are noted for the manner in which they can be subdivided by hair cracks and fissures, particularly as they dry out. Slickensides are smooth, almost polished, shear surfaces found in clays as a result of differential movements or expansions. Concretions are accumulations of carbonates and ion compounds which further disrupt the continuity of the soil mass, thereby significantly affecting their engineering properties.

Forces in soil systems

The forces operating in soils which exert major influences on their stability are, as can be surmised from the previous discussions, very numerous and complex. For the sake of simplicity, however, such forces can be grouped under the following headings: adsorption, cohesion, surface tension, internal friction and inertia. Soil stability is the product of these forces acting in unison, but their relative importance in any given instance will vary with the kind of soil. The following is a brief summary of the more important of these forces.

Adsorption. Adsorption is the phenomenon responsible for the attraction of water molecules and ions to the surfaces of soil particles. As has been previously mentioned, this attractive force is chemical in nature; the soil surfaces are predominantly negatively-charged, the water molecules are dipolar, and the ions in the soil water that are attracted to the particle surfaces are predominantly positively-charged. All soils have adsorptive powers which increase with decreasing particle size and increasing soil surface area. The mineralogical/chemical composition of a particle also influences its adsorptive capacity. Positively-charged ions display adsorption characteristics, and these are reflected in their ability to hold films of water. This adsorptive attraction of cations for water molecules generally increases with decreasing cation size, and this is reflected in the fact that calcium ions, which are actually much larger than sodium ions, carry thinner water films, so that the resultant 'effective' diameter of a Ca^{++} ion is much smaller than that of a Na^+ ion.

Adsorption plays its most important role in the fine fraction of a soil, particularly with the clay and colloidal-sized particles. It is in this size-range that the conditions of surface area and mineralogical/chemical composition that foster high adsorptive forces are usually present. As a result clay particles never exist in nature without adsorbed films of water. In fact, an individual colloidal particle may be capable of attracting and holding a water film that is much thicker than its own diameter. However, the thickness of an adsorbed film can vary considerably, depending on such factors as the availability of soil water, soil temperature and the nature of the adsorbed cation, e.g. calcium ions tend to result in adsorbed water films on the soil particles that are much thinner than when sodium ions are adsorbed.

No discussion of the adsorption phenomenon is complete without mention being made of the effect of having organic matter or humus present in the soil. Since they are predominantly negatively-charged, the humus particles display similar, but greater, adsorptive properties to clay minerals. Humus has an extremely high adsorptive capacity and is therefore capable of attracting and holding very thick films of water.

Cohesion. Cohesion can be defined as the attraction between like molecules, and in the soil system it is exemplified by the attraction of one water molecule for another. In ordinary tap water placed in a container and apparently at rest, the individual water molecules are actually moving too fast to be able to cohere or stick together. When, however, they are present in a soil, the water molecules come under the influence of strong particle surface charges; hence they are considerably slowed down so that they become more or less orientated in the force fields of the soil particles. Under such conditions the dipolar molecules of water are attracted to each other through their oppositely charged ends, in much the same manner as tiny bar magnets are attracted to each other.

Cohesion plays a most important role in cementing clay particles together in soil. An over-simplified way of picturing the mechanism of this cementation is to visualize the orientated water molecules as links in a chain connecting up the soil particle-cation-soil particle system. When soil particles are connected by films of water which are only a few molecules thick the cementation is very strong—in fact, it may be stronger than the cementation in Portland cement concrete. However, as the films become thicker, due to additional water being allowed into the soil, the forces of cohesion decrease, since the water molecules which are most distant from the charged surfaces are less strongly orientated and their attraction for each other is accordingly weakened. At the point where the forces of cohesion disappear, the soil-water system takes on the properties of a liquid.

From this, and the preceding discussion, it is apparent that adsorption and cohesion are the principal cementitious forces operating in clayey soils, and that conditions are most favourable for strong inter-particle cementation when the particles are flaky in shape and the water films between the flat surfaces are very thin.

Surface tension. Water exhibits the force of surface tension due to the molecular attraction which exists between its molecules at the air-water surface. When surface tension combines with the attraction between water and the surfaces of soil particles (adsorption) to constitute a force that is opposed to gravity and tends to draw or retain water above the water table, the combination of forces is called capillarity. Capillarity has an important effect on soil stability since it contributes to excess moisture in soils and also to frost action.

In small voids which are partially filled with water, moisture films tend to 'bridge' the spaces between the soil particles; this is especially noticeable in fine-grained granular soils. The surface tension in the films of water tends to pull the particles together and this gives rise to the phenomenon of *apparent cohesion* which may often be observed in damp sands and silts. The cohesion is called apparent because immersion or drying will destroy these films, thereby eliminating the tendency of these soil particles to stick together. Thus the contribution of apparent cohesion to stability is not permanent unless moisture conditions can be very carefully controlled.

Internal friction. As the name implies, this is a friction phenomenon which in soils is reflected in the resistance to displacement caused by interlocking of the bulky-shaped particles. The factors which have important effects on internal friction are particle shape and texture, the gradation of particle sizes, and the pressure of the particles against each other. The more angular the particles the greater is the frictional resistance of the soil. Similarly, the better the gradation and the greater the dry unit weight of a soil, the greater the internal friction.

Generally speaking, coarse-grained soils such as gravels and sands derive their shear strengths almost entirely from inter-granular friction, while the shearing resistances of clays are due to the forces of adsorption and cohesion. Many soils, however, derive their shear strength from a combination of all of these forces.

SOIL CLASSIFICATION

The previous discussions have shown that not only are soils very different but their behaviour can be variable also. Because of these many differentiating factors, soil investigators have tended to group soils together on the basis of some common characteristics which are of particular interest to their areas of study. As a result very many soil classification systems are in use throughout the world, and within different sciences, so it is necessary for the engineer to understand the fundamentals of the more important of these systems.

Casagrande system

Undoubtedly one of the most widely used, if not the most complete, engineering classification system based on common soil characteristics is that which was developed by Professor A. Casagrande of Harvard University in the early 1940's. It was developed as a result of difficulties experienced in the use of the other systems then in existence for the design and construction of military airfields during the war, and hence it was first known as the Airfield Classification System.[6] Subsequently, it was modified to a minor extent and today there are two 'outgrowth' systems in use which have many common characteristics. One of these is known as the Unified Classification System; it is used extensively in the United States where it was adopted in 1952 by both the U.S. Corps of Engineers and the Bureau of Reclamation. The second, the Extended Casagrande Classification System, is primarily used in Britain.[7]

A comprehensive summary of the Extended Casagrande system is given in Table 4.4. In this table, the first four columns describe the system, while

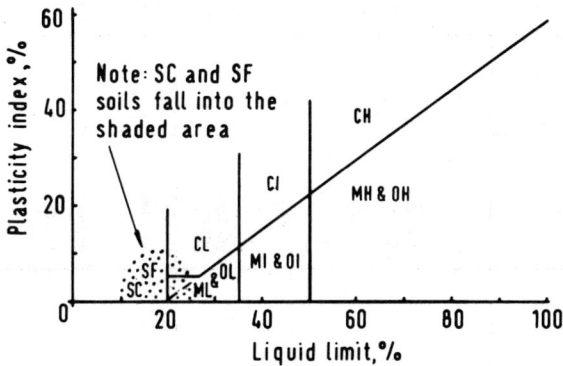

Fig. 4.9.   Plasticity chart for the Extended Casagrande classification system

the remaining columns provide information regarding particular properties and characteristics of each group. A plasticity chart is used in conjunction with Table 4.4 when fine-grained soils are being classified; this chart is shown in Fig. 4.9.

Although this system has more groupings than any other engineering classification system, it has been widely accepted because of the logical and orderly fashion in which these groupings are arranged. The system recognizes three main soil groups—the coarse-grained, fine-grained and fibrous soils—within which there are many subdivisions. These groupings, and the classification prefixes used, are as follows;

A. Coarse-grained soils.
    1.   Gravels and gravelly soils, prefix G.
    2.   Sands and sandy soils, prefix S.
    The gravels and sands are subdivided into the following:
    (a) Well graded soils with little or no fines, suffix W.

TABLE 4.4. Extended Casagrande classification system and its interpretation[7]

Material	Major divisions	Subgroups	Casagrande group symbol	Drainage characteristics	Potential frost action	Shrinkage or swelling properties	Value as a road foundation when not subject to frost action	Bulk unit weight before excavation, kg/m³		Coefficient of bulking, %
								Dry or moist	Saturated	
	Boulders and cobbles	Boulder gravels	—	Good	None to very slight	Almost none	Good to excellent	—	—	
		Well-graded gravel and gravel-sand mixtures, little or no fines	GW	Excellent	None to very slight	Almost none	Excellent	1920–2165	2080–2325	
		Well-graded gravel-sand mixtures with excellent clay binder	GC	Practically impervious	Medium	Very slight	Excellent	2000–2245	2160–2405	
	Gravels and gravelly soils	Uniform gravel with little or no fines	GU	Excellent	None	Almost none	Good	1520–1765	1840–2085	10–20
		Poorly-graded gravel and gravel-sand mixtures with little or no fines	GP	Excellent	None to very slight	Almost none	Good to excellent	1600–1845	1760–2005	
Coarse-grained soils		Gravel with fines, silty gravel, clayey gravel, poorly-graded gravel-sand-clay mixtures	GF	Fair to practically impervious	Slight to medium	Almost none to slight	Good to excellent	1760–1925	1920–2085	
		Well-graded sands and gravelly sands, little or no fines	SW	Excellent	None to very slight	Almost none	Excellent to good	1840–2005	2000–2165	
		Well-graded sand with excellent clay binder	SC	Practically impervious	Medium	Very slight	Excellent to good	1920–2085	1920–2325	
	Sands and sandy soils	Uniform sands with little or no fines	SU	Excellent	None to very slight	Almost none	Fair	1520–1845	1840–2165	5–15
		Poorly-graded sands with little or no fines	SP	Excellent	None to very slight	Almost none	Fair to good	1440–1685	1520–1765	
		Sands with fines, silty sands, clayey sands, poorly-graded sand-clay mixtures	SF	Fair to practically impervious	Slight to high	Almost none to medium	Fair to good	1520–1765	1760–2005	

Category	Description	Symbol						
Fine-grained soils — Soils having low compressibility	Silts (inorganic) and very fine sands, rock flour, silty or clayey fine sands with slight plasticity	ML	Fair to poor	Medium to very high	Slight to medium	Fair to poor	1520–1765	1600–1765 · 20–40
	Clayey silts (inorganic)	CL	Practically impervious	Medium to high	Medium	Fair to poor	1600–1765	1760–1925
	Organic silts of low plasticity	OL	Poor	Medium to high	Medium to high	Poor	1440–1685	1520–1765
Fine-grained soils — Soils having medium compressibility	Silty and sandy clays (inorganic) of medium plasticity	MI	Fair to poor	Medium	Medium to high	Fair to poor	1520–1765	1600–1765 · —
	Clays (inorganic) of medium plasticity	CI	Fair to practically impervious	Slight	High	Fair to poor	1600–1765	1765–1925
	Organic clays of medium plasticity	OI	Fair to practically impervious	Slight	High	Poor	1440–1685	1520–1765
Fine-grained soils — Soils having high compressibility	Micaceous or diatomaceous fine sandy and silty soils, elastic silts	MH	Poor	Medium to high	High	Poor	<1680	<1925 · —
	Clays (inorganic) of high plasticity, fat clays	CH	Practically impervious	Very slight	High	Poor to very poor	<1925	<1925
	Organic clays of high plasticity	OH	Practically impervious	Very slight	High	Very poor	<1765	<1925
Fibrous organic soils with high compressibility	Peat and other highly organic swamp soils	Pt	Fair to poor	Slight	Very high	Extremely poor	<1765	<1765 · —

(b) Well graded soils with suitable clay binder, suffix C.
(c) Uniformly graded soils, with little or no fines, suffix U.
(d) Poorly graded soils with little or no fines, suffix P.
(e) Poorly graded soils with appreciable fines, or well graded soils with excess fines, suffix F.

B. Fine-grained soils.
1. Inorganic silt soils, prefix M.
2. Inorganic clay soils, prefix C.
3. Organic silts and clays, prefix O.

There are three subdivisions to the fine-grained soils:
(a) Soils whose liquid limits are less than 35 per cent, and which have low compressibility, suffix L.
(b) Soils whose liquid limits are between 35 and 50 per cent and which have medium compressibility, suffix I.
(c) Soils whose liquid limits are greater than 50 per cent, suffix H.

C. Fibrous soils.
1. Peat and swamp soils, prefix Pt.

There are no subdivisions of the fibrous soils.

Description of soil groups. The following summarily describes some particular features of the soils in the major groups.

Coarse-grained soils. The determination of the classification groupings of the coarse-grained soils is primarily based on particle-size distribution. Gravels and gravelly soils (G) have more than 50 per cent by weight of the particles retained on the 2·36 mm BS sieve, while the sands and sandy soils (S) contain a preponderance of particles passing the 2·36 mm sieve and retained on the 75 μm sieve.

GW and SW soils. These are well graded soils with less than 5 to 10 per cent passing the 75 μm BS sieve.

GC and SC soils. These are also well graded but contain sufficient clay or other natural cementing binder to provide cohesion without causing undue swelling or shrinkage.

GU and SU soils. These are uniformly or closely graded soils which have a particle-size distribution extending over a very limited range of particle sizes, e.g. poorly graded and having an excess of only one small range of particle sizes and a deficiency of all others.

GP and SP soils. These are poorly graded soils but here the poor gradation is a result of a gap gradation. A typical example would be a soil composed of a mixture of coarse material and fine sand with the intermediate sizes absent.

GF and SF soils. These groups contain all coarse-grained soils which contain excess fines and which are poorly graded.

Fine-grained soils. The fine-grained soils are differentiated on the basis of their plasticity characteristics. The chart shown in Fig. 4.9 is used for this purpose, each soil being grouped according to the area of the chart in which the plotted point lies. An empirical boundary line whose equation is

$$\text{Plasticity index} = 0.73 \, (\text{Liquid limit} - 20)$$

generally separates the inorganic clays from the silty and organic soils; this boundary is known as the A-line. Vertical lines at liquid limits of 35 and 50 per cent are assumed to subdivide the fine-grained soils into those of low, medium and high plasticity. The suffixes for these subdivisions are L, I and H respectively.

At values of the liquid limit below 25 per cent there is considerable overlapping of classification groupings. The inorganic silt soils (ML) cross above the A-line until they are limited by a horizontal boundary line. Similarly, the lean clays (CL) overlap horizontally into the shaded area enclosing the coarse-grained soils containing binder material. In these cases a soil may be classified by rigidly accepting that the transition between the coarse-grained and fine-grained soils is where more than 50 per cent of the material is finer than 0.1 mm. More often, however, these soils are described by boundary line designations such as CL-SC, CL-SF, SF-ML or SF-SC.

Most inorganic clays lie fairly close to the A-line but some may lie well above it. Kaolin clays usually lie below the A-line. They behave in a similar manner to inorganic silts and so are classified as either ML, MI or MH. Some volcanic clays and bentonite (montmorillonitic) clays lie just above the A-line at liquid limits of several hundred per cent.

Fibrous soils. These are highly organic soils which are unsuitable as subgrade materials because of their compressibility characteristics. Peat, humus and marshy soils are typical examples of this group.

Laboratory classification of a soil can easily be carried out by following the procedure suggested in Fig. 4.10.

Highway Research Board system

In 1929 the U.S. Bureau of Public Roads, then known as the Public Roads Administration, presented a system of soil classification which divided soils into eight groups designated from A-1 to A-8 inclusive. Soils were placed into particular groups according to a number of their physical characteristics. Thus a well-graded soil composed primarily of gravel and sand, but with some clay binder material present, was designated an A-1 soil, while other soils of lesser stability as highway foundations were placed by means of their physical characteristics in higher numbered groups.

As the usefulness of this classification system became apparent, various engineering organizations modified it to meet their own particular needs and soon a number of subdivisions of the various groups came into rather widespread use. In 1945 the Public Roads system was extensively revised by a special committee of the Highway Research Board,[8] and the result of their deliberations is given in Table 4.5. This system, generally called the HRB system, is. also known as the Revised Public Roads system and, since its

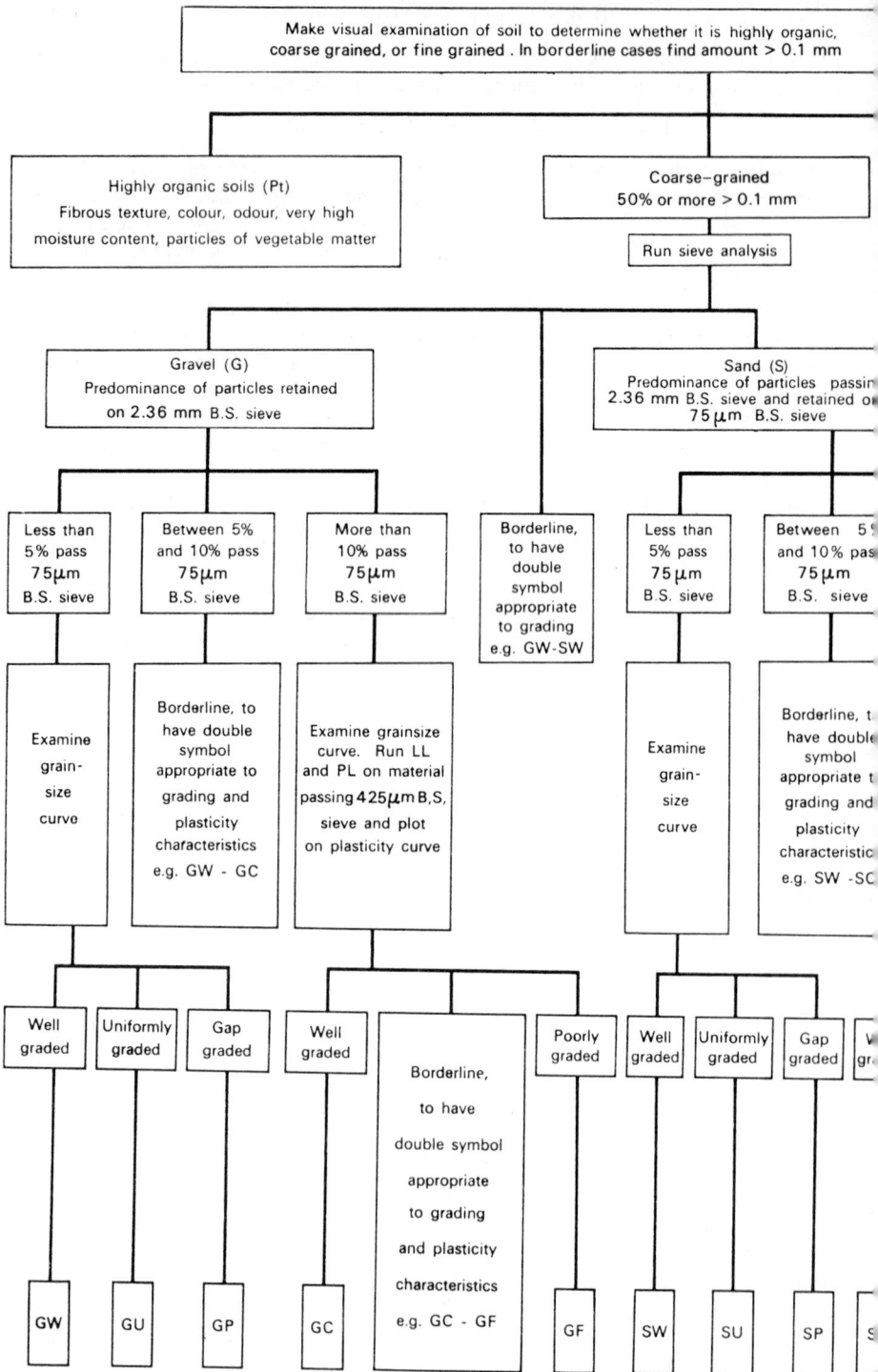

Fig. 4.10. Suggested outline of laboratory procedure

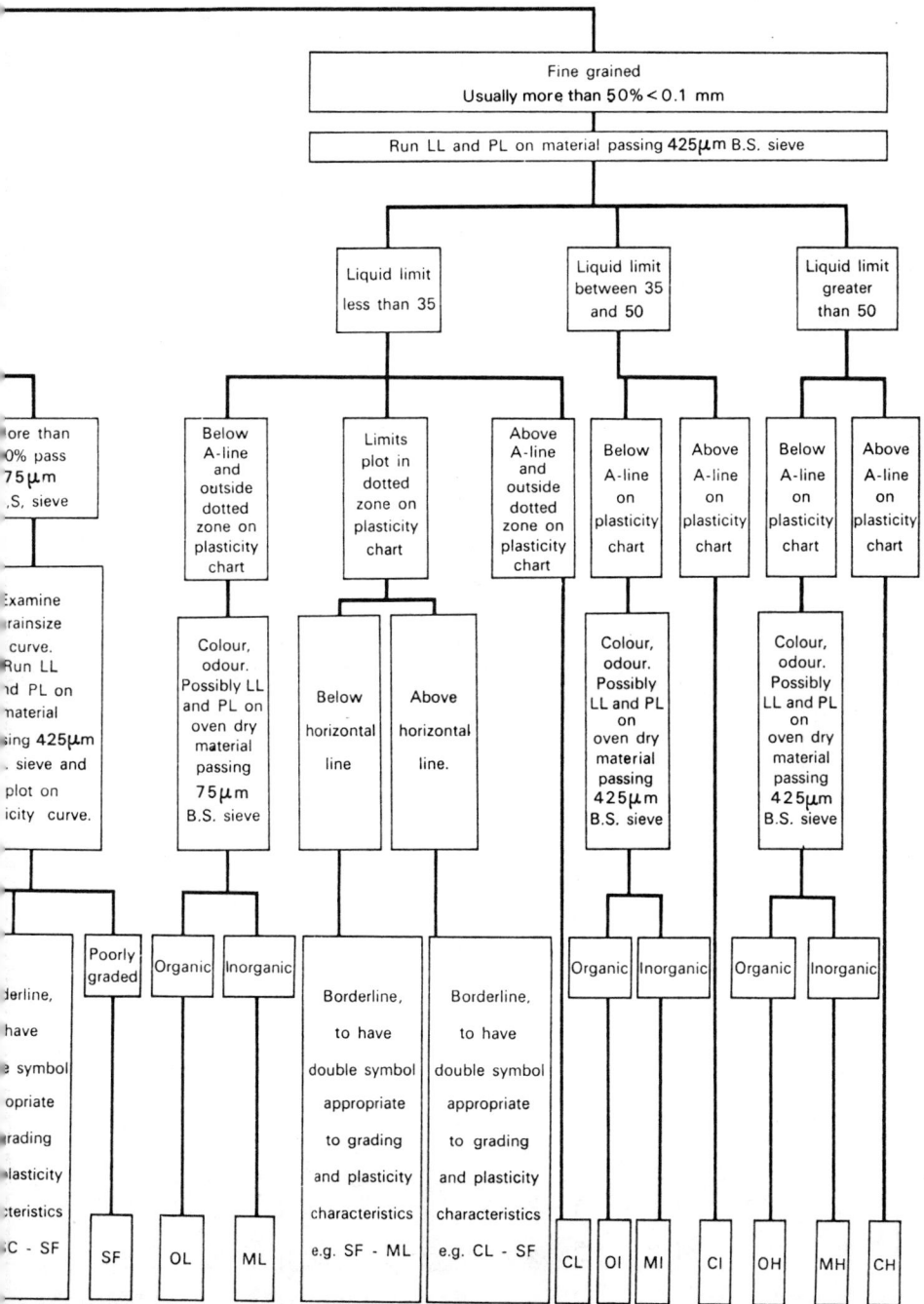

Fine grained
Usually more than 50% < 0.1 mm

Run LL and PL on material passing 425μm B.S. sieve

Liquid limit less than 35

Liquid limit between 35 and 50

Liquid limit greater than 50

More than 0% pass 75μm ,S, sieve

Examine grainsize curve. Run LL nd PL on material sing 425μm sieve and plot on icity curve.

Below A-line and outside dotted zone on plasticity chart

Limits plot in dotted zone on plasticity chart

Above A-line and outside dotted zone on plasticity chart

Below A-line on plasticity chart

Above A-line on plasticity chart

Below A-line on plasticity chart

Above A-line on plasticity chart

Colour, odour. Possibly LL and PL on oven dry material passing 75μm B.S. sieve

Below horizontal line

Above horizontal line.

Colour, odour. Possibly LL and PL on oven dry material passing 425μm B.S. sieve

Colour, odour. Possibly LL and PL on oven dry material passing 425μm B.S. sieve

Poorly graded

Organic

Inorganic

Organic

Inorganic

Organic

Inorganic

derline, have e symbol opriate rading lasticity cteristics C - SF

Borderline, to have double symbol appropriate to grading and plasticity characteristics e.g. SF - ML

Borderline, to have double symbol appropriate to grading and plasticity characteristics e.g. CL - SF

SF

OL

ML

CL

OI

MI

CI

OH

MH

CH

classifying soils within the Extended Casagrande system.

TABLE 4.5. *Highway Research Board classification system*

General classification	Granular materials (35% or less passing 75 µm sieve)							Silt-clay materials (More than 35% passing 75 µm sieve)			
	A-1		A-3	A-2				A-4	A-5	A-6	A-7
Group classification	A-1-a	A-1-b		A-2-4	A-2-5	A-2-6	A-2-7				A-7-5 A-7-6
Sieve analysis, % passing											
A.S.T.M. BS sieve sieve											
No. 10 2·0 mm	50 max.										
No. 40 425 µm	30 max.	50 max.	51 max.								
No. 200 75 µm	15 max.	25 max.	10 max.	35 max.	35 max.	35 max.	35 max.	36 min.	36 min.	36 min.	36 min.
Characteristics of fraction passing A.S.T.M. No. 40 (BS 425 µm)											
Liquid limit				40 max.	41 min.	40 max.	41 min.	40 max.	41 min.	40 max.	41 min.
Plasticity index	6 max.		N.P.	10 max.	10 max.	11 min.	11 min.	10 max.	10 max.	11 min.	11 min.
Group index	0		0	0		4 max.		8 max.	12 max.	16 max.	20 max.
Usual significant materials	Stone fragments, gravel and sand		Fine sand	Silty or clayey gravel and sand				Silty soils		Clayey soils	
General rating as subgrade	Excellent to good							Fair to poor			

Note. The A-7 group is subdivided according a plastic index criterion. When a soil has a **PI** equal to or less than the liquid limit minus 30, it is placed in the A-7-5 subgroup. When the PI is greater than LL–30, the soil falls into the A-7-6 subgroup.

adoption in 1950 by the American Association of State Highway Officials, as the AASHO system.

The HRB system is similar to the old Bureau system in that the lower the classification number of a soil, the better it is from the stability point of view. The new system differs, however, in that the number of classification groups was reduced from eight to seven, twelve subgroups were officially introduced, and the number of physical properties required to place a soil was reduced to three, i.e. mechanical analysis, liquid limit and plastic index. In addition, the concept of a 'Group Index' was developed in order to subdivide the fine-grained soils.

Description of soil groups. The soil materials included in the various groups of the Highway Research Board system are divided into two major classes. These are the granular materials containing 35 per cent or less material passing the 75 μm BS sieve, and the silt-clay materials containing more than 35 per cent passing the 75 μm sieve. The granular materials are composed of the A-1, A-2 and A-3 soils, while the silt-clay soils fall into the A-4, A-5, A-6 and A-7 soils.

Group A-1. The typical material of this group is a well-grained mixture of stone fragments or gravel, coarse sand, fine sand and a non-plastic or feebly plastic soil binder. However, it also includes stone fragments, gravel, coarse sand, volcanic cinders, etc., without soil binder. Subgroup A-1-a includes these soils which consist predominantly of stone fragments or gravel either with or without a well-graded binder of fine material. The A-1-b subgroup includes those which consist predominantly of coarse sand with or without a well-graded soil binder.

Group A-2. This group includes a wide variety of granular materials which are borderline between the materials falling in groups A-1 and A-3, and the silt-clay materials of groups A-4, A-5, A-6 and A-7. It includes all materials containing 35 per cent or less passing the 75 μm sieve which cannot be classified as A-1 or A-3, due to having a fines content or plasticity or both in excess of the limitations for these groups.

The subgroups A-2-4 and A-2-5 include various granular materials containing 35 per cent or less passing the 75 μm sieve and with the minus 425 μm portion having the characteristics of the A-4 and A-5 groups. These groups include such materials as gravel and coarse sand with silt contents or plasticity indices in excess of the limitations of group A-1, and fine sand with non-plastic silt content in excess of the limitations of group A-3.

The subgroups A-2-6 and A-2-7 include materials similar to those described under subgroups A-2-4 and A-2-5, except that the fine portion contains plastic clay having the characteristics of the A-6 or A-7 group. The approximate combined effects of plasticity indices in excess of 10 and percentages passing the 75 μm sieve in excess of 15 is reflected by group index values of 0 to 4.

Group A-3. The typical material of this group is fine beach sand or fine desert blow-sand without silty or clay fines or with a very small amount of non-plastic silt. The group includes stream-deposited mixtures of poorly-graded fine sand and limited amounts of coarse sand and clay.

Group A-4. The typical material of this group is a non-plastic or moderately plastic silty soil usually having 75 per cent or more passing the 75 μm sieve. The group includes mixtures of fine silty soil and up to 64 per cent of sand and gravel retained on the 75 μm sieve. The group index values range from 1 to 8, with increasing percentages of coarse material being reflected by decreasing group index values.

Group A-5. The typical material of this group is similar to that described under group A-4, except that it is usually of micaceous character and may be highly elastic as indicated by a high liquid limit. The group index values range from 1 to 12, with increasing values indicating the combined effect of increasing liquid limits and decreasing percentages of coarse material.

Group A-6. The typical material of this group is a plastic clay soil usually having 75 per cent or more passing the 75 μm sieve. The group includes mixtures of fine clayey soils and up to 64 per cent of sand and gravel retained on the 75 μm sieve. Materials of this group usually have high volume changes between the wet and dry states. The group index values range from 1 to 16, with increasing values indicating the combined effect of increasing plastic indices and decreasing percentages of coarse material.

Group A-7. The typical material of this group is similar to that described under group A-6, except that it has the high liquid limits characteristic of the A-5 group and may be elastic as well as subject to high volume change. The range of group index values is 1 to 20, with increasing values indicating the combined effect of increasing liquid limits and plasticity indices and decreasing percentages of coarse material.

The subgroup A-7-5 includes those materials with moderate plasticity indices in relation to liquid limit, which may be highly elastic as well as subject to considerable volume change. The subgroup A-7-6 includes those materials with high plastic indices in relation to liquid limit, which are subject to extremely high volume changes.

Group index. A new feature of the HRB system was the introduction of a subsidiary rating system as a means of placing a soil containing appreciable amounts of fine-grained material within its group or subgroup. The system chosen utilizes what is called the group index of a soil. This is simply a number between 0 and 20 which is dependent on the percentage of material passing the 75 μm sieve, the liquid limit and the plasticity index. A low group index is a reflection of high subgrade stability, while high group indices reflect the poor stability conditions associated with high liquid limits, high plasticity indices and low granular material contents.

The following empirical formula is used to obtain the group index of a soil:

$$G.I. = 0 \cdot 2a + 0 \cdot 005ac + 0 \cdot 01bd$$

where $a =$ that portion of the percentage passing the 75 μm sieve greater than 35 and not exceeding 75, expressed as a positive whole number from 1 to 40,

b = that portion of the percentage passing the 75 μm sieve greater than 15 and not exceeding 55, expressed as a positive whole number from 1 to 40,

c = that proportion of the numerical liquid limit greater than 40 and not exceeding 60, expressed as a positive whole number from 1 to 20,

and d = that portion of the numerical plasticity index greater than 10 and not exceeding 30, expressed as a positive whole number between 1 and 20.

The above formula is weighted so that the maximum influence of each of the three variables is in the ratio of 8 for the percentage passing the 75 μm sieve, 4 for liquid limit, and 8 for plasticity index. This weighting and the adopted critical ranges are based on the study of average relative evaluations placed on subgrade materials by several highway organizations in the United States which use the tests involved in this classification system.

The group index of a soil is expressed to the nearest whole number and is written in parenthesis after the group or subgroup designation, e.g. A-2-6 (3). Even if the determined value is zero, it is still given, as this indicates that the HRB rather than the original Public Roads system was used to classify the soil.

Pedological system

Pedology is defined as the science that treats of soils, including their nature, properties, formation, functioning, behaviour and response to use and management.[9] The development of this soil science gains its impetus from its relationship with agriculture, where its influence on the advancement of scientific agricultural practices has been of great significance. Unfortunately the value of pedology in engineering studies, particularly those relating to highway engineering, has yet to be fully recognized. It is only relatively recently that highway engineers have begun to make use of the great wealth of data and knowledge accumulated over the years by the pedologists. Of particular value in helping solve highway engineering problems are the agricultural soil maps which utilize the pedological soil classification system in their preparation.

The pedological system classifies soils according to their morphology and genesis. In other words, it is based on the premise that the profile of a soil is the result of the combined influence of all the past events and environment in the history of the soil. Five factors are assumed to act in combination in forming a soil's structure, form and properties; these are climate, vegetation, parent material, time and topography. A difference in kind or degree of any one of these factors results in a different soil profile being formed.

Description of system. In the pedological system, which is world-wide in application and scope, six main categories of soils are described. The so-called 'higher' categories are called the order, suborder, great soil group and family, while the 'lower' categories—which are the ones of most direct interest to the engineer—are known by the terms soil series and soil type.

Higher categories. The first category divides soils into three orders known as zonal, intrazonal and azonal. The zonal order includes those soils which have well-developed profiles that reflect the dominating influence of climate and vegetation. The zonal soils can be divided according to whether they are of the forested or treeless zones. Within these two subgroups there are a number of suborders which depend, in the main, on rainfall and temperature, or on the presence or absence of acid humus or soluble salts, etc. The zonal suborders are again subdivided on the basis of particular combined effects of climate, vegetation and topography into what are known as the great soil groups.

In contrast to the zonal soils, which are found over wide land areas, and primarily reflect the soil-forming influences of climate and vegetation, the intrazonal soils are found only in small areas and their characteristics are reflections of the passive factors of soil formation, i.e. time, topography and parent material. They are usually poorly-drained and have more or less well-developed profiles. The intrazonal soils are subdivided into suborders and great soil groups which depend mainly on rainfall and the degree of drainage imperfection. The bog soils are typical hydromorphic intrazonal soils of the humid regions, while the saline soils are representative halomorphic soils of the imperfectly drained arid regions. The calomorphic suborder is a special one devoted to well-drained soils formed from parent material rich in calcium. The main difference between the two members of this suborder's great soil groups is that one was formed under forest cover while the other was developed under grassland.

The azonal order includes those young and immature soils which have not had the opportunity to develop profile characteristics distinct from their parent materials. There are no suborders to this order, but instead there are three great soil groups: lithosols—these are thin surface soils formed *in situ* from soft or hard rock—alluvial or flood-plain soils, and dune sands.

All of the great soil groups are divided into *soil families*. The members of a soil family have similar profiles. Each of the families is named after one of the areas in which the particular profiles were first extensively studied.

Lower categories.   If a vertical section is taken down from the ground surface it will be found that there is a natural succession of strata or horizons which represent alterations in the original soil material brought about by the weathering process. In easily-drained soils with well-developed profiles there are normally three distinct layers known as the A, B and C horizons.

The uppermost layer, the A horizon, is often called the zone of eluviation because much of the original ultrafine colloidal material and the soluble mineral salts have been leached out from it by downward percolating water. The agriculturist is most concerned about this layer because it is rich in the humus and organic plant residues so vital to good crop growth. These characteristics render it a highly undesirable construction layer from the highway engineer's point of view because it often exhibits high compressibility and elasticity, high resistance to compaction, and variable plasticity.

The B horizon is also known as the zone of illuviation since it is in this layer that the material accumulates which is washed down from the A horizon just above it. This horizon is usually more compact than the horizons above or below it, contains more fine-grained particles, is less permeable, and is usually more surface-chemically active and unstable. These features render the B horizon a most important layer from the point of view of the highway engineer. With sandy soils it may be possible to utilize excavated B horizon material to improve the gradation of the A horizon. On the other hand, with fine-grained soils, the extra accumulation of fine particles in the B horizon may make it so unsuitable as a subgrade material that it has to be removed and wasted.

The layer in which the engineer is most interested is the C horizon. This contains the unchanged material from which the A and B horizons were developed. It is in exactly the same physical and chemical state as when it was first deposited in the geological cycle by water, wind or ice. It is this material which is normally utilized as fill material in the greater part of embankment construction.

Although not a normal occurrence, there may sometimes be found a further horizon, namely the D horizon, beneath the C horizon. This is not usually noted by the pedologist unless it is within about 1–1·5 m of the surface. Within this depth the D horizon has a significant effect on the characteristics of the overlying soil, thereby influencing the profile development.

As illustrated in Fig. 4.11, the A, B and C horizons may be subdivided into a number of sub-horizons. It must be emphasized, however, that the soil profile illustrated in this figure is an idealized one. More often than not many of the sub-horizons, and indeed horizons, may be missing *in situ*, depending both on the influential soil-forming factors and the erosional features present.

Soil series. By carefully observing such soil profile features as number, colour, texture, structure, relative arrangement, chemical composition, and thickness of the soil horizons, as well as the geology of the soil material, the pedologists have been able to subdivide the great soil groups into what are known as soil series. It is these subdivisions which, of all the pedological categories, are of most significance to the highway engineer, as it is these units which appear on the agricultural soil maps.

A series consists of a group of soils having the same mode of origin. Members of a particular series have the same number and arrangement of horizons, are developed from the same kind of parent material under the same climatic and vegetative conditions for the same period of time, and have the same slope on the same type of topography. In other words, the soils within a particular series are essentially homogeneous; only minor differences, primarily in texture of the A horizon material, are tolerated between the various members of a series.

Soil series are given names which are usually the proper names of some

Description (left)	Horizon	Description (right)
Organic debris lodged on the surface of the soil, usually absent from grassland soils	A_{00}	Loose leaves and organic debris, largely undecomposed
	A_0	Organic debris, partially decomposed or matted
	A_1	A dark-coloured horizon with a high content of organic matter mixed with mineral matter
	A_2	A light-coloured horizon representing the zone of maximum leaching. It is prominent in woodland soils and faintly developed or absent in grassland soils
The solum; composed of A and B horizons; soil developed by the soil forming processes	A_3	Transitional to B, but more like A than B. It is sometimes absent
	B_1	Transitional to A, but more like B than A. It is sometimes absent
	B_2	A usually deeper-coloured horizon representing the zone of maximum accumulation and maximum development of blocky or prismatic structure. In grassland soils, it has comparatively little accumulated material and represents a transition between A and C
	B_3	Transitional to C
Parent material; composes the C horizon	G	Horizon G, represents gleyed layer; found in hydromorphic soils
	Cca	Horizons Cca and Ccs; represent layers of accumulated calcium carbonate and calcium sulphate found in some soils
	Ccs	
Any stratum underneath the soil, such as hard rock or a layer of clay or sand that is not parent material but which may affect the overlying soil	D	Underlying stratum

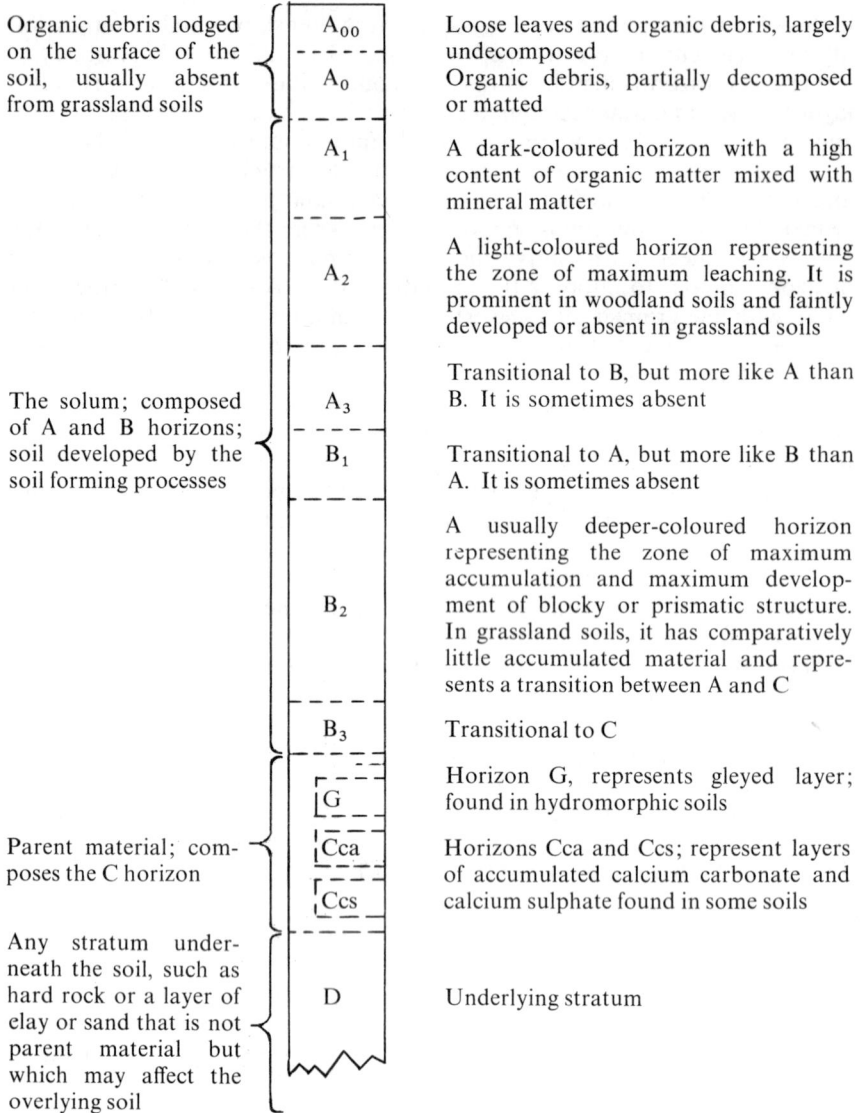

Fig. 4.11.   Hypothetical soil profile

places, lakes, streams, or other geographical entities. The name of a particular series is usually taken from near the spot where the prototype soil profile was identified and scientifically defined. Thus, after a soil series is first found and named, that name is used to describe all other soils which fit the same definition no matter where they are found.

Soil types. The lowest formal category of the pedological classification system is known as the soil type. The definition of this classification unit is identical with that of the soil series except that the textures of the A horizons of all soils within each soil type are not significantly different. Thus the name of a soil type consists of the name of the series to which it belongs together with a suffix describing the texture of the A horizon material, e.g. Brooke clay loam.

Textural classification system

The term 'texture' is a very difficult one to define. Perhaps it might be said that a soil is composed of particles of various sizes and shapes which give to it a distinctive appearance and 'feel' and the term which describes these features is the texture of the soil. Thus, for instance, if a soil is described as texturally harsh, then the impression is that soil is composed primarily of particles which are sharp and angular as against ones which are made up of flat or round particles.

It is very difficult, however, to have a common understanding of what is

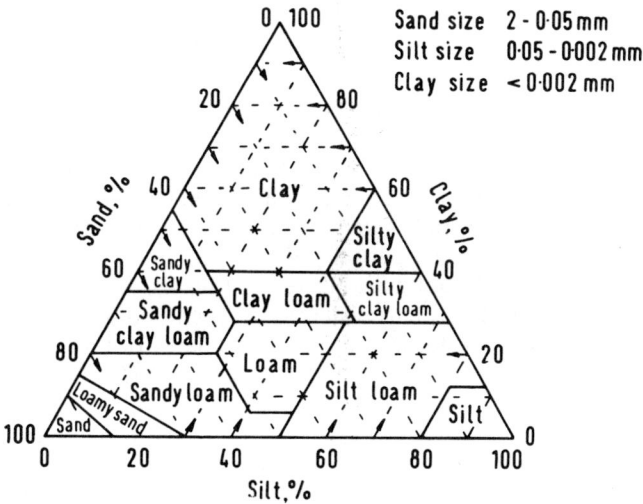

Fig. 4.12. Textural classification chart

meant by such terms as harsh or light, and so most textural classification systems utilize descriptions of particle-size distributions as indirect means of reflecting soil texture. Textural charts such as the one illustrated in Fig. 4.12 are then used to determine the proper term for the soil in question.

The terms used to indicate the texture of a soil are purely arbitrary in nature, and their meanings vary from organization to organization and from country to country. It is most important therefore that, when using a particular system, special care should be taken to specify the particle-size scale on which it is based. The triangular chart shown in Fig. 4.12 is that used by the Soil Survey of Great Britain and by the U.S. Department of Agriculture Soil Conservation Service; it is based on the particle-size classification used by those organizations.

Illustrative examples

The manner in which the different classification systems are utilized in order to classify a soil can be illustrated by taking a typical problem. Let it be assumed that it is intended to classify a soil about which the following information is available:

BS Sieve size, mm	Percentage passing
2·36	88
2·00	85
0·425	54
0·106	30
0·075	28
<0·050	25
<0·002	8
L.L. = 38	
P.I. = 16	

Extended Casagrande system. Following the procedure outlined in Fig. 4.10, it is seen that, since 70 per cent of the soil is retained on the 106 μm BS sieve, it falls into the coarse-grained category. Since 58 per cent of the particles passes the 2·36 mm sieve but is retained on the 106 μm sieve, the soil falls into the sand group and is given the suffix S. More than 10 per cent passes the 75 μm sieve so the particle-size curve should be examined and the plasticity index and liquid limit plotted in Fig. 4.9. Although not shown here, an examination of this particle-size curve will indicate that the soil is well-graded and that the plot in the plasticity chart does not fall inside the dotted zone. Thus the soil is given the classification SC.

Highway Research Board system. This soil is classified by entering the Table 4.5. It is first noted that, since this particular soil has less than 35 per cent passing the 75 μm sieve, it falls into the coarse-grained category. Next, beginning at the left-hand column of the chart and moving to the right, it is seen that the soil cannot be placed in the A-1 group as the maximum gradation and plasticity specifications are exceeded. For the same reasons it cannot be considered in the A-3 group. The gradation does meet the requirements of the A-2 group, so this can be further examined. Since the soil has a liquid limit of 38 and plasticity index of 16, it can only meet the requirements of the A-2-6 subgroup. The group index of the soil is next calculated as follows:

$$\text{G.I.} = 0.2a + 0.005ac + 0.01bd$$

$$= 0.2(0) + 0.005(0)(0) + 0.01(13)(6)$$

$$= 1, \text{ approx.}$$

The correct HRB classification for this soil is therefore A-2-6(1).

Pedological system. The data supplied are not suitable for classification of the soil within the pedological system.

Textural system. The textural chart shown in Fig. 4.12 will be used in this instance to illustrate how soils are classified within the textural systems.

Textural systems using triangular charts can only classify on the basis of the amount of sand, silt and clay present in the soil. When a soil contains a substantial amount of coarse material (usually taken as more than 10 per cent) such that it merits the description of a stony or gravelly soil, its textural class has to be determined on the basis of the percentages of sand, silt and clay adjusted to allow for the gravel or stone which must be 'excluded' in order to utilize the chart.

In this particular example, the soil contains 15 per cent of its material retained on the 2·00 mm BS sieve. (This is normally taken as the lower limit for gravel in the USDA system.) Hence, it will be classified as a gravelly soil of some kind. In order to use the triangular chart it is necessary to multiply the percentages of the sizes smaller than gravel by the ratio 100/85 so as to express these percentages in terms of the weight of the material exclusive of the gravel. Thus,

$$\text{Sand (2 to 0·05 mm)} \qquad 60 \times \frac{100}{85} = 71\%$$

$$\text{Silt (0·05 to 0·002 mm)} \qquad 17 \times \frac{100}{85} = 20\%$$

$$\text{Clay (}<0·002\text{ mm)} \qquad 8 \times \frac{100}{85} = 9\%$$

With these adjusted values it is now possible to enter the triangular chart, and it is seen that the textural class of the adjusted soil is that of a sandy loam. The complete designation of the textural classification of this soil is therefore 'gravelly sandy loam'.

SOIL MOISTURE

No special engineering knowledge is required to be aware of the fact that the engineering behaviour of a soil is greatly influenced by changes in its moisture content. It is common knowledge that a fine-grained soil which is hard and dry during the Summer may become soft and slippery, perhaps even a morass, after a period of heavy rain. On the other hand, the layman also realizes that the dry beach sand is a poor material on which to walk, but that it can be quite stable and forms a good walking surface when dampened by sea water.

These two simple examples of the influence of moisture on a soil suggest the need for understanding the basic factors relating to moisture movement in soils.

Types of soil water

Many attempts have been made by soil scientists to classify soil moisture into various categories. One of the simplest of the classification systems evolved was the one proposed in 1897 by a soil physicist named Briggs. He divided soil water into three types:

1. *Gravitational water.* This is not held by the soil, but drains under the influence of gravity.

2. *Capillary water.* This is held by surface tension forces as a continuous film about the particles and in the capillary pores.

3. *Hygroscopic water.* This is adsorbed from an atmosphere of water vapour as a result of attractive forces within the surfaces of the particles.

Gravitational water. When moisture enters a soil after rain, the tendency is for it to migrate downwards under the influence of gravity. The rate at which this occurs is mainly dependent on the soil's texture and structure. In general, however, the more coarse-grained the soil, the more quickly the water percolates down. Fine-grained soils, while they may have considerable porosity, are noted for the slow rate at which gravitational water moves through them; this is mainly due to the longer percolating passages and the higher side-wall frictions incurred by the water attempting to seep downwards.

Water table. Downward percolation of gravitational water continues until a depth is reached below which the soil pores are saturated with water. The water in this saturated zone is given the name of groundwater, and the upper surface of the zone is called the water table. Contrary to what is often believed by the layman, the water table is not a horizontal plane, but in reality is a surface whose elevation is as constantly changing as the topography above it. Put in a simple way, it tends to be a subsurface replica of the ground surface.

Because of the elevation differences of the water table, groundwater is not at rest in a soil but tends to flow laterally and perhaps emerge eventually either as ground seepage or to feed streams and lakes. In dry weather, this means that the water table may be considerably lowered. On the other hand, in wet weather, the water table rises towards the ground surface.

Two practical factors arise as a result of these movements. First of all, it is most important to note the time of the year at which the measurement of the depth of a water table is taken, so that, if necessary, a seasonal variation can be taken into account at a later date. Secondly, when estimating the flow of water into drains installed to intercept or lower a high water table, the laboratory tests to determine the coefficient of permeability of the soil should be carried out on undisturbed soil samples cut in the direction in which the moisture flow will actually occur. This is most important since, for instance, the coefficient of permeability in a horizontal plane of a sand and gravel material may be as much as 10 times greater than the vertical value, while in uniformly graded sand soils the coefficient of permeability in the horizontal direction may be up to 4 times greater than the vertical one.

Capillary water. If the forces of gravity were the only ones acting on the water in a soil, then the pores in the material above the water table would be dry at all times other than when gravitational water was flowing through. In fact, however, this is far from being so. If a cross-section were to be taken through the ground, it would be found that a zone exists above the water table within which the pores are wholly or partially filled with moisture. This zone, often termed the 'capillary fringe', is typically 2–5 cm thick in coarse sands, 12–35 cm in medium sands, 35–70 cm in fine sands, 70–150 cm in silts, and 200–400 cm or greater (after considerable periods of time) in clays.

Two slightly different viewpoints exist in the literature as to the manner in which capillary water rises in the soil above the water table. The older of these concepts relates the movement and retention of water in the capillary fringe of a soil to the rise and retention of water in a capillary tube. In a capillary tube of uniform diameter, the capillary rise from a free water surface, i.e. one at atmospheric pressure, is given by the well-known formula

$$h = \frac{2T}{r\rho_w g} . \cos \alpha$$

where h = height of rise of water, cm

T = surface tension, dynes/cm

r = radius of circular capillary tube, cm

ρ_w = density of water, g/cm^3

g = acceleration due to gravity, cm/sec^2

and α = angle of contact between the meniscus of the column of water and the wall of the tube.

From this equation it follows that the capillary rise is increased as the radius of the tube is decreased. Now, if the pores present in the soil above the water table are considered to form an inter-related mass of irregular capillary tubes, it can be hypothesized that the size of the pores controls the height of the capillary fringe, i.e. the more finely grained the soil, the higher the moisture will rise.

The second and more modern viewpoint explains the rise of water above the water table by thermodynamic reasoning. This approach has caused a gradual displacement of the capillary tube hypothesis and is now the more generally held. It is not feasible in the limited space here to discuss the thermodynamic concepts involved in this theory, so the reader is referred to the literature, where details and references relating to both the thermodynamic and the capillary concepts are readily available.[10, 11, 12] It should be clear, however, that essentially there is no conflict between the two points of view; rather they can be considered as two different approaches towards explaining the same phenomenon.

There are several important features associated with the presence of capillary moisture in a soil. One, for instance, is that, although the capillary fringe is usually considered to be vertically above the water table, capillary flow can and does occur in any direction. The main tendency is for the moisture to find its way from a zone of saturation to one with a low degree of saturation irrespective of whether this movement is a lateral or vertical one.

Capillary movement is also associated with what engineers term embankment 'pumping'. This refers to the phenomenon observed in the construction of embankments when a fine-grained soil is over-compacted. In this case, even though the water table may be at some depth, the repeated kneading of the soil has the effect of stimulating the upward capillary movement of water.

The stability of fine-grained soils, particularly clays, is very heavily dependent on the moisture content. If a soil is above the water table and within the capillary fringe, the cohesion of the material is mainly dependent on the amount of capillary water present; this moisture is held between the adjacent soil particles. If the soil is dried out, the amount of water between the particles will obviously decrease and the water will recede into the interstices. This causes the particles to be pulled closer together and this is why a clay soil hardens when it dries out.

When a soil is compacted, the soil particles are pushed closer together under the influence of the compaction forces. If two particles which are bound by capillary moisture are pushed together, the water will tend to be pushed out from between them and the radius of the meniscus is increased. If sufficient compaction is applied, the soil pores can become filled with water, the radii of the menisci becomes infinite, and the cohesion between the particles can be lost.

Hygroscopic moisture. As has been discussed before, the moisture forming part of the colloidal particles and their associated ions is termed adsorbed moisture, and it has been explained that some of this moisture is held more tightly than other parts. When the soil is dried in the air, say in a laboratory, moisture is removed by evaporation until the water remaining in the soil is in equilibrium with the moisture vapour in the air. If a sample of the soil is then dried at 105–110°C and the moisture content driven off is determined, the value obtained is called the hygroscopic moisture content.

Knowledge of the hygroscopic moisture content is very useful when laboratory studies are being carried out which involve the addition of controlled amounts of moisture to the soil. It must be remembered, however, that the hygroscopic content can be a variable quantity for a given soil, as its value depends on the temperature and humidity of the air over the soil at the time the determination is carried out.

FROST ACTION IN SOILS

Frost action can be defined as any action resulting from freezing and thawing which alters the moisture content, porosity or structure of soils, or affects their capacity to support loads. In order to take these factors into consideration, it is most important to have an understanding of frost action and its consequences. Accordingly, this discussion will be confined to the mechanics of frost action and its consequences and to the recognition of frost-susceptible soils. Methods to alleviate the effects of frost action from the standpoint of initial prevention and protection of frost-susceptible roads are presented in the chapter on Flexible Pavements.

Mechanism of frost heaving

Although much basic research has been carried out in the field of frost heaving, the mechanism by which it occurs is still not completely understood. Some of the recorded literature, prior to 1900, showed that earlier engineers believed that frost heaving was attributable only to the volume increases of soil water upon freezing. This theory was discarded as soon as it was generally realized that the density of ice at 0°C was 0·91674, and that of water at the same temperature was 0·99987. This means that the increase in volume when water changes to ice can only be 0·0907 cm^3 or just over 9 per cent; this increase is normally well exceeded in practice.

Most cases of severe frost heave in soils are attributed to the formation of ice lenses. Figure 4.13 illustrates the manner in which layers of segregated ice are formed in the ground where excessive heaving occurs. These ice

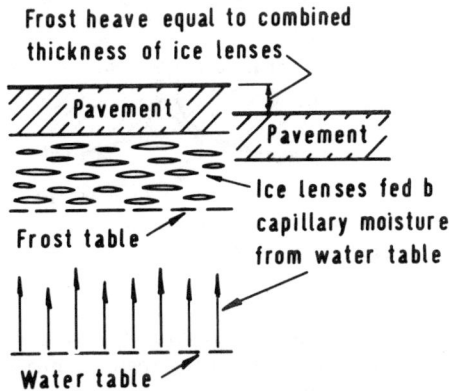

Fig. 4.13. Ice lens formation in a soil

lenses, which normally grow parallel to the ground surface and perpendicular to the direction of heat flow, may vary from small hairline lenses to centimetres thick, depending on many factors which will be discussed later.

For ice lenses to develop, three conditions must be met: sufficiently cold temperatures to freeze *some* of the soil water, a supply of water available to the freezing zone and, finally, means by which the moisture supply can be moved to the freezing zone, i.e. a frost-susceptible soil. Ice lensing does not occur when any of these conditions is missing.

Under conditions favourable to ice lens growth, the moisture necessary for the lens to form can usually only be supplied by capillary action and by vapour movement from the water table. It has been shown[13] that moisture diffusion in the vapour phase takes place in soils with large pore sizes rather than fine ones, and that the amounts of moisture transferred upward in the vapour phase towards the freezing ice lenses are negligible and hence can be ignored. It being clear therefore that the necessary moisture can only be fed to the ice lens by means of the capillary (suction) process, it is now possible to discuss a mechanism by which this may occur.

Figure 4.14 is an enlarged schematic diagram showing a section of the ice lens with respect to the soil particle and soil pore. Surrounding each particle is an adsorbed layer of water, and this separates the growing ice crystal from the mineral particle. In the lens-growing process, molecules of water from this adsorbed layer become attached to the bottom of the growing crystal, thus tending to reduce the thickness of the adsorbed layer. This layer seeks to maintain equilibrium thickness and so attracts moisture from the soil pores. As equilibrium thickness is restored, the ice lens and the soil above it is pushed upwards and the result is 'frost heave'. The pore moist-

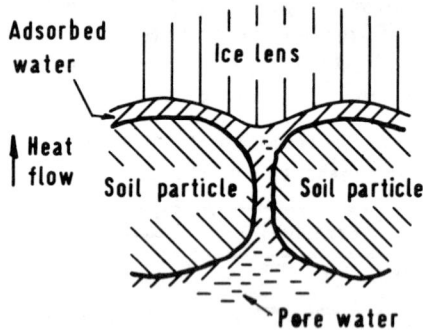

Fig. 4.14.   Enlarged schematic diagram showing a section of the ice lens with respect to the soil particle and soil pore

ure, in turn, is replaced by water taken from the water table. In other words, ice lens growth is really no more than the result of moisture film adjustment around the particle which continues until the temperature drops sufficiently for the ice front to propagate through the soil pore.

The theory underlying this conception can be briefly explained as follows.[14] The relationship between the size of a stable spherical crystal in its own melt and the absolute temperature can be shown by

$$\Delta T = \frac{2T\sigma_{iw}}{r\rho_i Q_f}$$

where ΔT = freezing point depression,

T = temperature of melting at zero curvature of the soil-liquid interface,

σ_{iw} = interfacial energy, ergs/cm^2

r = radius of the crystal, cm

ρ_i = density of the ice, g/cm^3

and Q_f = latent heat of fusion, ergs/g.

This equation states that the product of the curvature radius and freezing-point depression is a constant determined by the interfacial tension and the bulk properties of the substance. Now, assuming that the size of the soil pore determines the size of the advancing ice crystal, the quantity r in this equation can be interpreted as the radius of the pore. A part of the soil pore is occupied by the adsorbed phase, and this probably places additional size limitations on the ice crystal. This mechanism appears to be responsible for the growing of the ice phase at the supercooled temperatures necessary to release the energy required for the work involved in heaving and moisture movement. Accordingly, the smaller the pore radius r, the lower the temperature needed for the ice front to advance, since ΔT is proportional to $1/r$. In other words, the advance of the ice water interface will occur at a lower temperature in a soil containing mostly small pores than in one in which large pores predominate. This means that super-cooled water (i.e. moisture in the liquid state at a temperature below normal freezing point) exists in the small pores and is in contact with the adsorbed layer of the particle.

Referring to Fig. 4.14, the ice front is temporarily prevented from propagating downward until the temperature has been lowered sufficiently to satisfy the conditions of the equation. Before this can occur, part of the adsorbed layer will freeze, and this moisture in turn is replaced by the supercooled moisture as described before. From this it can be seen that the existence of supercooled water, and of the adsorbed phase, are both vital links in the movement of moisture from the water supply to the ice phase. Similarly, it should be noted that if the propagation of ice occurred at the freezing temperature of bulk water, 0°C, as it does in the case of large pores, the evolution of energy necessary to cause heaving and moisture movement would not occur.

It has been shown that the rate of frost heaving is for practical purposes independent of the rate of frost penetration.[15] If freezing occurs rapidly, the water sucked up is frozen into many thin layers; if it occurs slowly, the layers are thick. Thus the ice stratification of the frozen soil is an index of the rate of freezing, with thick ice indicating slow, and thin ice fast, frost penetration. From a practical point of view, it is more desirable to have a rapid early freeze than a slow one since, by influencing the thickness of the ice lenses, both the distribution of the extra moisture and the consequent reduction in capacity are also influenced. This rapid early freezing leaves the critical upper layers relatively barren of ice and thus when the thaw begins there is less moisture accumulation in the soil at a given time and this benefits stability.

Depth of frost penetration

The engineer is very much interested in the depth to which the freezing temperatures will penetrate into the soil beneath the highway pavement. The importance to be attached to this determination can be gauged from the fact that the most complete protection is afforded to a pavement by providing non-frost-susceptible material to a depth below that of the frost penetration, and thus preventing any underlying frost-susceptible soil from freezing.

Great care should be taken when estimating the depth of frost penetration in the case of a new road, as it is not always advisable to base the prediction on records of previously observed frost depths in the raw soil, as new road pavement surface temperatures are usually lower than the surrounding natural surfaces during the winter months. In addition, since the pavement itself is likely to be well-drained, it is likely to exist at a lower moisture content than the natural soil in the surrounding area, and this can also contribute toward a relatively large temperature differential.

The depth of frost penetration at any given location, and the nature and severity of the freezing to which the various elements of the road structure are subjected, therefore depends primarily on the thickness of the layers and the ingredient material in each layer, as well as on the prevailing air temperature. Nevertheless, it can be said that it is rarely that the freezing isotherm would ever penetrate more than 51 cm beneath the carriageway surface in Britain.[16]

Moisture and frost action

The greatest potential trouble from frost heave arises when the water table is relatively close to the ground surface and just below the freezing zone. It may, however, be concluded that a potentially troublesome water supply for ice segregation is present when the highest water table level at any time of the year is within, say, 1–1·5 m of the proposed subgrade surface or the bottom of any frost-susceptible material used in the roadway. When the depth to the water table is in excess of 3 m throughout the year, a source of water for substantial ice segregation is usually not present.

The existence of a deep water table does not necessarily ensure that the detrimental effects associated with frost action will not occur. For instance, the equilibrium moisture content which a homogeneous clay subgrade will develop under the road pavement after its construction is usually sufficient to provide water for at least limited ice lens formation, even though the water table be at a great depth. It can be assumed, however, that appreciable ice lensing will not normally occur in frost-susceptible soils with a remote water table when the degree of saturation of soils in the zone subject to freezing is less than about 70 per cent.

For a given water table position, heaving tends to be less for heavy clays than in lean clays and silt soils. This is due to moisture being able to move only relatively slowly over short distances during the freezing process because of the much lower permeability of the heavy clays. On the other hand, in relatively pervious but frost-susceptible sands and gravels, differential frost heave may be more intense than in clay subgrades. In such cases water may move to the growing ice lenses from substantial distances, both laterally and vertically.

Frost-susceptible soils

Investigators of frost problems in soils discovered early that certain types of soil are more susceptible to frost action than others. The most expedient rule-of-thumb means of identifying, without benefit of laboratory testing procedures, soils in which damaging frost action may occur were suggested by Professor A. Casagrande in 1932.[17] He stated that, under natural freezing conditions and with a sufficient water supply, considerable ice lensing could be expected in any non-uniform soil with more than 3 per cent of its particles smaller than 0·02 mm and in very uniform soils with more than 10 per cent smaller than 0·02 mm. According to the study on which these recommendations are based, no significant ice lensing need be expected in soils containing less than 1 per cent of particles smaller than 0·02 mm, even if the ground water level is as high as the frost table. In using these recommendations, it should be realized that the values of 3 and 10 per cent are not intended to represent points at which no ice lensing will take place, but rather levels below which frost heave will not usually exceed tolerable limits in ordinary roadway applications.

TABLE 4.6.   *U.S. Corps of Engineers' groupings of frost-susceptible soils*

Group	Description
F1	Gravelly soils containing between 3% and 20% finer than 0·02 mm
F2	Sands containing between 3% and 15% finer than 0·02 mm
F3	(a) Gravelly soils containing more than 20% finer than 0·02 mm
	(b) Sands, except very fine silty sand, containing more than 15% finer than 0·02 mm
	(c) Clays with plasticity indices greater than 12
	(d) Varved clays, existing with uniform subgrade conditions
F4	(a) All silts including sandy silts
	(b) Very fine silty sands containing more than 15% finer than 0·02 mm
	(c) Clays with plasticity indices less than 12
	(d) Varved clays existing with non-uniform subgrade conditions

Because of their extreme simplification the Casagrande recommendations cannot be expected to cover all soils and materials or be applicable to all situations. It is important to realize that the intensity of ice lens formation is dependent not only on the percentage finer than 0·02 mm, but also on the particle size distribution and/or the physico-chemical properties of these fines. Other contributing factors are unit weight, initial moisture content, and permeability. The influence of unit weight for instance is indicated in Fig. 4.15, which illustrates that, for a given increase in dry unit weight, the rate of frost heave may increase or decrease depending upon the soil. This figure is indeed a reminder that the most obvious present solution for guaranteeing the built-in stability of high-unit weight roadbases and sub-bases is to use only free-draining non-frost-susceptible materials within the freezing zone.

The soils shown in Fig. 4.15 are classified according to the Unified classification system; this is very similar to the Casagrande system used in Britain. Using the Unified system as a basis, the U.S. Corps of Engineers has proposed an adaptation specifically for use with frost-susceptible soils. This adaptation is given in Table 4.16. As can be seen, the system states that the more frost-susceptible a soil, the higher its F-number. The soils in group F4 are especially frost-susceptible. The manner in which this last grouping can be catered for in highway pavement design is discussed in the chapter on Flexible Pavements.

British experience[16] is that cohesive soils can be regarded as non-frost

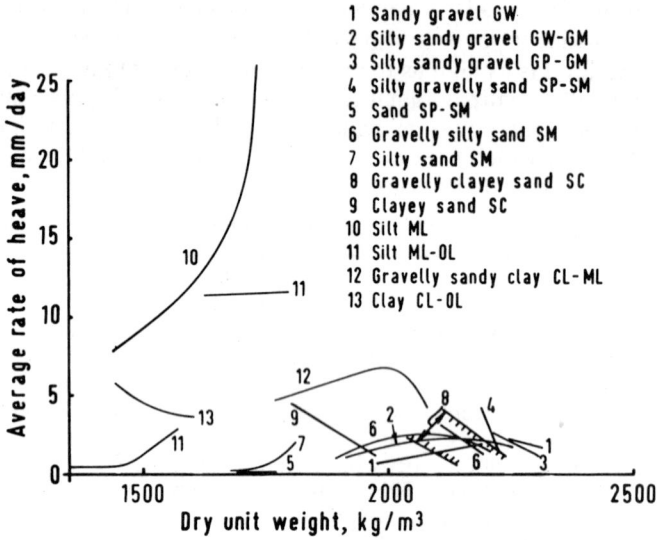

```
1  Sandy gravel GW
2  Silty sandy gravel GW-GM
3  Silty sandy gravel GP-GM
4  Silty gravelly sand SP-SM
5  Sand SP-SM
6  Gravelly silty sand SM
7  Silty sand SM
8  Gravelly clayey sand SC
9  Clayey sand SC
10 Silt ML
11 Silt ML-OL
12 Gravelly sandy clay CL-ML
13 Clay CL-OL
```

Fig. 4.15.   Effect of dry unit weight on average rate of frost heave[18]

susceptible when the plasticity index is greater than 15 per cent for well-drained soils, or 20 per cent for soils which are poorly drained, e.g. with the water table within about 0·6 m of the formation level. Furthermore, the liability of these soils to frost heave decreases with increasing state of compaction. Non-cohesive soils (except limestone gravels) can be regarded as non-frost-susceptible if the amount of material passing the 75 micron BS sieve is 10 per cent or less; limestone gravels with a particle saturation moisture content in excess of 2 per cent should be considered as potentially frost-susceptible. The state of compaction of these non-cohesive soils does not significantly affect their liability to frost heave.

METHODS OF TESTING SOILS

Because of the great upsurge in highway making in Britain, young and often untried engineers are having to supervise highway construction projects and assume responsibilities which previously they would only have been allowed to do after years of apprenticeship and experience. In order to bridge the knowledge gap brought about by inexperience, the engineer has to rely on the results of physical tests for guidance when designing and constructing roadways. However, if the results are to be of maximum value, then the tests themselves must not only be carried out correctly but the data obtained from them must be analyzed intelligently. The highway engineer must therefore not only know how to carry out these tests but also understand their significance so that the proper interpretations can be made.

The following is a brief discussion of what might be termed the fundamental physical tests carried out on soils for highway purposes. All of these tests can be considered routine ones, and information regarding the exact procedures is readily available in the literature.[10, 19, 20] The emphasis therefore in this discussion is placed on the purpose for which a particular test is carried out rather than on the test procedure itself.

Particle-size analysis

This method of test is a determination of the percentage of individual particle sizes present in a soil. It is usually carried out in two parts. The first part, called the sieve analysis, determines the amount of coarse material with the aid of sieves or screens of graded mesh. The second part consists of a study of the fine-grained fraction of the soil and utilizes a sedimentation test.

The sieve analysis is a basic test which consists of sieving a measured quantity of soil through a series of successively smaller sieves. The weight retained on each sieve is then expressed as a percentage of the total sample.

The analysis for the fine-grained portion consists of a sedimentation test on the material passing the $75 \, \mu$m BS sieve. Sedimentation tests are based on the assumption that soil can be dispersed uniformly through a liquid and that the individual particles of different size have settling velocities in accordance with Stokes' Law, i.e. the liquid at a given depth after a given period of time only contains those particles whose velocities have been sufficient to carry them there. The computations may include corrections for temperature, viscosity of the liquid and the specific gravity of the soil particles. The results are first expressed as a percentage of the fine-grained sample and then, if there is a coarse-grained fraction, converted to percentages of the total soil sample. The sedimentation test is well known to be subject to many errors, and hence is frequently omitted for soils having a combined silt and clay content of less than about 25 per cent.[21]

Typical test results. The results of a particle-size analysis can be presented in either of two ways. The first of these is a table in which is listed the percentage of the total sample that passes a given sieve size or is smaller than a specified particle diameter. The second way is to make a plot of the sieve or particle size versus the percentage passing the given sieve (or smaller than the given diameter). Because of the wide range of values which are possible, the usual procedure is to plot the particle size to a logarithmic scale while the percentages finer are plotted on an arithmetic scale.

Some typical particle-size distributions for different soils are shown in Fig. 4.16. Apart from being examples of different soils, some other very interesting factors are illustrated. For instance, the sandy clay curve is relatively flat, indicating that this soil contains a wide range of particle sizes.

Sieve sizes, mm

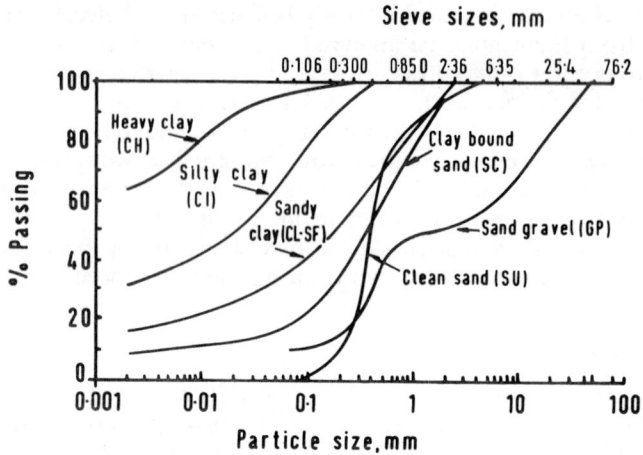

Fig. 4.16.   Typical particle-size distribution curves for different British soils[10]

Such a soil might well be termed a fairly well-graded material. On the other hand, the steepness of the clean sand curve indicates that this soil contains a large number of particles which are essentially the same size; this soil would therefore be called a uniform soil or a poorly-graded one. A gradation curve having a horizontal or near-horizontal 'hump' in it, such as the sand gravel soil, may mean that the soil either has had a large amount of its intermediate sizes removed or is composed of a mixture of two soils of near normal particle distribution.

Significance of test.   The results of this test are of most value when used for soil classification purposes, and further use of the gradation should be discouraged unless verification by studies of performance or experience permit empirical relationships to be formulated.[22] Although it is often found that the larger the particle size the better the engineering properties of the soil, nevertheless only rough approximations of strength or resistance properties should ever be attempted. It is a known fact that detrimental capillarity and frost damage are not a problem when a soil is coarse-grained, whereas fine-grained silts and clays can be very dangerous. As discussed elsewhere, empirical relationships have been developed for determining the susceptibility of a soil to frost action. Highway specifications for roadbase and sub-base construction use the particle-size analysis as a means of quality control for soil materials.

When stabilizing soils by either mechanical or chemical means, use is frequently made of particle-size analyses for mix design or control purposes. For instance, one American design procedure allows the percentage of cement to be used in coarse-grained soil-cement mixtures to be estimated on the basis of particle size. For mechanical stabilization, the results of gradation analyses are used to determine the size and percentage of coarse or fine materials needed to obtain a dense impermeable road pavement.

On occasions, it is possible to make general estimates of the permeability of a soil on the basis of its gradation. Broadly, it can be said that the coarse-grained soils will more readily permit the flow of water than the finer-grained ones, e.g. sands are more permeable than silts, and silts are more permeable than clays. On the other hand, a well-graded granular material can be virtually impermeable whereas a poorly-graded one is usually the opposite.

Consistency tests

By consistency is meant that property of a soil which is manifested by its resistance to flow. As such it is a reflection of the cohesive resistance properties of the soil rather than of the intergranular ones. These properties are considerably affected by the moisture content of the soil.

About 1910 the Swedish agricultural scientist Atterberg, suggested that soil consistency should be described by arbitrarily dividing a soil's cohesive range into six stages and expressing the limits of each range in terms of moisture content. The limits he chose are as follows:

1. Upper limit of viscous flow above which the mixture of soil and water flows like a liquid.

2. Liquid limit, or lower limit of viscous flow, above which the soil and water mixture flows as a viscous liquid and below which the mixture is plastic.

3. Sticky limit, above which the mixture of soil and water will adhere or stick to a steel spatula or other such object that is wetted by water.

4. Cohesion limit, which is at the water content at which crumbs of the soil cease to adhere when placed in contact with each other.

5. Plastic limit, or lower limit of the plastic range, which is at the water content at which the soil starts to crumble when rolled into a thread under the palm of the hand.

6. Shrinkage limit, or lower limit of volume change, at which there is no further decrease in volume as water is evaporated from the soil.

Two of these limits, the liquid and plastic limits, have won wide acceptance by highway and soil engineers and so they are briefly discussed here.

Liquid limit. The liquid limit of a soil is the moisture content, expressed as a percentage by weight of the oven dry soil, at the boundary between the liquid and solid states. This boundary was arbitrarily defined by Atterberg as being the moisture content which caused the soil to begin to flow when lightly jarred against the heel of the hand. Obviously this procedure involves considerable possibility of human error, so a mechanical liquid limit apparatus was devised in order to eliminate the personal factor. This device, which was developed by Professor A. Casagrande of Harvard University, consists of a shallow, circular brass cup resing on a hard rubber or bakelite base. In the course of the test, a soil pat consisting of a mixture of water and soil passing the $425 \mu m$ BS sieve is placed in the cup and divided with a special wedge-shaped cutting tool. By means of a cam on a shaft turned by a handcrank, the brass cup is raised through a height of 1 cm and then dropped sharply on the hard base. The device has been calibrated so that 25

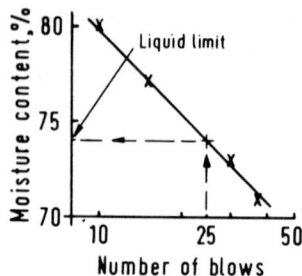

Fig. 4.17.   Liquid limit flow curve

drops of the cup at the rate of two drops per second is equivalent to 10 light jars as specified by the Atterberg hand method. The liquid limit of the soil is then taken as the moisture content at which the groove cut in the soil pat closes for a length of 13 mm after 25 blows.

Typical test results.   In the standard method of test, at least three moisture contents are determined; one of these is below the liquid limit, one at or near the liquid limit and the third above the liquid limit. A 'flow curve' such as is illustrated in Fig. 4.17 is then plotted on semi-logarithmic graph paper, with the moisture contents on the arithmetical scale and the number of impacts on the logarithmic scale. The results should be approximately a straight line and from this the liquid limit can be taken as the moisture content corresponding to the intersection of the flow curve with the 25-shock point on the ordinate scale.

Liquid limits vary widely but values of 40 to 60 per cent are typical for clay soils. For silty soils values of 25 to 50 can be expected. Sandy soils do not have liquid limits and are reported as non-plastic.

The liquid limit test is usually carried out on material which has been air-dried only. Whereas the values obtained from oven-dried, inorganic soil samples are little different from those obtained with air-dried samples, the liquid (and plastic) limits of organic clays determined from oven-dried samples are much lower than the limits obtained when the same soils have not been oven-dried before testing.

When carrying out the liquid limit test, the most common sources of error have been found to be as follows:

1. Inaccurate height of drop of the cup.
2. A worn cup due to scratching the bottom with the grooving tool.
3. Too thick a soil pat.
4. Variation in the rate of dropping the cup.
5. The human element in deciding when the groove has closed exactly 13 mm.

Plastic limit. The plastic limit of a soil may be defined as the moisture content of the soil at the boundary between the plastic and semi-solid states. This boundary was originally defined by Atterberg as the moisture content at which a sample of soil begins to crumble when rolled into a thread under the palm of the hand. As now arbitrarily defined, the plastic limit of a soil is the moisture content, expressed as a percentage of the oven-dried sample, of the air-dried material passing the 420 μm BS sieve at the time that the soil-water mixture has been rolled into a thread 3 mm in diameter and it begins to crumble at this size.

Typical test results. Pure sands cannot be rolled into a thread and so these soils are termed 'non-plastic'. The plastic limits of silts and clays do not vary too widely; the normal values range from 5 to 30, with the silty soils having the lower plastic limits.

When carrying out the test, it is left to the operator to decide how much pressure to apply to the soil thread when it is rolled on a glass plate. This means that there is a tendency for a heavy-handed operator to get a consistently higher value than one with a lighter touch.

TABLE 4.7.   *Some soil characteristics indicated by the consistency tests*

Characteristic	Comparing soils of equal L.L., with P.I. increasing	Comparing soils of equal P.I., with L.L. increasing
Compressibility	About the same	Increases
Permeability	Decreases	Increases
Rate of volume change	Increases	—
Dry strength	Increases	Decreases

Significance of consistency tests. A value usually used in conjunction with the liquid and plastic limits is the plasticity index. The plasticity index (or P.I.) of a soil is the arithmetic difference between the liquid and plastic limits; in other words, it is the range of moisture content over which the soil is in the plastic state.

The most common application of the plasticity test results is for the purpose of soil classification. The P.I. can also be used (in Britain) to directly estimate the strength of a subgrade soil, as reflected by its California Bearing Ratio value. In addition, it is possible to draw some general conclusions about soils with particular plasticity results. For instance, soils with high liquid limits also have poor engineering properties. A low plasticity index indicates a granular soil with little or no cohesion. Some characteristics indicated directly by the consistency tests are summarized in Table 4.7. Another important use of the consistency tests' results is therefore as a means of excluding from pavements and embankments those materials with undesirable plastic qualities.

Specific gravity test

The specific gravity of a soil is the ratio of the weight in air of a given volume of soil particles to the weight in air of an equal volume of distilled water at a stated temperature. Unless otherwise stated, specific gravity in Britain is usually referred to distilled water at 20°C.

The test requires that a series of weighings be carried out after the soil has been placed in a special density bottle. The specific gravity G_s of the material is then calculated from the following formula:

$$G_s = \frac{W_2 - W_1}{(W_4 - W_1) - (W_3 - W_2)}$$

where W_1 = weight of density bottle,
$\qquad W_2$ = weight of bottle and dry soil,
$\qquad W_3$ = weight of bottle, soil and water,
and $\qquad W_4$ = weight of bottle when full of water only.

In many soils the presence of a number of minerals, each of which has a different specific gravity, may present measurement difficulties. This is why testing is divided into two parts when a soil has material retained on the 2·36 mm BS sieve. The specific gravity of the whole soil may then be estimated from the weighted average of the values for the coarse and the fine particles using the following formula:

$$G_{cf} = \frac{1}{\dfrac{p_c}{g_c} + \dfrac{p_f}{g_f}}$$

where $\quad G_{cf}$ = combined specific gravity,
$\qquad p_c$ $\ $ = ratio of weight of coarse fraction to total weight of sample,
$\qquad g_c$ $\ $ = specific gravity of coarse portion,
$\qquad p_f$ $\ $ = ratio of weight of fine fraction to total weight of sample,
and $\qquad g_f$ $\ $ = specific gravity of fine portion.

Typical test results. The true specific gravity of a soil.is actually the weighted average of the specific gravities of all the mineral particles present in the soil. Since about 1000 minerals have been identified as present in rocks, and since soil is derived from rocks, very many of these minerals will normally be present in any given soil. In practice, however, it is found that because of the natural preponderance of quartz and quartz-like minerals ($G = 2·6$ to $2·7$) the specific gravities of most soils range between 2·55 and 2·75, with the average being about 2·65. Nevertheless, it must be remembered that the values can readily range from about 2·0 for organic or porous particle soils to over 3·0 for soils containing heavy minerals.

Accurate results depend upon extreme care in removing the air from the soil after the addition of water. A small error can be quite significant and result in a low specific gravity value.

Significance of test. The specific gravity is used in the computations of many laboratory tests on soils. In particular, it is required in the calculation of the void ratios of soil specimens, in the determination of the moisture content of a soil by the pycnometer method, and in the particle-size analysis (sedimentation test).

Prior to 1969, .the compaction of earthworks and most sub-base and roadbase materials in Britain was specified in terms of air content, for which knowledge of the specific gravity was essential. The introduction of *Method Specification* for compaction[23] has caused a decline in the use of this test, although it still needs to be carried out if for any reason the state of compaction achieved needs to be ascertained in terms of the air content of the product.

Moisture-unit weight test

In highway soil engineering practice at the present time, the term 'density' is commonly used instead of unit weight, the same meaning being however applied. It should be pointed out that this usage is at variance with the definition of density as applied in other scientific areas. For instance, in physics density is defined as mass per unit volume; this is equal to weight per unit volume divided by the acceleration due to gravity. In this text the more proper 'unit weight' will be used instead of the term density.

The term dry unit weight is used to define the weight of the dry soil particles per unit volume, while the weight of the wet solids plus the water contained in the void spaces per unit volume is called the wet (bulk) unit weight. Thus it can be seen that, depending on the moisture condition of the soil, the value of the wet unit may range from that at saturation to the dry unit weight when there is no moisture at all present in the soil.

In 1933, R. R. Proctor of the Los Angeles Bureau of Waterworks and Supply published what might be described as one of the first scientific approaches to the study of soil compaction; his work was carried out during the construction of several large earth dams in the United States. He devised a laboratory method of test, now commonly known as the Proctor test, in order to ensure that soil compaction would be carried out in such a way that unit weights would be attained which would give the desired impermeability and stability to the earth dams. By compacting over 200 different soils at various moisture contents, and using laboratory compacting energies which were considered equivalent to those produced by field compaction, he discovered that as moisture contents were increased the unit weights increased to maximum values, after which they decreased. The moisture content at which the *Maximum Dry Unit Weight* was attained under a given compactive effort was termed the *Optimum Moisture Content*. Following these early studies of Proctor, standardized laboratory procedures were developed and it is these which are in use today.

The moisture-unit weight test is normally carried out by compacting a prepared soil sample in a given number of layers into a metal cylindrical mould of volume equal to $944 \, cm^3$. In the *standard* Proctor test, a 2·5 kg metal rammer is dropped from a height of 30·5 cm on to each layer of soil in the mould; a total of 25 blows is used to compact each of three layers. In the *modified* Proctor test, a 4·5 kg rammer falls 45·8 cm, and a total of 25 blows is applied to each of five layers. Following compaction the weight of the soil in the mould is determined and, with the volume of the mould known, the wet unit weight is calculated by dividing the weight by the volume. A determination is then made of the moisture present in the soil sample in the mould, after which the soil is removed from the mould. A new soil sample is then prepared, an additional increment of water added, and the procedure repeated. The test is continued until the weight of a compacted sample in the mould is less than that obtained in the preceding measurement.

As stated above, it is necessary to carry out a moisture content determination on every sample prepared in the course of the test. The moisture content referred to in this case is the total amount of water contained in the compacted sample; this includes free water, capillary water and hygroscopic moisture. It is obtained by weighing a representative sample of the soil before and after drying it in an oven at a temperature between 105°C and 110°C. The difference in weights represents the amount of water in the sample and, when this is expressed as a percentage of the dried sample, the moisture content of the soil is given.

Knowing the wet unit weight and moisture content of each sample, it is then possible to calculate the dry unit weights. The following formula is used for this purpose:

$$\psi_d = \frac{100\psi}{100 + m}$$

where ψ_d = dry unit weight, kg/m^3
       ψ = wet unit weight, kg/m^3
and m = moisture content, %

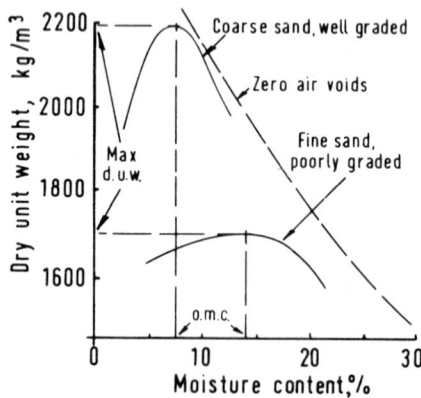

Fig. 4.18.   Typical moisture-unit weight curves

When the dry unit weights are plotted against their corresponding moisture contents, and a smooth curve is drawn through the data points, a curve similar to those shown in Fig. 4.18 is obtained.

Both of the Proctor tests can be criticized as providing only a poor guide for specifications on the site compaction of highly permeable soils, e.g. fine-grained clean gravels, or on uniformly graded and coarse clean sands, and so a new test, the *Vibrating Hammer* test, has been developed in Britain[19, 20] for use with these soils. With this test a 3·18 kg electric-powered vibrating hammer is used to compact three layers of moist soil (each layer for 60 sec) in a California Bearing Ratio test mould; the total downward load during compaction, including the weight of the hammer and tamper, must lie between 32 and 41 kg. The wet unit weight of the soil (kg/m^3) can be calculated from the formula

$$= \frac{W}{0 \cdot 158h}$$

where W = weight of wet soil, g
and h = height of compacted specimen, cm

The dry unit weight is then easily calculated using the same formula as for the Proctor tests. By varying the moisture content of the soil, a moisture content-dry unit weight relationship can then be derived as before.

Typical test results. The maximum dry unit weight that can theoretically be attained with a soil having a specific gravity of 2·65 is 2650 kg/m^3. In fact, this unit weight could only be obtained if all the soil particles were fitted against each other exactly. Since soil particles come in various shapes and sizes, there are always air voids in the soil and so this theoretical unit weight can never be achieved. They dry unit weight of soils *in situ* is often about 50 to 60 per cent of the theoretical maximum value. Soils such as peat and muck have natural unit weights in the order of 15 per cent of the theoretically possible. On the other hand, well-graded, dense gravel mixtures can be compacted to dry unit weights of perhaps 90 per cent of the theoretical limit.

The maximum dry unit weight which may be obtained with any of the laboratory tests is considerably influenced by the gradation of the soil and the amount of energy imparted to it in the course of compaction. The influence of gradation is shown in Fig. 4.18, where it can be seen that a considerably higher dry unit weight is obtained with the well-graded soil. This figure also illustrates that for poorly-graded sandy (and gravelly) soils, there is normally little significant change in the unit weight with the addition of water.

Using the standard Proctor method of compaction, the following is a list of the range of values that may be anticipated for the moisture-unit weight test:[22]

| Clays | Maximum dry unit weight | 1440–1685 kg/m^3 |
| | Optimum moisture content | 20–30% |

Silty clays	Maximum dry unit weight	$1600-1845\,\mathrm{kg/m^3}$
	Optimum moisture content	$15-25\%$
Sandy clays	Maximum dry unit weight	$1760-2165\,\mathrm{kg/m^3}$
	Optimum moisture content	$8-15\%$

Changing the amount of energy applied to a soil by varying the number of blows and/or the weight of the compaction hammer also has a considerable influence on the dry unit weight. The net effect of increasing the amount of compactive energy is to increase the maximum unit weight and decrease the optimum moisture content of a given soil. If a number of such curves are compared, it is seen that, when the air voids content is small, the effect of increasing the compaction is negligible; when the air voids are large, the effect of increasing the compaction is considerable.

As previously mentioned, two laboratory methods of test are most commonly used in Britain: the *Standard Proctor Test* and the *Modified Proctor Test*. The results obtained when both of these tests were carried out on five soils ranging from a well-graded gravel-sand-clay mixture to a heavy clay are summarized in Table 4.8. The increase in the amount of compaction had the greatest effect on the heavy and silty clays, resulting in large increases in the maximum dry unit weights and correspondingly large decreases in optimum moisture contents. This table also illustrates that for a given compaction effort the particle size and gradation of the soil are of considerable importance.

Significance of test.   The moisture-unit weight test is designed specifically to aid in the field compaction of soils, so that their best engineering properties may be developed. The assumption underlying the test is that the strength or shearing resistance of a soil increases with increasing dry unit weight.

The laboratory tests just described use compactive efforts which are generally assumed to be similar in effect to that of the construction equipment available in the field. This assumption is of course one which is very much open to criticism. However, as long as the specifications require these tests, the arguments are academic only and are not relevant considerations to the construction forces in their routine operations.

A most important factor to be noted about this test is that the presence of a certain amount of water is needed in order to get the desired dry unit weights. For simplicity, this water can be regarded as acting as a lubricant which enables the soil particles to slide over each other freely in the course of compaction. However, if too much moisture is present there is a tendency for the particles to be forced apart, due to pressures which are set up as the soil is compacted, with the result that higher unit weights cannot be obtained. Thus the laboratory test not only defines the dry unit weight that should be obtained in the field, but it also delineates how much water should be used during the compaction if this unit weight is to be achieved.

Given the maximum dry unit weight and the optimum moisture content of a soil, the construction forces can compact the soil into the practical design condition. As a check, field forces employ a unit weight test to determine the unit weight obtained by the construction equipment; if the result

TABLE 4.8.   *Comparison of results of the standard and modified compaction tests*[10]

Type of soil	Average results of standard Proctor compaction test		Average effect of modified Proctor compaction test	
	Maximum dry unit weight kg/m^3	Optimum moisture content %	Maximum dry unit weight, kg/m^3	Optimum moisture content %
Heavy clay	1555	28	increased by 320	decreased by 10
Silty clay	1670	21	increased by 275	decreased by 9
Sandy clay	1845	14	increased by 210	decreased by 3
Sand	1940	11	increased by 145	decreased by 2
Gravel-sand-clay	2070	9	increased by 130	decreased by 1

obtained is lower than the value permitted by the specifications, the contractor is required to recompact the soil. Normally relative compaction specifications require that a certain percentage of the maximum dry unit weight be achieved. This percentage varies from 90 to 95 per cent for the more granular materials, and 95 to 100 per cent for the fine-grained silts and clays.

Unconfined compressive strength test

This test can only be carried out on cohesive soils or soils stabilized with an additive which binds the particles together. With raw soils the test is best carried out on cylindrical specimens having a diameter of about 38 mm and a height to diameter ratio of 2:1. Cylindrical specimens are also used to test all stabilized soils in many countries, although in Britain it is recommended[20] that 15·2 cm cubes be used for coarse-grained (passing the 40 mm BS sieve) and medium-grained (passing the 20 mm BS sieve) soils; for fine-grained (passing the 5 mm BS sieve) soils, and, alternatively, for medium-grained soils, cylindrical specimens of 5·1 cm dia by 10·2 cm high, and 10·2 cm dia by 20·3 cm high, respectively, are recommended.

Stabilized specimens may be prepared by either static or dynamic compaction methods so that if desired the final dry unit weights are the maximum values obtained during the moisture-unit weight compaction test. In the course of the compression test, the load is normally applied at a predetermined uniform rate of deformation: the maximum load exerted by the compression machine is recorded as the unconfined compressive strength.

Typical test results.   Although a load-versus-deflection curve can be plotted, it is the ultimate compressive strength which is most frequently used in highway engineering studies, particularly those relating to soil stabilization. For mixtures that produce a reasonably rigid material, the load-deflection curve may be a straight line over a relatively long range, indicating fairly elastic materials.

The ultimate compressive strengths of soils and stabilized soil mixtures vary over a wide range. Results obtained are considerably influenced by such factors as the amount and type of stabilizing additive, the method and length of curing of the test specimens, and whether or not the specimens are saturated before testing.

Significance of test.   The unconfined compressive strength test of a natural soil may be classed as a shear test, since it is essentially a triaxial shear test with zero lateral pressure. Reference is made to the discussion on shear tests for further details on this aspect.

For soil stabilization work, the test serves much the same purpose as for concrete work. Particular uses of the test are to determine the suitability of the soil for treatment with a given additive and to compare different mixtures, to specify the additive content to be used in construction, and to provide a standard by which the quality of the field processing can be assessed. The measured strength value is not used for design purposes, nor is the modulus of elasticity that is available from the load-deflection curve. Rather the unconfined compressive strength data are principally significant for control purposes.

Shear tests

All laboratory shear tests require specimens that are as similar to field conditions as possible. In addition the loads applied to the specimens in the laboratory should approximate the actual field loading conditions. In the case of natural soil which is unworked, this requires an 'undisturbed' sample which must be carefully removed from the ground with as little disruption to its structure as possible. When the soil has to undergo compaction or other manipulation, the sample for laboratory testing should be prepared in anticipation of the expected future condition of the soil.

Shear tests may be described as 'quick', 'consolidated quick' and 'slow' tests, depending on the loading conditions expected and desired. For quick tests the sample is loaded relatively rapidly and the moisture content does not change during the test. In the slow tests, on the other hand, water is permitted to drain freely from the samples during all stages of testing. For the consolidated quick test, the specimen is preloaded and allowed to drain freely prior to the actual shearing test being carried out. The variations in these three types of test result from the need to approximate the actual conditions to which the soil will be subjected.

The principal types of shear test are the direct shear and the triaxial shear tests. The primary difference between the two lies in the equipment and technique used. The following is a brief description of each of these tests.

Direct shear test. Specimens used in the direct shear test may be saturated prior to testing if it is desired to represent the most critical condition for the given soil structure. To carry out the test the specimen is placed in the lower (holding) place of the shear-box testing device, the upper plate is then placed in position and a shearing load is applied perpendicular to the axis of the sample. Measurements are made of the incremental loads applied and the lateral displacements as a result of these loads. The test is conducted on a minimum of two samples, for each of which a normal or compressive load is applied in a direction parallel to the axis of the soil cylinder; this normal force is different for each sample.

Triaxial shear test. In this test, the specimen is subjected to three compressive stresses at right angles to each other, and one of these stresses is increased until the specimen fails in shear. The triaxial test differs from the direct shear test in that the plane of shear failure is not predetermined. It is similar, however, in that a number of identical specimens must be tested, and the test conditions are selected to correspond as closely as possible with the field conditions.

The triaxial cell consists of a transparent plastic cylinder which is much larger than the soil sample; this enables the specimen to be observed during the test. The cylinder is capped at the top and bottom with removable metal plates. The soil sample is placed inside a very thin rubber membrane which is slightly smaller than the specimen, capped at the top and bottom with a porous disc and placed in the cylindrical cell. In the course of the test a compressive load is applied by a piston arrangement through the top metal plate, while at the same time a uniform lateral pressure is obtained by means of a liquid placed in the cell and about the rubber enclosed specimen.

A series of tests is run, using a different lateral pressure for each test. Measurements are made in each case of the compressive loads applied and of the deformations of the samples in the direction parallel to the axis of the cylinder.

Typical test results. Whatever the type of test, the shear test results are analysed in a similar fashion. Load-versus-displacement curves are plotted, and a critical point on each curve is determined; this point is related to the allowable displacement, the slope of the load-displacement curve, and other factors.

To find the actual shear value is not the purpose of the test. Rather, the shearing resistance of the soil is generally assumed to result from two components, friction and cohesion, and so the test is aimed at determining the value of these two constituent characteristics. Since frictional resistance is increased by normal or compressive loads, the shearing resistance of a soil in the field at any instant is influenced by the loads perpendicular to the plane of shear. Therefore, laboratory shear tests determine the 'angle of internal friction', commonly designated by ϕ, and cohesion, usually termed c. These two values are considered constant for both laboratory and field conditions, and the actual shearing resistance can then be calculated.

Typical values for friction and cohesion are as follows:

Sandy soils $\phi = 28°$ to $45°$

           $c = 0$ to $2·06 \, MN/m^2$

   Clay soils $\phi = 0°$ to $15°$

           $c = 0·7$ to $13·8 \, MN/m^2$

Generally, sandy soils develop their shearing resistance through friction, with little or no cohesion. The opposite is true for clay soils, the bulk of their resistance coming from cohesion. When an unconfined compression test is carried out on a clay soil, ϕ can be assumed to be zero, and then the value of c is one-half the compressive strength.

Significance of test.   Shear tests are used in highway engineering to determine experimentally the shear characteristics of the soil. Normally the test values obtained are applied in the form of a stability equation which attempts to determine the probability of a soil failing in shear under the loads and conditions imposed in the field.

Not only do pavements, embankments, and bridge structures rest directly on soil, but most embankments are composed of soil materials, and so the shearing resistance offered by a soil is an important factor in a great many highway problems. Shearing of the foundation soil can result in complete pavement disintegration, collapse of a bridge, or the loss of an embankment through sliding. While the prediction of the unit shearing resistance is the objective of laboratory shear tests, the application of these results requires extensive soil mechanics' experience.

Consolidation test

The consolidation test is used to estimate both the rate of settlement and the total amount of settlement of a soil layer under an applied load. The procedure and analysis is restricted to problems involving saturated soil masses, principally clays and other such soils of low permeability. Prediction of the settlement of a structure by the use of a laboratory consolidation test requires that the sample used should be as nearly identical and representative of the soil mass as possible.

To carry out the test, a soil sample is cut and trimmed so as to fit into a special metal ring provided for the test; this sample is normally a disc 75–100 mm in diameter by 25 mm thick. Porous discs are placed on top and beneath the specimen, and the assembled sample discs and ring are placed in a loading unit. A compressive load is applied and the changes in thickness of the sample are read at set time intervals. After settlement is complete—this is usually taken as having occurred after 24 hours—the applied unit load is doubled and readings are taken as before. This procedure is generally followed through four to six loading increments, with the magnitude of the final increment depending on the actual loadings expected in the field.

Typical test results.   The two principal values obtained from the consolidation test are the compressive index C_c and the coefficient of consolidation C_v. These values are calculated from the test data and are used to estimate the rate as well as the total settlement under a given load condition.

The compressive index is used in the analysis of total settlement. It is a dimensionless factor which normally ranges in value from 0·1 to 0·3 for silty clays, and from 0·2 to 1·0 for clays.[22] The total amount of settlement is also related to the thickness of the layer under consolidation and the applied load. Settlements of 150–450 mm are not at all unusual beneath large structures.

The values for the coefficient of consolidation range from 2–0·2 cm²/sec for silty clays, and 0·1–0·02 cm²/sec for clays.[22] The coefficient is used to estimate the amount of settlement for a given period of time under a given increment of load.

The settlement for the time period is compared with the total settlement for the load, and the ratio of the two, usually expressed as a percentage, is called the degree of consolidation. Thus, in the construction of a 10 m embankment, if the work is discontinued after a height of 3 m is reached, a time-lapse of three weeks may produce a degree of consolidation of 90 per cent. Subsequently another 3 m may be added, and time for settlement again allowed, and so on. Stage construction of this form permits the underlying, saturated soil mass to safely eliminate pore water and develop shearing resistance.

Significance of test.   When a load is applied to a soil mass, the immediate tendency is for the soil particles to be pushed closer together. However, when the soil mass is saturated, the water, being incompressible, must initially carry part of the applied load. This results in the production of an initial pressure, commonly termed the pore-water pressure, which continues as water proceeds to drain from the soil. During the drainage period, which in very impermeable soils may require many years to complete, the soil particles are forced closer together, thereby producing the volume change known as settlement.

The laboratory consolidation test attempts to determine in an accelerated manner both the rate of settlement and the total amount to be expected under the total load applied. These values can be most important for analyses of highway embankment settlements and slope stabilities during construction stages. As described in the chapter on Flexible Pavements, vertical sand drains can be used to accelerate settlement by providing readily available passageways for the escape of pore water. Settlement can then proceed more quickly because of the shorter distance which the water has to travel in getting out of the soil. Thus the road pavement can be placed in safety on the embankment when it is known that the bulk of the settlement is complete.

It is important to remember that the shearing resistance of a soil is lower during periods of high pore pressures. Thus the consolidation test can also be used to determine how rapidly the height of an embankment can be increased during construction without a shear failure being produced in the soil.

SELECTED BIBLIOGRAPHY

1. JOFFE, J. S. *Pedology*. New Brunswick, N.J., Rutgers University Press, 1936.
2. BS 892:1967. *Glossary of Highway Engineering Terms*. London, British Standards Institution, 1967.

3. LEONARDS, G. A. Engineering properties of soils. Ch. 3. In LEONARDS, G. A. (Ed.) *Foundation Engineering*. London, McGraw-Hill, 1962.
4. MILLAR, C. E., L. M. TURK and H. D. FOTH. *Fundamentals of Soil Science*. New York, John Wiley, 1962.
5. GRIM, R. E. *Clay Mineralogy*. New York and Maidenhead, McGraw-Hill, 1953.
6. CASAGRANDE, A. Classification and identification of soils, *Proc. Amer. Soc. Civ. Engrs*, Part 1, 1947, **73**, No. 6, 783–810.
7. C.P. 2003:1959. *Earthworks*. London, British Standards Institution, 1959.
8. ALLEN, H. Report of committee on classification of materials for sub-grades and granular-type roads, *Proc. Highway Research Board*, 1945, **25**, 375–392.
9. U.S. DEPARTMENT OF AGRICULTURE. *Soil Survey Manual*. U.S.D.A. Handbook No. 18. Washington, D.C., U.S. Government Printing Office, 1951.
10. ROAD RESEARCH LABORATORY. *Soil Mechanics for Road Engineers*. London, H.M.S.O., 1952.
11. SPANGLER, M. G. *Soil Engineering*. Scranton, Pa., International Textbook Co., 1960.
12. WOOLTORTON, F. L. D. *The Scientific Basis of Road Design*. London, Arnold, 1954.
13. JUMIKIS, A. R. Soil moisture transfer in the vapour phase upon freezing, *Highway Research Board Bull*. 168, 1957, 96–114.
14. PENNER, E. The mechanism of frost heaving in soils, *Highway Research Board Bull*. 225, 1959, 1–13.
15. ALDRICH, H. P., Jnr. Frost penetration below highway and airfield pavements, *Highway Research Board Bull*. 135, 1956, 124–149.
16. CRONEY, D. and JACOBS, J. C. The Frost Susceptibility of Soils and Road Materials. *RRL Report* LR 90. Crowthorne, Berks., The Road Research Laboratory, 1967.
17. CASAGRANDE, A. Discussion on frost heaving, *Proc. Highway Research Board*, 1932, **11**, pt. 1, 167–172.
18. LINELL, K. A. and C. W. KAPLAR. The factor of soil and material type in frost action, *Highway Research Board Bull*. 225, 1959, 81–126.
19. BS 1377:1967. *Methods of Testing Soils for Civil Engineering Purposes*. London, British Standards Institution, 1968.
20. BS 1924:1967. *Methods of Test for Stabilized Soils*. London, British Standards Institution, 1967.
21. SHERWOOD, P. T. The Reproducibility of the Results of Soil Classification and Compaction Tests. *RRL Report* LR 339. Crowthorne, Berks., The Road Research Laboratory, 1970.
22. COMMITTEE ON SIGNIFICANCE OF TESTS FOR HIGHWAY MATERIALS. Significance of tests for highway materials—basic tests, *Proc. Amer. Soc. Civ. Engrs, Journal of the Highway Division*, Sept. 1957, **83**, No. HW4.
23. MINISTRY OF TRANSPORT, et al. *Specifications for Road and Bridge Works*. London H.M.S.O., 1969.

5 Soil stabilization

To the road engineer, the definition of soil stabilization which is perhaps of most interest is that ascribed to the chairman of the American Highway Research Board in 1938, viz:
'A stabilized fill, subgrade, road surface or roadbase is one that will stay put, and stabilizing is the process by which it has been made that way.'
This definition has the beauty of emphasizing that highway soil stabilization is concerned with any process by which a soil may be improved and made more stable. Normally, the construction of roadbases and surface courses with already stable materials, e.g. crushed rock, concrete, etc., is not considered as falling within this definition, but other than that any treatment used to improve the strength of a soil by reducing its susceptibility to the influence of water and traffic is soil stabilization, whether the process is performed *in situ* or applied to the soil before or after it is placed in the roadway or embankment.

In practice the methods by which soils may be stabilized for highway purposes can be divided into the following main groups:

1. *Mechanical stabilization.* This, by far the most widely used method of stabilization, relies for stability on the inherent properties of the soil material. If a soil cannot be made stable simply by compaction, then additional soil or other aggregate materials may be admixed to produce a mixture having the required stability characteristics. Although not mechanical stabilization in the true sense of the word, the use of additives such as chlorides, calcium lignosulphonate (lignin), and molasses is usually associated with this process. This is because only soils which already have some mechanical stability can be satisfactorily improved by these chemical additives. Other methods sometimes included in the mechanical stabilization category are the thermal procedures involving freezing and heating of the soil.

2. *Cement stabilization.* This is a process in which cement is mixed with the soil to cause it to harden into a compact mass. A properly designed cement-stabilized soil will not soften in the presence of water and will withstand detrimental forces resulting from frost action.

3. *Lime and lime-pozzolan stabilization.* In this method of stabilization the functions of the additives are twofold. They may be used to modify the soil properties—principally by chemically changing the soil gradation—or they may also cause the soil to harden into a compact mass having properties and uses similar to those of a cement-stabilized soil.

4. *Bituminous stabilization.* In this process, bituminous materials—

221

bitumens or road tars—are mixed with soil so as to waterproof the particles and/or provide the additional cohesion necessary for stabilization.

5. *Other experimental stabilization methods.* Due to the paucity of good roadmaking materials in many parts of the world, considerable research has been carried out in an effort to find a cheap 'magic chemical' which, when added in small quantities to a soil, will quickly result in the formation of a highly stable pavement mixture. Included in this group is the cementing inorganic stabilizer sodium silicate, the organic cationic chemicals which act primarily as waterproofers of the soil, and the natural and synthetic resinous materials which cement and/or waterproof the soil particles.

MECHANICAL STABILIZATION

When a granular structure, such as a roadbase or surfacing, has the property of resistance to lateral displacement under load, it is said to be mechanically stable.[1] In mechanically stabilized soils this resistance is provided by the natural forces of cohesion and internal friction in the soil. Cohesion is mainly associated with the silt and clay content of the material, while internal friction is a characteristic of the coarser particles.

Usage

Mechanical stabilization is still the stabilization method most widely used in road construction throughout the world. Its popularity is based on the fact that it makes possible the maximum usage of locally available materials in highway embankments, sub-bases, roadbases and surface courses. Mechanically stabilized roads may range from simple earth tracks which have been cleared of vegetation and compacted by traffic—but which still meet the stability criteria demanded by their traffic—to highways which utilize highly sophisticated blends of different materials in one or more of their component layers.

In countries with well-developed economies, pavements formed from mechanically stabilized locally-available materials are primarily used in rural areas, where the critical need is to provide access to lands and farms but the traffic demands are not sufficient to necessitate the construction of more expensive, higher-type pavements. In a given situation it may well be that the estimated future traffic demand justifies a high type of pavement but sufficient funds are just not available with which to construct it—in which case a mechanically stabilized pavement may lend itself to stage-construction. By stage-construction is meant the step-by-step improvement of the road as the expenditure is justified by the demands of increased traffic; its object is to provide, in an efficient and economic manner, adequate service to the traffic throughout the development of the highway. Thus, at an early stage the pavement would be formed entirely of the mechanically stabilized material, but at the final stage this same material might only form part or all of the roadbase and/or sub-base of a bituminous-surfaced or concrete road designed to carry heavy traffic.

At this time mechanically stabilized soils are not used to any large extent in Britain as pavement construction materials. Their most extensive utilization is in the United States and so it is convenient to examine practice there with respect to usage under particular traffic conditions.[2] The general opinion amongst engineers in America is that properly constructed and designed mechanically stabilized pavements are entirely satisfactory when the annual average daily traffic is less than about 50 veh/day; below this volume the loss of surface material due to abrasion is not very great and the dust problem is not serious. When the traffic exceeds an ADT of about 100 veh/day, maintenance costs can be considerably reduced by the application to to the surface of some type of dust palliative such as a liquid bituminous material or calcium chloride. Once, however, the traffic reaches between 200 and 300 veh/day, the mechanically stabilized surface can no longer be maintained economically, and so a higher type of surface will need to be constructed; however, the existing surface is very often used in this construction, to form part of the roadbase for the higher type of road.

Types of mechanical stabilization. Again there are differences amongst authorities as to what exactly is meant by mechanically stabilizing a soil. Some go so far as to regard the drying out of a soil after a rainstorm as a form of mechanical stabilization, while others confine it to a process whereby stabilization is brought about by altering the mechanical analysis of a soil. Since it is this author's view that mechanical stabilization does not merely entail alteration of a single criterion such as gradation, but requires a consideration of many of the fundamental phenomena of soil (and earthworks) engineering, the discussion in this text will deliberately take the following broad form:

A. Mechanical stabilization by treatments:
    1. Compaction
    2. Consolidation
    3. Electrical and thermal methods
B. Mechanical stabilization with the aid of additives:
    4. Soil and aggregate
    5. Chlorides
    6. Lignin
    7. Molasses

Stabilization by compaction

The need for an adequate state of compaction of the soil in highway pavements, subgrades and embankments is a fundamental consideration which is accepted by most road engineers. There are many instances of road settlement and unstable embankment slopes which can be traced to poor compaction. Increasing the state of compaction of a soil stabilizes it by increasing its strength, reducing the possibility of settlement and, usually, minimizing changes in moisture content. Thus, if economical designs of pavements, embankments and subgrades are to be obtained, and subsequent maintenance costs reduced, it is essential that the soil used in these road components be brought to a satisfactory stable state of compaction during the construction work.

Definition of compaction. The term compaction is often, and wrongly, used interchangeably with the terms 'compression' and 'consolidation', and it is important that the reader should recognize and appreciate the distinctions between them. By compaction is meant the artificial increase in unit weight of a material which is brought about by mechanical means, e.g. rolling; it is normally accomplished by expelling air from the soil mass, thereby decreasing the void ratio. Compressibility and consolidation on the other hand are associated with the increase in unit weight due to the gradual expulsion of water from the soil pores. Whereas compaction is an 'instantaneous' process, consolidation occurs over much longer periods of time; very often many years will pass before consolidation is complete and the final settlement is known.

In natural soils consolidation is greatly influenced by the soil structure and the past stress history of the material. For instance, in-situ soils which are the result of a sedimentation process are more compressible than their residual or aeolian counterparts. Even though a soil may be compacted, consolidation may still take place if the equilibrium conditions are upset, e.g. the lowering of the water table beneath an embankment or subgrade which is 'adequately' compacted may result in an effective increase in soil stress such that a volume change will take place within compressible layers below the original water table; this leads to settlement of compacted material at the surface. In other instances, volume changes may take place as a result of shrinkage and swell phenomena. These phenomena, which are most noticeable in fine-grained soils, result from a build-up and release of capillary tensile stresses within a soil's pore moisture and from the varying degree of affinity for moisture which particular clay minerals have.

Factors affecting compaction in the field. The compaction which can be obtained in a soil or soil-aggregate mixture in the course of construction is dependent on the following factors: 1. The moisture content at which compaction is carried out. 2. The soil type. 3. The means by which compaction is carried out.

The first two of these factors are treated in the discussions on moisture-unit weight in the chapters on Soil Engineering for Highways and on Flexible Pavements, so the following is confined to the means by which the desired compaction can be obtained in the field.

The selection of the proper compaction equipment and method is vital to the production of the long-term economy expected and desired from a highway. Essentially there are four ways by which soil may be compacted by mechanical means. These methods, and some examples of equipment used for each, are:

1. Using heavy weights to press the particles together. Smooth-wheeled rollers are examples of compaction equipment which operate on this principle.

2. Kneading of soil while at the same time applying pressure. This type of compaction is applied by the sheepsfoot roller. The pneumatic-tyred roller has a compaction action which is a cross between that obtained with a smooth-wheeled roller and a sheepsfoot roller.

3. Vibrating the soil so that the particles are shaken together into a compact mass. The vibrating rollers utilize this procedure.

4. Pounding the soil so that the particles are forced to move closer together. This is perhaps the oldest type of compaction; hand-tamping as a means of obtaining high unit weights is literally ages old. Mechanical tampers have now been devised which make this method of compaction a most practical one in particular instances.

Smooth-wheeled rollers.   Smooth steel-wheel rollers have long played an important part in road construction and are acknowledged to be the oldest form of mechanical compaction. The early steam roller consisted of three steel wheels on a body which was self-propelled by a steam engine. The wheels were arranged so that the central track of the front wheel was not overlapped by the two rear wheels. In present roller models the steam engines are replaced by petrol or diesel engines and have either two or three rolls in line behind each other, so that each roll follows in the track of the front one. The rolls are actually hollow steel drums which are filled with water or damp sand to obtain the desired pressures.

While satisfactory results can be obtained with smooth-wheeled rollers on most soils, experience has shown that they are most suitable for compacting gravels, sand and suchlike materials where a crushing action is needed. When used on moist plastic soils, smooth-wheeled rollers have difficulty in maintaining traction. Another disadvantage is that they frequently create a crust on the surface which can prevent the proper compaction of the soil beneath; this 'bridging' effect is most noticeable with heavy plastic soils.

The principal characteristics of a smooth-wheeled roller which affect its compacting performance are the load per unit width under the compaction rolls, and the width and diameter of the rolls themselves. The load per unit width and the roll diameter control the pressure near the surface of the soil, while the gross weight of the rolls affect the rate with which this pressure decreases with depth. Thus, for compaction work, it is important that the load per unit width as well as the gross weight of the roller be specified. The depth of layer which may be satisfactorily compacted depends on the type of soil, the weight of the roller and the purpose of the work; in Britain it varies from a low of about 115 mm for pavements to a high of 250 mm in embankments.

Sheepsfoot rollers.   A roll on this type of roller consists of a steel cylindrical drum with steel projections or 'feet' extending in a radial direction outward from the surface of the cylinder; the drums are manufactured in diameters ranging from 1–1·8 m and weigh, when loaded, from 2·7–27 tonnes. They may be self-propelled, but more commonly they are towed either by track or pneumatic-tyred tractors, up to three rollers being towed in tandem at one time.

It is said that the concept of a roller of this type came to a contractor who noted the compaction caused by a flock of sheep crossing an earth road in California just after the turn of the century. Whatever the validity of this story, there is no doubt that the sheepsfoot roller is the most commonly used compactor in the United States today whereas it is rarely used in Britain.

As opposed to other rollers, the sheepsfoot compacts from the bottom of the lift up. It operates in the following manner. During the first pass over a lift of loose soil, the feet penetrate to near the base of the layer so that the bottom material is well compacted. As additional passes are made, the feet penetrate to lesser depths as the unit weight, and thus the bearing capacity, of the soil is increased. Eventually the soil will be so well compacted that the feet of the roller will only penetrate into the ground to a depth of a few centimetres, in which case the roller is said to have 'walked-out' of the layer.

The sheepsfoot roller is designed to break-up the surface 'crust' on a soil by compacting small areas at high load concentrations. The principal characteristics affecting its performance are the contact pressure and the coverage of ground obtained per pass. The contact pressure applied by each foot is a product of the roller weight divided by the total area of one row of tamping feet. This pressure should be as large as possible, but should not exceed the bearing capacity of the soil being compacted. If this occurs, the roller will sink into the ground until enough adjacent feet, or the drum itself, come into contact with the ground and the contact pressure is that which the soil's bearing capacity can sustain.

Since compaction is obtained by the feet penetrating and applying a high vertical pressure while at the same time providing lateral pressure, the sheepsfoot roller is most efficient in compacting fine-grained soils. Unfortunately, however, it produces a larger amount of air voids in a soil than do smooth-wheel or pneumatic-tyred rollers; this can be very detrimental in the case of subgrades in which a large air content increases their liability to absorb moisture. A further factor which limits the usage of this type of roller in Britain is the fact that, when high dry unit weights are required, the optimum moisture contents required by these rollers are usually less than those found in fine-grained soils in the field.

Pneumatic-tyred rollers. For many years, it was common practice for earthworks to be compacted by directing the regular construction haul units to travel over freshly placed material until it was quite dense. This led eventually to the development of special pneumatic-tyred rollers to compact all types of soil.

Pneumatic rollers can be divided into three types or classes called medium, heavy and super-heavy weight. The medium class includes both tow-type and self-propelled units up to about 12 t rolling capacity. The heavy class covers the range beyond this up to the 50-tonne rollers, while the super-weights—which are mostly used for compacting airport runways and taxi-ways, and for proof rolling of highway subgrades—include rollers up to 200 t capacity. In all types of pneumatic roller, ballast boxes are carried which can be filled with suitable materials to provide the necessary compaction load.

The pneumatic wheels on the rollers are arranged so as to distribute equally and uniformly the compaction load, regardless of the ground profile. Full ground coverage is obtained by having the wide-face tyres arranged so that there is always overlap between the front and rear wheels. In addition, many pneumatic-tyred rollers have a wobble-wheel motion which provides a type of kneading action to the soil. As is discussed in more detail in the chapter on Flexible Pavements, the characteristics of a pneumatic-tyred roller which most affect its performance are the tyre inflation pressures and the load carried by each wheel. Controlled compaction studies have shown that there appears to be little point in employing tyre-inflation pressures much in excess of 275–345 kN/m^2 with wheel loads of about 5 tonnes when compacting clayey soils.[3] For the compaction of granular soils, there appears to be some advantage in employing the highest tyre-inflation pressure and the heaviest wheel load practicable, consistent of course with avoiding over-stressing of the soils. Pneumatic rollers are generally unsuitable for compacting uniformly graded materials.

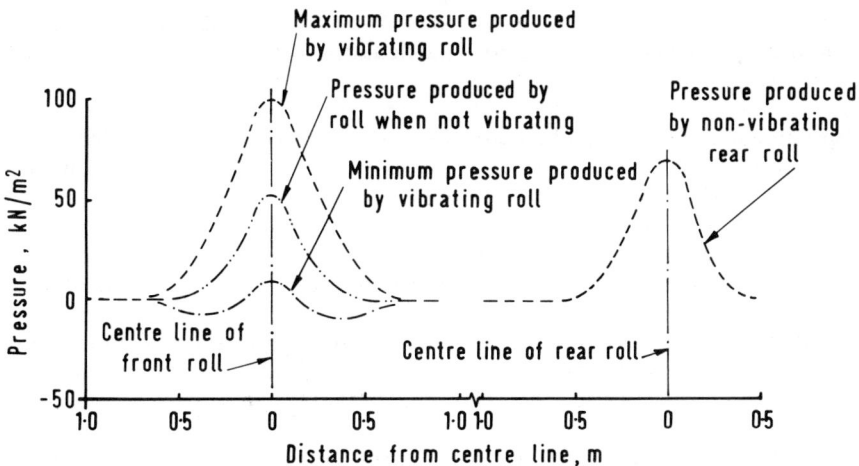

Fig. 5.1. Pressures produced at a depth of 200 mm in a silty clay soil by a $2\frac{1}{2}$-tonne vibrating roller

Vibratory compactors. These consist of a vibrating unit of either the out-of-balance weight type or a pulsating hydraulic type mounted on a screed, plate or roller in such a way that the net effect is an up-and-down vibratory movement of the compactor at a frequency of about 20 to 30 cycles per second. This frequency is in the range of the natural frequency of soils, particularly coarse-grained ones, and so the particles are literally shaken about when the compactor moves over the ground surface. When coarse-grained soils are being compacted, the result is that the particles are continually rotated until they slip over each other and a compact mass is obtained. With pure sands and gravels, the particles are usually most easily vibrated into position when the soils are saturated. Not only does the excess water provide the lubrication necessary to allow the particles to slide over each other, but it breaks the apparent 'cohesive' bonds between particles which are formed when the material is damp.

Even with non-coarse-grained soils, vibratory rollers can be quite effective. Fig. 5.1, illustrates the results of some pressure measurements taken beneath a $2\frac{1}{2}$-tonne vibrating roller, with the front roll not vibrating, as well as vibrating, while a silty clay soil was being compacted. With the front roll not vibrating the results were those expected from a basic smooth-wheeled roller, whereas when the front roll was vibrating the maximum pressure produced under the roll was doubled.

Impact compaction. Obviously the methods of compaction which have been presented so far all involve fairly substantial pieces of equipment which require large manoeuvring space for satisfactory results to be obtained. When trenches have to be back-filled or soil compacted near abutments, walls or columns, it will very often not be possible to use large compactors, and so recourse is made to smaller, portable impact compactors such as the internal combustion-type frog-rammers. These can be quite effective in compacting relatively small areas, particularly when the materials are non-cohesive.

Specification and control of compaction. 'Stabilization' by means of compaction constitutes one of the major items of cost in the construction of a highway, and it is important therefore that the state of compaction specified and obtained should not be greater than is really necessary. From a long-term cost aspect, it is more important for a desirable state of compaction to be obtained, as failure to do so can lead to settlement and damage to the highway structure.

It is the practice in the United States to specify the state of compaction required as a percentage of the maximum dry unit weight obtained with one of the standard Proctor tests. This 'end-product' method of specification can be criticized on the following grounds:

1. If the moisture content selected for compaction in the field is much lower than the optimum moisture content obtained with the laboratory compaction test, a poor state of compaction, measured in terms of air-voids content of the soil, will result even if a high relative compaction is actually obtained.

2. If the natural moisture content of the soil in the field is much higher than the laboratory optimum moisture content (a common situation in Britain), then the specified relative compaction may well be unattainable.

To overcome these disadvantages the practice in Britain in recent years has been for the moisture content specified for compacting purposes to be that which is likely to be naturally obtained in the field, and for the state of compaction to be controlled by specifying the amount of compactive effort to be applied to a given type of material ('method' specification). Details as to how different types of equipment may be used to compact British soils (and other materials) are available in the literature.[4]

Stabilization by consolidation

By this is meant the preloading of a soil—usually a clay or peat—prior to the construction of an earth embankment, with time allowed for the soil to be partially or completely consolidated under the increased load. Properly carried out, it results in an increase in the bearing capacity of the underlying soil, decreases the amount of subsequent settlement and, in certain instances, eliminates the need for supporting piles beneath the pavement. Methods of preloading are by means of the direct construction of an embankment—with or without the use of vertical sand drains—lowering of the ground water level, stage construction, and consolidation by atmospheric pressure.

Preloading by direct construction of an embankment is by far the most common way of prestressing a poor soil. On large construction schemes it is usually preceded by a thorough soil engineering investigation to determine the height to which the fill should be constructed and the length of time it should be left in place. If at all possible it is advisable to compact the fill as it is being constructed, and then that material can be used to form the final embankment; it is desirable, also, to use a preloading fill having a greater height and weight than the desired embankment, and then to 'cut back' to the desired formation level. In the course of construction vertical sand drains may be constructed, to accelerate the rate of consolidation. Details of the preloading method of construction are given in Chapter 6 on Flexible Pavements.

Lowering of the ground water level is also a method of consolidating a foundation soil, and drainage methods for so doing are described in Chapter 2.

Consolidation by means of stage construction is simply another means of preloading a soil by direct embankment construction. It is a long-term process which results in the consolidation of a weak underlying soil without displacing it. The embankment is first built to a height which is limited by the shearing strength of the foundation, and further construction is delayed until the excess pore-water stress induced in the weak layer by the weight of the embankment has dissipated over the course of the consolidation process. When this pressure has been reduced to a safe value, an additional height of fill is placed and the process repeated until the embankment is finally completed.

A most interesting low-cost method of stabilizing a soil by consolidation has been developed by the Royal Swedish Geotechnical Institute; it uses the atmospheric pressure as a temporary surcharge and vertical sand drains to accelerate consolidation.[5] In the field the soil area which is being stabilized is covered with a sand filter 400–500 mm thick and on top of this a tight impermeable membrane, i.e. sheet plastic or a layer of clay or a bituminous covering, is placed. Outside the width of the filter the covering membrane is connected tightly to the underlying soft clay, which is free from cracks and root channels and is beneath the normally dry soil at the very surface of the ground. A suction pipe is then carried through the membrane and into the filter and a pump starts sucking air out of the filter and, gradually, out of the clay. In any horizontal plane the total normal pressure, i.e. the atmospheric pressure plus the weight of the overlying soil, remains constant and therefore the grain pressure increases as the pore pressure decreases. Thus the pump sucks the water out of the clay, which is forced to consolidate accordingly.

Stabilization by electrical and thermal methods

A method of soil stabilization which, at the moment, would appear to be very limited in scope in terms of highway construction is that of solidification by *freezing*. This process relies on the fact that freezing of the water in the pore spaces of a soil gives a very high strength to the material. Soil-freezing has been used very successfully for the temporary stabilization of the foundations of buildings, sinking of shafts, enlarging of a railway tunnel, etc., but other than on these isolated projects little work has been done in this area.

What would appear to be much more promising is soil stabilization by *heating*. Again this is a method which has been pioneered in non-highway engineering schemes; however, research has shown it to be practical in those areas where there is an extreme shortage of natural aggregates.[6] The method relies on the fact that when a clay soil is heated it loses weight, this weight loss being due to the removal of adsorbed and interlayer moisture held by the clay minerals and, above certain temperatures, to changes in the actual structure of the clay minerals. This phenomenon, which has been observed by dehydration curves, differential thermal analysis and X-ray diffraction studies, has been used to obtain road aggregates in British Guiana, the Sudan and Australia, and as a means of stabilizing foundation soils in Rumania and the U.S.S.R.

These changes in structure brought about by heating clay soils result in hard durable road materials which, although of relatively low quality when compared with the normal types of stone used in road construction, can be used to advantage where high-quality materials are not available. Heating causes the plasticity index to decrease, while the attraction for water is considerably reduced. Permeability generally increases due to shrinkage of the clays by desiccation and the formation of cracks and fissures. Compressibility, both soaked and unsoaked, is reduced due to the decrease in water sensitivity.

Soil-aggregate stabilization

Stabilization by altering the gradation of a soil constitutes soil-aggregate stabilization. A soil-aggregate stabilized mass is analogous to Portland cement concrete in that the coarse aggregate content plays a role similar to the aggregate in concrete while the fines act as a mortar; the general aim therefore in changing the gradation is to obtain the proper proportions of particle-size fractions that will give a dense homogenous mass. The main use of this form of stabilization is for sub-base construction for high-type roads and in the roadbases and surface courses for secondary-type roads.

General requirements of surfaces and roadbases. It is necessary for the engineer to have a clear understanding of the functions of the various layers in a stabilized soil roadway in order to be able to decide which soil-aggregate materials may or may not be included in each layer. Apart from the common requirement of sufficient stability to meet the loads placed upon them, each layer has different material needs. These can be illustrated by discussing the requirements of the two most important elements, the roadbase and surface course, of a road pavement in which stabilized soil materials are to be used.

Granular stabilized soil surface courses should only be used in rural areas where traffic is quite light. The material requirements for these layers have been enunciated as follows:[7]

1. Stability to support the weight of traffic.

2. Resistance to abrasive action of traffic.

3. Ability to shed a large portion of the rain which falls on the surface, since a large amount of water penetrating the surface might cause loss of stability in the surface course or softening of the subgrade.

4. Capillary properties to replace the moisture lost by surface evaporation and thus maintain the desirable damp conditions in which the soil-aggregate particles are firmly bound together.

Consider these surface course requirements. From the point of view of stability and resistance to abrasion, the materials used must consist of hard, tough, durable pebbles or fragments of stone or gravel, and a filler of sand or other finely divided mineral matter, together with sufficient binding material to permit the consolidation and formation of a tightly bonded layer. Experience has shown that it is undesirable to use aggregate larger than 25 mm size in a surface course as it is very liable to be torn from the surface by a motor grader carrying out maintenance. Best aggregate interlock is obtained by using angular particles; for this reason glacial gravels and sands are preferred to river materials. A certain amount of clay binder must also be present to help prevent the smaller particles from being whipped away under the action of traffic. Again, from the point of view of stability, it is desirable that as little water as possible should penetrate the surface course. Rainwater must be shed from the surface very quickly, and for this reason A-type crowns are better with granular-stabilized surfaces than the more conventional parabolic ones. The clay content can also play an important role in preventing moisture infiltration, as during wet weather the clay particles will swell and plug the soil pores. In this case, however, care must be taken to ensure that the clay content is only just sufficient to perform this function, and that it will not swell and cause dislocation in the granular materials. In hot climates, the clay fraction will help retain against evaporation the moisture content necessary for stability and, if there happens to be an underlying source of available moisture, it may be able to replenish evaporated moisture by capillary attraction.

The requirements of a roadbase are very different from those of a surface course; this is often the reason why roadbases containing 'unscientifically-converted' surface courses have failed. The primary requirement of a soil-aggregate roadbase is that it be able to withstand the stresses imposed upon it. This it is able to do if it is composed of a dense mixture of coarse and fine soil materials which has high shear strength. Dense mixtures may be obtained when their particle size distributions follow Fuller's law which is expressed as

$$p = 100 \sqrt{\frac{d}{D}}$$

where p = percent by weight of the total mixture passing any given sieve size,

d = aperture size of that sieve,

and D = size of the largest particle in the mixture.

The clay content of a soil-aggregate stabilized roadbase must be very carefully controlled. Whereas in a surface course, a non-porous layer is required in order to prevent water infiltration, the very opposite is normally true in a roadbase, particularly one which is covered by, e.g., an impermeable bituminous surfacing. The presence of a significant quantity of clay in the roadbase will attract water and, since it cannot escape by evaporation, cause it to accumulate within the layer; this in turn will cause a softening of the roadbase and may eventually lead to the complete destruction of the pavement.

Usage. The most comprehensive usage of soil-aggregate roads is in the United States. Scientific investigation into their design was initiated as early as 1906 by Dr. C. M. Strachan of South Carolina who endeavoured to correlate the performances of sand-clay roads with their mechanical analyses. Interest in this work developed rapidly and since then a considerable amount of theoretical and practical research has been carried out into the factors influencing the desirable service qualities of soil-aggregate roads. Typical of the gradations which have been found to give satisfactory service are those recommended by the American Association of State Highway Officials.

Soil-aggregate pavements are relatively rarely constructed in Britain, particularly those in which part of the stabilized material forms the surface course. Experience in this country has resulted in the specification given in Table 5.1 for the use of sands and gravels in the sub-base construction of new major roads. This specification also requires that the material passing the 425 μm BS sieve should have a plasticity index of less than 6, and that it should be laid and compacted at a moisture content within the range 1 per cent above to 2 per cent below the optimum percentage determined in accordance with the Vibrating Hammer method of test.

TABLE 5.1.   *British recommendations for the gradation of granular stabilized sub-bases*[4]

BS sieve, mm	Percentage by weight passing
76·0	100
38·1	85–100
9·52	45–100
4·76	25–85
0·600	8–45
0·075	0–10

Blending soils and aggregate. In some localities deposits of naturally occurring soils will be found which meet the soil-aggregate specifications. More often than not, however, two or more soils will have to be blended in proper proportions to produce mixtures meeting these requirements. Generally this involves either adding aggregate to the *in situ* soil, or adding fine material to *in situ* granular material that is not sufficiently cohesive, or proportioning mixtures of fine soil and aggregate for plant mixing, or adding non-cohesive material of the proper composition to a surface course which is to be rebuilt for use as a roadbase. Methods for determining the proportions to blend are still essentially by trial and error. They involve either estimating trial gradations, preparing test mixtures and testing for gradation and plasticity, or else estimating the proportions which will give the desired plasticity index, preparing trial mixtures, and then testing for gradation and plasticity. Some of the more commonly used guide-procedures for blending soils are as follows:

1. *Blending two materials by gradation.* The graphical procedure outlined in the chapter on Pavement Materials is normally quite adequate for this purpose.

2. *Blending two materials by plasticity index.* The proportions to be blended to give the required plasticity index can be estimated from the following equations:

$$a = \frac{100S_B(P-P_B)}{S_B(P-P_B)-S_A(P-P_A)}$$

$$b = 100 - a$$

where a = percentage of material A in final mixture,
      b = percentage of material B in final mixture,
      P = required plasticity index of final mixture,
      P_A = plasticity index of material A,
      P_B = plasticity index of material B,
      S_A = percentage passing the 425 μm sieve in material A,
      S_B = percentage passing the 425 μm sieve in material B.

3. *Blending three materials by gradation.* Figure 5.2 illustrates a triangular chart procedure which is perhaps the most commonly used method of blending three different soil materials. Let it be assumed that three materials

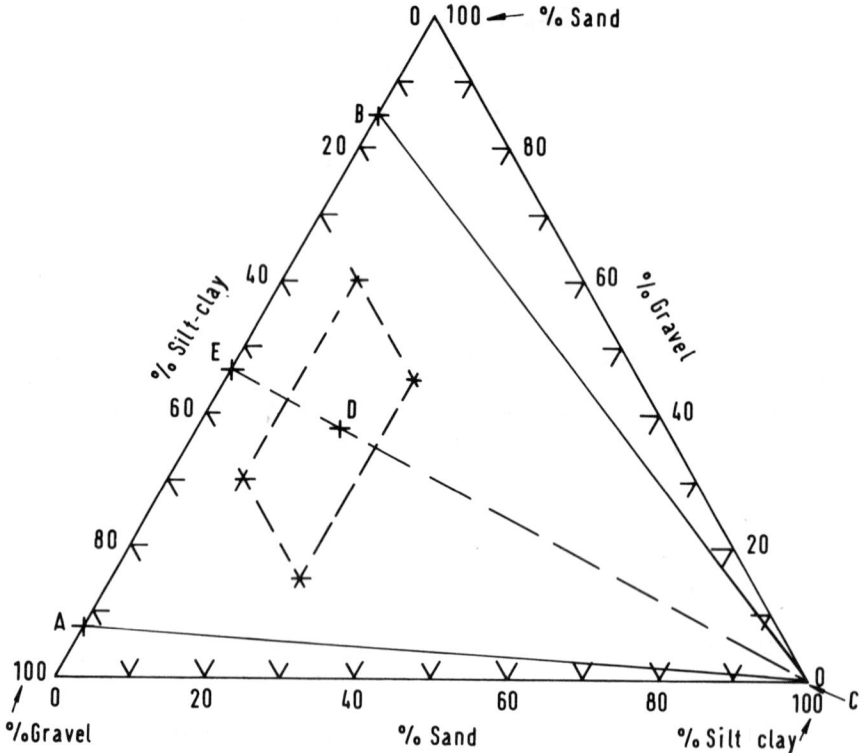

Fig. 5.2. Chart for combining three soil materials to meet a specified gradation

which are to be blended are designated A, B and C, and that their gradings are as shown in Table 5.2; their component proportions of gravel, sand and silt clay are shown in the same table. The blending procedure is then as follows:

Step 1. On the triangular chart plot the points A, B and C to represent the component fractions of each material. Join A to B to C. Note that the triangle ABC includes all possible soil combinations from the three different sources.

TABLE 5.2. *Soil gradations used in example blending problem*

Sieve size mm	Percentage passing			
	Soil A	Soil B	Soil C	Specification
25·4	95	100	100	100
19·0	70	100	100	85–100
9·52	21	100	100	65–100
4·76	11	100	100	55–85
2·36	7	85	100	40–70
0·425	2	55	100	25–45
0·075	0	0	100	10–25

(*a*) Gradations

Soil	Size fraction, per cent		
	Gravel	Sand	Silt-clay
A	93	7	0
B	15	85	0
C	0	0	100

(*b*) Component proportions

Step 2. Plot the specification limits on the chart. The parallelogram obtained includes the combinations of the three materials which will meet the specifications. Select point D at about the centre of this area and join C to D and continue it onwards until it intersects AB at point E.

Step 3. Using any convenient scale determine the measured lengths of various lines as follows:

Line	Length, mm	Line	Length, mm
AB	79·8	EC	88·9
AE	39·9	ED	16·5
EB	39·9	DC	72·4

Step 4. Determine the following ratios:

$$\frac{AE}{AB} = 0·5 \qquad \frac{EB}{AB} = 0·5 \qquad \frac{ED}{EC} = 0·186$$

Note that in this example point E represents a possible combination of materials A and B in the ratio of 50% to 50%, while point D represents a possible combination of material C with the already combined materials A and B in the ratio of about 19% C to 81% A and B.

Step 5. Determine the proportions in the final mixture as follows:

Material A $0.50 \times 80.9 = 40.5\%$
Material B $0.50 \times 80.9 = 40.5\%$
Material C $= 19\%$

Step 6. Check that the gradation of the mixture meets the specification. Thus

Sieve size mm	$A = 0.405$	$B = 0.405$	$C = 0.19$	Percentage passing	
				Combination	Specification
25·4	38·4	40·5	19	98	100
19·0	28·4	40·5	19	88	85–100
9·52	8·5	40·5	19	68	65–100
4·76	4·5	40·5	19	64	55–85
2·36	2·8	34·4	19	56	40–70
0·425	0·8	22·2	19	42	25–45
0·075	—	—	19	18	10–25

Step 7. Prepare a trial mix at this combination and determine its gradation. It may then be found necessary to adjust the proportions slightly.

Chloride stabilization

The first modern usage of chloride for stabilizing roads precedes the year 1910 and probably occurred when sea water was used as a dust palliative on roadways near the ocean. It was noticed that, by adding sea water, dust prevention was provided for much longer periods of time than when fresh water was used for this purpose, and this led to investigations which resulted in the use of chlorides as additives to fresh water in inland areas.

Types of chloride. The chlorides used to stabilize soils are calcium chloride and sodium chloride. Sodium chloride occurs naturally as rock salt in the ocean, in salt lakes and in brines present in cavities in rock formations; of these the most important source by far is rock salt which may be 99 per cent pure NaCl. Calcium chloride is obtained from these same natural sources, but in addition great quantities are obtained by refining the $CaCl_2$-rich waste liquors from the soda-ash industry; soda-ash, which utilizes salt, lime and ammonia as raw materials in its preparation, is a necessary ingredient in the manufacture of many chemicals, such as caustic soda and sodium carbonate.

Mechanism of stabilization. Chlorides stabilize roads by a number of mechanisms, and under certain conditions one or a number of these play more important parts than others. These may be outlined as follows:

1. Chlorides have excellent hygroscopic and deliquescent properties. The hygroscopicity of a material is a measure of its ability to absorb moisture from the air while a deliquescent material is a hygroscopic one which will

dissolve in moisture. Their relative deliquescence properties are indicated by the fact that calcium chloride will dissolve in moisture absorbed from the air at 25°C when the relative humidity is only 29 per cent, whereas sodium chloride dissolves in absorbed moisture at the same temperature when the relative humidity is 76 per cent. Calcium chloride is also more hygroscopic than sodium chloride; at 25°C and a relative humidity of 36 per cent, 1 kg of $CaCl_2$ will take up 1 kg of water, while at 95 per cent R.H. it is capable of absorbing 3·75 kg of water.

These two properties of the chlorides are perhaps their most important ones. They mean that in dry climates moisture which is evaporated from the roadway during the day may be replaced by moisture from the atmosphere during the night. Thus the mechanical stability given to the soil material by cohesive water bonds can be maintained, while other advantageous features resulting from having chloride brine distributed throughout the mixture can also be preserved.

2. Chlorides have the ability to lower the vapour pressure of the water in which they are dissolved. Since the rate of evaporation of a liquid is dependent on its vapour pressure, this means that moisture evaporation from chloride-treated soils takes place at a slower rate than if pure water were mixed with the soil.

3. Chlorides present in water-soil mixtures cause an increase in the surface tension of the water. This has two important effects. First of all, by increasing the surface tension of the pore water, the rate of evaporation is reduced. Some evaporation will take place, however, so the pore moisture content is reduced. This in turn causes the second effect, which is that the surface tension is further increased, and the water films may then cause the compacted soil particles to 'tighten' together, thereby increasing the unit weight and stability of the material. Allied with this latter property is the possibility of the appearance of fine cracks in the roadway, but their effects are apparently negligible in comparison with the benefits gained.

4. The addition of either of the chlorides to a soil usually results in an increase in the dry unit weight obtainable with a given compactive effort. Chloride solutions are good soil lubricants, with the result that a given soil dry unit weight can be obtained with less compactive effort than if water only is used. Calcium chloride is a better lubricant than sodium chloride.

5. Compaction in dry weather can be carried on for a longer period of time at a given moisture content, since evaporation takes place more slowly.

6. In dry weather also sodium chloride will crystallize in the surface of the roadway, thereby cementing it into a dense hard crust which is capable of withstanding considerable abrasion from traffic.

7. Perhaps the most widely known property of the chlorides is their ability to lower the freezing temperature of water. The extent to which they protect a soil against being frozen is dependent on the type of chloride used and the concentration present in solution in the pores. For instance, if the pore solution contains a 10 per cent $CaCl_2$ concentration, then freezing of the water may begin at a temperature of about $-4·5°C$. As soon as the water freezes, however, the chloride content of the remainder increases with the

result that the temperature has to be lowered before more water will freeze. Complete freezing will not occur until the $CaCl_2$ concentration of the unfrozen solution reaches 30 per cent—this is called the *eutectic* point of the solution—and the temperature has been lowered to $-51°C$. In the case of sodium chloride solutions the eutectic solution is 25 per cent and complete freezing occurs at $-19·5°C$.

8. The admixing of chlorides to a soil may change its plasticity characteristics. When calcium chloride is added, the exchange reaction which takes place may result in a reduction of the plasticity index, due to the calcium cations being preferentially adsorbed on to the surfaces of the soil particles. If sodium chloride is the admixing material, the result may be an increase in the plasticity index, due to the ability of the Na^+ ions to become highly hydrated during moist weather. In temperate climates this can mean that it may not be possible to use sodium chloride to stabilize certain soil-aggregate roadbases.

9. Although not too much is known about the extent to which this actually occurs, the presence of sodium chloride is believed to increase the solubility of limestone particles in water. Eventually the calcium ions precipitate out of solution as minute amounts of calcium carbonate. The calcium carbonate then acts as a weak cement which, in a well-graded soil-aggregate mixture, may help to bind the particles together.

Usage. A successful chloride-stabilized pavement is of necessity a mechanically stable one; chlorides are not good cementing agents and therefore the material being stabilized must already have high stability. The best results are obtained when well-graded materials are compacted to dense impervious masses, as these reduce considerably the ease with which the chlorides may be leached downwards from the upper portions of the pavement.

The maintenance of soil-aggregate surface courses in warm climates is considerably eased by the addition of chlorides, since they ensure sufficient moisture to maintain the initial stability of the mass, and prevent ravelling and the creation of 'wash-board' surfaces. In general calcium chloride is more successful at this—although initially it is more expensive than sodium chloride. Most economical usage is obtained by periodic applications of calcium chloride and shaping of the surface at regular intervals.

Use of calcium chloride in the soil-aggregate roadbases is also advantageous. If the surface course is composed of an impermeable bituminous material, then the effectiveness of adding $CaCl_2$ is only limited by the amount of ground-water movement and the length of time it takes to leach the chloride from the roadbase. If the surface course is permeable and contains $CaCl_2$, and the climate is a temperate one, then calcium chloride leached from the roadbase will be replenished to a certain extent by the downward movement of moisture through the surface course. In either case the calcium chloride in the roadbase will reduce the deleterious effects resulting from frost action in the pavement.

Details regarding chloride stabilization are available in the literature.[8, 9]

Lignin stabilization

Wood consists of cellulose and several cementing materials which bind the cellulose fibres together. In the sulphite paper-making industry, the paper mill retains the fibres and the cellulose, while the cementing materials are wasted from the process in a water solution called 'spent sulphite liquor', which is actually calcium ligno-sulphonic acid. The constituents of this liquor which are used to further stabilize mechanically stabilized roads are salts which are referred to as calcium lignosulphonates or lignin sulphonates, or simply lignins. A representative analysis of lignin will normally average about 61 per cent carbon, 28 per cent oxygen, 6 per cent hydrogen, 2 per cent sulphur, and 3 per cent calcium.

Mechanism of stabilization. Although it is used as a stabilizing agent in forest roads in Sweden, Canada and parts of the United States, not too much is known of the exact mechanism by which lignin stabilizes soils. Work that has been done, however, indicates that the addition of lignin to soils results in higher dry unit weights and decreased permeabilities. As regards the latter, sulphonates are excellent clay-dispersing agents and thus, when it rains on to a lignin-stabilized surface course, the dispersed clay particles swell and plug the pores, thereby reducing water penetration. After the lignin has cured, it becomes cementitious and binds the soil particles together; unfortunately the cement formed is water-soluble and, if not protected, may disappear with the onset of wet weather. Frost-action problems are also reduced by having lignin present in a soil; this is due both to a slight lowering of the freezing point of the soil moisture and to the increased impermeability of the lignin-stabilized roadbase. Lignins also have hygroscopic properties due to the sugars which are present. However, the moisture retention properties are limited as bacteria will attack these sugars.

Usage. As with chlorides, the most advantageous usage of lignin is with soil-aggregate pavements which are already stable. Best results are obtained when these pavements are very dense, since this prevents rapid leaching of the lignin and promotes good cementation. Although lignin will also increase the stability of poorly-graded soils, they are apt to be leached out very rapidly; in areas where lignin is readily available, however, this may make little difference. At the present time it appears that effective use of lignin is confined to hot climates and on roadways adjacent to sources of the material.

Molasses stabilization

Molasses used for highway soil stabilization purposes is a waste residue known as 'black strap molasses' which is obtained as a by-product of the manufacture of sugar from sugar cane. It is a very thick syrupy liquid which contains resinous and some inorganic constituents which render it unfit for human consumption.

Mechanism of stabilization. Black strap molasses is a hygroscopic material and this enables it to take up moisture from the air and to control the evaporation of water from the soil-aggregate pavement as it is being compacted. It is also a cementing agent; unfortunately, however, the cement formed is water-soluble, but if water can be kept away the binding action is very strong indeed.

Usage. At this time the use of molasses as a stabilizing agent is confined to tropical localities where cane sugar is produced. In these areas it is primarily used as a dust palliative and surface binder, usually in the form of surface treatments to roads which initially are mechanically stable. The outstanding limitation to its use is that it is readily soluble in water and will easily leach out of the pavement. However, if an impermeable material is used in the surface course, there would appear to be no reason why molasses could not be used to improve the stabilities of soil-aggregate roadbases.

CEMENT STABILIZATION

Of all the methods of soil stabilization now in use, that which utilizes cement as the stabilizing agent is second only to mechanical stabilization in importance and usage. Modern usage of cement for stabilization purposes can perhaps be said to date back to 1917 when a Dr. J. H. Amies took out a patent in the United States on soil and cement mixtures labelled 'Soilamines'. While usage in Britain can be said to have begun in 1920 with experiments involving the addition of cement to chalk soil in earth cottage construction near Amesbury, systematic use was not initiated until about 1939 when the (now) Transport and Road Research Laboratory followed up a study of American experience with a programme of laboratory tests and small-scale field trials on cement-soil mixtures.

Factors which have helped to make the use of Portland cement so popular as a soil stabilizer in nearly every other country in the world, are as follows:

1. Cement is readily available in most countries as a home product.

2. Cement is manufactured on such a large scale for concrete construction that its price is comparatively low.

3. Use of cement generally involves less care and control than many other methods of stabilization.

4. More information is generally available on cement-treated soil mixtures than on other types of soil stabilization.

5. Almost any soil can be stabilized with Portland cement if enough cement is used in combination with the right amount of water and proper compaction and curing.

Types of cement-treated soil mixtures

Cement-treated soil is a simple, intimate mixture of pulverized soil, Portland cement and water. All cement-treated soil mixtures are very often, and erroneously so, called soil cement mixtures. In fact, there are three different types of cement-treated soil mixtures of which soil-cement is just one. The other two are termed plastic soil-cement and cement-modified soil. It is important that the engineer should distinguish between the three so that he will know how and when each may be used most advantageously.

Soil cement. This is a hardened material formed by curing a mechanically compacted intimate mixture of pulverized soil, Portland cement and water. In other words, sufficient cement is added to the soil to harden it, and the moisture content of the mixture is adequate for compaction purposes and for hydrating the cement. Inherent in the definition is the assumption that a soil cement mixture is capable of meeting particular criteria relating to stability and/or durability.

Soil cement has a number of uses of which by far the most important is as a roadbase and/or sub-base material in roadways and parking areas, airport runways, taxiways and aprons. Another important use has been as a foundation material for large structures. It has also been used in the construction of low-cost buildings, particularly in arid climates. Soil cement is not used in road surfacings as it has poor resistance to abrasion. Hence, another superimposed material, e.g. a concrete pavement or bituminous surfacing, must be used to protect it from the wheels of traffic.

Cement-modified soil. Some soils cannot be *economically* stabilized with cement. For instance, the amount of cement that may be required to stabilize a clay soil so that it can be used as a roadbase for a major highway could well be in the order of 20 to 30 per cent. Because of the high cost which this would involve, it might be more economical to haul in some foreign roadbase material, and only add sufficient cement to modify the physical properties of the in situ soil sufficiently for it to be used as a sub-base material. Alternatively, a normally substandard granular material having relatively high plasticity and low bearing value may be made acceptable for pavement construction purposes by the addition of sufficient cement to change its properties.

In both of the above examples, the amount of cement added will be insufficient to harden the soil enough to meet soil-cement stability and durability criteria; the resultant material is *intended* to fragment under the action of traffic. In other words, the addition of the cement is primarily for the purpose of improving the physical characteristics of the soil. Usually the soils chosen for modification purposes have high water-holding capacities and volume-change characteristics which may cause excessive distortion of pavements; hence they have low supporting values, as their moisture contents vary with the weather.

Plastic soil-cement. This is a hardened material formed by curing an intimate mixture of pulverized soil, cement and enough water to produce a mortar-like consistency at the time of mixing and placing. By comparison, soil-cement is mixed and placed with only enough moisture to permit adequate compaction and cement hydration. Soil cement and plastic soil cement are similar in that they are both capable of meeting specified criteria of durability and/or stability.

Plastic soil-cement is used primarily as an erosion-control material by paving steep, irregular or confined areas, and as lining on the sides of ditches and canals. It has application also for levee and earth dam facings where the use of soil-cement might present construction problems. Plastic soil-cement has also been used to cast construction blocks for building purposes in underdeveloped areas.

In general, plastic soil-cement mixtures require about 4 percentage points more cement than soil-cement mixtures in order to meet the same criteria.[10]

Mechanism of cement stabilization

When water is added to neat cement, the major hydration products are basic calcium silicate hydrates, calcium aluminate hydrates and hydrated lime. The first two of these products constitute the major cementitious components, while the lime is deposited as a separate crystalline solid phase. When the cement is present in a granular soil, the cementation is probably very similar to that in concrete, except that the cement paste does not fill the voids between the soil particles. In other words, cementation is primarily by means of mechanical bonding of the calcium silicate and aluminate hydrates to the rough mineral surfaces. When cement is used to stabilize a fine-grained soil, cementation becomes a combination of mechanical bonding and chemical bonding which involves a reaction between the cement and the surfaces of the soil particles.

As the silt and clay content of a soil is increased, the particle surface area is also much increased, and so the opportunity is much greater for additional chemical reactions to take place between the hydrated cement and the particle surfaces. The free lime given off during hydration is believed to play a prominent part in these reactions. First of all, it is probable that there is an exchange reaction with cations already adsorbed on to the particles. In addition, the lime reacts with the soil silica (and alumina) to form secondary hydrous calcium silicates (and aluminates) on the silica particle surfaces. (These reactions are explained in more detail in the discussion on lime and lime-pozzolan stabilization.) The effect of the cation exchange reaction is to change the plasticity properties of the soil. The secondary silicates are cementitious and further contribute to inter-particle bonding. The greater the fines content of the soil and the more reactive the clay minerals, the more important are both of these secondary reactions.

The manner in which cement stabilizes fine-grained soils can perhaps be demonstrated by visualizing the grains of cement as a nucleus to which the fine soil particles adhere. As the cement in the soil is increased, the quantity of free silt and clay is progressively reduced and a coarse-grained material of lower water-holding capacity and increased volume stability and supporting value is obtained; this is a cement-modified soil. As more and more cement is added, the quantity of coarse-grained material is increased until the point is reached where all the soil grains remain in a solid mass as befits a structural material; this is soil cement.

Factors affecting cement-soil mixtures

It is literally impossible to list and discuss all the factors which affect the physical properties of cement-treated soil mixtures, for there are an infinite number of soils and an infinite number of combinations. While the literature on this subject is tremendous, there is one American reference[11] which is outstanding in summarizing data in this area prior to 1960, and another British one[12] which summarizes work in this country prior to 1968. The following is concerned with those factors which primarily affect cement-soil mixture characteristics. They are:

1. Nature and type of soil.   4. Mixing and compaction.

2. Amount and type of cement.   5. Curing conditions and length.

3. Moisture content.   6. Use of chemical additives.

Soil. The nature and type of soil is perhaps the single most important factor affecting the stability of a cement-treated soil. Any soil can be stabilized if sufficient cement is added, but the cement content required to achieve a given stability must normally be increased as the silt and clay content is increased, other factors remaining constant. This means that it may not be economically possible to stabilize with cement certain soils such as heavy clays; whether it is or is not so is dependent on an economic analysis of this and other, alternative, methods of construction. Excessive amounts of clay-size particles may cause considerable problems when pulverizing, mixing and compacting cement-soil mixtures. American experience indicates that sandy and gravelly soils with about 10 to 35 per cent silt and clay have the most favourable characteristics for soil cement construction. Experience in Britain is that soils with liquid limits greater than 45 and plasticity indices over 20 cannot normally be economically utilized in soil-cement construction. The upper grading limit for soils considered suitable for cement stabilization is shown in Fig. 5.3; materials coarser than this are unsuitable because of possible damage to mixers during he mixing process, and because they present construction problems such as their liability to segregate and the difficulty of controlling levels. (Also shown in Fig. 5.3 are the gradation limits for what are known in Britain as Lean Concrete and Cement Bound Granular Materials; these particular soil cement mixtures are discussed separately in some greater detail in the chapter on Flexible Pavements.) Soils that are poorly graded may require uneconomically high cement contents, and to overcome this it is practice in Britain to specify that a soil to be stabilized with cement should be well-graded and have a uniformity coefficient of not less than 5, i.e. the ratio of the particle size for which 60 per cent is finer to the particle size for which 10 per cent is finer ⩾ 5.

The type of clay mineral is also of considerable importance in soil stabilization in that the expansive types of clay are more difficult to stabilize. This particular aspect is discussed in greater detail in relation to lime stabilization.

Organic matter is almost invariably present in the surface layers of soil and may extend to a depth of 1·5 m in well-drained sandy soils. Its presence very often renders a soil unsuitable for stabilization with cement — which is one reason why it is necessary to strip *at least* the top soil before attempting this form of stabilization. Soils with B and C horizons which can fall into this

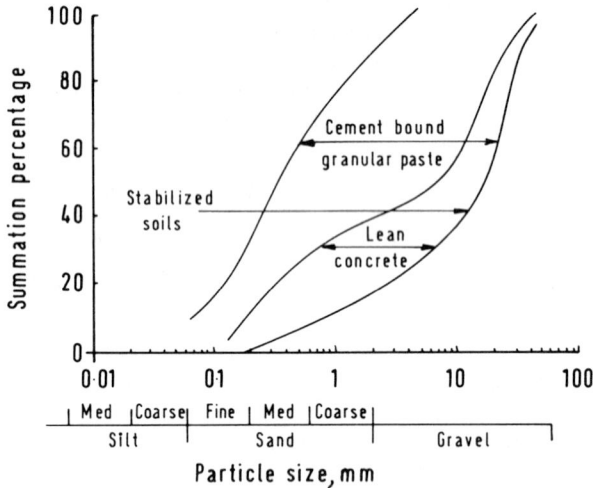

Fig. 5.3.   Grading limits for different types of cement-bound materials used in Britain[12]

category tend to have acidic podzolic profiles and were formed under wooded vegetative conditions. They can be easily detected by making up a cement-soil paste using 10 per cent Portland cement and determining the pH value of the paste 1 hour after the addition of water. If the pH is less than 12·1, the soil should be rejected, as this indicates the presence of organic matter capable of hindering the hardening of the cement.[12]

The presence of calcium sulphate in the soil water also influences its suitability for cement stabilization[13]. Research has shown that a soil cannot be economically stabilized with cement if it contains a sulphate content in excess of about 0·25 per cent. In Great Britain, sulphates are rarely found in freely draining soils since any that did exist would be removed by leaching; they can be frequently encountered in clay soils at depths (due to leaching) in excess of 0·75–1 m. Practically the hazard arises from the possibility that the sulphates may be transported by the seasonal movement of the ground water from an area below the water table into stabilized soil above it. Unsuitability for stabilization is due to the ability of the sulphate to combine with the tricalcium aluminate in the hydrated cement and produce calcium sulpho-aluminate. This new mineral expands to occupy a greater volume than the reactants from which it is formed, thereby breaking the cementitious bonds in the soil cement.

Cement. Any type of cement may be used to stabilize soils, but normally ordinary Portland cement is used. A high-early strength cement may be used if high strengths are required quickly, or if the soil is organic contaminated (it gives higher strengths). Sulphate resistant cement can be of limited value in the stabilization of materials containing sulphates since, if clay is present, reaction can still occur between the sulphates, clay minerals and lime.

It is now usual practice in Britain to specify the desired stabilities of most soil-cement mixtures in terms of minimum unconfined strengths and not, as formerly, in terms of minimum cement contents. In other words, it is up to the contractor to add as much cement to the soil as is necessary to achieve

TABLE 5.3. *American recommendations with respect to the amount of cement required to stabilize B and C horizon soils*[14]

A.A.S.H.O. soil group	Usual range in cement requirement	
	% by vol.	% by wt.
A-1-a	5 to 7	3 to 5
A-1-b	7 to 9	5 to 8
A-2	7 to 10	5 to 9
A-3	8 to 12	7 to 11
A-4	8 to 12	7 to 12
A-5	8 to 12	8 to 13
A-6	10 to 14	9 to 15
A-7	10 to 14	10 to 16

the specified strength at some minimum state of compaction. The most recent specifications[4] for soil cement require a minimum 7-day value of $2 \cdot 76 \, MN/m^2$ for moist-cured cylindrical specimens having a height/diameter ratio of 2:1 and $3 \cdot 45 \, MN/m^2$ for cubical specimens.

In the United States the desired cement content is normally selected to meet durability, i.e. the implied assumption is that strength needs will automatically be met, and not, as in Britain, to meet stability criteria, i.e. the implied assumption is that durability needs will also be met. The American requirements often result in cement contents of the order shown in Table 5.3.

It is interesting to note that the use of a minimum compressive strength to select a suitable cement content was considered and discarded by the

Americans on the grounds that cement contents obtained in this way are materially greater than those which are actually required to give satisfactory results for most soils in their country.[10] This in turn materially increases construction costs—the cement constitutes about one-half of the total cost of soil cement construction in the U.S.—and thereby defeats the primary aim of this work, which is low cost. In contrast, the British approach is reported to result in a stabilized material having more reliable properties, which cause it to perform more satisfactorily under heavy traffic conditions. In addition, specification on a strength basis is stated to result in some saving in the amount of cement required for a given mixture, due to the contractor having considerable freedom as to the manner in which the construction process may be carried out, as a result of which he is able to experiment with and use improved processing methods and equipment.[15]

Moisture. It is necessary for water to be present in a cement-soil mixture in order to hydrate the cement, to improve workability and to facilitate compaction. This water should be relatively clean and free from harmful amounts of alkalies, acids or organic matter.

The amount of moisture present in the mixture has a considerable effect on the strengths and unit weights obtained with a soil-cement mixture. For example, fully hydrated cement takes up about 20 per cent of its own weight of water. At high soil moisture contents, the cement has no difficulty in obtaining this water, but if the water content is decreased the cement has to compete with the soil for moisture; if the material has a high suction potential, it may have a greater affinity for the moisture, the cement will not fully hydrate and the strength will be reduced from what it should be.

As with natural soils, cement-treated soils exhibit the same type of moisture-unit weight relationships, i.e. for a given compactive effort there is one moisture content which will give a maximum dry unit weight. It is important to realize that the optimum moisture content for maximum dry unit weight is not necessarily the same as that for strength. This is illustrated in Fig. 5.4, which further shows that the optimum value for strength may vary, depending on the manner and length of time the test specimens are cured prior to testing. In general the optimum moisture content for maximum strength tends to be on the dry side of the optimum for maximum dry unit weight for sandy soils, and on the wet side for clayey soils. For the clayey soils, the location of the optimum value for strength is dependent not only on the amount of clay present in the soil, but also on the type of clay mineral.

Mixing and compacting. In general, the more efficiently cement, water and soil are mixed, the greater the stability and durability of the soil-cement product. Alternatively, high mixing efficiencies will result in lower cement contents in order to achieve a given soil cement field strength.

Increasing the moist mixing time and/or delaying compaction after ending the moist mixing generally results in an increase in the optimum

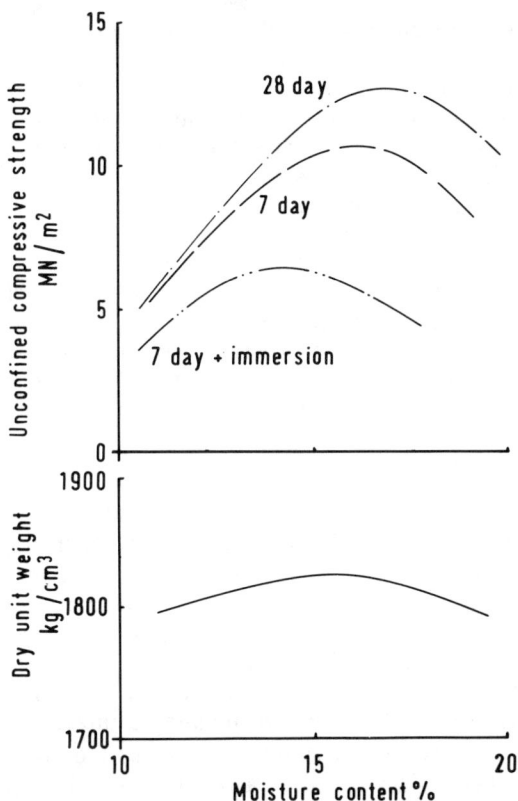

Fig. 5.4. Influence of moisture content on dry unit weights and strengths of soil cement mixtures composed of 20 per cent cement and a 50/50 mixture of dune sand and montmorillonitic clay[16]

moisture content for maximum dry unit weight of the cement-soil mixture, while its durability and unconfined compressive strength are decreased. The extent to which each occurs in any particular instance is dependent on the type of soil, the period of mixing and/or delay, and the amount of cement present. The reduction in unit weight can be attributed to the bonding of particles as the cement hydrates during mixing, with the result that the particles cannot slide over each other as easily during compaction. The reductions in strength and durability are undoubtedly due to the rupturing of these same bonds during the mixing and compacting processes. For these reasons, many specifications place an upper limit on the length of time between when moisture is added to the mixture and when compaction is complete. For instance, in Britain the current specifications require that compaction be completed within 2 hours of mixing being initiated.

For a given soil-water-cement mixture it can be generally said that the greater the unit weight, the more stable and durable is the hardened structural material.

Curing conditions. The environmental conditions under which curing takes place have a considerable effect also on the extent to which a soil may be stabilized with cement. As illustrated in Fig. 5.4, the strength of soil cement increases with age; this strength gain is similar to that which is achieved with concrete and continues for many years. Like concrete also, soil cement must be moist-cured during the initial stages of its life so that moisture sufficient to meet the hydration needs of the cement can be maintained in the mixture.

Unlike with concrete, the temperature of curing has a considerable effect and higher strength results are obtained at higher curing temperatures. This factor—which is due to a pozzolanic reaction between the clay and the lime released by the cement during hydration—can be of economic importance in warm-weather countries when the design criterion is based simply on compressive strength. For instance, in a country where the compacted soil-cement is able to cure at an average temperature of, say, 38°C, lower cement contents will be required to achieve a given strength over a period of time than in Britain where the average summer temperature is about 15°C.

Additives. Work on the evaluation of chemical admixtures to soil-cement mixtures has been carried out for many years and, depending on the reasons for which they are used, several very successful additives have been discovered. For instance, the detrimental effects associated with organic soils can very often be negated by the pre-admixing of up to 2 per cent hydrated lime or 0·5 per cent calcium chloride. With highly plastic soils it is common practice in this and other countries to pre-mix a small quantity of lime to facilitate pulverization of the soil and the later admixture of cement.

Trace chemicals have been found which when added to soil cement mixtures can dramatically increase the early strengths obtainable. These can make it possible in particular instances to reduce the amount of cement required to obtain a given design strength. Where extra high early strength is not the principal objective, a significant reduction in cement content may be obtained without loss of long-term strength by the addition of a reactive fly ash to the cement-soil mixture.[17]

Construction

The objective in soil cement construction is to obtain an intimate mixture of the specified amounts of soil, cement and water, and then to compact this mixture so as to achieve a high dry unit weight. The main types of equipment used to mix the materials are travelling mixing machines, rotary mix-in-place equipment and stationary mixing plants.

There are two types of travelling plant machines. Both types, known as the windrow type and the flat type, perform the mixing of cement, soil and water in a single operation. With the *windrow* type, the prepared soil—best results are obtained with a friable material—is bladed into a windrow, i.e. a low accumulation of material of perhaps 200–300 m long with a trapezoidal cross-section. After cement has been added on top of the soil, the travelling plant passes over the windrow, picks up both materials, mixes them dry, adds water, mixes them again, and then drops the mixture to form a wind-row of the moist-mixed material behind it as it moves along. The mixture is then spread and compacted. With the *flat* type of travelling plant, the procedure is the same except that the soil is not pre-accumulated in a windrow, but instead the travelling plant picks up the cement and *in-situ* soil, mixes it, adds water, completes mixing and drops the moist mixture behind it. Both types of travelling plant machine have the particular advantage that the soil cement material is mixed and placed before much cementation can take place.

Rotary mix-in-place mixers are different from the flat type of travelling plant mixers in that they are used to carry out the mixing in a number of operations instead of just one. First of all the rotary mixer is used to dry-mix the *in-situ* soil and cement. Moisture is then added in a number of incre-ments, usually by a water pressure distributor, until the desired moisture content is obtained. After each increment is added the cement-soil-water mixture is again mixed by the rotary mixer. When enough moisture has been added and the cement, soil and water are thoroughly mixed throughout the full depth and width of treatment, the material is compacted and finished. This method of mixing has the advantage that placement of the water is more easily controlled, it is easier to compensate for the moisture lost through evaporation, and a more uniform mixture is obtained. The principal disadvantage is that the moist mixing may take a long period of time, with consequent detrimental effects to the soil cement.

Stationary mixing plants are the types which are probably most used in Britain. Mixing in these plants and then transporting the mixtures to the sites results in a relatively expensive form of construction, and so stationary mixing plants are best used on projects which involve the use of coarse-grained borrow soils and where the haul time to the construction site is short. At stationary plants, the proportioned amounts of cement, soil and water can be mixed in either batch-type or continuous-flow mixers.

Procedure steps. Regardless of the type of equipment used, the *mix-in-place* construction procedure generally involves the following step-by-step operations:

Step 1. Carry out initial preparation. Before actual construction can begin the subgrade or sub-base must be compacted and shaped to the proper crown and grade. This is very important as the shape existing at the start of construction determines to a large extent the final shape of the carriageway.

Step 2. Pulverize the soil. The first actual step in construction is to scarify and pulverize the soil until at least 80 per cent passes the 4·76 mm sieve. (This step may or may not be required, depending on the mixing equipment used and/or the soil type.) When the loose soil is considered to be in a suitable state, it is bladed and shaped to an approximate crown and grade.

Step 3. Spread the cement. The cement may be spread either by mechani-cal spreaders or by hand. The mechanical machines are used to uniformly spread bulk cement, while manual means are used to place and then spread cement in bags. If rain falls during the spreading operation, the spreading should be stopped and the cement cut into the soil mass immediately.

Step 4. 'Dry-mix' the soil and cement. The object of this exercise is simply to distribute the cement throughout the soil mass. The primary aim at this stage is not to achieve complete mixing but only to mix enough to prevent cement balls from forming when more water is applied. The soil need not be absolutely dry for dry-mixing to take place. In fact sandy soils can be relatively easily dry-mixed when the moisture content is as much as 2 per cent above the standard Proctor optimum moisture. Clayey soils will not mix readily with cement when the moisture content is much greater than 3 per cent below the same optimum moisture. Normally mixing takes place to a depth sufficient to provide a soil-cement layer thickness of 75–200 mm.

Step 5. Add water and wet-mix. If the moisture content of the cement-treated soil is less than the design value at the end of dry-mixing, additional moisture should be added and the material remixed as necessary.

Step 6. Compact and finish. Compaction should commence as soon as possible after mixing is ended, and it should be complete within as short a period as possible from when the cement is added to the soil. British practice, which is aimed at achieving satisfactory compaction, is to require the output of the plant to be such that a minimum rate of 18·25 m/h of completely stabilized material is maintained. Where it is necessary to employ more than one layer of stabilized material, each successive layer should be placed and compacted within an equally short time of the completion of compaction of the layer below. Special care should be taken to ensure that full compaction is obtained at the pavement edges and in the vicinity of all transverse and longitudinal construction joints.

Depending on the compaction equipment used, the finishing process involves to a greater or less degree the removal of all surface irregularities and the final shaping process of the compacted layer.

Step 7. Protect and cure. Once finishing is completed, the soil-cement mixture must be protected during the curing period. Curing can be achieved by:

(a) Covering the layer with an impermeable plastic sheeting.

(b) Spraying the surface of the layer with a quick-breaking emulsion. This method has the particular advantage that not only is sufficient moisture retained in the layer to help hydration of the cement, but in addition the bituminous binder acts as a prime course (in the case of a roadbase) for the bituminous surfacing which is placed later.

(c) On very small schemes, the cement-treated layer can be covered with at least a 76 mm thick layer of suitable soil, which is kept in a damp condition by periodic spraying with water during the curing period, and then completely removed on completion of curing.

During cold frosty weather, the soil-cement mixture should be protected by a layer of straw or soil which is at least 76 mm thick and covered with an impermeable sheeting which should be maintained in position during the cold period.

LIME AND LIME-POZZOLAN STABILIZATION

Lime

Strictly defined, lime is calcium oxide (CaO), but in practice the term is also used to describe both the oxides and hydroxides of calcium and calcium magnesium, i.e. quicklimes and hydrated limes formed from calcitic and dolomitic materials. As this implies, there are many varieties of lime.

Types of lime. Lime is most often produced by calcining limestones, but chalk and oyster-shells are also calcined for this purpose. If the limestone being treated is a pure or near-pure calcium carbonate ($CaCO_3$), the lime produced is described as being a *high-calcium* or *calcitic lime*, or simply as a 'fat' lime. If the stone being calcined is a dolomitic limestone containing a high proportion of magnesium carbonate ($MgCO_3$), the lime products are termed *dolomitic* or *high-magnesium limes*. While the British Standard specification on building limes[18] does not differentiate between calcitic and dolomitic limes, it is generally accepted that high-calcium limes contain less than about 5 per cent magnesium oxide, while most dolomitic quicklimes contain between 25 and 45 per cent MgO.

Hydraulic limes—these can be of either the calcitic or the dolomitic variety, and may or may not be hydrated—can be considered as being intermediate between the above 'normal' limes and ordinary Portland cements. They are obtained by calcining impure limestones containing up to about 35 per cent clay. If the rock has a significant clay content, but less than 15 per cent, then the lime produced is considered to be a semi-hydraulic material. If the clay content is greater than 15 per cent, the lime is termed an eminently hydraulic or, more simply, a hydraulic lime. During the normal calcining of argillaceous or siliceous limestones, the calcium carbonate is caused to dissociate by the heat, following which solid reactions take place between the calcium oxide which is formed and the silica, alumina and ferric oxide components of the clay. This leads primarily to the formation of the dicalcium silicate and tricalcium aluminate compounds which give to the hydraulic and semi-hydraulic limes their cement-like properties, while they still retain the plastic properties of limes.

Quicklimes. These are normally produced by direct calcination of crushed limestones in either shaft (vertical) kilns or rotary (near-horizontal) kilns. In the case of pure calcite, the following reaction takes place

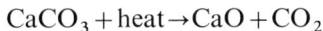

$$CaCO_3 + heat \rightarrow CaO + CO_2$$

At atmospheric pressure and at a heat of 900°C, the lime product of the above reaction is termed calcitic or high-calcium quicklime.

If a dolomitic limestone is calcined, dolomitic quicklime is produced in the following manner. At atmospheric pressure and a heat of 730°C the dolomitic rock decomposes to form magnesium oxide, carbon dioxide and calcium carbonate. At 900°C, the $CaCO_3$ also decomposes and the end-product is MgO and CaO, which is dolomitic quicklime.

A significant feature of quicklime which retards its usage in soil stabilization is that its highly exothermic hydration reaction and causticity can cause destruction to living tissue. Hence, particularly in windy weather, construction workers without adequate protective clothing may severely damage their eyes and skin when working with quicklime. Quicklime has, however, the economic advantage over hydrated lime of being cheaper while at the same time containing as much as 25 per cent more CaO or MgO: this saving may be, however, at least partially offset by the increased cost of handling the more active quicklime.

Hydrated limes. The term 'slaking' is used in the lime industry to refer to the combining of quicklime and an *excess* amount of water in order to produce a lime slurry or a lime putty of varying degrees of consistency. In contrast to this, the term 'hydration' is used to describe the addition to quicklime of *sufficient* moisture to satisfy its affinity for water under the hydration conditions; the net result is a dry powdered lime product.

Before quicklime is hydrated the lumps are usually ground to particles less than 13 mm in size. The high-calcium quicklimes are then able to react readily with water to produce calcitic hydrated lime in which all the calcium oxide is converted to calcium hydroxide with the evolution of heat. The MgO component of dolomitic quicklime, on the other hand, does not hydrate so readily at the temperatures, atmospheric pressure and short retention times used in the ordinary lime hydration process. Thus, dolomitic monohydrate lime is obtained after the following reaction has taken place:

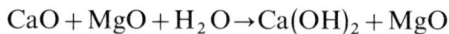

$$CaO + MgO + H_2O \rightarrow Ca(OH)_2 + MgO$$

Since about 1940 it has been found possible to hydrate dolomitic quicklime more fully under special conditions incorporating higher temperatures and pressure and longer retention times. With the MgO being hydrated to $Mg(OH)_2$, these pressure-hydrated limes are known as dolomitic dihydrate limes. These are not suitable for use in road construction, and so will not be further discussed here.

Although hydrated lime is considerably less dangerous to handle than quicklime, it is still a caustic material and can cause some damage to the eyes and skin. Both hydrated limes and quicklimes release heat upon contact with water, but quicklimes release much more heat because of the highly exothermic hydration reactions.

Pozzolans

A pozzolana or pozzolanic material is a siliceous or siliceous and aluminous material which in itself possesses little or no cementitious value, but will, when in finely divided form and in the presence of moisture, chemically react with lime at ordinary temperatures to form compounds possessing cementitious properties. The term pozzolana is a derivation from Pozzuoli, the name of a port near Naples in Italy. It was there that the early Romans found that a local volcanic ash formed a stabilized product when mixed with sand and lime.

The pozzolanic materials can be roughly divided into natural and artificial pozzolanas. The most active natural pozzolanas are clayey or siliceous materials of volcanic origin such as tuff and trass, which are characteristically glassy or non-crystalline in structure. While these form important road materials in volcanic countries, they are of no significance in Britain. As will be described later, certain fine-grained soils are capable of reacting with lime in the presence of moisture and so these can also be classed as natural pozzolanas. What may be of particular importance in this country are the artificial pozzolanas, which are predominantly waste products of industry. While ground bricks, pulverized blast-furnace slag and burnt shale all fall into the category of artificial pozzolanas, the pozzolanic material which has potentially the greatest possibility as a road construction material is pulverized fuel ash (P.F.A.).

Pulverized fuel ash. This may be defined as solid material extracted by electrical or mechanical means from the flue gases of boilers fired with pulverized coal.[19] In the scientific literature this material is also referred to as ' fly ash '.

Fly ash is a waste product which accounts for approximately three-fourths of the residue obtained from burning pulverized coal in the generation of electricity. During recent years considerable research work has been initiated in order to find economic outlets for this material. So far, in Britain, the main outlets for fly ash have been in the making of concrete, in the construction of highway embankments and in the patent processes for manufacturing lightweight sintered aggregates and building blocks. It has been found, for instance, that fly ash of an acceptable chemical and physical composition can be used as a partial replacement—up to 50 per cent has been used successfully—for ordinary Portland cement in the production of less expensive concrete. The facility with which compacted P.F.A. will harden with age to produce a material of significant bearing capacity, its resistance to settlement, and its free-drainage and frost-resisting characteristics tend to make fly-ash fill superior to many of the more conventional embankment materials. However, the use of fly ash alone, lime and fly ash, and cement and fly ash as cementing agents in the stabilization of soils is another economic outlet which so far has had little consideration in Britain.

Properties of P.F.A. One of the major drawbacks to the use of fly ash in highway construction has been the variability in the properties, and hence behaviour, of the fly ashes obtained from different power stations. These variations are now being minimized, however, and there is a growing uniformity of fly ash supplies as production is concentrated upon more efficient ' base-load ' power stations.[20] Basically the primary ingredients of fly ash are finely divided, glassy spheres of silica and alumina. Residual unburned carbon is usually also present, together with inclusions of unfused ash. Magnetic and non-magnetic iron compounds, some alkali, and limited amounts of water-soluble materials are also present. Calcium oxide derived from the burning of limestone present in the original coal may also occur alone or in combination with the other ingredients in the ash. Differences between fly ashes usually reflect the extent to which these various ingredients are present, and the physical sizes and forms of the ash components.

The quantity of unburnt carbon present in a fly ash depends on the efficiency with which the coal is burned. The carbon exists in the fly ash as irregular, porous, coke-like particles. A high carbon content increases the moisture requirement of any mixture in which the ash is used, results in lower dry unit weights, reduces the amount of reactive surface area, and physically limits the contacts of the cementitious materials. Hence it is generally accepted that fly ashes with high carbon contents are poor quality construction materials; an often quoted criterion is that a good fly ash does not contain more than 10 per cent by weight of carbon.

The importance of particle size is indicated by the fact that chemical analysis of different size-fractions of fly ashes have indicated that the reactive non-combustible materials, i.e. silica (SiO_2), alumina (Al_2O_3), and haematite (Fe_2O_3), tend to be concentrated in the finer fractions of the ashes. As a result, it is accepted by many authorities as a rule-of-thumb that in a good quality fly ash at least 80 per cent by weight of the particles should be smaller than 42 microns.

Sometimes fly ashes may be obtained which have retained latent electrical charges which cause the particles to repel each other, thereby causing bulking of the materials. These charges are difficult to dispel and fly ashes retaining them may have to be rejected for use in road construction, due to the very considerable difficulties encountered when compacting them.

Mechanism of stabilization

When lime or a mixture of lime and fly ash is added to a soil, a number of reactions begin taking place. All of these have not yet been isolated, and so the following discussion is confined to the ones which are generally considered to be most understood.[21]

Particle aggregation. When lime and a moist clayey soil are mixed in a loose condition, an almost immediate flocculation effect is observed; this is reflected in a lowering of the liquid limit of the soil, the raising of the plastic and shrinkage limits and a lowering of maximum dry unit weight. This alteration of soil character is due to most probably, a number of reactions. Firstly, a cation-exchange reaction occurs with the bi-valent calcium ions supplied by the lime replacing less firmly attached monovalent ions in the double layer surrounding the clay particles. This tends to decrease the thickness of the double layer, and to depress the zeta potential (a measure of the effectiveness of the particle negative charges in repelling a second particle), thereby causing flocculation. As the clay flocculates under this action, water which has become separated from the particles concentrates in the cavities between the coagulated particles. A clay flocculated by the action of excess calcium ions will thus absorb more water in the capillaries between the particle aggregates before swelling occurs than will a clay with predominantly monovalent ions.

In their natural condition many clays, of course, are already saturated with adsorbed calcium ions and thus observed flocculation effects following lime treatment could not result from an exchange mechanism. Under these conditions it has been postulated that flocculation must be due to a 'crowding' of additional cations onto the clay particles, thereby again depressing the zeta potential.

At the same time, it is not unlikely that a pozzolanic reaction (discussed later) is initiated between the lime and the clay, and this also contributes to soil aggregation.

These reactions suggest that the modification of a clay soil by the addition of lime is clearly dependent upon the nature of the dominant cation originally adsorbed in the double layer, and also upon the type of clay. For example a sodium-dominated montmorrillonite (which has a high cation exchange capacity) will require a relatively high addition of lime to achieve calcium saturation and the full potential flocculation effect. In contrast, a kaolinitic clay will require a considerably lesser amount of lime to achieve the full potential effects.

Pozzolanic reaction. A second chemical reaction which occurs when lime is added to a moist cohesive soil is the one which results in the slow, long-term cementing together of the soil particles at their points of interaction. This reaction, which is still not fully understood, possibly commences simultaneously with the flocculation process. It is believed to be a pozzolanic reaction which involves the lime and the alumina and silica present in the soil in the formation of hydrous calcium aluminates and silicates which are similar to the reaction products of hydrated cement. Thus, strength gain due to cementation is generally very much dependent on the amount and type of clay mineral present in the soil. For instance, montmorillonitic soils, which contain large amounts of reactive silica, may require several years before the pozzolanic reaction is completed and equilibrium reached; on the other hand, soils such as coarse silts and sands, which contain low amounts of reactive silica and alumina, react very little with lime. In the latter, the cementitious strength gains are usually very low and inadequate for highway purposes, and hence lime is rarely used to stabilize granular soils.

The pozzolanic reaction between the lime and the soil is considerably influenced by the amount and type of lime used. In general, strength increases as the lime content is increased, but for a given curing period there may be an optimum lime content; the exact content in any instance is influenced by the amount and type of clay present in the soil. Dolomitic limes can be more effective than calcitic ones in producing high strengths with 'average' soils; the reason why this is so cannot yet be fully explained, but it is believed to be related to the ability of the unhydrated MgO to act as an accelerator of the pozzolanic reaction. Quicklimes are more effective than hydrated limes in producing strength gains. Hydraulic limes generally tend to give higher strengths than non-hydraulic limes. Again, while this reaction mechanism is not completely understood, it is probable that it is a combination of the gel-formation reaction which occurs when a cement is hydrated, and the pozzolanic reaction between the hydrated lime present and the clay particles.

The strength of a lime-soil mixture will normally increase with age until all the free lime is used up. High temperatures are very beneficial to the rate of strength increase; this is one of the major reasons why lime stabilization is used frequently in hot climates. The moisture content at which the soil and lime should be mixed can be very critical. Research has shown that the optimum moisture content for maximum strength is not necessarily the same as that for maximum dry unit weight. With clayey soils, the optimum moisture content for strength tends to be on the high side of the value for dry unit weight, while with silty soils the opposite may be true. Furthermore, not only may the dry unit weight and strength optimums differ significantly from each other, but the strength optimums may vary for different curing periods.

Reaction with pulverized fuel ash.   As has been noted above, the amount of cementation developed in a soil-lime mixture is related to the reactivity and amount of pozzolanic material which occurs naturally in the soil. If the reactive pozzolana content is low, as with sandy soils, both the rate of strength gain and the strength ultimately attained will be low. In such soils both the amount and rate of cementation can be improved by adding to the soil-lime mixture a highly pozzolanic material such as fly ash which will react preferentially with the free lime.

Pozzolanic materials are generally considered to have no cementitious properties of their own. In the case of fly ash, however, this assumption is not always entirely true. Exploratory studies have shown that, in fact, some fly ashes produce cementation when mixed with soil without the addition of lime. Further studies have indicated that this is due to some fly ashes containing significant quantities of water-soluble lime (CaO and MgO) and calcium sulphate which become involved in pozzolanic reactions.

The choice of which fly ash to use with a particular granular soil, in combination with lime, is a most difficult one to make because the cementation effects are difficult to predict from the physical and chemical characteristics of the ashes. A quick test has, however, been suggested for the rapid evaluation of a fly ash[22] to consider its suitability as a soil-stabilizing agent.

Carbonation. A third reaction is that of carbonation. Carbon dioxide from the air and rainwater converts the calcium and magnesium oxides and hydroxides back to their respective carbonates. These carbonates are weak cements which at one time were considered to be a principal source of strength in lime-stabilized soils; now, however, it is accepted that the strength gain due to carbonation is minimal in comparison with the detrimental effects resulting from it.

Carbonation is particularly noticeable in industrial areas where the carbon dioxide content of the air is considerably higher than in rural areas. Here it is found that lime-soil and lime-fly ash-soil mixtures are likely to develop lower strengths than might be predicted by the laboratory tests on mixture specimens cured under rigidly controlled and protected conditions. This is mainly because lime, which would normally take part in the pozzolanic reactions, is now unavailable for this purpose due to carbonation having taken place.

Practically, the carbonation problem means that lime should be specially protected while in storage and in shipment prior to being used in the field. During construction it is advisable to protect the mixture from prolonged exposure to the atmosphere, and so compaction should take place as soon as possible.

Usage

At the present time, usage of lime and lime-fly ash as road building materials is still only in the experimental stage in Britain. Its main usage is in the United States where it is employed for subgrade modification and subbase and roadbase construction purposes. It is interesting to note that roughly 80 per cent of lime-soil construction is in the warm southern states of Texas and Louisiana, thereby reflecting to certain extent the practical importance of temperature in the hardening of lime-soil mixtures; it should also be pointed out, however, that both of these States have highly reactive montmorillonitic soils. Usage of lime-fly ash as a stabilizing agent has been mainly confined to road and car park schemes within economical hauling distances of power stations from which suitable fly ashes can be obtained.

At this time there is no universally accepted design procedure which caters specifically for the design of lime and lime-fly ash stabilized bases. It is generally accepted, however, that a suitably stabilized lime-soil or lime-fly ash-soil mixture will, cm for cm, substitute for a mechanically stabilized mixture of equal stability as measured by the California Bearing Ratio test.

The procedures used in the construction of soil-lime and soil-lime-fly ash roadways are very similar to those utilized with soil-cement and so need not be repeated here. One point that is worth emphasizing, however, is that, since the lime-pozzolan reaction is a relatively slow one, the time allowed between the addition of the cementing agent and the compaction of the mixture is normally not as critical with soil-lime and soil-lime-fly ash as it is with soil-cement.

Perhaps the greatest future for lime in temperate climatic areas as in Britain is as a soil *modifier* where the emphasis would be on improving otherwise unusable soils. One recent study[23] indicated that the early strength and deformation properties (cone index, CBR, shear strength, and modulus of deformation) of uncured lime-soil mixtures compacted wet of optimum were substantially improved relative to the untreated natural soils. It can be therefore deduced that the effects of the immediate lime-soil reactions of cation exchange, flocculation, and agglomeration should be sufficient to greatly improve poor in-situ soil conditions for purposes such as expediting construction, providing increased subgrade support, and constructing temporary roads.

BITUMINOUS STABILIZATION

Bituminous materials are believed to have been first used for modern stabilization purposes as dust palliatives on natural soil roads in Southern California in 1898. Interest in their usage as stabilizing agents grew particularly in the 1920's and 1930's, so that by the end of the Second World War a voluminous amount of literature was already available relative to bituminous soil stabilization knowledge and practice.[24] In spite of this wealth of early knowledge and some intensive research which has been carried out since then, use of bituminous-stabilized soil is limited at the present time to the construction of lightly trafficked roadways in dry areas where coarse mineral aggregates are very scarce. One major reason why usage is still confined is that, in spite of the many investigations which have been carried out to-date, there is still a tremendous lack of knowledge regarding the exact influences of the many variables involved in stabilizing a soil with a bituminous material.

Mechanism of stabilization

When a bituminous material is dispersed throughout a soil, there are two beneficial features which may occur. First, and perhaps most important, it will waterproof the soil, thereby maintaining already existing strength. Secondly, it may act as a cementing agent by binding soil particles together. Obviously, in many instances, a combination of these two mechanisms will occur.

Waterproofing action. How a bituminous material waterproofs a soil may be explained by a combination of two theories. According to the *Plug Theory*, the bituminous globules literally act as plugs or stoppers in the soil void-spaces, thereby removing the flow channels along which surface water might enter and mix water might leave the soil. The *Membrane Theory* visualizes thin bituminous films as coating individual soil particles or aggregations of particles, thereby producing the same result of waterproofing the soil, but in a somewhat different manner than visualized by the plug theory.

It is likely that what happens in practice is a mixture of both of these concepts. Whatever is the case, the fundamental feature to be emphasized is that the function of the bituminous material is to protect strength inherent in the soil from the destructive effects of moisture changes. This means that, from a purely waterproofing aspect, the greater the amount of bituminous material present and the more thorough the mixing procedure, the more waterproofed is the soil, since thick bituminous films and plugs are obviously more impermeable than thin ones. However, perfect waterproofing cannot be obtained without introducing so much additive that the soil will become too lubricated, so that compaction cannot be carried out in a satisfactory manner and hence stability will drop. Thus it must be remembered that the stability of a waterproofed product does not necessarily improve with the amount of bituminous material added, but declines in quality once a certain optimum amount is reached.

Cementation action. According to the *Intimate Mix* theory, the effectiveness of the cementing action of a bituminous material can be mainly explained on the basis of the adhesion which takes place between the binder and soil particle. This adhesion is the combined effect of the action of surface tension, adsorption, and other properties of the solid and liquid surfaces. Since water has a much lower viscosity than any bituminous binder, it has a greater wetting power, so true adhesion (or cementation) will only be achieved when the binder displaces any water on the surface. If, for instance, due to the lower surface tension of the binder, it does not displace the water, then coverage of the particle *and* water will occur, but adhesion will not take place between the particle proper and the binder, and cementation will be reduced accordingly. Adhesion is, however, promoted if the soil particle and the binder are of different polarity.

Once adhesion has been established between the bituminous material and the soil particles, the binder literally acts as a 'bridge' holding the particles together. It is important to understand its exact role in this process. In the case of, e.g., a cohesionless soil the addition of a bituminous binder helps considerably by providing cohesive strength. Thus the emphasis is on thoroughly mixing the binder so as to coat each and every particle. However, if too much binder is added and the film coatings about the particles become too thick, then contact between the particles will be prevented and strength will be lowered, since the bituminous cohesion will be essentially the only component contributing to the final stability of the mixture. Thus, as a general rule, it can be said that the thinner the bituminous film, the more stable the bituminous-coarse soil mixture.

Usage

The term 'bituminous stabilization' has been loosely referred to a variety of treatments, ranging from the application of unknown quantities of bituminous materials to undisturbed soil, to the construction of high-stability pavements containing controlled and intimately mixed amounts of soil and bitumen and/or tar. For descriptive purposes, the processes commonly employed may be divided into the following four main types: 1. Sand-bituminous mixtures. 2. Sand-gravel-bituminous mixtures. 3. Soil-bituminous mixtures. 4. Oiled-earth treatments.

Sand-bituminous mixtures. In these mixtures the bituminous material is expected to provide cementitious strength to such cohesionless materials as loose beach, river, dune or other types of clean sand. Proper application of the binder should provide cohesive strength with little interference to already existing stability resulting from friction between the particles. Proper application implies using an optimum binder content so that the overall stability is increased.

Sand-bituminous mixtures may be used for a variety of purposes, the exact usage in a given situation depending primarily on the quality of mixture-material desired, and also on experience with and availability of the bituminous binders. These binders may be any one of the following:

(a) *Penetration-grade bitumens or road tars*. American experience—which has been mostly with bitumens—suggests that the most suitable bitumen grades are those with 85–100 and 120–150 penetration, but higher grades are also used. The use of these grades of material requires that the sands be preheated, and that both mixing and construction take place at high temperatures. In recent years in the United States these mixtures have been designed by methods similar to those used for designing dense bituminous surfacings in that country; usually the binder contents obtained by these methods are between 5 and 10 per cent. In the course of construction the sand-bituminous layers are laid so that each is usually not more than 50 mm thick, and they are compacted to at least 95 per cent of the laboratory designed unit weight.

(b) *Cutbacks*. In American practice, the more liquid rapid-curing cutbacks are most commonly used, but the exact grade selected in any particular instance is dependent on the climatic and mixing conditions prevailing. In Britain, medium-curing cutbacks are mostly used. The nearest equivalent British procedure[25] utilizes the pre-addition of a small amount of lime or cement to damp sand together with a quantity of a special tar or cutback bitumen containing an amount of organic acid. The acid is necessary if the binder is to react with the hydrated lime to produce secondary products which will promote adhesion to the sand particles. Table 5.4 shows the recommended mixtures for this process, which is known as the 'wet-aggregate (hydrated lime) process'. Mixing of the ingredient materials takes place at ambient temperatures in paddle mixers, after which the mixture is spread either by hand or machine in layers of up to about 40 mm thick. In this country, the wet-aggregate process is used for surface course construction. However, since the sand carpets formed in this way have relatively poor resistance to abrasion, they are usually covered with a surface dressing at the first sign of fretting.

(c) *Emulsions*. Emulsions are not used at all in Britain to stabilize soils. In the southern United States, particularly in Florida, they have been used very successfully for the past 30 years on roads carrying up to 1500 vehicles per day. Generally, fully stable emulsions must be used for sand stabilization, since the labile and semi-labile types tend to break too quickly on application to the soil. Usually mix-in-place types of mixers and blade graders are used to mix the binder with the soil. After mixing, the mixture is normally allowed to aerate until the liquid content is down to about 5 per cent and then compaction is carried out.

Sand-gravel-bituminous mixtures. These mixtures are generally used under similar conditions and circumstances as the sand-bituminous mixtures. Normally a sand-gravel material falling into this category has a fairly good gradation, possesses good frictional characteristics, but has little cohesion.

The function of the added bituminous material is to act both as a binding and as a waterproofing agent. An example of British practice with respect to sand-gravel mixtures used for road surfacing purposes is also given in Table 5.4.

TABLE 5.4.   *Composition of sand and sand-gravel bituminous mixtures used in the wet-aggregate (hydrated lime) process*

	Sand			Sandy gravel
	Fine	Medium	Coarse	
Typical gradings of aggregate, % passing given sieve:				
19 mm				90–100
9·52				70–85
3·18	100	95–100	90–100	55–75
1·18				50–70
0·300	60–90	25–60	5–35	10–35
0·075	3–10	0–5	0–5	0–4
Clay content, % by weight	<2	<2	<2	<1
Moisture content of aggregate, %	4–12	3–10	3–8	3–8
Agent:				
Hydrated lime, % by weight	2	2	4	2
Portland cement, % by weight	3	3	5	3
Binder, % by weight of total mix, excluding water				
1. On roads carrying over 3000 tonne/day of traffic:				
(a) Cutback bitumen	4·5–6·0	4·25–4·75	4·25–4·75	3·5–4·5
(b) Tar	5·5–6·5	4·5–5·0	4·5–5·0	3·75–4·75
2. On roads carrying under 3000 tonne/day of traffic:				
(a) Cutback bitumen	5·0–6·5	4·75–5·5	4·75–5·5	4·0–5·0
(b) Tar	6·0–7·5	5·5–6·5	5·5–6·5	4·75–6·0

Soil-bituminous mixtures. This is a catch-all term used to describe those mixtures where a bituminous material is used to stabilize the moisture contents of cohesive fine-grained soils. American experience indicates that soils that can be satisfactorily stabilized in this way generally have not more that about 30 per cent passing the 75μm sieve, liquid limits less than 30 and plastic indices less than 18. Soils of greater plasticity cannot be used because of the difficulties experienced in dispersing the bitumen throughout the system. Soils being stabilized should contain little or no acid organic matter as this can be detrimental to the stability of a soil-bituminous mixture. Fine-grained soils which are very alkaline—these are usually found in arid and semi-arid regions—are also particularly difficult to stabilize with bituminous materials.

Only liquid bituminous materials are suitable for stabilizing cohesive soils; best results are obtained with the more fluid of the medium-curing cutbacks. When, however, the soil is relatively highly plastic, slow-curing cutbacks may have to be used to allow time for the thorough dispersal of the bituminous waterproofer throughout the soil. The amount required to stabilize a soil usually varies between 4 and 8 per cent. At the present time there is no satisfactory method of determining what content should be used in a given situation.

The clay mineral present in the soil has quite an effect on the stability of a soil-bituminous mixture. Kaolinitic soils are easy to stabilize but montmorillonites can be difficult due to mixing problems. The greater the silica content of the clay minerals, the greater the amount of stabilizer required. Research[26] has shown, however, that the addition of small amounts of certain antistripping agents and reactive chemicals, such as phosphorus pentoxide (P_2O_5), hold promise for improving and broadening the effectiveness of bituminous binders as stabilizing agents for cohesive soils. Considerable care must be exercised, however, in both the selection of additives and their method of incorporation in a soil, if the maximum benefits are to be realized.

Moisture content. It is appropriate here to discuss the role which moisture plays as an ingredient of soil-bituminous mixtures. At first glance it might appear that, if it is desired to stabilize a soil by waterproofing it, the very last thing that is wanted is the addition of extra water to the mixture. In practice, however, this is actually what may happen. Not only are cohesive soils practically impossible to pulverize adequately without the presence of water, but water is an essential ingredient during the actual mixing and compacting. During mixing, water facilitates the even distribution of the bituminous material throughout the soil mass. It is known, for example, that the amount of moisture required for the thorough distribution of cutback bitumen increases as the amount of fine material in the soil increases; in any given instance, the most uniform distribution is obtained at a moisture content somewhere in the neighbourhood of the liquid limit of the soil. In compaction, the amount of water is important mainly because of its effect on dry unit weight. Again, it is known that the moisture content for maximum dry unit weight of the soil-bituminous mixture is different from that required for the soil alone.

While it is generally recognized that water is necessary in the mixture, there is little agreement as to what exact content should be utilized in any particular situation. The different water contents that have been suggested are, at the very least, controversial. For instance, a value of moisture content which has gained a certain acceptance is called the *Fluff-point* of the soil. In itself the term fluff-point is a misnomer, since it is actually a range of moisture-content points which gets narrower as the plasticity index of the soil increases. This range, which is usually between one-third and one-half of the standard Proctor optimum content for the soil, brackets the water contents at which the soil will 'fluff' to its maximum extent. The logic here is that, since this is the condition where the greatest amount of voids are apparent in the soil-water mixture, this should be the best condition at which to admix the bituminous material.

The effects resulting from the bituminous volatiles which are present during the compaction of soil-bituminous mixtures are also not clearly understood. For example, during the course of stabilization with cutback bitumens, it is usual to include a period of aeration between mixing and compaction to allow the volatiles to escape; this is because bitumen-volatile loss is associated with increases in strength of compacted mixtures. Again, however, there are considerable differences in the recommended aeration periods. These have ranged from that required for a reduction of at least one-fifth to that of one-half the original combined percentage of water and bitumen volatiles.

Thus it can be seen that there is still considerable confusion over the role which moisture plays in the stabilization of soil with bituminous materials. Perhaps all that can be said categorically at this time is that the percentages of mixing water required to produce maximum strength, maximum dry unit weight, minimum absorption during immersion, and minimum swelling of soil-bituminous mixtures are different for each property mentioned.[27] Thus in any particular situation the engineer should carry out a preliminary investigation to determine the optimum moisture values for these conditions. Then, based on knowledge which he has available regarding the critical local design criteria, he should choose the most appropriate moisture value. This may be a compromise moisture content at which the variance of the aforementioned properties from the optimum conditions is a minimum; this may be fairly close to the optimum moisture (water + volatiles) content for maximum dry unit weight of the mixture.

Oiled-earth treatments. The term oiled-earth is used when a bituminous material of low viscosity is applied to the exposed surface of a moist, densely compacted soil. While in Sweden the same term is applied to surfacings composed of pre-mixed mixtures of gravel and heated road oils,[28] it is normally reserved for processes in which the bituminous material is applied to the surface of the *in-situ* soil and allowed to penetrate downwards under the force of gravity. Since its main functions are simply to protect the underlying material from the deleterious effects brought about by changes in moisture content and to minimize dust nuisance, the process may be used with any type of soil. Normally, however, it is utilized only when use of the other higher forms of bituminous stabilization is not feasible.

Most effective results are obtained with the oiled-earth process when slow-curing cutbacks are applied to the clean surfaces of soils which are mechanically stable, well-compacted and well-drained. Use of slow-curing cutbacks is to be preferred for a number of reasons. Not only do the volatiles not escape so rapidly, thereby allowing more time for the bituminous material to seep into the soil, but it is also found that, when seepage is complete, fractionation of the cutback has occurred with the heavier bitumen remaining at the top of the stabilized layer. This has the additional advantage that the lower and more liquid constituents of the bituminous material act to replace water which might otherwise have evaporated in hot dry weather, thereby helping to prevent the ravelling of the surface which would occur if the soil was allowed to dry out.

OTHER CHEMICAL STABILIZATION METHODS

Before discussing some of these newer stabilization methods, it must be emphasized that they are all still in the experimental stage. In fact the reader may wonder why they are included at all since most can be regarded as 'failures' at this time. Perhaps this can be explained by the following:[29]

'The alchemist of ancient times sought the philosopher's stone, which was believed to have the power to transmute the baser metals into gold. The philosopher's stone that intrigues the imagination of the highway engineer is the thing or method that will have the power to transmute cheaply any kind of soil into a material that will resist abrasion and displacement under traffic in all kinds of weather, and that will retain these properties indefinitely.'

The following is intended to indicate just a few lines along which laboratory research has sought this 'magic' chemical.

Sodium silicate

Sodium silicate is available in various forms, e.g. sodium ortho-silicate $(Na_2 SiO_7)$, sodium sesquisilicate $(Na_6 Si_2 O_7)$, sodium metasilicate $(Na_2 SiO_3)$, and sodium disilicate $(Na_2 Si_2 O_5)$. Because of the various uses to which they can be put—e.g. as adhesives, cements, detergents, deflocculants, rust inhibitors, catalyst bases, etc.—sodium silicates are readily available and easily obtained for commercial use.

The best known sodium silicate stabilization process is that originally developed by the Dutch engineer, Dr. Hugo Joosten, for use as a grout into deep foundations. He found that an injection of sodium silicate followed by an injection of calcium chloride resulted in the instantaneous formation of an insoluble gel of calcium silicates which filled the soil voids and provided considerable strength while at the same time preventing the seepage of water. Another injection method of interest is that developed by Charles Langer of France in 1934, in which the precipitation reaction is much slower than in the Joosten two-shot method. This process enables the admixture of the second chemical to the sodium silicate solution to take place prior to injection, and the combination is then injected as a single shot. Unfortunately, the silica gel formed by the addition of the most common second admixture, sodium bicarbonate, neither attains the strength of the two-shot process nor is insoluble, and so it gradually deteriorates under attack from ground-water; hence the single-shot can only be regarded as a temporary process.

The use of sodium silicate as a soil stabilizing agent on its own is due to its ability to react with soluble calcium salts in water to form cementing agents composed of insoluble gelatinous calcium silicates. From the relatively limited evidence[30] that is available it would appear that sodium silicate alone or in combination with certain chemical precipitants can stabil-ize sandy soils and produce a roadbase or sub-base that could well retain its beneficial effects in mild climates. At the present time, however, there is no field evidence to suggest that the strength and durability of relatively non-plastic fine-grained soils can be sufficiently increased to withstand the freezing and thawing cycles typical of temperate climates. Furthermore, the total cost of the sodium silicate additives necessary to achieve stability is probably greater than that of stabilizers such as Portland cement with which much more experience is available.

Sodium silicates with or without precipitants are of little value in dust-proofing or waterproofing fine-grained soils. On the basis of laboratory tests, various sodium silicates used as secondary additives appear to improve the strength and durability of non-plastic soils stabilized with Portland cement, lime, or lime-fly ash. They appear to be especially useful in increasing the resistance of cement-stabilized soils to sulphate attack.

Organic cationic stabilizers

Organic cationic chemicals are those compounds, organic in nature, which dissociate in water to produce organic cations which may have ex-ceedingly complex structures. Compared with the inorganic cations such as calcium, magnesium, hydrogen or sodium, the organic cations are very large, hence the term 'large organic cations' which is frequently used in the litera-ture when discussing these stabilizers.

Numerous organic cationic chemicals have been tested for effectiveness as soil stabilizing agents[31] and several have been shown to be economically feasible for road construction. Of these, a quarternary ammonium chloride will be described here as an illustration of how an organic cationic chemical may stabilize a soil.

A quarternary ammonium chloride may be thought of as an organic counterpart of ammonium chloride ($NH_4 Cl$) which has all its hydrogen atoms replaced by organic groups. One of the more effective of these salts is a di-hydrogenated tallow, dimethyl ammonium chloride, which is obtained in the United States under the trade names of Arquad 2HT, DDAC, and Aliquat H226. When one of these is added to a soil, it is hypothesized that a rapid cation exchange reaction takes place between the organic cationic compound and the inorganic cations on the clay surfaces. This has the effect of tending to flocculate the clay particles and lower the plasticity index by reducing the surface charges of the particles, thereby lessening the ability of the clay to take up moisture. In the case of montmorillonitic soils, the organic cations are also adsorbed between layers of the expandable lattice minerals, thereby retarding changes in the thickness of the water film between these layers and reducing swelling and shrinking. In addition, the organic cations partially coat the clay surfaces with thin hydrophobic films which have little or no affinity for water; thus the soil particles are waterproofed. Hence it can be seen that the water stability of an aggregated clayey soil can be increased by the waterproofing actions of the organic cationic compound.

When however the soil is 'dry', the addition of the organic cationic compound will reduce strength, since the natural stability which results from the binding caused by the thin water films in the soil is decreased due to the partial coating and waterproofing action of the compound.

Resinous stabilizers

A resin may be defined as a solid or semi-solid complex amorphous mixture or organic substance which has no definite melting point and shows no tendency to crystallize. It is characterized by such physical properties as a typical lustre and a conchoidal fracture rather than by a definite chemical composition. Common plastics are composed of synthetic resins and filler material. Thus it is a useful analogy to consider soil stabilized with resin as essentially a mixture in which the soil particles act as filler material in a kind of plastic.

Vinsol resin and *rosin* are two natural resinous agents which have been very carefully studied[32] for their waterproofing characteristics. Vinsol resin $(C_{27}H_{30}O_5)$ is made from the residues obtained after the distillation of pine tree stumps in the manufacture of turpentine. Rosin is also obtained as a by-product of turpentine extraction and its chief constituent is abietic acid $(C_{19}H_{29}COOH)$. Resins are believed to waterproof a soil by forming thin coatings on the moisture at the water-air interfaces in soil pores; then when an initial amount of moisture has been absorbed by the soil the air spaces become filled and the air-water interface diminishes in area. This in turn causes the resin film coating to be crowded into a smaller surface area. The resistance of the film to compression reduces the surface tension of the water to zero, thereby opposing the further entry of water.

The maximum waterproofing effect of these resinous materials is developed within about three days after mixing takes place. There appears to be an optimum amount of admixture for a given soil, as when too much is added the water absorption of the mixture increases. Stability is dependent on the original strength characteristics of the untreated soil, which the resins then simply help to preserve.

Calcium acrylate is a white powder which has the special advantage of being about 30 per cent soluble in water at normal temperatures; this means that it is very easily added to and dispersed throughout a soil. When a catalyst (ammonium persulphate) and an activator (sodium thiosulphate) are added to the solution, polymerization takes place and results in the formation of a water-insoluble gel which imparts unique rubbery properties to the soil within a few hours. When dried so that the polymer contains less than about 25 per cent water, cross-linkages are formed, which results in a hard rigid stabilized soil which in this state is an excellent pavement material. Unfortunately, this rigid state is not permanent, for when it is again subjected to excessive moisture it reverts to its former rubbery condition.

American field studies have shown that between 4 and 6 per cent of this chemical can stabilize a loess soil sufficiently in four to six hours to allow temporary lorry traffic over it. However, because of its reversible characteristics—as well as its high cost—it can only be considered as an emergency stop-gap method where the rate of hardening is critical. In this sense, it is important to note that the polymerization rate is easily controlled by adjusting the catalyst-activator concentration; a greater concentration will reduce the polymerization rate.

SELECTED BIBLIOGRAPHY

1. ROAD RESEARCH LABORATORY. *Soil Mechanics for Road Engineers.* London, H.M.S.O., 1952.
2. HUANG, E. Y. *Manual of Current Practice for Design, Construction and Maintenance of Soil-Aggregate Roads.* Urbana, Ill., University of Illinois Engineering Experiment Station, 1959.
3. LEWIS, W. A. Investigation of the performance of pneumatic-tyred rollers in the compaction of soil, *Road Research Technical Paper No. 45.* London, H.M.S.O., 1959.
4. MINISTRY OF TRANSPORT. *Specification for Road and Bridge Works.* London, H.M.S.O., 1969.
5. KJELLMAN, W. Consolidation of clay soil by means of atmospheric pressure, *Proc. Conference on Soil Stabilization,* Massachusetts Institute of Technology, 1952, 258–274.
6. GRAINGER, G. D. A study of burnt clay as a roadmaking aggregate, *Recherches et Essais sur les Structures en Terre Cuite, Symposium R.I.L.E.M.,* Milan 1962, Rome 1965.
7. WILLIS, E. A. Design requirements for graded mixtures suitable for road surfaces and base courses, *Proc. Highway Research Board,* 1938, **18,** Pt. II, 206–208.
8. DAVIDSON, D. T. and R. L. HANDY. Soil stabilization with chlorides and lignin derivatives. In WOODS, K. B. *Highway Engineering Handbook.* New York and Maidenhead, McGraw-Hill, 1960.
9. THORNBURN, T. H. and MURA, R. Stabilization of soils with inorganic salts and bases: A review of the literature, *Highway Research Record* No. 294, 1–22, 1969.
10. CATTON, M. D. Early soil-cement research and development. *Proc. Amer. Soc. Civ. Engrs, Journal of the Highway Division,* Jan. 1959, **85,** No. HW1, Paper No. 1899.
11. COMMITTEE ON SOIL-PORTLAND CEMENT STABILIZATION. Soil stabilization with Portland cement, *Highway Research Board Bull.* 292, 1961.
12. SHERWOOD, P. T. The Properties of Cement Stabilized Materials. *RRL Report* LR 205, Crowthorne, Berks., The Road Research Laboratory, 1968.
13. SHERWOOD, P. T. Effect of sulphates on cement and lime-stabilised soils, *Highway Research Board Bull.* 353, 1962, 98–107.
14. PORTLAND CEMENT ASSOCIATION. *Soil-Cement Laboratory Handbook.* Chicago, Ill., The Association, 1956.
15. MACLEAN, D. J. and W. A. LEWIS. British practice in the design and specification of cement-stabilized bases and sub-bases for roads, *Highway Research Record* No. 36, 1963, 56–76.
16. DAVIDSON, D. T., G. L. PITRE, M. MATEOS and K. P. GEORGE. Moisture-density, moisture-strength, and compaction characteristics of cement-treated soil mixtures. *Paper presented at the 41st Annual Meeting of the Highway Research Board,* Washington, D.C., Jan. 1962.
17. O'FLAHERTY, C. A., M. MATEOS and D. T. DAVIDSON. Fly-ash and sodium carbonate as additives to soil-cement mixtures, *Highway Research Board Bull.* 353, 1962, 108–123.

18. BS 890 *Building Limes*. London, British Standards Institution, 1966.

19. BS 3892:1965. *Pulverized-Fuel Ash for Use in Concrete*. London, British Standards Institution, 1965.

20. DYER, M. R. and O'FLAHERTY, C. A. Some variable properties of pulverized fuel ash, *Chemistry and Industry*, pp. 8–15, Jan. 2, 1971.

21. DYER, M. R. and O'FLAHERTY, C. A. Lime-fly ash stabilization of soil: Development of the Technology, *Highways Design and Construction*, 1972, **40**, 1754, 10–14, and 1755, 24–28.

22. MATEOS, M. and D. T. DAVIDSON. Steam curing and X-ray studies of fly ashes, *Proc. A.S.T.M.*, 1962, **62**, 1008–1018.

23. NEUBAUER, C. H. and THOMPSON, M. R. Stability properties of uncured lime-treated fine-grained soils, *Highway Research Record* 381, 1972, 20–26.

24. HIGHWAY RESEARCH BOARD. Soil bituminous roads, *Current Road Problems* No. 12. Washington, D.C., 1946.

25. ROAD RESEARCH LABORATORY. Bituminous surfacings made by the wet-aggregate (hydrated lime) process, *Road Note* No. 16. London, H.M.S.O., 1953.

26. MICHAELS, A. S. and V. PUZINAUSKAS. Additives as aids to asphalt stabilization of fine-grained soils, *Highway Research Board Bull*. 129, 1956, 26–49.

27. KATTI, R. K., D. T. DAVIDSON and J. B. SHEELER. Water in cutback asphalt stabilization of soil, *Highway Research Board Bull*. 241, 1960, 17–49.

28. HALLBERG, S. A brief account of Swedish experiments with oil treatment of gravel roads, *Roads and Road Construction*, 1958, **36**, 421.

29. MEYERS, B. Iowa studies surfacing problems, *Better Roads*, 1947, **17**, 29–31.

30. HURLEY, C. H. and THORNBURN, T. H. Sodium silicate stabilization of soils: A review of the literature, *Highway Research Record* 381, 1972, 46–79.

31. HOOVER, J. M. and D. T. DAVIDSON. Organic cationic chemicals as stabilizing agents for Iowa loess, *Highway Research Board Bull*. 129, 1956, 10–25.

32. CLARE, K. E. The waterproofing of soil by resinous materials, *J. Soc. Chem. Ind.*, 1949, **68**, 69–76.

6 Flexible pavements

A highway pavement is a structure consisting of superimposed layers of selected and processed materials whose primary function is to distribute the applied vehicle loads to the subgrade. The ultimate aim is to ensure that the transmitted stresses are sufficiently reduced that they will not exceed the supporting capacity of the subgrade. Two types of pavement are generally recognized as serving this purpose—flexible pavements and rigid pavements.

The distinguishing feature of a flexible pavement lies in its structural mechanics and the fact that the pressure is transmitted to the subgrade through the lateral distribution of the applied load with depth, rather than by beam and slab action as with a concrete slab. Thus a flexible pavement can be most easily defined by contrasting it with a rigid Portland cement concrete pavement.

When the subgrade deflects beneath a rigid pavement, the concrete slab is able to bridge over localized failures and areas of inadequate support because of its structural capabilities. Thus its thickness is relatively little affected by the quality of the subgrade as long as it meets certain minimum criteria.

In direct contrast to this, the strength of the subgrade is the main factor controlling the design of a flexible pavement. When the subgrade deflects, the overlying flexible pavement is expected to deform to a similar shape and extent. Thus the basic design criterion is the depth of pavement required to distribute the applied surface load to the subgrade; the subgrade must not be overstressed and caused to deform to a greater extent than the pavement itself can deform without damaging its own structural integrity.

In its simplest form a flexible pavement is generally considered to be any pavement other than a concrete one. It is this definition that is accepted by the great majority of practising engineers and so it is the one that will be utilized in this textbook. It should be clearly understood, however, that the term is simply one of convenience and does not truly reflect the characteristics of the many different and composite types of construction masquerading as 'flexible' pavements.

Elements of a flexible pavement

Before discussing in detail the various features of a flexible road, it is necessary to mention briefly some terminology. As illustrated in Fig. 6.1, the cross-section of a flexible road is composed of a Pavement superimposed

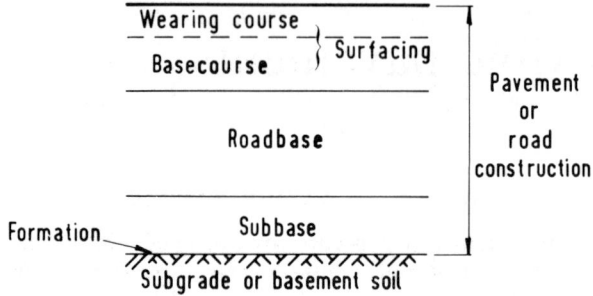

Fig. 6.1.   Diagrammatic illustration of the structural elements of a flexible pavement

on the basement soil or Subgrade. The intersection of the subgrade and the pavement is known as the Formation.

The subgrade is normally considered to be the in-situ soil over which the highway is being constructed. It should be quite clear, however, that the term subgrade is also applied to all native soil materials exposed by excavation and to excavated soil that may be artificially deposited to form a compacted embankment. In the latter case the added material is not considered to be part of the road structure itself but part of the foundation of the road.

Surface course. Whether it be a pavement for an expensive motorway or a simple country road, the basic structural cross-section is essentially that illustrated in Fig. 6.1; it is composed of several distinct layers superimposed on the subgrade in the manner indicated.

The uppermost layer of a flexible pavement—and probably the most important one since it is the one which the motorist sees and rides upon—is called the surface course. The primary function of this layer is to provide a safe and comfortable riding surface for the traffic. In addition it is expected to protect the layers beneath from the effects of the natural elements, as well as from the disintegrating effects caused by vehicles skidding and braking on the roadway.

It used to be that for design purposes, the surface course was not normally considered to have any beam strength. As will be discussed later this assumption usually errs on the side of safety.

The materials in pavement surface courses can vary from loose mixtures of earth and gravel to the very highest type of bituminous mixtures. The type utilized in any particular situation depends on the quality of service that is required of the roadway.

If a bituminous surfacing is used it may consist of either a single hom-ogeneous layer or, in the higher types of road, of two distinct sub-layers known as a wearing course and a basecourse. The wearing course provides the actual surface on which the traffic runs. If an unsuitable material is used in the wearing course it can be a major cause of skidding accidents. The basecourse acts as a regulating layer to provide the wearing course with a better riding quality, as well as adding to the structural integrity of the pavement. It is not as impervious a material as the wearing course.

Roadbase. This layer must not be confused with the basecourse, which is an integral part of the surface course. One is a sub-layer within the bituminous surfacing while the other is normally the thickest element of the flexible pavement on which the surfacing rests. From a structural aspect the road-base is the most important layer of a flexible pavement. It is expected to bear the burden of distributing the applied surface loads and to ensure that the bearing capacity of the subgrade is not exceeded. Hence the material used in the roadbase must always be of a high quality.

Sub-base. In its simplest sense this layer can be considered merely as an extension of the roadbase; in fact it may or may not be present in the pavement as a separate layer. Whether or not it is utilized in a pavement depends on the purpose for which it is to be used. Its function can be examined from a number of aspects:

1. As a structural member of the pavement the sub-base helps to distrib-ute the applied loads to the subgrade. Since the stresses which the sub-base is required to transmit are significantly less than those required of the road-base its constituent materials can often be of much lower quality and cost. The sub-base material must, however, always be significantly stronger than the subgrade and capable of resisting within itself the stresses transmitted to it by the roadbase.

2. In certain instances a coarse-grained material may be used in the sub-base to act as a drainage layer. The primary purpose of this is to pass to the highway drainage system any moisture which falls during construction or which enters the pavement after construction. The quality of the material used must be such that the free-draining criterion of the sub-base is always met.

3. On fine-grained subgrade soils a granular sub-base may be provided (a) to carry constructional traffic and act as a platform on which subsequent layers can be constructed, (b) to act as a cut-off blanket and prevent moisture from migrating upward from the sub-grade, (c) to act as a cut-off blanket to prevent the infiltration of subgrade material into the pavement.

The type of material used in any of these designs depends on the purpose for which it is being used and the grading of the subgrade soil.

STRESS CONSIDERATIONS

As compared with many other engineering design procedures, the design of flexible pavements is in the embryonic stage. The development of a rational design procedure is at present being investigated from two aspects. Firstly, studies are being carried out by means of theoretical and laboratory analyses of the static and dynamic stresses induced by traffic, the stress-deformation characteristics of the layers in the road under various forms of loading, and the variation of these properties with time. The second approach leans to the construction and observation of full-scale experimental test roads, of which the most notable are the WASHO[1] and AASHO[2] roads in the United States and the Boroughbridge and Alconbury Hill test roads[3, 4, 5] in Britain.

Although very promising, these approaches have not yet reached the stage where a fundamental method of flexible pavement design can be evolved, and reliance is still placed on empirical design procedures and on existing pavement performance results. The following discussion on stress considerations presents some of the variables which must be evaluated in developing a rational design procedure.

Stress distribution

The vertical pressure on any horizontal plane in or beneath a pavement, due to the wheel load applied at the surface, is distributed over an area that is considerably larger than the area of contact between the tyre and the carriageway. While the total vertical pressure on the horizontal plane is equal to the applied load at the surface, the greatest vertical stress occurs directly beneath the centre of the tyre contact area; it becomes less and less as the horizontal distance away from the centre becomes larger.

Thinking in terms of a 3-dimensional graph which has two horizontal axes drawn to linear scales, while the remaining (vertical) axis has force units, then the pressure distribution on any horizontal plane in the pavement or subgrade can be considered as represented by a helmet-shaped surface. The base of this pressure-helmet tends to have an elliptical shape because of the approximately elliptical outline of the contact area between the tyre and the carriageway. The height of the helmet, which represents the maximum pressure, varies according to the depth of the horizontal plane beneath the surface; its height decreases and the helmet-shaped stress surface flattens out as the depth increases.

Boussinesq theory. If the material through which the applied stresses are being transmitted is considered to be an idealized elastic homogeneous and isotropic mass which extends infinite distances both laterally and vertically downward from where the load is applied to a level surface, it is possible to determine the stresses at any point within the mass by means of the theoretical equations developed by the French elastician Boussinesq. These equations, which completely define the normal stresses at various co-ordinate distances from the point of application of a concentrated load P, are given in Fig. 6.2.

In this figure the normal and shear stresses which keep a minute cube of material in equilibrium are denoted by σ and τ. Their subscripts denote the orientation of the lines of action; thus σ_z is a normal stress acting on the plane normal to the z-axis and τ_{xz} indicates the shearing stress acting in a plane normal to the x-axis and on a line of action parallel to the z-axis.

These equations have been integrated to determine the stresses beneath a uniformly loaded circular area. The vertical and horizontal normal stresses at any point directly below the centre of the loaded area are given by:

$$\sigma_z = p\left[1 - \frac{z^3}{(a^2+z^2)^{3/2}}\right]$$

and
$$\sigma_x = \sigma_y = \frac{p}{2}\left[1 + 2\mu - \frac{2z(1+\mu)}{(a^2+z^2)^{1/2}} + \frac{z^3}{(a^2+z^2)^{3/2}}\right]$$

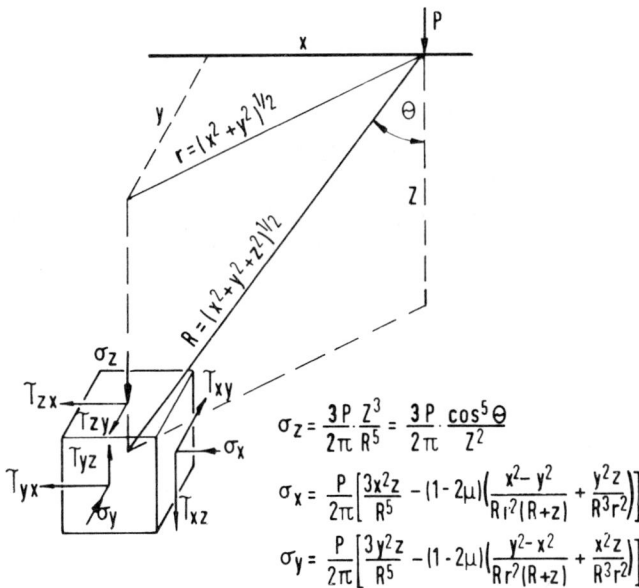

$$\sigma_z = \frac{3P}{2\pi}\cdot\frac{z^3}{R^5} = \frac{3P}{2\pi}\cdot\frac{\cos^5\theta}{z^2}$$

$$\sigma_x = \frac{P}{2\pi}\left[\frac{3x^2z}{R^5} - (1-2\mu)\left(\frac{x^2-y^2}{Rr^2(R+z)} + \frac{y^2z}{R^3r^2}\right)\right]$$

$$\sigma_y = \frac{P}{2\pi}\left[\frac{3y^2z}{R^5} - (1-2\mu)\left(\frac{y^2-x^2}{Rr^2(R+z)} + \frac{x^2z}{R^3r^2}\right)\right]$$

Fig. 6.2. Stresses acting on an element due to a point load according to Boussinesq

In these equations p is the unit load applied over a circular area of radius a, and μ is Poisson's ratio. The other symbols are as designated in Fig. 6.2.

It should be noted here that both of these equations are independent of the soil modulus of elasticity. In addition, the vertical stress equation is independent of Poisson's ratio as well.

Poisson's ratio.   Poisson's ratio is defined as the ratio of the strain normal to the applied stress to the strain parallel to the applied stress. It is an inherent property of an elastic material and ranges between zero and 0·5. A material such as cork has a μ-value approximately equal to zero. An incompressible material on the other hand will have a maximum value of $\mu = 0.5$; this occurs when the decrease in volume of a vertical cylinder under vertical compression is approximately equal to the increase in volume due to lateral expansion. Although experimenters have the greatest difficulty in measuring Poisson's ratio for soils, it has been shown conclusively that it lies closer to 0·5 than to zero. For this reason, as well as the fact that it greatly simplifies later calculations, a value of $\mu = 0.5$ is most often used in order to simplify the Boussinesq equations.

Deflection equations.   Accepting the Boussinesq vertical and horizontal stress equations, it can be shown that the total elastic deformation at the surface at the centre of the applied uniform load is given by

$$\Delta = \frac{2pa}{E}(1 - \mu^2)$$

where Δ = total deflection,
$\quad\quad\; p$ = applied uniform load,
$\quad\quad\; a$ = radius of circular contact area,
$\quad\quad\; E$ = modulus of elasticity of material,
and μ = Poisson's ratio.

If a pavement of depth z is interposed between the subgrade and the applied load at the surface, then the total elastic deformation in the subgrade, that is from depth z to infinity, is given by

$$\Delta = \frac{p}{E}\left[(2 - 2\mu^2)(a^2 + z^2)^{1/2} - \frac{z^2(1 + \mu)}{(a^2 + z^2)^{1/2}} + z(\mu + 2\mu^2 - 1)\right]$$

When Poisson's ratio is taken as equal to 0·5 the two above deflection equations become

$$\Delta = \frac{1\cdot5pa}{E}$$

and
$$\Delta = \frac{1\cdot5pa^2}{E(a^2 + z^2)^{1/2}} \text{ respectively.}$$

It should be noted that the deflections obtained from these two equations become the same when z approaches zero, and it is clear that the deflections in the pavement itself are not taken into account by the equations.

Validity of Boussinesq equations.   The values obtained by the Boussinesq equations can be considered to be relatively valid for a very uniform soil in which the boundary conditions are compatible with the theoretical assumptions and provided that the modulus of elasticity of the soil is a constant.

Frequently, however, the environmental conditions are different from the theoretical assumptions and as a result the theoretical values are considerably different from actual ones. In particular, in the case of a flexible pavement superimposed upon a subgrade soil the equations do not obviously hold true. For instance, the deflection equation takes no account of the stress distribution properties of the materials in the pavement and assumes that the deflection in the subgrade is simply a function of the radial distance a, the applied surface load p and the modulus of elasticity of the subgrade E. As a result the theoretical deflections in the subgrade are usually greater than the actual values beneath a road pavement.

The reliance on the modulus of elasticity of the soil is also subject to discussion. Soil is not truly elastic and does not exhibit a constant stress/strain relationship. Instead of being a straight line, the relationship is more likely to take the form of a convex curve such that the stress to strain ratio decreases as the strain increases. Thus the E-value used in the equations must always be chosen for some subjectively selected strain value.

To avoid confusion that might arise through using a term which ought to be only used when referring to a material which is truly elastic, the term Modulus of Deformation should be used in reference to the value E; this is in line with current practice in the research literature.

Other elastic theories. While the Boussinesq equations are not directly applicable to roadway design conditions, they have been presented here because of the very considerable importance that is attached to elastic theory as a means of estimating the stresses and deflections in and beneath the pavement. A measure of its importance can be gained by considering Table 6.1 which lists various multi-layer elastic theories that have been proposed. Detailed examination of these theories is beyond the scope of this text and so discussion will be confined to a few general comments.

Although the term 'multi-layer' is used in Table 6.1, these theoretical studies have been limited to two-and three-layer systems because of the very complex difficulties associated with the analyses of the systems. Flexible roads are normally examples of at least a three-layer system, the surface course being the upper layer, the roadbase the intermediate layer and the soil subgrade the semi-infinite lower layer. When there is a sub-base in the pavement it is difficult to fit the construction into the three-layer system unless the sub-base has properties similar to those of either the roadbase or subgrade.

As illustrated in Table 6.1, there is disagreement among investigators regarding the properties to be assigned to the layers above the subgrade. The uppermost layer is regarded by some as an elastic plate, by others as an elastic layer. The difference between the two approaches is that an elastic plate is considered to undergo bending deformation but no vertical deformation under direct stress, whereas in an elastic layer all the stresses are considered and no restrictions are placed on the deflections. The elastic layer

TABLE 6.1.   *Elastic theories for two- and three-layered systems*[6]

Type of system	Author	Assumed conditions of the layers	Assumed conditions of the interface	Form of loading	Parameters used	Stress and deflections determined	Comments
Two-layered	Westergaard (1926)	Elastic plate on a subgrade acting as a series of parallel vertical independent springs	Discontinuous	Uniform, circular loading	Modulus of subgrade reaction (i.e. spring constant)	Tensile stresses at three points on the surface of the plate	The subgrade is unrealistic and stresses in the subgrade are not considered
	Hogg (1938)	Elastic plate on a semi-infinite elastic subgrade	Discontinuous	Uniform, circular loading	Young's modulus and Poisson's ratio for the subgrade	Tensile stresses in the plate	Extension of Westergaard's theory
	Burmister (1943)	Elastic layer on a semi-infinite elastic subgrade	(a) Continuous (b) Discontinuous	Bessel function loading	a/h, E_1/E_2	w, zz, rr, rz, in the pavement and subgrade	Equations derived but not evaluated for specific cases
	Fox (1948)	ditto	ditto	Uniform, circular loading	ditto	All stresses in the pavement and subgrade	Extension and numerical analysis of Burmister's work
	Hank and Scrivner (1948)	ditto	ditto	ditto	ditto	Stresses at both sides of the interface	ditto
	Pickett and Ai (1948)	Elastic plate on a semi-infinite elastic subgrade	(a) Discontinuous (b) No horizontal displacement at the interface	ditto	a/h and fnctions of E_1/E_2, μ_1 and μ_2	All stresses and deflection in the subgrade	Mainly applicable to concrete pavements. Results were modified to allow for shear in the pavement on deflection, and horizontal shear at the interface, by making them agree with Burmister's results (approx.)
	Odemark (1949)	ditto	Discontinuous	ditto	ditto	Deflections at the surface and radial stress at the base of the upper layer	
Three-layered	Burmister (1945)	Two elastic layers on a semi-indefinite elastic subgrade	(a) Continuous (b) Discontinuous	Bessel function loading	a/h_2, h_1/h_2, E_1/E_2, E_2/E_3	Deflection at the second interface	A method of calculating all the stresses is given, but no numerical evaluation of the deflection or stresses
	Hank and Scrivner (1948)	ditto	Continuous	Uniform, circular loading	ditto	All the stresses at the first surface	Extension and numerical evaluation of Burmister's equations for the first interface
	Acum and Fox (1951)	ditto	Continuous	ditto	ditto	Vertical and radial stresses at both interfaces	Extension and numerical evaluation of Burmister's equations for vertical and radial stresses
	Jeuffroy and Bachalez	Elastic plate on an elastic layer on a semi-infinite elastic	Discontinuous at the first interface, continuous at the	ditto	h_2/a, $[h_1(E_1)^{1/3}]/[a(6E_2)]$-	w_1, rr_1 and zz_2	

concept is thus a more general treatment of the problem. Actual measurements show that transient compression of the upper layer does occur, indicating that it behaves as an elastic layer and not as a plate. Some data illustrating the validity of this statement are presented in Fig. 6.3.

Table 6.1 also indicates some differences of opinion regarding continuity conditions between the different layers. Some theories assume the interfaces between the layers to be perfectly smooth and without friction, while others assume them to be rough so that the strains are completely transmitted across them. The interfaces occurring in practice are unlikely to be ideally rough, but they are certainly far from being smooth. Because the theories can only deal with the two extreme conditions, it seems preferable to consider the interfaces as rough, since it is closer to the condition which actually exists in the field.

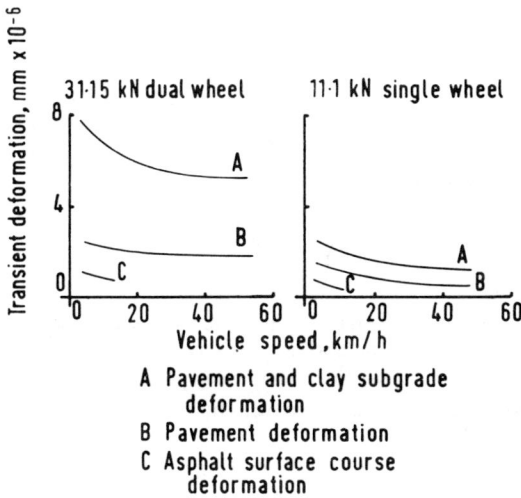

Fig. 6.3.   Variation of transient deformation with vehicle speed[6]

Burmister theories.   Basic theories which utilize assumptions close to actual conditions in a flexible road are those proposed by Burmister. Stress and deformation values obtained by these theories are dependent on the moduli of deformation of the different layers. Since typical flexible roads are normally composed of layers whose moduli decrease with depth, the net effect is that the Burmister equations predict stresses and deflections in the subgrade that are considerably less than those obtained from the Boussinesq equations. This is illustrated for the two-layer system in Fig. 6.4, which compares results obtained by the Boussinesq theory and the two-layer theory.[7]

On the left-hand side of Fig. 6.4 the applied wheel load can be considered as resting directly on a subgrade with a modulus of deformation E_2. On the right-hand side of the figure, a pavement of thickness h_1 and modulus E_1 is inserted and the stresses calculated taking this into account. The data are plotted in the form of bulbs of pressure for the two systems. A bulb of pressure is a surface obtained by connecting points of equal stress on the various horizontal planes at various depths. The pressure at any one point on the surface of a bulb is the same as that at any other point. Pressures at

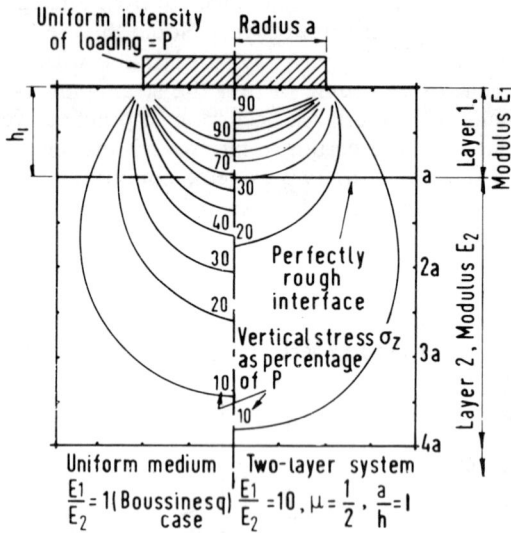

Fig. 6.4.   Theoretical comparison between the vertical stress distribution in a uniform material and in a two-layer system

points inside any given bulb are greater than those on the bulb, while pressures outside the bulb are less.

While Fig. 6.4 is shown here primarily to illustrate the differences between the values obtained from the two theories, it also illustrates admirably the main function of a pavement, which is to reduce to an acceptable level the pressures applied to the subgrade. As can be seen, the stresses in the subgrade at a depth h_1 are considerably influenced by the insertion of the stronger pavement material. With the pavement inserted, the vertical stress at the interface and directly below the centre of the applied load is estimated to be approximately 30 per cent of p, whereas without the pavement the stress at a depth h_1 is approximately 70 per cent of the applied unit load.

Effects of various factors

Tyre pressure. Solid tyres are not entirely obsolete at this time—even in modern countries such as Britain they are still used for special purposes. Designed to carry abnormal and heavy loads at low speeds they can be encountered on certain routes although, it must be pointed out, not in any great numbers. In the majority of under-developed countries solid tyres are standard fittings on bullock carts and the like type of rolling vehicle.

There are very few published data relative to the stress transmitted to a road by this type of tyre. One 1909 paper which presented the results of an investigation into the loading imposed on a road surface by the solid rubber tyres of the motor omnibus of that period gave the results shown in Table 6.2. In this study the force on each front wheel was approximately 8·36 kN, and about 7·27 kN on each of the back four wheels.

While the values given in Table 6.2 are of little practical use, they serve to illustrate the pressures that can be exerted by solid tyres on the pavement

TABLE 6.2. *Stresses on a road surface due to a vehicle with solid rubber tyres*[8]

Wheel location	Tyre condition	Elliptical contact area dimensions, mm		Applied pressure kN/m^2	
		Short axis	Long axis	Maximum	Average
Front	New	63·5	175·3	1586	965
	Worn	86·4	144·8	1379	889
Back (outer	New	55·9	193·0	1379	889
twin tyre)	Worn	80·3	144·8	1241	786
Back (inner	New	53·3	194·3	1379	889
twin tyre)	Worn	85·1	139·7	1234	779

surface, under relatively light wheel loads. These pressures are significantly greater than the pressures that pneumatic tyres will normally transmit to the surface under similar wheel loads. Since the solid-tyred vehicles now in use carry loads well in excess of those represented in Table 6.2, it can be expected that in many cases they transmit much greater pressures to the pavement.

For all practical purposes it is not normally necessary to pay special attention to the requirements of solid-tyred vehicles when designing highway pavements in Britain. The design procedure in use in this country at this time is based on results obtained from examining pavement performances under existing traffic loads, and hence the occasional solid-tyred vehicle is automatically taken into account. On the other hand, for pavements in this country and abroad on which excessive numbers of solid-tyred vehicles can be expected, care should be taken when using the routine design procedures, since they may not adequately meet the more stringent requirements of the solid-tyred vehicles.

Pneumatic tyres. Most tyres on private cars have recommended inflation pressures of less than 205 kN/m². Inflation pressures of 450–585 kN/m² are normal on medium-sized lorries of the 5-tonne class, whereas the normal type of heavier commercial vehicle may use tyres inflated to well over 700 kN/m². Special vehicles have tyre-inflation pressures considerably greater than these.

The relationship between the tyre inflation pressure and the stresses induced in a pavement is more easily discussed by referring to Fig. 6.5. The data in this figure illustrate the effect of changing the inflation pressure in a tyre when the applied wheel load is held constant.

A pneumatic tyre pressing on a road surface forms a contact area that is approximately elliptical. Figure 6.5(a) indicates how, for a given wheel load, the contact area decreases as the inflation pressure is increased. The extent of the decrease in any given situation will of course depend on the initial wheel load and the quality of the tyre itself.

Figure 6.5(b) indicates the manner in which the actual pressure transmitted to the surface increases in an apparently near-linear fashion as the inflation pressure is increased. Note that at any given time the applied surface pressure is always considerably greater than the tyre-inflation pressure. For the data in this figure the vertical pressure on the pavement surface appears to average about 200 per cent of the inflation pressure.

The data in Fig. 6.5(c) indicates the change in the vertical pressure measured at the pavement-subgrade interface, i.e. the formation level, as a result of changing the tyre pressure. Note that the measured stress at the formation is only increased significantly for very substantial increases in the inflation pressure. Theoretical studies would seem to indicate that the stress at the formation is proportional to $a^{1 \cdot 9}$ where a is the equivalent radius of the tyre contact area.

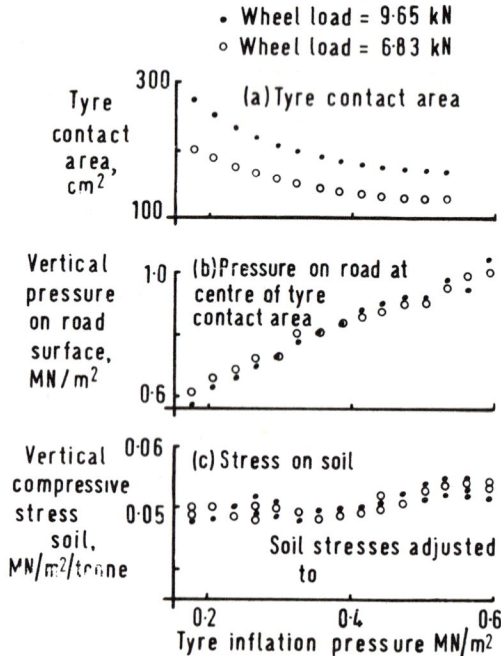

Fig. 6.5.   Effects of changing the tyre-inflation pressure[6]

Figure 6.5(c) also indirectly reflects the role of the tyre pressure in inducing stresses in the pavement. The effects of high inflation pressures are most pronounced in the upper layers of a pavement and have relatively little differential effects at greater depths. In other words, for a given wheel load the tyre-inflation pressure has little effect on the depth of pavement required above the subgrade, but it is this pressure which controls the quality of the materials used in the upper layers.

Wheel load. While the tyre-inflation pressure designates the materials to be used in the pavement, it is the total applied wheel load which determines the depth of pavement required to ensure that the subgrade is not over-stressed. The extent of this influence is illustrated in Fig. 6.6, which shows how the stresses at the top and bottom of a pavement were changed when the tyre-inflation pressure was kept constant at $414\,kN/m^2$ while the load applied to

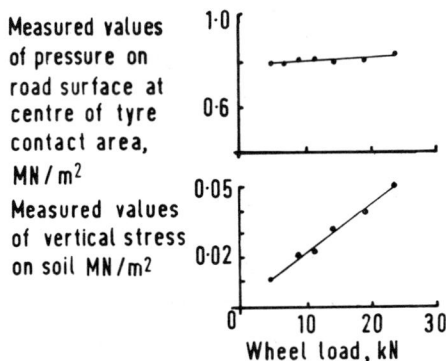

Fig. 6.6.   Effect of changing the applied wheel load[6]

the smooth-treaded tyre of the testing vehicle was progressively increased from 4·45 to 22·24 kN.

In this illustration it is seen that as the wheel load is increased the tyre deflects and the contact area is increased; as a result the peak unit pressure applied to the carriageway shows only a very small increase although the total load is increased by 400 per cent. The additional wheel load had the effect of causing the vertical stress at the pavement-subgrade interface to be increased in direct proportion to the extra load. Thus it is clear in this case that as the wheel load is increased the depth of pavement must also be increased so that the allowable subgrade stress is not exceeded.

Wheel configuration. Many commercial vehicles have dual rear-wheels and theoretically, these can also influence the stress-distribution and deflections within and below the highway pavement. The most definitive investigations into the effect of various wheel arrangements have been carried out on airport pavements, where they are of significant importance because of the greater wheel loads, and little research has been done on their effect on highway structures.

Theoretically it can be shown that the single-wheel load required to reproduce the same maximum stresses in a homogeneous material as are given by a dual-tyred assembly is

$$P_E = P + \frac{Pz_5}{(z^2 + S^2)^{5/2}}$$

where P_E = equivalent single wheel load,
         P = load on each dual-tyre,
         z = depth to the plane being stressed,
and S = distance between the centres of the individual tyres.

This relationship clearly illustrates the two most important features of the dual-tyred assembly. Firstly, the calculated stresses at the pavement surface (when $z = 0$) are due only to the individual wheels of the assembly and there are no interacting effects. Secondly, the distance between the tyre centres plays an important part in the stress distribution beneath the surface. At greater depths, however, where the S-value is small in comparison with depth, the stresses due to the dual-tyres become near-additive.

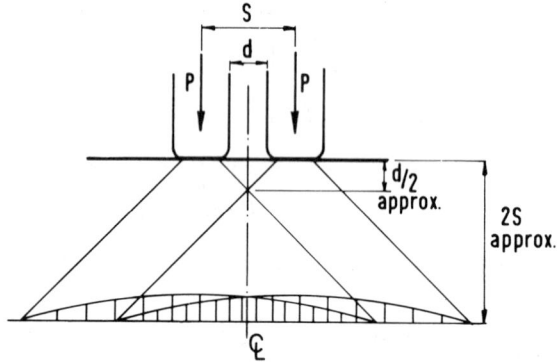

Fig. 6.7.   Schematic diagram of vertical stresses under a dual-tyred wheel assembly

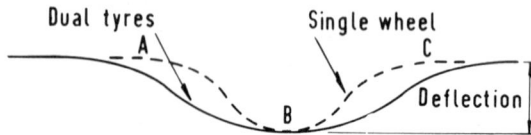

Fig. 6.8.   Schematic diagram of deflections under single- and dual-tyred wheel assemblies

Studies carried out by the U.S. Corps of Engineers in the course of airport pavement investigations indicate that the maximum significant depth at which each tyre of a dual-tyred assembly acts as an independent unit happens to be equal to half the distance d between the inner faces of the two tyres. At depths greater than $d/2$ the pressure helmets on a given plane begin to overlap; at a depth $2S$, which is twice the distance between the centre lines of the tyres, the dual-tyred assembly acts essentially as a single unit carrying a load equal to $2P$. These factors are illustrated schematically in Fig. 6.7.

These airport studies also showed that at shallow depths the maximum deflections take place beneath the centres of the individual tyres, whereas at a depth of 2S the maximum deflection shifts to beneath the centre-line of the assembly. At intermediate depths the maximum deflections occur at lateral points between the centre-line of the assembly and the centre-line of each tyre of the duals.

It was also shown that a dual-tyred assembly which caused the same maximum deflections in a material as a single-wheel load actually produced less strain in the material. This is best explained by referring to Fig. 6.8, in which the dotted line indicates the deflection under an equivalent dual-tyred assembly. In each the depth of the maximum deformation is the same, but the shapes of the deformed areas are different. This is most significant, since it is not so much the amount of the deflection which causes a flexible pavement to crack, but rather the degree of curvature of the deformed surface. Thus it can be seen that the pavement will be much more severely stressed at points A, B and C under the single-wheel load than under the dual-tyred assembly.

Number of axles. The heavier types of commercial vehicle have multiple axles and these also have a pronounced effect on stress distributions and deflections. Results from the WASHO Road Test[1] clearly indicate that the

TABLE 6.3. *Equivalent single- and tandem-axle loads*[1]

Surfacing*	Single-axle load, kN	Equivalent tandem-axle load, kN	
		Based on deflection	Based on distress
5 mm ashphalt	80·1	155·7	125·9
	99·6	195·7	161·9
10 mm ashphalt	80·1	135·7	125·9
	99·6	181·9	149·5

* Surface plus roadbase thickness is 152·5 mm.

relative effects cannot be dealt with by simple summation. Thus, for example, Table 6.3 shows that in no case were the results obtained with an equivalent tandem-axle equal to twice that of a single-axle. This table also shows that, irrespective of the surfacing material or the measurement criterion, the load that a pavement can safely carry may be considerably increased if the vehicles have tandem axles.

Static versus moving loads. The effects of transient loads on stresses and deflections are indicated in Figs. 6.9 and 6.3. For the data illustrated in both of these figures the stresses and deflections tend to decrease as the vehicle speed increases from creep speed to about 24 km/h. Above 24 km/h the values tend to be constant. In the study from which these data were abstracted it was found that the speed effect was much greater when the roadbase consisted of bituminous-bound instead of cement-bound materials. These differences were considered to be due to the moduli of deformation changing when the rate of loading was changed.

In practice the vehicle speed effect is most noticeable on particular sections of roadway. For instance it means that, for a given volume of traffic, greater thicknesses and/or quality of paving materials may be required for pavements in urban areas than for those in rural areas because of the lower

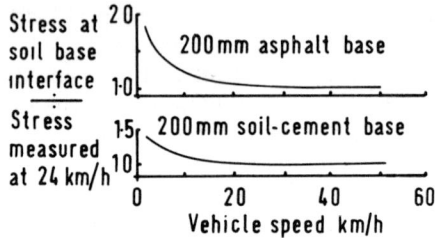

Fig. 6.9.   Variation of vertical stress at the subgrade-pavement interface with vehicle speed[6]

average speeds in urban areas. Similarly, pavement requirements for uphill gradients may be more demanding than for downhill gradients; there is little doubt that the increased distress shown by uphill traffic lanes can at least be partly attributed to the vehicle-speed effect.

Repetition of loads. Although the effect of material fatigue on highway pavement behaviour is little understood at this time, there is little doubt that it plays a critical role in pavement failures. On highways with bituminous surfacings a flexible pavement which fails by fatigue may be indicated by alligator or 'map' cracking of the surface. This form of cracking may be a

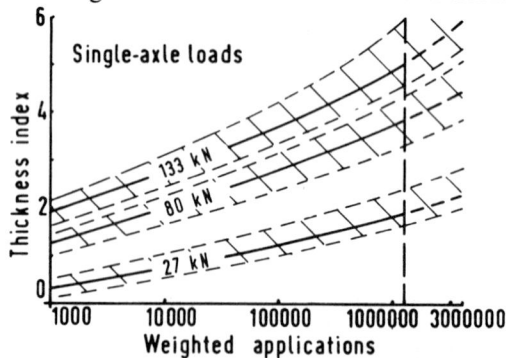

Fig. 6.10.   Relation between the number of axle applications and pavement thicknesses required to provide a given terminal level of service[2]

result of the fatigue characteristics of the bituminous material itself or it may reflect the effect of the repeated loading on the roadbase, sub-base and/or subgrade materials. In this latter respect roadbase aggregate materials may be broken down under the action of repeated loads, just as soil materials in the sub-base or subgrade may be caused to work their way upwards under the kneading action of traffic.

The effect of the number of repeated loads is directly indicated in Fig. 6.10 and indirectly illustrated in Fig. 6.11. Fig. 6.10, obtained during the AASHO Road Test, suggests that for a given axle load, the pavement thickness required to provide a given terminal level of service is proportional to the logarithm of the number of repetitions of the axle load. Figure 6.11, which is based on British test road data, clearly indicates that pavement deformation is a function of time and hence the number of load applications.

Pavement thickness and material. As already stated, the basic concept underlying flexible pavement design is that if the subgrade is overstressed and deflects then the pavement will deflect. Thus most design procedures attempt to evaluate the stability of the subgrade in some fashion, so that the thickness of the overlying material needed to safely distribute the applied wheel load to it can be estimated. In so doing many thickness design

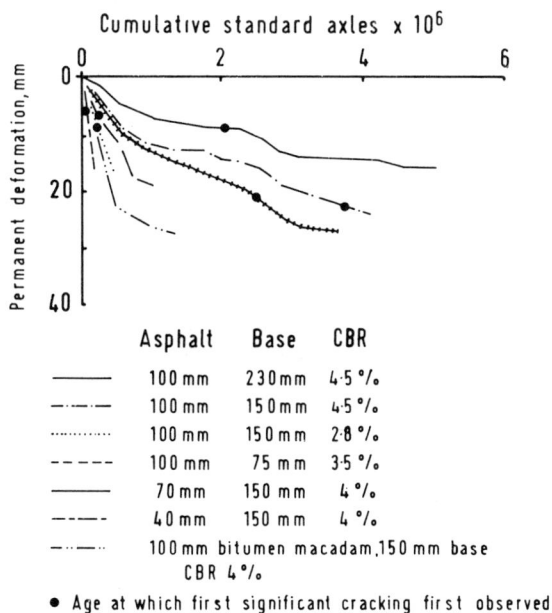

Asphalt	Base	CBR
100 mm	230 mm	4·5 %
100 mm	150 mm	4·5 %
100 mm	150 mm	2·8 %
100 mm	75 mm	3·5 %
70 mm	150 mm	4 %
40 mm	150 mm	4 %
100 mm bitumen macadam,150 mm base CBR 4 %		

● Age at which first significant cracking first observed

Fig. 6.11. Comparison of deformations for road sections with pavements having different surfacings and wetmix roadbase thicknesses on subgrades of varying C.B.R. strength[5]

procedures pay relatively little attention to the quality of the pavement materials and usually merely specify that they should meet certain minimum criteria.

It is here that the great weakness of the empirical methods is reflected since, as shown in Figs. 6.11 and 6.12, it is not only the pavement thickness but also the qualities of the pavement materials which determine stress distribution and resultant deformation. Figure 6.12 presents the results of some dynamic stress measurements taken at a heavily trafficked road on a uniform clay subgrade; the stresses shown were measured before the road was opened to traffic. The stresses under the rolled asphalt roadbase were lower than those under the other sections, lean concrete being next, while the soil-cement and crushed stone wet-mix were least successful in spreading wheel loads. Measurements were also made at intervals during several years after the highway was open to traffic and it was found that the stress under the rolled asphalt roadbase did not increase with time, whereas the stress under the lean concrete doubled during the first year of traffic and increased still further during the second year. The deterioration of the load-spreading

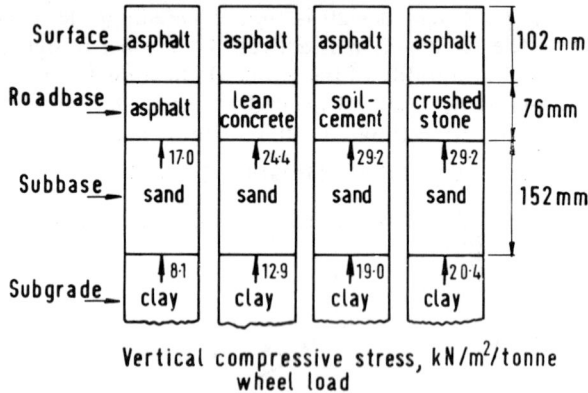

Vertical compressive stress, kN/m²/tonne wheel load

Fig. 6.12.   Effect of type of roadbase on subgrade stress[5]

properties of the lean concrete was considered to be due to excess cracking arising from the tensile stresses generated within it.

The stresses beneath the soil-cement base increased by about 90 per cent during the first year of traffic, after which they decreased again. The initial increase was probably due to the tensile stresses generated within the soil-cement layer. The later fall of stress was attributed to wetting of the subgrade through the cracks which developed in the surfacing; this decreased the modulus of deformation of the soil, thereby increasing the stresses in the upper layers of the road.

Although not illustrated in any of the diagrams shown in this chapter, studies have also been carried out to measure the effect of the road temperature on the load-spreading properties of flexible pavement materials. Basically, these studies have shown the necessity for taking pavement behaviour under extreme temperature conditions into account when designing a flexible pavement, particularly when high wheel loads are involved.[9] It was found that the load-spreading properties of bituminous roads can deteriorate markedly with rising temperatures. This is due to the reduction of the viscosity of the bituminous material, which in turn causes a reduction of the dynamic elastic moduli of the layers forming the road. These experiments also emphasize the necessity of recording temperatures when measuring the dynamic stresses or deflections of pavements under traffic.

The above comments re the effect of temperature behaviour on pavement stability emphasises the need for taking great care when deciding whether or not to utilize in a given environmental area design methods which have been developed elsewhere. For example, it is common practice in tropical zones to utilize pavement design procedures that have been empirically derived in shallow water-table temperate climate conditions. This can lead to considerable over-design where the water-table is deep, a common situation in much of the tropics.[10]

Edge loading. One of the most striking conclusions of the WASHO Road Test was the effect that the lateral positioning of the vehicles on the carriageway had upon pavement distress. For instance, Fig. 6.13, which is based on British test road data, shows that the most deformed portion of the pavement is about 1 m from the edge of the carriageway, i.e. beneath the nearside vehicle-wheel track in the slow outer traffic lane. This lane normally carries the great majority of the commercial vehicles. The farside traffic lane, which primarily serves only overtaking vehicles (mostly cars), is relatively little distressed.

Deformations near the edges of a road can be explained on the basis that they are due to the discontinuity of pavement and the lack of impervious covering. Because of the lack of impervious surfacing beyond the carriageway edges, moisture can find easy entrance into, and exit from, the pavement and subgrade, thus rendering them more susceptible to deformation. Because of the discontinuity of pavement, shear failure and lateral displacement of the pavement and/or subgrade can more easily occur.

The latter explanation is based on the concept of pavement failure which assumes that, as a result of the applied force at the surface, the particles in any layer of the pavement or subgrade are displaced along a curved path and develop an upward force against the underside of the layer above. The resistance to the subgrade displacement is provided by the interparticle friction and cohesion in the subgrade, the weight of the overlying pavement, and the flexural strengths of the overlying layers. Now, if the pavement is not continuous, as is so if there is no hard shoulder along the edge of the carriageway, the resisting force is very severely reduced and failure may take place.

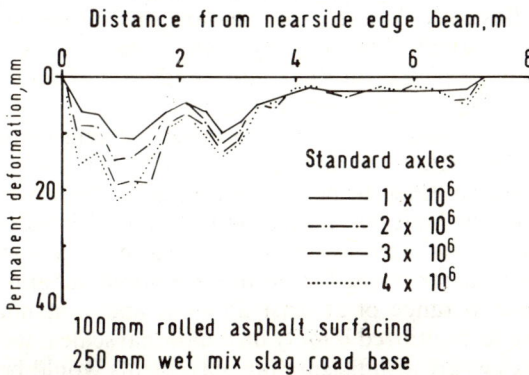

Fig. 6.13.   Typical transverse deformation profiles on the same pavement section at four different levels of traffic flow[5]

METHODS OF DESIGN

One of the more important applications of soil engineering is in the formulation of methods for calculating the minimum thickness and strength of highway pavement which must be placed on a subgrade in order to provide a stable riding surface for a specified traffic load. Prior to the early 1940's little attention had been directed towards the problems of pavement design, primarily because the methods of construction then in use gave satisfactory results. With the onset of the war, however, scientific attention became focused in this direction because of the urgent need to construct great mileages of roads and airport runways for heavy traffic, as quickly and as economically as possible. Since then the subject of pavement design has continued to receive much attention and, as regards roads in particular, it has been accepted that reliable methods of design can lead to important economies by avoiding costly failures and wastage due to overdesign.

It must be made clear at this stage that no 'correct' way of designing a flexible pavement has yet been developed. Indeed it can even be said that there is no universal agreement amongst engineers as to which design procedure is the 'best'. There are at least 25 design procedures in use throughout the world today; some of these differ only in their method of application while others are based on entirely different procedures. While they differ considerably in their reliability, they all have the common feature that they depend on the experience and academic knowledge of the engineer utilizing them.

It is not practicable in the space available here to discuss all the design methods that are now in use; nor is it necessary, since most of these methods can be gathered into five classification groups. These groups and typical examples from them will now be discussed in detail.

Methods based on precedent

Design methods of this group might also be called design procedures by rule-of-thumb. Basically they rely on standard thicknesses of pavement for particular types of road. Thus in very many cases the same thickness of construction may be placed on weak clay subgrades as on strong granular ones, as long as the highway type designates it.

Many satisfactory highways have been, and are being, constructed on the basis of precedent experience and there is no reason why this method of design should not continue to be used on a purely local basis where prior experience definitely indicates that economic and reliable results can be expected. However, the highway engineer should be very wary of designing on this basis over larger areas. For instance, normal subgrade strengths in Britain vary over a range of at least 25 to 1, and if standard pavement thicknesses were to be utilized blindly on a national scale it would inevitably result in instances where insufficient, thin, pavements would be laid on weak subgrades and unnecessarily thick pavements on strong subgrades.

Empirical methods using a dispersion angle for load distribution

Design methods falling into this group assume that the wheel load applied to the surface is distributed at a fixed angle in a downward direction. The thickness of pavement necessary to ensure that the subgrade is not overstressed is then determined on the basis of this assumption. The first and best known thickness formula utilizing this assumption is the Massachusetts' formula.

Massachusetts' and similar methods. First presented about 1901, the Massachusetts' formula assumes that the applied wheel load acts through a point on the surface and is distributed downwards in the form of a 45 degree isosceles pyramid. This results in a stress pyramid having a base that is square, and side dimensions equal to twice the height.

If the depth of pavement as determined by the Massachusetts' formula is t, the applied wheel load is P, and the allowable subgrade pressure is q, then the formula derived is as follows:

$$P = q(2t)^2$$

and

$$t = 0{\cdot}5(P/q)^{1/2}$$

No method of determining the subgrade bearing capacity was recommended for use with this formula.

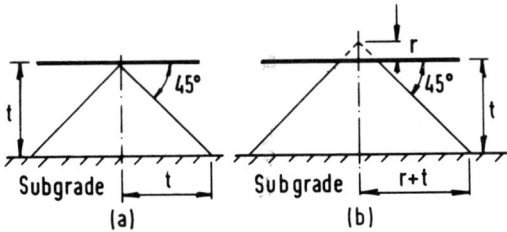

Fig. 6.14. Schematic diagrams of the manner in which the applied wheel load may be transmitted to the subgrade according to the cone theory

A relatively early modification of this formula assumed that the applied wheel load still acts through a point but is distributed downward in the form of a 45 degree cone instead of a pyramid. This assumption, which is illustrated schematically in Fig. 6.14(a), results in the thickness formula

$$t = 0{\cdot}564 \ (P/q)^{1/2}$$

Eventually it was recognized that the wheel load applied to the surface is not a point load, but is actually distributed over a small area. This gave rise to the concept that the load was applied over an equivalent circular area of radius r, and that it was then distributed through the pavement in the form of a conic frustum. This concept is illustrated in Fig. 6.14(b). For a 45°

transmission it means that both the height of the cone and the radius of its base are equal to $(r+t)$. Then

$$P = \pi q(r+t)^2$$

and

$$t = 0.564\,(P/q)^{1/2} - r$$

This formula, which was first introduced about 1934, is known as the Asphalt Institute design formula.

The usefulness of procedures of this type is very limited. While this approach represents a certain advancement over design by rule-of-thumb, its basic concepts lead to built-in error. Not only does the stress distribution not follow the simplified assumption, but there is no accepted method for determining the allowable subgrade stress to be used in the equations.

Empirical methods using soil classification tests

Pavement thicknesses as determined by procedures falling into this group rely upon previous experience of the thicknesses required for various wheel loads on different types of soil. Soil classification tests utilizing particle-size analyses, and liquid limit and plastic limit tests, may be used to describe the soils.

The two most well-known and representative of these procedures are the Group Index method and the Michigan State Highway Department method based on the pedological classification system.

Group Index method. In the chapter on Soil Engineering for Highways there was presented a soil classification system known as the HRB system. Within this classification system groups of soils are differentiated according to a particle analysis, liquid limit and plasticity index. Soils within each group are differentiated by means of a qualifying group index number obtained from the following formula:

$$\text{G.I.} = 0.2a + 0.005ac + 0.01bd$$

where a = that portion of the percentage passing the no. 200 sieve that is greater than 35 per cent and does not exceed 75 per cent, expressed as a positive whole number (1 to 40),

           b = that portion of the percentage passing the no. 200 sieve that is greater than 15 per cent and does not exceed 55 per cent, expressed as a positive whole number (1 to 40),

           c = that portion of the numerical liquid limit that is greater than 40 and which does not exceed 60, expressed as a positive whole number (1 to 20),

and d = that portion of the numerical plasticity index that is greater than 10 and which does not exceed 30, expressed as a positive whole number (1 to 20).

General evaluation of subgrade	Group Index range of subgrade	Daily volume of comm. vehs.			
		Light < 50	Medium 50-300	Heavy > 300	
Excellent (A-1-a)					Surface & roadbase thicknesses vary with volume of commercial traffic
Good	0-1	152mm	229mm	305mm	
Fair	2-4	102	102	102	Selected subbase thicknesses vary with subgrade characteristics
Poor	5-9	203mm	203mm	203mm	
V.poor	10-20	305mm	305mm	305mm	

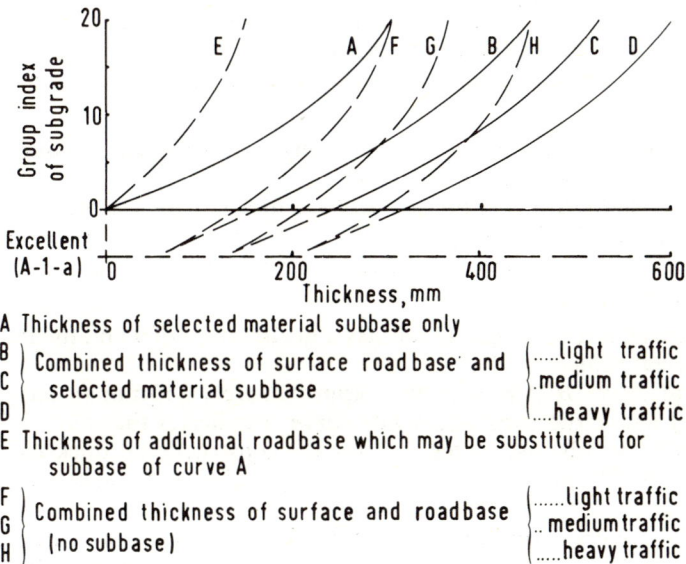

A Thickness of selected material subbase only

B } Combined thickness of surface roadbase and { ....light traffic
C } selected material subbase { .medium traffic
D } { ....heavy traffic

E Thickness of additional roadbase which may be substituted for subbase of curve A

F } Combined thickness of surface and roadbase { .....light traffic
G } (no subbase) { .. medium traffic
H } { .....heavy traffic

Fig. 6.15. Original Group Index design curves

When the HRB classification system was first presented it was accompanied by a discussion presenting a simplified pavement thickness design method based on the group index of the subgrade soil.[11] The original design curves associated with this method are shown in Fig. 6.15. These curves have since been modified by various highway agencies for their own purposes.

As can be seen from Fig. 6.15, two of the main variables that the Group Index design procedure attempts to take into account are the intensity of traffic and the characteristics of the subgrade.

The number of commercial vehicles per day is taken as the traffic criterion and it is assumed that the amount of daily wheel loads greater than 40 kN does not exceed 15 per cent. Three traffic classifications are used; light traffic is less than 50 commercial vehicles per day, between 50 and 300 per day is considered to be medium traffic, and heavy traffic is more than 300 per day.

Subgrades are divided into the following five groups: (*a*) Excellent soils are those classified according to the HRB designation A-1-a. (*b*) Good soils are those with group indices between 0 and 1. (*c*) Fair soils have group indices between 2 and 4. (*d*) Poor soils have group indices between 5 and 9. (*e*) Very poor soils have group indices between 10 and 20.

The manner in which the charts in Fig. 6.15 can be used for design purposes is illustrated by the following examples:

EXAMPLE 1. The subgrade soil is A-1-a and the number of commercial vehicles is 40 per day.

In this case the subgrade soil is composed of an excellent roadbase material in its own right, and so the addition of another roadbase is not necessary. For the traffic conditions expected, it will be quite sufficient to cover the subgrade with a 50 mm surface course; this is usually a bituminous surfacing.

EXAMPLE 2. The subgrade soil has a group index equal to 1 and the daily number of commercial vehicles is 40.

From curve *B* a combined roadbase and surface course thickness of 152 mm is obtained. The pavement will then be composed of at least a 50 mm bituminous surface course and a granular or crushed stone roadbase of a thickness of 102 mm.

EXAMPLE 3. The subgrade soil has a group index of 9 and the number of commercial vehicles is 40.

From curve *B* it is seen that approximately 355 mm of pavement is required above the subgrade, while curve *A* indicates that only 203 mm of this can be of sub-base material. Thus the constructional thickness may be made up of 50 mm of bituminous surfacing, 102 mm of granular or crushed stone roadbase, and 203 mm of sub-base material.

An alternative solution to Example 3 would be to omit the sub-base material entirely and, using curve *F*, have a total pavement thickness of 250 mm. In this case the thickness may be composed of 50 mm of bituminous surfacing on top of 203 mm of granular or crushed stone roadbase material.

Assumptions. The curves presented in Fig. 6.15 are expected to provide adequate pavement thicknesses for most climatic conditions. In areas of severe frost penetration, however, it is recommended that the values given by the curves should be qualified by local knowledge based on experience.

The thicknesses recommended by this method of design are based on the assumptions that the top 152 mm of the subgrade are compacted to at least 95 per cent of standard Proctor dry unit weight and that the sub-base and roadbase materials are compacted to not less than 100 per cent of the standard Proctor values. They also assume that the water-table is maintained at least 1–1·25 m below the carriageway.

Reliability. The group index method of pavement design has the particular advantage that it is a most simple design procedure. The data utilized directly are those normally collected as part of the standard soil classification procedure and thus no extra tests are required. It is scientific in that it attempts to allow for both the traffic conditions and the characteristics of the subgrade soil. Rather than attempting to evaluate the subgrade material by means of some empirical test, it attempts to take into account the effects of the various constituents which make up the soil. Thus a clayey soil, which would normally have a high group index, requires a substantial thickness of pavement above it, while a granular soil requires a relatively thin coverage.

In general, however, the pavement thicknesses recommended by the group index design procedure tend to err on the safe side; while this has certain advantages, it has the disadvantage that uneconomical pavement thicknesses may often be constructed. The most obvious reason for this lies in the fact that no attempt is made to take into account the different load-spreading abilities of pavement materials. A further important disadvantage of the procedure is that its usage is entirely dependent on particular conditions of subgrade moisture and compaction: if these conditions are not met in the field, the design curves are not applicable.

Michigan pedological method. This design procedure might in one sense be termed design by scientific experience. As has already been discussed in the chapter on Soil Engineering for Highways, soil scientists have for many years been studying soils with regard to their importance to agriculture. This form of scientific study is known as Pedology and the soil classification system utilized in it is called the Pedological classification system. It is this system that forms the basis of a design procedure used in Michigan in the United States.[12]

The design procedure is a simple one. Since the method is primarily concerned with the treatment of the subgrade soil, this is thoroughly investigated and eventually classified according to the pedological system, i.e. by soil series and soil type. The analyzed records of nearly 40 years of road construction on similarly classified soils are then consulted and from these it is possible to state what treatment must be applied to the subgrade in order to resist frost action damage, which is the most critical highway problem in that state. Treatment as used in this context normally refers to the need for drainage facilities, the necessity for stripping the top soil and/or constructing an embankment and, if an embankment is required, the manner in which it should be constructed.

Once the subgrade treatment has been selected, an appropriate standard pavement cross-section is chosen. This selection is based on an evaluation of the volumes of traffic and the frequency of particular highway loads that may be expected on the new facility.

Empirical methods using a soil strength test

These procedures normally utilize particular forms of penetration or bearing tests that are only applicable to their associated design methods.

Each procedure then recommends the necessary pavement thickness on the basis of past experience or, as in Britain, on test road results.

The best known of these methods is the California Bearing Ratio procedure, which is based upon a standard penetration test. This will be discussed here in some detail, since not only is it the procedure that is most widely used throughout the world today but it forms the first step in the pavement design method in use in Britain at this time.

C.B.R. test. The California Bearing Ratio (abbreviated to C.B.R.) design procedure was, as its name suggests, originally devised by engineers of the California Division of Highways following an intensive investigation into flexible pavement failures in California in the nine years prior to 1938.[13] It was taken up by the U.S. Corps of Engineers during the war and adapted for use in the design of airport pavements. As a result of the extensive investigations undertaken by the Corps[14] other organizations began to take an active interest in the design method and its acceptance became widespread after the war.

Although the state of California itself has now stopped using the design method, there is no doubt that the C.B.R. procedure is the most universally accepted pavement design method in use today.

To use the method in designing a pavement it is necessary to carry out a standard penetration-type load-deformation test and then, using values obtained from the test, an empirical design chart is entered and the pavement and/or layer thickness required is read off the proper curve. The original test procedure and design curves developed by the California investigators have since been modified by various agencies for their own purposes. While some of these modifications are minor in form and others represent fairly extensive changes, they all have the same basic outline in common, so that the engineer who has carried out the complete procedure according to one set of specifications can quite readily carry it out according to a different agency's requirements.

For this reason no attempt will be made here to discuss in detail the original C.B.R. procedure or to describe the various modifications to it. Instead emphasis will be placed on the method as used in Britain; the development and particular features associated with this procedure will be discussed in detail.

Laboratory test. The C.B.R. penetration test is carried out in the laboratory in the following manner:

A sample of soil is placed in a steel mould with attached collar that is 152 mm in diameter by a total of 178 mm high. This soil is then compacted, in a manner which will be discussed later, until a specimen 152 mm in diameter by 127 mm high is obtained that is at the moisture and unit weight conditions which the site investigation and consideration of available constructional methods and equipment indicate to be appropriate.

Following compaction the sample, still in the mould, is subjected to a surcharge weight equivalent to the estimated weight of flexible pavement expected above the layer from which the test soil is selected. Annular weights, each equivalent to 63·5 mm of flexible pavement, are used for this purpose. Small errors in estimating the surcharge load are of minor importance except in tests on cohesionless soils.

As indicated in Fig. 6.16, the specimen is next placed in a testing machine and a cylindrical plunger having an end area of 19·35 cm^2 is caused to penetrate the specimen at a standard uniform rate of 1·3 mm/min. The penetration of the plunger is measured by a dial gauge; at penetrations of 0·63, 1·27, 1·91, 2·54, 3·81, 5·08, 6·35 and 7·62 mm the applied loads are noted and permanently recorded.

After the test a load versus penetration curve is drawn for the data just obtained. Two examples of load-penetration curves are shown in Fig. 12.17(a). Since it was necessary to establish a scale for the assessment of soil strength, the original California investigators determined the pressures

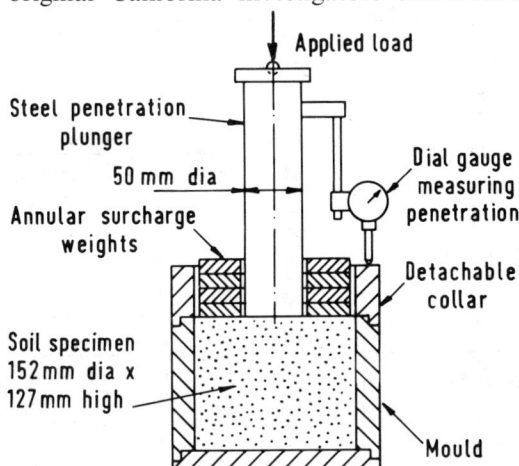

Fig. 6.16.   Laboratory California Bearing Ratio test

necessary to obtain the same penetrations in a compacted specimen of crushed limestone. The load penetration curve developed for this standard material is also shown in Fig. 6.17(a); this is the standard scale used in the procedure utilized in this country.

In practice an initial seating load is applied to the plunger before the loading and penetration gauges are set to zero and this seating load is then neglected in subsequent calculations. It sometimes happens, however, that the plunger is still not perfectly bedded in the soil and, as a result of this and other factors, a load-penetration curve with a shape similar to that of curve B in Fig. 6.17(a) may be obtained instead of the more normal shaped curve illustrated by curve A. When this happens the curve must be corrected by drawing a tangent at the point of greatest slope and then transposing the axis of load so that zero penetration is taken as the point where the tangent cuts the axis of penetration. The corrected load-penetration curve is the tangent from the new origin to the point of tangency on the re-sited curve, and then the curve itself. This correction is illustrated in Fig. 6.17(a).

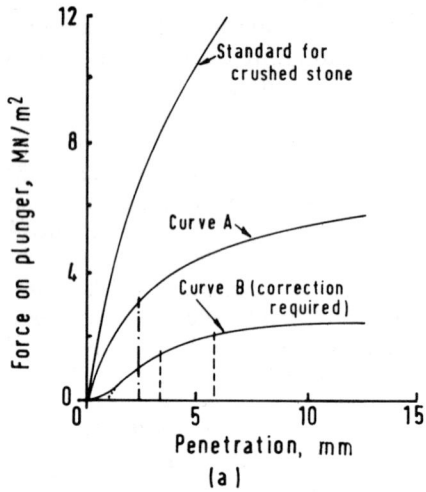

Fig. 6.17. California Bearing Ratio test data
(a) Examples of load-penetration curves.
(b) Typical laboratory C.B.R. values for British soils compacted at their natural moisture contents.

Type of soil	Plasticity index,%	CBR,%	
		Well drained*	Poorly drained
Heavy clay	70 60 50 40	2 2 2·5 3	1 1·5 2 2
Silty clay	30	5	3
Sandy clay	20 10	6 7	4 5
Silt	–	2	1
Sand(poorly-graded) Sand(well-graded)	Non-plastic ''	20 40	10 15
Well graded sandy gravel	''	60	20

*Water table at least 0·6m below formation level

(b)

The C.B.R. of a soil is then obtained by reading off from the curve the load which causes a penetration of 2·54 mm and dividing this value by the load required to produce the same penetration in the standard crushed stone mixture. This standard load is 6·895 MN/m². At the same time the load causing a penetration of 5·08 mm is determined and this is divided by 10·343 MN/m². In each case the answer obtained is multiplied by 100 so that the values are expressed in percentage form.

The two percentage values are compared and if, as is normal, the 2·54 mm value is greater than the 5·08 mm value, it is used for design purposes and is called the C.B.R. of the soil. If on the other hand the 5·08 mm value is larger, the test is entirely repeated on a fresh specimen. If the new percentage value at 5·08 mm penetration is still greater, then this is called the C.B.R. of the soil and is the value used for design purposes. Typical C.B.R. values for common British soils are given in Fig. 6.17(b).

In-situ test. Apparatus is also available which can be used to measure the C.B.R. of a soil in its natural habitat. This test is carried out in exactly the same way as the laboratory test except, of course, that the soil is not contained in any form of mould, and the load is applied to the plunger by means of a jack mounted on a suitable vehicle.

In-situ and laboratory C.B.R. tests only give equivalent results when fine-grained soils are being evaluated. Hence, since the actual design curves are based on the results of the standard laboratory test, the in-place test should never be used with granular soils.

The in-situ C.B.R. test can validly be used on subgrade soils that cannot be improved by further compaction, i.e. heavy clays, or when no further compaction of the subgrade is expected during the course of construction. In such cases the test must be carried out while the soil is in the worst condition of moisture content and unit weight that are likely to arise during the life of the road. This usually means testing the soil under the wet conditions of Winter.

Factors affecting C.B.R. test results. The C.B.R. test is entirely empirical and cannot be considered as even attempting to measure any basic property of the soil. Hence the C.B.R. of a soil can only be considered as an undefinable index of its strength which, for any particular soil, is dependent on the condition of the material at the time of testing. This means that the soil must be tested in a condition that is critical to the design. To ensure that proper values are obtained it is necessary for the designer to have a clear understanding of the effects of the method of compaction, the unit weight, and the moisture content on the C.B.R. of a soil sample.

Method of compaction. The highway designer is concerned with the strength of a soil when it is at a particular condition of moisture content and unit weight. The C.B.R. specimen may be brought to this condition by two methods of compaction.

In the first method, which is that favoured in Britain, the soil is mixed with the amount of water needed to bring it to the design moisture content and then a calculated amount of the mixture is placed in the steel C.B.R. mould with attached collar. The wet soil is ' rodded' continuously into the mould and then a 51 mm displacer disc is placed in the collar and the specimen is compressed in a compression machine until the top of the displacer disc is flush with the top of the collar. The soil specimen will then be exactly 127 mm high, and ready for testing.

The main advantage of this method lies in the fact that the desired unit weight is obtained directly without the need for any preliminary testing. In addition it requires less physical effort than other methods. Its principal disadvantage is that a vertical compaction gradient may be obtained within the specimen. Thus if dry unit weight measurements are made of the top, middle and bottom thirds of the specimen, significantly different values may be obtained.

The second method of compaction requires the specimen to be prepared dynamically. With this method a predetermined amount of wet soil is placed in the mould in five equal layers, each layer being compacted by applying a required number of blows with the standard or modified AASHO compaction hammer. The number of blows needed to compact the calculated amount of soil to the required height of 127 mm is determined on the basis of necessary preliminary testing.

This latter method has the advantage that a more uniformly compacted mixture is obtained in the specimen. In addition, it produces a soil structure that is believed by many authorities to be closer to that which is actually obtained under roller compaction in the field. Its main disadvantage is that some preliminary experimentation is necessary to find the thickness of each layer and the number of blows per layer that will produce the desired unit weight.

Unit weight. Generally a soil consists of mineral matter, water and air. At any given moisture content its strength will be increased if its dry unit weight is increased, i.e. if the air content of the soil is decreased. Figure 6.18 indicates the manner in which the C.B.R. of a heavy clay decreases as the

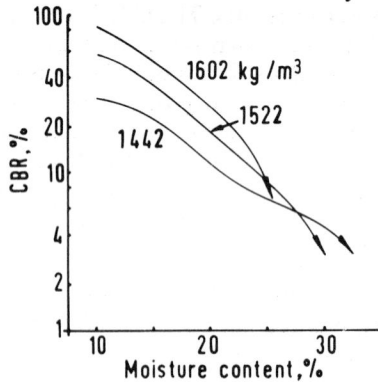

Fig. 6.18. Effect of moisture content and unit weight on the C.B.R. of a heavy clay

unit weight is decreased, the moisture content of the soil being kept constant. For this reason great emphasis must be placed on the concept of selecting a design dry unit weight which corresponds to the minimum state of compaction which it is anticipated will be achieved in the field at the time of construction.

Moisture content. Figure 6.18 also shows that the strength of a soil varies with its moisture content. Hence it is most important that the design C.B.R.

should be assessed at the highest moisture content that the soil is liable to have subsequent to the completion of roadworks. The selection of this moisture content requires an understanding of the way in which moisture moves within a soil.

Before a soil is covered by a road pavement, its moisture content will fluctuate seasonally; this is particularly noticeable when the soil is covered by a dense vegetation. At any given time the uncovered soil may be drier or wetter than the ultimate condition that will be reached when an impervious road structure is constructed over it. For instance, a gradient of moisture within the soil profile that is determined during the summer might look like line A in Fig. 6.19, while another taken in the winter might look like line B. Sometime after the construction of the pavement, the soil moisture conditions will become stabilized and subject only to changes due to fluctuations

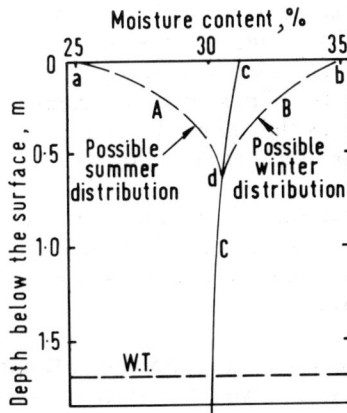

Fig. 6.19.   Diagrammatic illustration of some seasonal variations in the moisture content of a heavy subgrade soil

in the level of the water table. When this has happened the moisture gradient might look like line C.

Examination of Fig. 6.19 leads to the following conclusions regarding the moisture content at which the C.B.R. specimen should be tested if it is to be used to determine design values.

1. If the pavement is to be constructed during the winter, the soil should be tested at whatever is the moisture content at point b. This high moisture content must be used since the road may be opened to traffic before the moisture content can decrease to c per cent.

2. If the road is to be built during the summer the soil should be tested at the moisture content at point c, since this will ultimately be its stable moisture content.

A method has been developed by which the ultimate moisture content at c can be estimated for a given soil when the water table is at a given depth.[15] This requires laboratory suction tests on samples taken from the proposed formation level using a suction corresponding to the most adverse anticipated water-table conditions.

If facilities for conducting suction tests are not available, an estimate of the ultimate moisture content can be obtained from measurements taken at a point below the zone of seasonal fluctuation but at least 0·3 m above the water table. This is at or below point d in Fig. 6.19. In cohesive soils in Britain, particularly those supporting dense vegetation, the moisture content at a depth of 1–1·25 m will normally provide a good estimate of the ultimate moisture content. Sandy soils, on the other hand, do not normally support a dense cover of deep-rooted grasses and therefore seasonal fluctuations of moisture content are confined close to the surface. Thus the moisture content at a depth of 0·3–0·6 m will usually be adequate for estimating the C.B.R. test moisture content for this type of soil.

If a cohesive soil is to be used as an embankment material and then compacted, the disturbance to the soil structure is likely to cause the ultimate moisture content to be increased above the normal ultimate value for the undisturbed soil. To take this into account the moisture content measured at a depth of 1–1·25 m should be increased by about 2 per cent if a heavy clay is to be used as embankment material. For lighter silty and sandy clays used in embankments a corresponding addition of 1 per cent should be made.

In all cases, of course, the soil at the depth at which the moisture content is estimated must be the same as the soil which is to be compacted, otherwise large errors will be made in the estimation of the correct C.B.R. value to be used for design purposes.

British design procedure. What has happened in Britain relative to the application of the C.B.R. test as a basis for pavement design is in itself a useful illustration of the fact that the highway engineer must continually keep abreast of research developments, that knowledge never stands still. It is useful therefore to briefly trace the development of the design procedure now used in this country.

The original California investigators concluded that soils having the same C.B.R. required the same thickness of overlying flexible construction in order to prevent plastic deformation. Two design curves were originally prepared relating pavement thickness with the C.B.R. of the underlying materials, one for heavy traffic and the other for light traffic. An implied assumption underlying each of these curves was that all kinds of flexible construction of the macadam type spread the applied wheel load to the same extent.

The use of a range of wheel loads to identify the different design curves was never accepted in Britain, as it was considered to frequently lead to overdesign through catering so fully for the heaviest vehicles likely to use the road. Instead, curves were developed (see Figure 6.20) which took direct account of the total number of commercial vehicles having unladen weights in excess of 1·53 tonnes that were carried in the course of a busy 24-hour day. (Commercial vehicles were chosen because it is they that cause the main damage to roads.) The number of vehicles was determined from traffic counts in both directions if on a single carriageway, or on both carriageways if on a dual carriageway road.

Note that each design curve shown in Fig. 6.20 assumed a relation between the C.B.R. of a soil material and the thickness of flexible construction to be placed above the layer from which the soil material was taken, so that it would not be overstressed when loaded. If the material under evaluation was the subgrade soil, then the design curve gave the total thickness of pavement required. If the test was carried out on sub-base material, then the

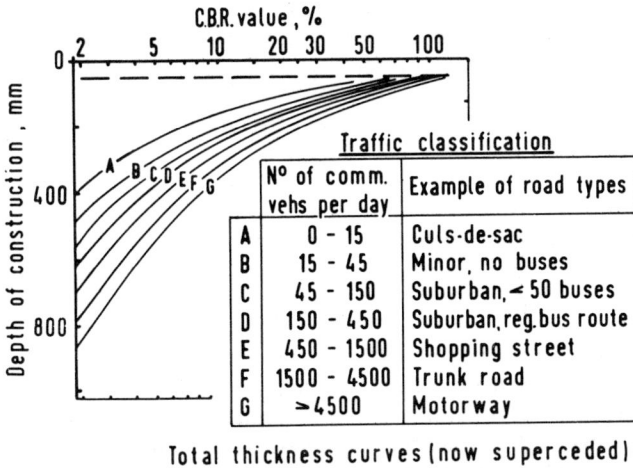

Total thickness curves (now superceded)

Fig. 6.20. The first stage in the development of the current British pavement design curves

curves were used to estimate what combined thickness of roadbase and surfacing was required, *irrespective of what materials were used in these layers.*

The design curves shown in Fig. 6.20 were superseded in 1960 by a new series of design charts. The overall thickness requirements recommended by these charts were based on failure investigations of roads which had generally been in service for at least 20 years after construction or major maintenance and, hence, these new designs were based on the traffic intensity expected after 20 years and not on the initial traffic.[16] Unless other data were available a growth rate of 4 per cent per year was assumed in calculating this future traffic.

By 1965 sufficient information had been gathered from the British test roads to differentiate between the performance characteristics of different roadbase materials, and so the curves were again amended to take this into account. Thus, for example, for the heaviest traffic category (>4500 comm. veh/day), the roadbase thickness could be reduced from 25·4 to 17·75 cm when asphalt was used as the roadbase material, and from 25·4 to 20·3 cm for dense-coated macadam roadbases. An equivalent increase in sub-base thickness was, however, made to maintain the total thickness requirements.

The design curves currently in use in Great Britain are given in Fig. 6.21. It should be noted that only one series of curves utilizes the C.B.R. procedure, i.e. the sub-base curves. Furthermore, the remaining curves (Figs. 6.21(b)–(e)) specify independently the thicknesses of surfacing and roadbase to be used with given materials for varying conditions of traffic flow, i.e. for different design lives. This is a major development as it enables the design engineer to see at a glance the implications in relation to pavement life, of increasing or decreasing the initial pavement thickness—and this in turn enables him to examine more fully the economics of stage construction. This latter point is quite important because additional surfacings are nowadays so often required throughout the design life of a pavement, in order to restore its skidding characteristics. Another feature of the current curves is that they are also immediately applicable to the design of private roads carrying specialized vehicles of known axle loadings, e.g. refinery and mineral roads. It may also be noted that the traffic intensities reflect the numbers of commercial vehicles per day in each direction.

Steps in design.   The British design procedure is perhaps best explained as a series of design steps, viz:

Step 1. Determine the number of commercial vehicles expected to use the roadway on the day it is opened, i.e. the 'present' or 'construction' traffic.

This information is usually obtained from traffic survey data. For certain types of roads, e.g. residential and associated developments, it is quite common for accurate data not to be available, and a detailed traffic survey may not be justified. In such instances, the Transport and Road Research Laboratory recommend that the following initial values be assumed:

Type of road	*Est. traffic flow in each direction, comm. veh/day*
1. Cul-de-sacs and minor residential roads	10
2. Through roads and roads carrying regular bus routes involving up to 25 p.s.v./day in each direction	75
3. Major through roads carrying regular bus routes involving 25–50 p.s.v./day in each direction	175
4. Main shopping centre of a large development carrying goods deliveries and main through roads carrying more than 50 p.s.v./day in each direction	350

Step 2. Determine the cumulative number of commercial vehicles (one direction of traffic) expected to use the design lane over its design lifetime.

The curves shown in Fig. 6.22 are used for this purpose. Those in Figs. 6.22(a)–(c) refer to each lane of 2-lane single carriageway roads, or to the 'slow' (nearside) lanes of single carriageways with more than two traffic lanes or of dual carriageways with up to three lanes per carriageway. These

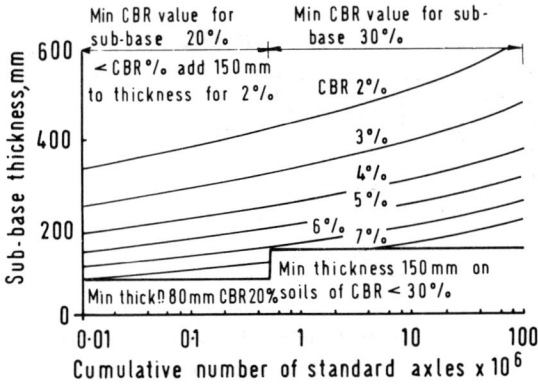

(a) Sub-base

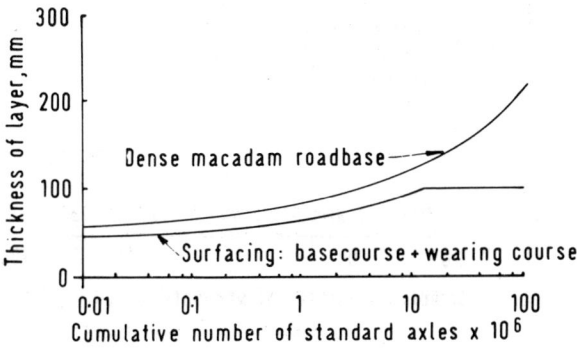

(b) Rolled asphalt roadbase

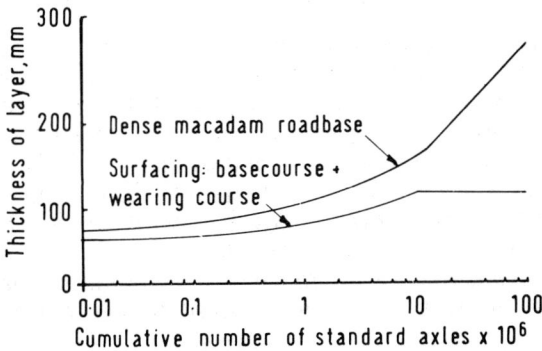

(c) Dense macadam roadbase

Fig. 6.21.   Flexible pavement design curves currently in use in Great Britain[17]
(See over for parts (d) and (e).)

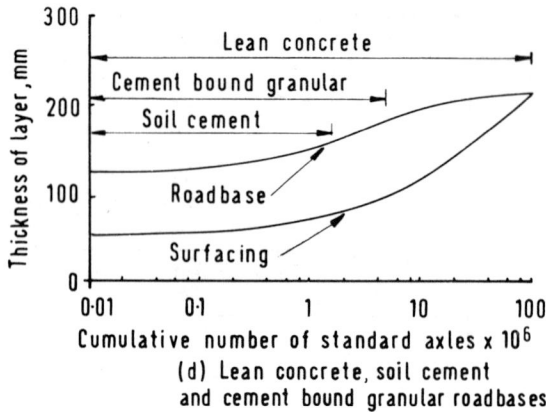

(d) Lean concrete, soil cement
and cement bound granular roadbases

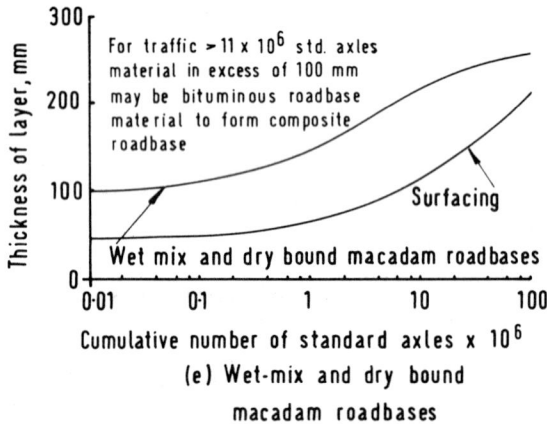

(e) Wet-mix and dry bound
macadam roadbases

Fig. 6.21.   Flexible pavement design curves currently in use in Great Britain[17]

curves also take into account the fact that the percentage of commercial
vehicles using the slow lane of a multilane dual carriageway decreases from
about 97 per cent for a carriageway carrying 3000 comm. veh/day to 70 per
cent for 13 000 comm. veh/day, while commercial vehicle usage of the adjacent lanes generally rises in a corresponding fashion.[18] The curves in Fig.
6.22(d) are used when it is desired to estimate the number of commercial
vehicles in the lane adjacent to the slow lane, with the ultimate intention of
utilizing a different pavement thickness in that lane.

For most flexible roads a design life of 20 years will normally be chosen,
i.e. because of the relative ease with which flexible roads can be given
overlays.

Step 3. Determine the equivalent 'life' number of standard axles to be
used for design purposes.

Having calculated the number of commercial vehicles expected for the
selected design year, the next step is to convert this number into an equiva-

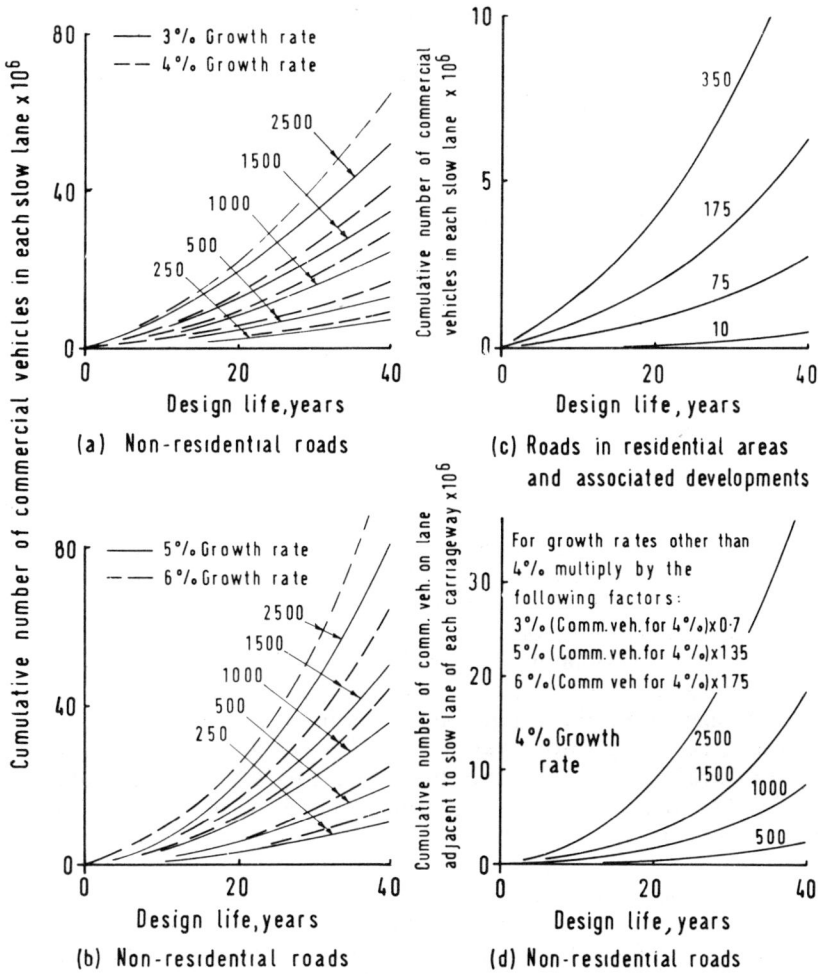

Fig. 6.22. Relationships between design life and cumulative number of commercial vehicles carried by each design lane. The numbers on each curve refer to the initial traffic expressed as the number of commercial vehicles per day in each direction[17]

lent axle value which can be used directly for design purposes. Studies carried out by the Transport and Road Research Laboratory[18] have shown that the number of axles on commercial vehicles used on different classes of British road in 1969 were slightly less than those shown in the first column of Table 6.4; however, as the number of axles per vehicle is continually rising, the Laboratory recommends that the equivalence values given in the table should be used for design purposes in Great Britain. As noted previously in relation to Table 6.3, it is possible to carry out empirical road tests to indicate different combinations of axle weight and configuration which will

TABLE 6.4.   *Conversion factors used to obtain the equivalent number of standard axles from the number of commercial vehicles.*

Type of road	No. of axles per comm. veh	No. of std axles/comm. axle	No. of std axles/comm. veh
1. Motorways and trunk roads designed to carry over 1000 comm. veh/day in each direction at the time of construction	2·7	0·4	1·08
2. Roads designed to carry 250–1000 comm. veh/day in each direction at the time of construction	2·4	0·3	0·72
3. All other public roads	2·25	0·2	0·45

cause given amounts of a damage to a pavement. Thus, as a result of extensive controlled road tests carried out under the auspices of the American Association of State Highway Officials, a series of 'equivalence factors' have been developed which enable the damaging powers of different axle loadings to be related to a common standard.[19] From these A.A.S.H.O. road test data, relationships were established whereby any given axle load could be equated with an 80·05 kN axle load in terms of its damaging effect on a pavement, e.g. a single-axle load of 177·95 kN was found to have a damaging effect equivalent to 22·8 times that of an 80·05 kN axle (see Table 6.5). Lacking similar British data, the Transport and Road Research Laboratory have accepted these relationships, and applied them to known distributions of axle loadings on British roads in order to obtain an average or 'standard' axle value which could be applied directly to the commercial vehicles on each of three major types of road (these are the values given in the middle column of Table 6.4). The product of the first two columns in the table then gives the average number of standard axles per commercial vehicle which, when multiplied by the cumulative number of commercial vehicles expected throughout the design life of the pavement, gives the value to be used for design purposes.

Step 4. Determine the sub-base thickness.

This is done by entering Fig. 6.21(a) with the appropriate cumulative number of standard axles and the subgrade C.B.R.-value. The C.B.R. of the subgrade is determined, preferably by means of a laboratory test at the estimated equilibrium moisture content and dry unit weight of the subgrade or, if the plasticity values of the soil and the depth to the water table are known, deduced from the data in Fig. 6.17(b). Note that for housing estate-type roads, i.e. those carrying $<0·7 \times 10^6$ standard axles, a lower quality of sub-base is permitted as compared with that recommended for the more heavily trafficked roads.

It should also be noted, however, that whilst all sub-base materials must be frost-resistant, there are certain circumstances when the frost-susceptibility of the subgrade may over-ride strength considerations in determining sub-base thickness.[20] Note also that if the C.B.R. of the subgrade is greater than the minimum requirement for the sub-base, then obviously no sub-base need be provided.

TABLE 6.5. *Equivalence factors and the damaging power of different axle loads*

Axle load, kgf(kN)	Equivalence factor	Axle load, kgf(kN)	Equivalence factor
910 (8·92)	·0002	9 980 (97·90)	2·3
1 810(17·76)	·0025	10 890(106·83)	3·2
2 720(26·68)	·01	11 790(115·66)	4·4
3 630(35·61)	·03	12 700(124·59)	5·8
4 540(44·54)	·09	13 610(133·51)	7·6
5 440(53·37)	·19	14 520(142·44)	9·7
6 350(62·29)	·35	15 420(151·27)	12·1
7 260(71·22)	·61	16 320(160·10)	15·0
8 160(80·05)	1·00	17 230(169·03)	18·6
9 070(88·98)	1·5	18 140(177·95)	22·8

Step 5. Determine the roadbase and surfacing thicknesses.

These thicknesses are obtained from Figs. 6.21(b)–(e). Note that for the very high quality roadbase materials, i.e. rolled asphalt and dense coated macadam, the thickness of surfacing increases up to 100 mm with increasing traffic; above 11×10^6 standard axles any further increase in pavement strength is obtained only by increasing the roadbase. In all instances, the quality of the surfacing material increases as the traffic to be catered for is increased.

Step 6. Consider whether there is an overall economic advantage to be gained from designing a tapered pavement cross section.

This normally applies only to roads with four or more lanes. It refers to the obvious fact that if the pavement design is based on the slow lane (which normally carries between 70 and 90 per cent of the commercial vehicles), then the use of a uniform thickness of construction must represent a significant over-design with respect to the other lanes. It is then a matter of determining whether the savings derived from designing a tapered cross-section are more than offset by the disbenefits from having to provide different crossfalls for the formation, sub-base and roadbase.

Illustrative examples. The manner in which the above steps are utilized is illustrated in the following two design examples.

Problem 1. A 3-lane dual carriageway motorway is required to carry 2500 comm. veh/day (in each direction) at the time of construction. This traffic is expected to grow at the rate of 4 per cent per annum. The soil is classified as CH under the Extended Casagrande system, and has a plastic index of 41 per

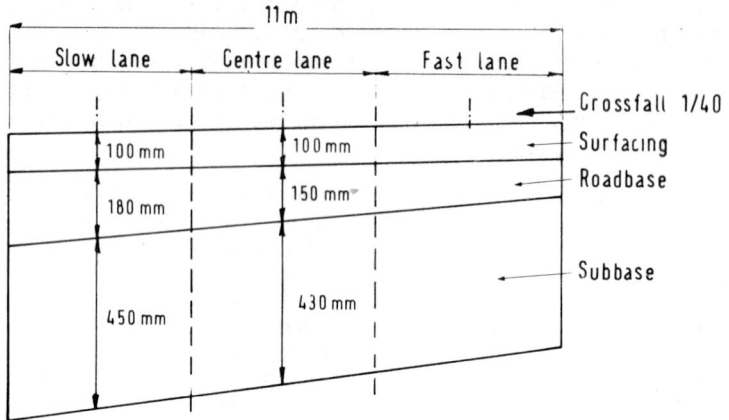

Fig. 6.23.   Example taper design cross-section

cent. The water-table is some 1550 mm below the proposed formation level. The design life is to be 30 years.

1. The present traffic is 2500 comm. veh/day per carriageway.

2. Fig. 6.22(a) indicates that during a 30-year design life, the slow lane will carry about 39×10^6 commercial vehicles.

3. The number of standard axles to be used for design purposes is 1·08 (from Table 6.4) multiplied by $39 \times 10^6 = 42 \times 10^6$ approx.

4. From Fig. 6.17b the estimated C.B.R. of the subgrade is 3 per cent. The required sub-base thickness is therefore 450 mm of material with a C.B.R. of greater than 30 per cent (see Fig. 6.21(a)).

5. From Fig. 6.21(b)—which, in this case, assumes a rolled asphalt roadbase—the thickness of roadbase required is about 180 mm. The surfacing thickness is 100 mm. (The materials and thicknesses to be used in the surfacing are described in Chapter 7.)

6. During a 30-year design life, the remaining lanes will carry 21×10^6 commercial vehicles (see Fig. 6.22(d)), or about $22·5 \times 10^6$ standard axles. This indicates sub-base roadbase and surfacing thicknesses of 430 mm, 150 mm (rolled asphalt), and 100 mm, respectively. This represents a reduction in the sub-base requirement of 4 per cent, and 17 per cent in the roadbase thickness, as compared with the slow lane design.

A suitable pavement cross-section would therefore be as indicated in Fig. 6.23. Note that this form of taper design does not normally allow the surfacing thickness to be lessened from lane to lane; any reduction in thickness must be confined to the sub-base and roadbase. Furthermore, discontinuities across the width of the carriageway are obviously not desirable, and so the calculated design thicknesses should be applied to the centrelines of the slow lane and middle lane; this in turn fixes the thickness design at both edges of the carriageway.

Problem 2.   It is intended to build a 2-lane road within an oil refinery along a route where the C.B.R. of the subgrade soil is 4 per cent. The daily traffic (each direction) on the day it is opened will likely be 100 passages of 4-axle vehicles each exerting a force of 89 kN through each of the two rear axles, 71 kN on the second axle, and 27 kN on the front axle; 200 passages of 3-axle vehicles with loads of 89 kN on each of the two rear axles and 18 kN on the front axle; and 100 passages of 2-axle vehicles with 80 kN on the rear axle and 27 kN on the front axle. The road is required to have a design life of 20 years.

    1. The traffic in each direction is composed of six hundred 89 kN axles, one hundred 80 kN axles, one hundred 27 kN axles, and two hundred 18 kN axles.

    2 and 3. The number of standard axles is calculated as follows:

Axle load, kN	No. during design life	Equivalence factor	No. of std. axles $\times 10^6$	Total $\times 10^6$
89	$600 \times 365 \times 20$	1·5	6·57	
80	$100 \times 365 \times 20$	1·0	0·73	7·77
71	$100 \times 365 \times 20$	0·61	0·45	
27	$200 \times 365 \times 20$	0·01	0·015	
18	$200 \times 365 \times 20$	0·0025·	0·004	

    4 and 5. Reference to Figs. 6.21(a)–(e) gives the following design:

Surfacing:	
With rolled asphalt or dense	
macadam roadbase	100 mm
With lean concrete or wet-mix	
roadbase	110 mm
Roadbase:	
Rolled asphalt	120 mm
Dense coated macadam	140 mm
Lean concrete	190 mm
Wet-mix	210 mm
Sub-base:	310 mm

    Before finally deciding on the pavement thickness, the width of road should be considered. If the carriageway is so narrow that vehicles moving in the two directions follow essentially the same tracks, the design will need to cater for, say, double the above traffic, assuming the vehicles are equally loaded in each direction.

Methods based partly on theory and partly on experience

    Procedures falling into this category estimate the strength or stress-strain properties of the subgrade by means of some form of shear or bearing test.

The values obtained from the test are then utilized in a simplified theory that has been shown to have experimental justification.

A noteworthy example of these methods is the California Stabilometer procedure devised by F. N. Hveem of the California Division of Highways. This procedure[21] replaced the California Bearing Ratio as the pavement design method in that state soon after the war.

Stabilometer design method.[22] Whereas most empirical design procedures attempt to estimate the thickness of pavement necessary to distribute the applied wheel stresses to the subgrade, the Stabilometer method requires the pavement to be designed to resist any lateral and upward movement of soil particles which may take place in the manner illustrated in Fig. 6.24. The three factors considered to be of major importance in developing the flexible pavement thickness and which are taken into account by this procedure are:

1. *The structural quality of the subgrade soil.* This quality is measured by means of stabilometer and expansion pressure tests; it is expressed in terms of a resistance value, R.

2. *The traffic conditions.* The design procedure considers that the required thickness of cover over any particular layer is proportional to the average tyre pressure, the square root of the effective tyre imprint area, and the logarithm of the number of load repetitions. In utilizing these considerations the magnitude and number of wheel loads estimated for the design year are converted into equivalent numbers of 49·05 kN wheel loads.

3. *The tensile strengths of the pavement layers.* The required thickness of a given layer or group of layers is considered to be inversely proportional to the 5th root of its tensile strength. This tensile strength is expressed in terms of a cohesive value.

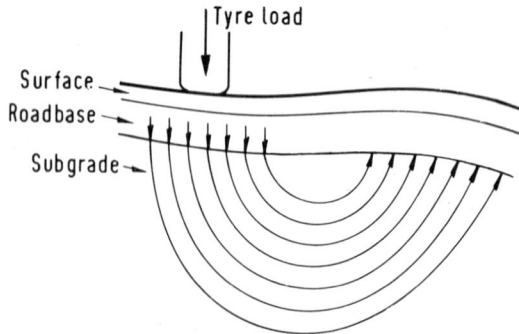

Fig. 6.24.   Diagrammatic example of subgrade failure

Preparation of test specimens.   Prior to using the actual design procedure it is necessary to carry out a series of tests in order to obtain basic relative data. All these tests require the preparation of soil specimens by means of a unique compaction procedure. The first step in this procedure requires the preparation of sufficient soil to make three test specimens, each 102 mm in diameter by 63·5 mm high. Soil used in these specimens is restricted to material passing the 19 mm sieve, since it is considered that the resistance to plastic flow is dependent on this portion; all larger particles are considered to be merely floating in a matrix of the passing 19 mm material.

The soil for the first test specimen is mixed with slightly more moisture than is expected to produce saturation of the specimen voids after compaction. The soil is then compacted in a standard mould by a special kneading-type compactor which compacts neither by straight compression nor damaging impact, but instead makes a series of individual impressions with a vertical roving ram or tamping foot which has its face shaped as a sector of a 102 mm diameter circle. This small-size tamping foot has a kneading action which is supposed to simulate that given to the road pavement by rollers. Compaction of the specimen is achieved by 100 applications of the tamping foot, each application being applied to a different sector of the surface.

Exudation-pressure test. The compacted specimen, still in its mould, is next subjected to loading over the whole of the specimen face at the rate of 8·9 kN/min. This compression is continued until free moisture is forced from the sample. The pressure at which this occurs is called the exudation pressure, and it should normally occur at a pressure of about 2·07 MN/m^2. The value of 2·07 MN/m^2 is desired because the California investigators found that soils underlying pavements in their State exuded moisture under pressure of about this value.

On the basis of the results from the first specimen, two other test specimens are prepared at higher or lower moisture contents so as to bracket the required exudation pressure. A special apparatus is then used to determine the pressure under which moisture exudes from each test specimen. One might say that the exudation-pressure test is simply another way of determining the design moisture content. This is assumed to be the point at which enough water is present in the soil to produce saturation.

Expansion-pressure test. After allowing sufficient time for any soil rebound to take place, the three exudation test specimens, still in their moulds, are placed in an expansion-pressure testing apparatus for 24 hours. This is a device in which each specimen is confined under a perforated disc covered with 200 ml of water. The tendency of the soil specimens is to absorb the water and expand, each to a different degree because of initial differences in moisture content and unit weight. This tendency is resisted and the force that is required to stop each specimen from expanding is called the expansion pressure.

This test is carried out to determine the force necessary to prevent a soil from swelling in the field when a supply of water is made available to it. This preventive force can be translated into the depth of cover material required to resist the expansion. If the test soil is from the subgrade, then the depth of cover is of course the total pavement thickness.

Stabilometer test. Upon completion of the expansion test each specimen is removed from its mould and tested in the Hveem stabilometer. A diagrammatic representation of a specimen in the stabilometer is shown in Fig. 6.25. In a restricted sense the stabilometer test is a form of triaxial test except that

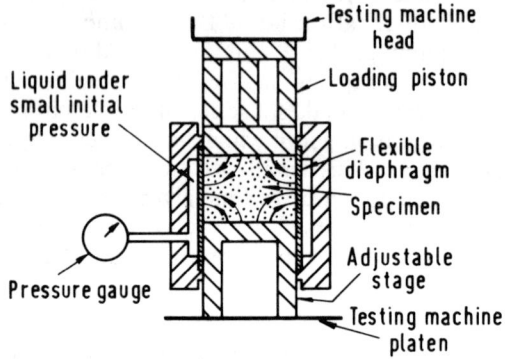

Fig. 6.25.   Diagrammatic sketch of Hveem stabilometer test

Note. The specimen is enclosed in a flexible diaphragm which transmits the horizontal pressure to the liquid. The magnitude of this pressure is then read on the gauge.

the results are influenced primarily by the resistance due to internal friction and only slightly by the portion of the total resistance that may be due to cohesion.

The strength of the soil as determined by the stabilometer test is called its resistance value R; it is expressed as a function of the ratio between an applied, arbitrarily chosen vertical pressure and the transmitted lateral pressure. The vertical force is applied slowly at a rate of 1·27 mm/min until it reaches a total of 4·45 MN or 1·1 MN/m². The R-value is then calculated from the following equation:

$$R = 100 - \frac{100}{\frac{2·5}{D}\left(\frac{P_v}{P_h} - 1\right) + 1}$$

where R = resistance value,

P_v = applied vertical pressure of 1·1 MN/m²,

P_h = horizontal pressure, MN/m² (obtained from stabilometer gauge when $P_v = 1·1$ MN/m²),

and D = displacement of stabilometer liquid due to increasing the horizontal pressure from 34·5 to 690 kN/m²; it is measured in terms of the number of revolutions of a calibrated pump handle. This is essentially a correction factor to allow for the surface roughness of the test specimen and the penetration of the flexible diaphragm into it.

Examination of this equation reveals that the R-value of a rigid material $(P_h = 0)$ is 100, whereas that of a liquid material $(P_h = 1·1 \text{ MN/m}^2)$ is zero.

Cohesiometer test. As indicated in Fig. 6.24, the subgrade material, when deforming, tends to bend the layers about it. Hence the top part of each overlying layer in the pavement must possess cohesive strength in order to resist tensile deformation. For this reason, in the stabilometer design procedure, it is necessary to take account of the tensile strength of each overlying layer.

Hveem devised a simple device called the Cohesiometer in order to measure the tensile strength of a compacted specimen. The tensile strength measurement obtained with this apparatus is expressed in grammes per square inch; it is approximately equal to the modulus of rupture multiplied by 45·4.

When the cover material consists of more than one layer, the cohesiometer value for the combined layer can be estimated by taking two layers at a time and determining an equivalent value from the following equation:

$$C_m = C_l + \left(\frac{t_2}{t_1 + t_2}\right)^2 (C_2 - C_1)$$

where C_m = equivalent cohesiometer value of the two layers,
　　　　C_l = cohesiometer value of the top layer,
　　　　C_2 = cohesiometer value of the next lower layer,
　　　　t_1 = thickness of the top layer,
and 　　t_2 = thickness of the next lower layer.

Design procedure. The actual design procedure is based on three concepts:

1. There must be sufficient thickness of overlying material to provide a weight which will prevent expansion of the underlying material after placement, thereby preventing it from taking up more moisture. This thickness estimate is determined on the basis of the expansion-pressure test results.

2. There must be a sufficient thickness of cover material to prevent the plastic deformation of the underlying material. This thickness is determined from the results of the stabilometer tests.

3. There must be a sufficient thickness of cover material to resist strength loss due to accumulation of moisture while the highway is in service. This cover thickness is determined from the exudation-pressure tests.

The use of these concepts is best illustrated by the following example:

PROBLEM. A subgrade soil is to be overlaid by a pavement composed of (it is estimated) 102 mm of bituminous surfacing and 203 mm of soil-cement roadbase. The equivalent cohesiometer value for the pavement is 375 g/in². The current daily traffic estimate of commercial vehicles is 774 with two axles, 212 with three axles, 68 with four axles, 118 with five axles, and 112 with six axles. It is expected that this traffic will increase by 50 per cent over the next ten years. Tests on the subgrade soil gave the following results:

Moisture content %	R-value	Expansion pressure kN/m^2	Exudation pressure MN/m^2
22·3	53	8·62	4·14
23·2	43	3·38	3·10
26·3	14	0	1·17

The average unit weight of the pavement materials is $2080\,kg/m^3$.

Determine whether the assumed pavement thickness is adequate for the design conditions.

SOLUTION. The first step is to estimate the design traffic index. This is expressed in terms of the total number of 22·24 kN equivalent wheel loads, abbreviated EWL, in one direction that is expected over the service life of the highway. Only commercial vehicles are considered in this analysis, the heavier vehicles being given greater EWL weightings determined on the basis of the pavement studies in California.

Thus the current traffic can be expressed in EWL repetitions as follows:

No. of axles	Current 1-way ADT		EWL weightings		Product
2	774	×	370	=	286 000
3	212	×	910	=	193 000
4	68	×	2000	=	136 000
5	118	×	3120	=	368 000
6	112	×	2200	=	246 000
Total current EWL repetitions				=	1 229 000

Therefore the design number of EWL repetitions in one direction expected over the course of the next 10 years is:

$$10 \times 1\,229\,000 \times \frac{1\cdot0 + 1\cdot5}{2} = 15\cdot4 \text{ million EWL}$$

This EWL design number is next converted into a design Traffic Index by means of the empirical formula:

$$\text{T.I.} = 1\cdot35 \text{ EWL}^{0\cdot11}$$

In this case,

$$\text{T.I.} = 1\cdot35 \times 15\,400\,000^{0\cdot11}$$

$$= 8\cdot9 \text{ (say 9)}$$

Knowing the equivalent cohesiometer value of the cover material and the traffic index for the road, the next step is to calculate the thickness of pavement as required by the R-values obtained from the subgrade soil-test specimens. These are calculated from the empirical equation:

$$T = \frac{2 \cdot 413(\text{T.I.})(90 - R)}{C^{1/5}}$$

where T = thickness of cover material, mm
  T.I. = traffic index,
  R = stabilometer resistance value,
  C = cohesiometer value,
and $2 \cdot 413$ = constant, which includes a factor for safety.

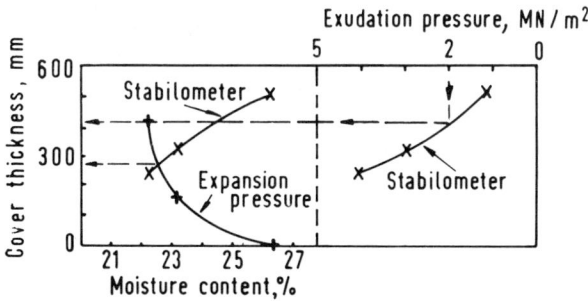

Fig. 6.26. Determination of pavement thickness by Hveem method

Using this equation cover thicknesses of 249, 318 and 510 mm are required for R-values of 53, 43 and 14 respectively. These thicknesses are then plotted against their respective moisture contents in Fig. 6.26.

The next step is to calculate the thickness of cover required to resist the expansion pressures. For each moisture content the thickness is obtained by dividing the tabulated expansion pressures by, in this instance, 20·4 kN/m^2 which is the force exerted per metre depth of pavement. These thicknesses are 0·422, 0·166 and 0 m respectively; they are also plotted in Fig. 6.26.

The thicknesses required by the stabilometer values are next plotted against their respective exudation pressures in Fig. 6.26.

From these three curves it is possible to deduce two factors. Firstly, the pavement thickness which will resist expansion of the underlying soil, while at the same time it prevents plastic deformation of the soil, is 267 mm. Secondly, the pavement thickness required for an exudation pressure of 2·07 MN/m^2 is 420 mm. Thus it would appear that, in order to resist all detrimental forces, the assumed pavement thickness of 305 mm is inadequate and the design procedure should now be repeated for a new pavement thickness of 420 mm.

Methods based primarily on theory

These methods are based on completely theoretical analyses of the stresses and strains within the pavement and the subgrade. At this time there is no doubt that none of the existing procedures falls truly within this design category. Most theoretical approaches attempting to fill this gap have had to make certain limiting and arbitrary assumptions which nullify the completely theoretical aspects.

One example of the modified theoretical approach is the official method of pavement design used in the U.S.S.R.[23] The SOJUSDORNII method of design, which came into use in 1943 is based on the following thesis:

1. The stress conditions in the road structure at the beginning of failure are characterized by a definite ultimate value of the deflection of the pavement.

2. The modulus of deformation E, which depends on the unit applied load p and the total relative deflection λ caused by the applied load, is the criterion for determining the resistance to deformation of the pavement. This relative deflection λ is a non-dimensional value obtained by dividing the absolute deflection Δ by the diameter r of a circle whose area is equivalent to the contact area of the dual-tyre wheels of a design vehicle. The λ-value used for design purposes is predetermined by the operating qualities desired of the roadway—in the instructions for thickness design of flexible pavements issued in 1960, the following limit values were given for λ:

Asphalt concrete	0·03	Unbound roadbase	0·05
Asphalt macadam	0·35–0·04	Subgrade	0·01–0·02

When determining the modulus of deformation, its value is calculated at the design figure chosen on the basis of experience as the relative deflection of the given pavement layer. Account is also taken of the moisture conditions expected and the manner in which they affect the modulus determination.

3. As a rule the pavement is a multi-layer construction, each layer having a different E-value. It is therefore designed so that the absolute deflection of each layer does not exceed the allowable design value under the action of the design loading, taking the nature and intensity of the road traffic into account.

The design load is obtained by multiplying the dual-type wheel load of a design vehicle by a constant K, where

$$K = 0\cdot5 + 0\cdot65 \log N$$

In this equation N is the number of design vehicles passing along a two-lane road in 24 hours in the season of maximum and most frequent weakening of the pavement. The coefficient K is determined on the basis of laboratory and site observations, and takes into account the influence of both the repeated loads and the disturbance caused by the dynamic actions of moving vehicles.

Vehicles of various types are converted to the design vehicle by an equivalent-action substitution. The U.S.S.R. studies have led their investigators to the conclusion that the effect produced on the pavement by vehicles of various types can be determined with sufficient accuracy by utilization of the ratio between the products of the unit pressure and the diameter of the contact area of the design vehicle and the vehicles actually on the road.

It is easiest to discuss the detailed design procedure by considering a basic two-layer system consisting of an upper layer of thickness h_1 and modulus of deformation E_1, resting on top of a semi-infinite subgrade layer with modulus E_2. A semi-empirical formula was developed which enables the pavement thickness h_1 to be replaced by an equivalent thickness of subgrade h_{equiv}, where

$$h_{equiv} = h_1 \left(\frac{E_1}{E_2}\right)^{1/2 \cdot 5}$$

The introduction of the equivalent layer concept allows the use of a theory of stress distribution in a uniform solid which is similar to Boussinesq's theory. According to this theory the vertical stress distribution beneath the pavement and along the centre line of the applied force is given by

$$\sigma_z = \frac{p}{1 + a \left(\dfrac{Z + h_{equiv}}{r}\right)^2}$$

where σ_z = stress in the bottom layer at a depth Z below the contact plane between both layers,

a = correction factor, taken in the given case as equal to 1, due to the higher concentration of stresses in real materials in comparison with the ideal elastic-isotropic semi-infinite layer. (Note that when $a = 2 \cdot 5$ the stresses obtained by this equation practically coincide with the values given by Boussinesq's theory),

and p, r and h_{equiv} are as defined before in this section.

The calculation of the amount of vertical deformation assumes a linear relationship to exist between stress and strain. The compression of an elementary layer of thickness dz located at a depth Z is given by

$$d\Delta = \frac{\sigma_z \, dz}{E}$$

and the total deflection by

$$\Delta = \int_0^\infty \frac{\sigma_z \, dz}{E}$$

This latter equation is integrated separately between the limits of 0 to h_1 (for the upper layer with modulus of deformation E_1) and from h_1 to ∞ (for the semi-infinite layer with a modulus E_2). As a result the total vertical deflection of a two-layer system is given by the following equation:

$$\Delta = \frac{pr}{E_2} \left[\frac{\pi}{2} - \left\{ 1 - \frac{1}{2 \cdot 5 \sqrt{(E_1/E_2)^{3 \cdot 5}}} \right\} \left\{ \arctan \frac{h_1}{r} \, ^{2 \cdot 5}\sqrt{E_1/E_2} \right\} \right]$$

When $E_2 = E_1 = E_{equiv}$, the total vertical deflection can be written as the deformation of an equivalent uniform solid, and

$$\Delta = \frac{\pi}{2} \frac{pr}{E_{equiv}}$$

Combining these two equations permits the derivation of an equation for the equivalent modulus of deformation for a two-layer system. Thus,

$$E_{equiv} = \frac{E_2}{1 - \frac{2}{\pi} \left[1 - \frac{1}{\sqrt[2.5]{(E_1/E_2)^{3.5}}} \right] \left[\arctan \frac{h_1}{r} \sqrt[2.5]{E_1/E_2} \right]}$$

If the allowable relative deformation is taken as λ, the following equation is derived which can determine the equivalent modulus of deformation, E_{equiv}, that is required so that the allowable λ-value will not be exceeded. Thus,

$$\lambda = \frac{\Delta}{r} = \frac{\pi}{2} \times \frac{pK}{E_{equiv}}$$

Therefore,

$$E_{equiv} = \frac{\pi}{2} \cdot \frac{p}{\lambda} (0.5 + 0.65 \log N).$$

Design steps. When designing flexible pavements using the SOJUSDOR-NII equations the steps to be followed are:

Step 1. Decide upon the level of service that is desired of the pavement. This is reflected in the values of λ selected for design purposes.

Step 2. Estimate the expected traffic conditions. This estimate is then expressed in terms of the number of design vehicles (N, p, r) that the pavement will have to carry.

Step 3. Calculate the equivalent modulus of deformation, E_{equiv}, that is required so that the allowable relative deformation λ will not be exceeded.

Step 4. Determine the subgrade modulus E_2 and the modulus of the upper layer E_1 (see ref. 23).

Step 5. Knowing E_{equiv}, E_2 and E_1, calculate the depth of pavement h_1 that is required above the subgrade.

Normally there are more than two layers to be considered when designing a flexible highway pavement. The procedure for this is similar to that described above except that E_{equiv} for the subgrade and the layer immediately above it is first determined, then E^1_{equiv} is determined for the system composed of the next upper layer and a semi-infinite layer with a modulus of deformation E_{equiv}; this is continued until the equivalent modulus for the entire multi-layer structure is determined.

Note on pavement design research

As will be gathered from the previous material, pavement design practice is, on the whole, still very much based on experience, coloured by results

DESIGN REQUIREMENTS

Service life
Commercial traffic
Geographical location
Failure criteria: Cracking and Deformation

Comparison

DESIGN DATA

Thicknesses and materials of
pavement layers.
Dynamic moduli and Poisson's ratio
of pavement layers and soil.
Material and pavement costs

Wheel load spectrum
No. of wheel loads in service life

Deformation and cracking of
pavement layers and soil under
repeated loading

CALCULATION

Multi-layer elastic theory used
to compute stresses and strains
in pavement for range of wheel
loads: Also transient surface
deflection under standard wheel
load

Estimates of cracking and
deformation for pavement
layers, and hence for whole
pavement

FULL OR PILOT SCALE TEST

Measurements of deformation and cracking
and transient surface deflection

Comparison
Comparison

ACTUAL ROAD

Measurement of transient
surface deflection

Comparison

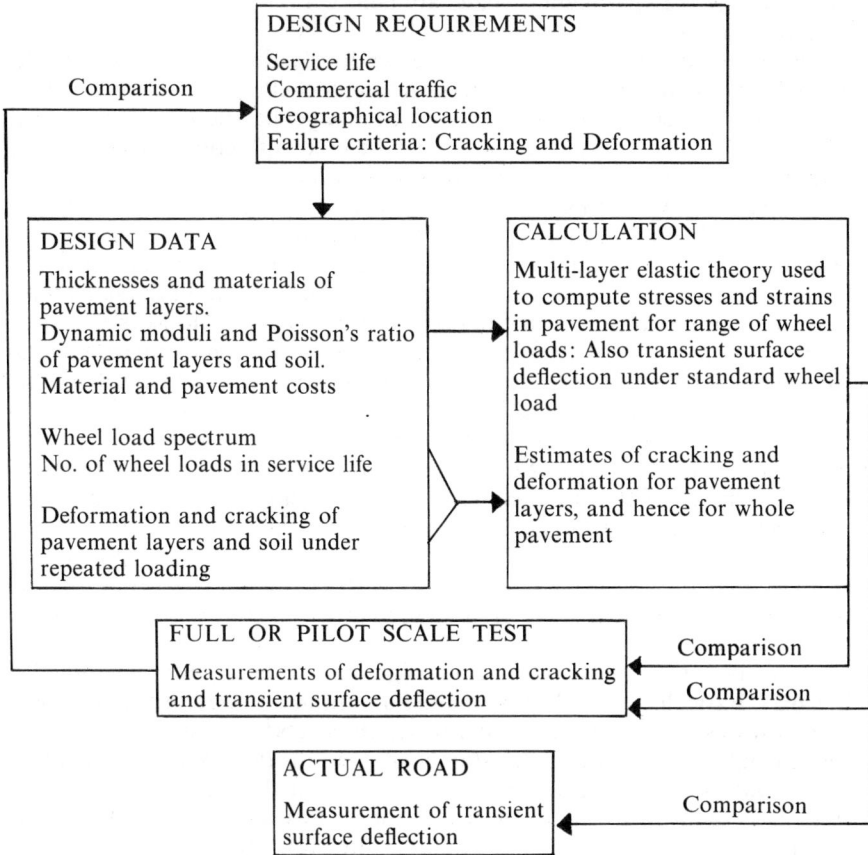

Fig. 6.27.   Analysis of a pavement design method[24]

obtained from full-scale experimental pavements. While this results in designs which will give satisfactory service for many years, the experimental
road approach does have a number of disadvantages:

(a) *The time taken to get results.*   A major experiment may require 2–3 years
preparation before the sections can be opened to traffic; thereafter many
years must normally elapse before useful results are obtained. A more rapid
procedure is needed to determine whether new materials or forms of construction are suitable for general use.

(b) *The empirical nature of the results.*   The sections in full-scale experiments, although numerous, cannot possibly include all the combinations of
material, thickness, environmental conditions, etc., that are of interest. A
procedure is needed to complete and extend the information obtained from
the full-scale experiments.

(c) *Their cost.* Experimental roads are considerably more expensive than conventional roads.

Pavement design has advanced tremendously during the past decade, particularly with respect to understanding the effects of using different materials in pavement layers. Even so, it cannot yet be said that a theoretical procedure has been developed which can be used with complete confidence within the design constraints. What is still needed is a reliable method of design based upon permissible stresses in the various layers, which will limit the pavement deformation or cracking to acceptable levels under given numbers of vehicles of various weights. Its preparation requires several major fields of research to be brought to an adequate stage at about the same time. Essential features of such a method have been defined as follows[24]:

1. A specification of vehicle loading. 2. A specification of environmental conditions particularly temperature and moisture. 3. A means of calculating the stresses and strains caused by traffic, which implies adequate theory and a knowledge of the relevant material properties. 4. A knowledge of the permissible levels for various materials, i.e. the values which, when repeated a large number of times under the appropriate environmental conditions, just keep the deformation or cracking of the pavement within the limiting values. 5. Previous demonstration that 1 to 4 can be used to predict the service behaviour of the pavement. 6. Preferably an accelerated-time method of testing the performance of a complete pavement under laboratory conditions.

A likely outline of the pavement design method of the future is shown in Fig. 6.27. Basically, it would first require an initial estimate of the thickness of the various pavement layers, after which the stresses and strains caused by traffic would be calculated by elastic theory, and the deformation and cracking caused by their repeated applications computed. The deformation and cracking would then be compared with the permissible values and, if different, the thicknesses of the layers would be altered to suit. The unit cost data would be used to compute the cost of the design, and alternatives compared in a search for the cheapest. Comparison of computed deformation, cracking and deflection would be made with test results when available.

FLEXIBLE PAVEMENT DESIGN IN FROST AREAS

The pavement design procedures that have so far been described are applicable only to what might be termed 'normal' soil conditions. When these conditions no longer apply, such as when constructing a roadway in an area where resistance to frost action is the controlling design criterion, then the normal procedures may not be adequate. The following discussion refers to the design and construction of flexible pavements under these extreme conditions.

Detrimental effects of frost action

There are two main detrimental effects associated with frost action in soil. The first of these is frost-heaving resulting primarily from the accumulation of moisture in the form of ice lenses at the freezing plane in the soil. (The mechanism by which this occurs has already been described in the chapter on Soil Engineering for Highways). Secondly, there is a decrease in supporting strength when thawing takes place. Other detrimental effects which may occur in conjunction with these are loss of compaction, development of permanent roughness, restriction of drainage by the frozen layers and the cracking and deterioration of the pavement surface.

Heave damage. Frost heaving itself would cause little damage if it occurred in a uniform manner. However, due to variations in soil composition, moisture content and other factors, severe heaving is seldom uniform in nature. Most damaging perhaps is the abrupt heave which can cause fractured pavements. These are usually transverse when they reflect changing ground conditions.

Another common type of damage to flexible pavements is the lifting of the crown of the road. Such heaving, which also is common with rigid pavements, is attributable to two factors. The first is the incomplete removal of snow, so that the snow along the edges of the pavement acts as an insulator and retards the frost penetration beneath it, while the cleared areas permit deeper frost penetration and hence greater uplift of the pavement along the centre line. More important from the British viewpoint is the second factor, i.e. the differential uplift caused by the difference in subgrade moisture between the edges and mid-portion of the pavement.

Thaw damage. Most serious to both flexible and rigid pavements is the effect of the softening of the roadbed due to the thawing of the frozen soil beneath and the resulting reduction in the road-carrying capacity of the pavement. As thawing begins, the ice melts primarily from the top down and, if the drainage is not adequate, melt water may be trapped beneath the surface above and the still frozen soil beneath. The weakened subgrade—or pavement, if composed of frost susceptible materials—permits greater deflections under traffic. These increased deflections further disturb the soil, reducing its load-carrying capacity so that cracking and eventual failure of the pavement occur.

If the ice content is high and the traffic conditions are sufficiently heavy to cause considerable reworking of the soil, a free-flowing mud may be formed which is forced out at the pavement edges or at breaks in the pavement. This type of action, termed 'frost boil', is common to both types of pavements, but especially to rigid ones.

The more usual damage to flexible pavements in the thaw season is in the form of a close network of cracks accompanied by distortion of the carriageway. This distortion may produce a rough riding surface and, in the more severe cases, deep rutting and complete breaking of the surface.

Frost-action design

The design of pavements from a frost-action point of view can be based on any of three basic concepts: (a) elimination of surface deformation resulting from frost action, (b) limiting to an acceptable amount the subgrade frost penetration and thus the possible detrimental effects, and (c) provision of adequate bearing capacity during the most critical period.

Complete protection. Under this method of design any surface deformation is completely eliminated by removing all frost-susceptible soil to a depth below the frost-penetration line and replacing it with clear granular non-frost-susceptible material. A variant on this recommendation is to excavate the soil to the frost line, place an impervious seal such as a bituminous membrane or a 100–150 mm layer of granular non-frost-susceptible material at this level to cut off the source of moisture, and then replace the original material. The pavement is then constructed using normal design criteria.

If a granular cut-off layer is used, great care should be taken in designing it, so that the cut-off will not become clogged as time progresses (see Chapter 2).

Limited subgrade frost penetration. This method, which was developed by the U.S. Corps of Engineers,[25] attempts to hold deformations to small acceptable values, instead of eliminating them entirely. It is applicable primarily to variable subgrade conditions which might produce differential heaving. Soils to which the procedure is applicable are (a) all silts, including sandy silts, (b) very fine silty sands containing more than 15 per cent finer than 0.02 mm, and (c) clays with plasticity indices less than 12.

The procedure requires the depth to which freezing will penetrate to be estimated, assuming that the roadbase is of unlimited depth. The initial frost penetration value determined in this way represents the thickness of surface and roadbase materials required to completely eliminate frost heave. If this total thickness is a, and the surface thickness is p, then the required roadbase thickness is $c = a - p$ (see Fig. 6.28).

It is not normally necessary to provide this full thickness, since a small amount of frost penetration into the subgrade can be tolerated during occasional winters. Therefore, to allow for this, Fig. 6.28 should be entered with the dimension c as the abscissa, and at the applicable value of r (in this case r is the ratio of water content of the subgrade to the water content of the base; if $r \geq 2$, use $r = 2$) find on the left-hand scale the design roadbase thickness b which will result in an allowable value of subgrade frost penetration s, as shown on the right-hand scale. The reason for limiting r to a maximum of 2.0 is that not all the moisture in fine-grained soils will actually freeze at the temperatures which will be reached in the portion of the subgrade into which freezing temperatures will penetrate.

This design procedure is stated to result in a sufficient thickness of material between the frost-susceptible subgrade and the surface to ensure that, for average field conditions, the subgrade frost penetration s will not cause

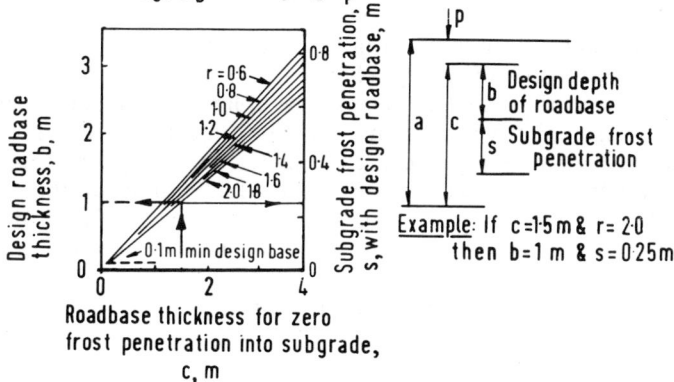

Fig. 6.28.   Design depth of non-frost-susceptible roadbase for limited subgrade frost penetration[25]

excessive differential heaving and cracking of the surface during the design freezing index year. The bottom 100 mm of the design roadbase of thickness *b* must be designed as a filter, such as that previously noted in relation to the cut-off blanket.

Reduced subgrade strength. This method of design is based on the reduction in strength of the subgrade which occurs during thawing of frost-affected soils. It usually permits less thickness of pavement than that needed for limited subgrade frost penetration. A procedure[25] has been developed by the Corps of Engineers which may be used for both flexible and rigid pavements on soils where the subgrade is horizontally uniform or only slightly variable such that significantly differential heaving will not occur.

It is worth noting that this design method does not make use of the California Bearing Ratio, subgrade modulus, or other types of field test performed during the frost melting period, since most of these tests do not give values that are representative of the weakening that occurs when (*a*) the subgrade is thawing and (*b*) sensitive subgrades are subjected to the kneading action of traffic. When a thickness determined by the reduced subgrade strength method exceeds that determined for limited or complete subgrade frost penetration, the applicable smaller value should be used, *provided that it is at least equal to the thickness required for non-frost conditions.*

Special considerations. When constructing the pavement the subgrade should be properly graded and compacted prior to the placement of the roadbase, and equipment-hauling materials should be so routed that rutting of the subgrade is avoided. If this is not done, undesirable mixing of the frost-susceptible subgrade with the roadbase may take place. The existence of a layer of fines on top of the roadbase should also be avoided. This may be a source of sufficient frost action to cause rapid deterioration of the pavement, particularly if it begins to crack after a number of years' service and so permits surface moisture to infiltrate.

Transition points between cut and fill require special treatment if differential heaving is to be avoided. The exact treatment varies in any particular case, but basically it amounts to interposing a transition zone which will mitigate the effects of any abrupt change.

Another potentially dangerous instance of frost action under pavements arises from what is known as 'chimney action' in culverts. This is especially noticeable in cross-road pipes of large diameter. Cold air circulating through the pipe can cause frost penetration into the material above the pipe. This will result in differential vertical and horizontal heaving if the soil is frost-susceptible and sufficient moisture is available. The best remedy for chimney action is to place non-frost-heaving material round the pipe for a thickness equal to the frost penetration around the pipe.

Additives. Many attempts have been made over the years by numerous investigators to reduce or eliminate ice-lens growth in soils by the use of additives. Some of the approaches made are as follows:

Plug soil voids. If the voids can be effectively plugged or sealed so that water cannot migrate, then ice lenses cannot grow. Bituminous materials have been mainly used in this method. The main drawback is that the percentage of additive needed approaches that which is used in pavement surfacings and this makes this procedure economically impracticable.

Cement soil particles. This approach is closely related to the plugging of soil voids. Cement of course is a very effective soil-binding agent but its use can be relatively costly.

Alter characteristics of the pore fluids. Salts may be added to lower the freezing point of the pore fluid in the soil. This reduces the depth of frost penetration under a given set of temperature conditions but does not affect the heave characteristics of the soil. The main disadvantage of this method is its non-permanency.

Alter soil properties by aggregation. The soil fines are primarily responsible for the frost susceptibility of soils. 'Dirty' gravels, for instance, can be rendered non-frost-susceptible by washing out the fines. When this is not feasible, the effective quantity of fines can be reduced by admixing additives, such as lime, that cause small particles to aggregate into larger units, thus effecting a 'cleaner' soil.

Alter soil properties by dispersion of fines. Additives that can cause an increase in the interparticle repulsion within the soil fines tend to disperse the soil aggregates. Particles that do not stick together can then be manipulated into a more orderly structure. Dispersants such as tetrasodium pyrophosphate are most effective in dirty sands and gravels. The reason why is not exactly known but it is suggested that disaggregating the fines permits them to pack into smaller spaces, thereby enlarging the voids among the gravel particles; the dispersed fines are then eventually leached out of the soil, thus cleaning it.

Alter characteristics of the surfaces of soil particles. Use of the proper additives enables mineral surfaces to be made hydrophobic or water-hating. A soil so waterproofed cannot be wetted and should have little or no absorbed moisture. Conversely, coating a fine-grained soil with additives that have high polar groups exposed to the soil moisture can increase the amount of absorbed moisture; these decrease the air void content, thereby reducing the permeability of the soil sufficiently to make it non-frost-susceptible.

PAVEMENTS ON COMPRESSIBLE SOILS

Probably the most difficult problem that can confront a highway engineer is that of constructing a roadway over a highly compressible soil. In this situation there is no method of 'design' that the engineer can turn to for guidance; rather he is faced with the problem of deciding the most economical method of 'constructing' the pavement so that it will not deform unduly over a given period of time. In the following discussion therefore an outline is given of the various methods that have been used in constructing roadways over compressible soils, with particular reference to construction over peat. This outline is based largely on a survey of the world literature on the subject, together with a more detailed examination of actual sites in Britain and Ireland.[26,27]

Elasticity, compressibility and plasticity of soils

Before going into the details relative to the construction of a roadway over soft sites it is necessary to define the basic properties of soils that are of particular interest to this discussion, viz. elasticity, compressibility and plasticity.

Elasticity. By elasticity is meant the property of a material which allows it to be compressed under a load and then to rebound to its original shape when the load has been removed. In other words, no permanent deformation takes place under the applied load or loads.

No soils can be said to be truly elastic, although some possess it to the extent that they can cause substantial fatigue damage to a pavement. These are primarily silts and clays having sizeable amounts of flat flakey particles such as mica, certain diatomaceous earths, and those containing large quantities of organic colloids. Peat can be included in this last category, the degree of elasticity being particularly associated with the more fibrous and less humified varieties of peat.

Highly elastic soils can be detected by a characteristic compression-rebound action-reaction under a moving wheel load. Repeated wheel loads on these pavements will cause a reworking of the soil which may well lead to subgrade failure, as well as contributing heavily to fatigue failure of the pavement itself.

Compressibility. By compressibility is meant the property of a material which enables it to occupy a smaller volume under an applied load. It is different from elasticity in that little or no recovery takes place when the load is removed. The term consolidation is used instead of compression when it is a saturated material which is being loaded for a long period of time.

The compressibility of a soil is due to a reduction in the volume of the voids rather than to the compressibility of the individual soil particles or the air and moisture present in the soil. In other words, when a load is applied the air and water present in the soil attempt to escape, thereby decreasing the pore sizes.

The problems associated with compressibility are best illustrated by considering a saturated soil that is suddenly loaded. When the load is applied, the particles are pressed about the saturated pores as the soil attempts to consolidate. Instantly the load is transferred to the incompressible water in the voids for an amount of time which in any particular case depends on the type and state of compaction of the soil. If the soil is a granular one having a high coefficient of permeability, the water in the pores is able to escape very quickly and compression takes place very rapidly. It is for this reason that compression of sands is rarely a problem, since whatever deformation is going to take place occurs while the load is being applied. If, on the other hand, the loaded soil is a saturated clay, the water will not be able to escape rapidly because of the relative impermeability of the material. As long as the load is applied, however, the increase in pore-water pressure will ensure that the water continues to try to escape and that compressibility will be an important factor for a long period of time.

Thus it can be seen that the compressibility of a soil is a most important consideration from two aspects. First of all the total amount of settlement or deformation is important. This is combined with the fact that more often than not, because of the very nature of soils, differential settlement is liable to occur along the line of a roadway. Secondly, this deformation may take place over a long period of time, the rate being relatively rapid at the beginning of settlement but decreasing with time. This decrease is due to the normal reduction in hydrostatic pressure due to the water escaping and the soil's structure gradually taking up the load. For this reason a heavy clay subgrade may take many years before the equilibrium settlement condition is reached. With peats, a further disquieting factor is involved. Whereas with a clayey soil the time taken for it to consolidate can be regarded as varying with the square of the length of the drainage path, the coefficient of permeability being kept constant, this is not so with peat. Because of the much greater change in the void content, the permeability of a peat decreases significantly with time and this, coupled with the normal decrease in pore-water pressure, means that the consolidation of peats may continue for periods of years, depending on the depth and type of peat being stressed.

Plasticity. Failure of a soil by plastic deformation can occur when a pavement is constructed on top of a saturated peat subgrade. When the pavement is constructed the pore-water pressure is increased and the moisture attempts to escape. As a result the already low resistance to shear is considerably reduced, the lines of least resistance are followed and the ground areas alongside the pavement, but not covered by it, are seen to bulge upwards. If ditches are adjacent to the pavement they will bulge inwards progressively with time and often close up completely.

Such movements are invariably reflected in the distortion of the road surface itself. The extent of the undulations depends on the traffic conditions, the type of road and the characteristics of the peat.

Methods of construction

When constructing a highway over highly compressible soils the traffic demand and the level of service required of the roadway largely determines the exact design and methods to be used. Existing methods of construction vary from direct construction on the compressible material, to carrying the road on piles, to complete removal of the soil and replacement by embankment material. In addition, there are a number of ancillary methods of providing support for roads over compressible soils.

Direct lightweight construction. It is always cheaper initially to construct a roadway directly on top of the compressible soil. None of the methods available, however, give any guarantee that large deformation of the road surface and high maintenance costs will not subsequently become a permanent feature of the highway. For this reason considerable thought should be given to the cost of future maintenance before a final decision is made about constructing a pavement directly upon a compressible soil.

In direct construction of this type it is obviously desirable to keep the applied load as small as possible. For this reason roads over peat which carry light volumes of traffic are sometimes composed of a foundation layer of lightweight material, acting as a raft, which is covered by a thin flexible pavement. This can be very expensive and in practice the raft is seldom thick enough to give much buoyant support to the pavement. Nevertheless, the raft does help in keeping settlements small; it is of particular value where a road has to be widened with as little disturbance as possible to the adjacent peat.

Fascines of brushwood or logs form the traditional raft construction. These, however, must be maintained below the water table or they will rot within a few years. The reasons for installing fascine are to provide a certain amount of buoyancy for supporting the pavement, to spread the weight of the road as evenly as possible and so reduce differential settlement, and to prevent the fill material from penetrating and sinking into the compressible material. A typical fascine mattress used for these purposes might consist of two frameworks of logs set about 750 mm apart, the space between them being filled with loose brushwood laid criss-cross in three layers. The frameworks can be constructed by cording together 125 mm and 175 mm diameter logs or fascines in a grid-patterned network.

A considerable amount of labour is involved in making and laying fascines. This makes the process a costly one and it is therefore dying out in most highly-developed countries. Other materials are continually being studied in order to produce alternative acceptable lightweight forms of construction. One such material is a mixture of sludge and granulated slag from blast furnaces. Another alternative requires the laying of 300 mm diameter bundles of straw beneath the pavement. Each bundle is made by wrapping straw around poles of 50–75 mm diameter and then securing the straw with thin wire and coating it with bitumen. Turfs cut from the heather mattress which commonly covers areas of peat have also been laid as a raft 'subgrade' foundation for the pavement.

Direct embankment construction. Where available funds do not permit the use of other suitable methods or where the depth of peat is very great, construction of an embankment usually becomes the final alternative when building a heavily-trafficked highway over a highly-compressible soil. The reason for this is not to displace the underlying soft material but rather to consolidate it. For instance, peat may have an in-place dry unit weight of the order of 80–160 kg/m^3 and a moisture content of over 1000 per cent by weight of dry peat; its shear strength is normally less than 27·5 kN/m^2. However, if an embankment is constructed *over* the peat it will be considerably compressed; this causes the moisture content to be decreased and the dry unit weight to be increased. If the filling is carried out properly the shear strength of the compressed peat can be sufficiently increased to resist deformation by plastic flow.

Two methods are in general use for constructing embankments on compressible soils. They are the surcharge and sand drain methods.

Surcharge method. The first step in this procedure consists of laying a construction layer of embankment material over the soft soil; this is to support the initial construction equipment. The embankment is then constructed in suitable compacted layers until its height approximates the total amount of settlement expected. A surcharge of uncompacted material is added on top to accelerate the outflow of water and the compaction of the underlying compressible material. After the desired settlement has been achieved the surcharge material is removed and the pavement constructed according to normal procedure.

The main problem associated with the construction of an embankment over a compressible soil is the determination of the rate at which the layers should be laid so that the development of hydrostatic excess pressure within the pore water is minimized. If this is not controlled, the shearing strength of the foundation soil may be exceeded during construction and it will be displaced sideways, with the resultant excessive settlement of the embankment. This will mean that not only will large quantities of fill material be wasted but unsatisfactory unstable conditions may develop which will be very difficult to put right.

The simplest way of controlling the rate of construction is to install a vertical piezometer or standpipe in the foundation material and observe the elevation to which the water rises as the embankment is constructed. The increase in the height of water in the piezometer, when compared to the normal elevation of the water table, is evidence of the hydrostatic excess pressure in the pore water and thus an indication of the reduction in shear strength. By installing pressure gauges of this type throughout the embankment a continuous check can be made to see that the excess pore-water pressures generated by placing a layer of fill material are largely dissipated before the next layer is added.

Sand-drain method. In essence this method is the same as the surcharge method except that vertical sand-drains are used to hasten the consolidation of the compressible foundation soil beneath the embankment. These drains are constructed by boring vertical holes through the first embankment layer and down through the compressible material until firm soil is reached beneath. These holes may be 500–750 mm in diameter and between 9 to 18 m deep; it is usually uneconomical to utilize them for depths less than 4·5 m. The holes are then filled with a clean uniform coarse sand, after which the

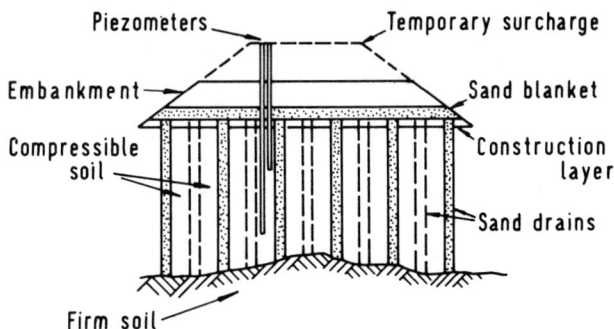

Fig. 6.29.   Diagrammatic representation of a vertical sand-drain installation

tops of the vertical drains and the initial construction layer are covered with a sand blanket of the same coarse-grained material to a depth of from 1–1·5 m. The normal embankment is then constructed on top of this horizontal sand blanket, and a surcharge is added as necessary.

An example of a typical sand-drain installation is illustrated in Fig. 6.29. The sand-drains shown will normally be spaced alternately about 3–6 m apart. Four different methods of installing them are in general use: by means of rotary drills, rotary jets, driven mandrels and jetted mandrels.

The *Rotary Drill plus Bucket* method is the standard straightforward one of drilling the drainage hole, excavating the compressible soil and then back-filling with sand. This has its limitations in unstable saturated soils, which are liable to squeeze into the holes before they can be back-filled.

The *Rotary Drill and Jet* method relies on jets of water to wash the drilled soil to the surface. This method in particular is most unsuitable in already saturated soils which are not capable of supporting the walls of the holes.

The *Driven Mandrel* method is the most widely used procedure. The mandrel consists simply of a hollow tube with a closed flap on the bottom; this is forced into the soil to the required depth by means of a pile-driver. The tube is then filled with the coarse sand and as it is withdrawn the bottom flap opens and the sand flows out and fills the hole. This is a very convenient and easily used method. Its main limitation is in very sticky soils which make it difficult to withdraw the mandrel.

The *Jetted Mandrel* method is similar to the driven ones except that jets of water are used to prepare the way for the tube as it is inserted into the soil. The sand is then allowed to flow through the tube and into the hole as the mandrel is withdrawn.

The purpose of sand-drains is to speed up the consolidation and stabilization of the compressible soil by speeding up the time taken for the pore-water to escape. By providing shorter drainage paths this time is considerably reduced, since the time of consolidation is approximately a function of the square of the length of the drainage path. In addition, when the water reaches the vertical drain it can rise more quickly because of the higher permeability of the sand and drain away through the horizontal sand blanket provided for this purpose.

Sand-drains of this type have been known to increase the rate of consolidation in clayey soils by as much as 25 times. They have their limitations, however, in the very highly compressible soils such as peat. For instance, one study actually showed that the installation of sand-drains actually *retarded* the rate of consolidation of the overlying embankment. The reason for this is not exactly known, but a possible explanation may be that the columns of sand forming the vertical drains literally acted as supporting piles, thus reducing the stresses set up in the peat and thereby lowering the rate of settlement.

Roads on piles. If local conditions, such as the presence of buildings founded on soft soil adjacent to the line of a road, or perhaps a multiplicity of services such as water mains and electric cables which are buried in the soil, limit movement of the foundation material, then consideration may have to be given to constructing that section of the roadway on piles. These piles, which are normally between 1·5 and 3 m apart, are used to transfer the weight of the road to the firm strata beneath the compressible soil.

Normally, this is considered a very expensive method of construction and, hence, is not generally recommended. However, instances can arise, e.g. due to a shortage of a suitable locally available fill material,[28] where pile construction may be a relatively economical method of road construction.

Removal and replacement. The only completely reliable method of constructing a roadway across a compressible soil is to remove the soil completely to the depth of firm strata beneath and then construct a stable embankment in its place. While this requires a large initial outlay, it can often be justified when considering the very large maintenance costs that may be saved.

The following are the four main methods of removing the soft soil and installing an embankment in its place.

Mechanical excavation and replacement. When the depth of the unstable material is less than about 3·5 m it is usually an economical proposition to use mechanical excavators and to remove the soil completely. The embankment is then constructed in the vacant space on the firm underlying stratum. Depending on the depth of compressible soil and the height of the water table, the excavators can either work from the underlying firm stratum or from the embankment itself as it is being constructed forward. The excavated material is usually wasted alongside the embankment, unless it happens to be in a built-up area.

Displacement by gravity. When the depth of unstable material is greater than about 3·5 m but less than about 15 m, and where the material itself is soft enough to be moved sideways by the weight of the embankment, simple displacement methods may be adequate. The two principal gravity displacement methods are by end-tipping and by side-tipping.

With the *End-tipping* method the embankment is slowly advanced along the line of the road by depositing fill at its head. At the same time a dragline sitting on the head of the embankment partially excavates the unstable soft material just ahead of it. The weight of the embankment then displaces the unstable material upon which it is resting forward into the excavated cavity and the embankment sinks to the level of the firm strata. When the embankment has settled, new filling is begun ahead again and the same procedure carried out.

The more usual method of *Side-tipping* requires a shallow ditch to be cut through the surface mat of the unstable soil to the depth of the water table and along the proposed centre line of the road. Its purpose is to form a plane of weakness within the compressible soil. Fill material is then added symmetrically on either side of this line until its weight displaces the underlying soil to the side.

The displacement method is straightforward but very slow. If the underlying material is too stiff to be displaced by the weight of the embankment alone, it may have to be softened by using jets to impregnate it with water. The usual method of jetting is to sink pipes of about 25 mm diameter through the embankment down to the bottom of the peat. Water under a pressure of as much as 1·7 MN/m² is then forced into the peat or soft clay as the pipes are slowly withdrawn. Alternatively the jets are used to inundate the fill with water so that the unstable material is displaced by the increased weight of the embankment.

A further disadvantage of the displacement method is that, if it is not very carefully controlled, large pockets of unstable material may be left beneath the embankment which may give rise to further settlement with time. In addition the symmetrical tipping methods may give rise to weak shoulders due to poor lateral penetration of the fill material.

In general, however, the gravity displacement method appears to be most useful at locations where the unstable material is very soft and where a plentiful supply of granular fill material is available.

Displacement by blasting.   Explosives have been used successfully, mainly in the United States, Germany and Ireland, to blast away peat either in front of or from beneath embankments. The principal methods of blasting are known as trench-shooting, toe-shooting, and underfill blasting.

In the *Trench-shooting* method the explosive charges are placed near the bottom of the peat, usually by jetting, in longitudinal rows in front of the embankment being constructed. This is illustrated in Fig. 6.30(a). The charges are placed apart at distances equal to one-half to two-thirds the thickness of the peat. When they are fired an open trench is formed into which further fill material is then placed. This method of bog blasting is used for depths of peat up to 6 m thick. It is especially suitable for the excavation of fairly stiff peat that is not liable to slip.

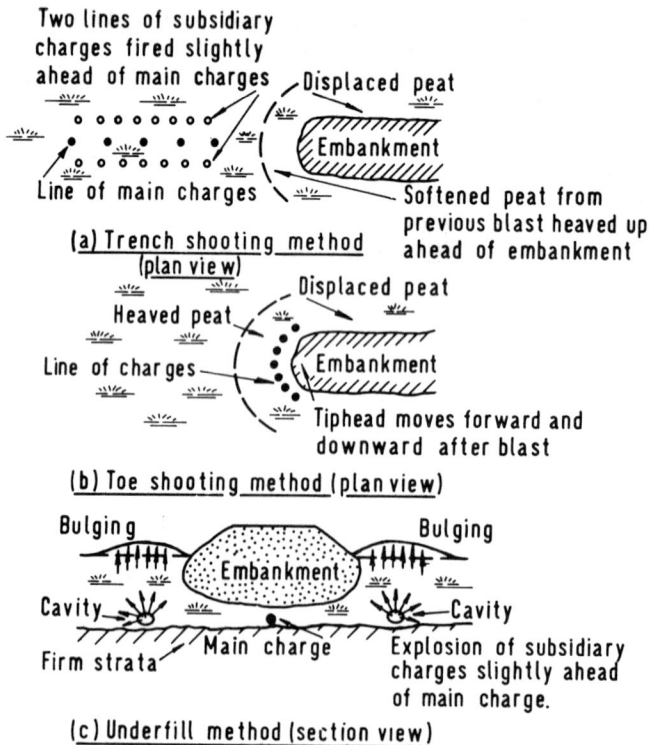

Fig. 6.30.   Methods of bog blasting

In the *Toe-shooting* method the end of the embankment is advanced and its head height raised until the peat is partially displaced from below and heaved up ahead of it. As illustrated in Fig. 6.30(b), charges are then placed at the bottom of the peat just immediately ahead of the embankment. When the charges are fired simultaneously, the peat is displaced and the embankment drops into the void left by the explosion.

Like trench-shooting, the toe-shooting method is used for moderately shallow depths of up to about 6 m. It is particularly suitable for fairly soft peats which can nearly flow by gravity displacement alone. For deeper deposits up to about 15 m thick, the peat can be displaced in the same way if *Torpedo-blasting* is used. This is the same as toe-shooting except that, instead of all the charges being set off at the bottom of the peat, they are attached at regular intervals to long posts which are placed vertically in the peat. The charges are then exploded simultaneously and the peat is displaced to the required depth.

The *Underfill* method of bog blasting is illustrated in Fig. 6.30(c). In this method the full desired width and height of the embankment is placed over the peat and charges are placed beneath it and near the bottom of the peat. The charges are usually in three rows; one row of main charges is placed under the centre of the embankment and a row of subsidiary charges are placed beneath each edge. The subsidiary charges are detonated just before the main charges so that the peat is blasted outwards in the most effective way. The embankment material then drops neatly into the continuous cavities left by the explosions.

The underfill method of bog blasting can be used for any depth of peat, although for the greater depths several series of blasts may be required. It is particularly useful when a sound material layer overlies the unstable soil. When the peat is over 9 m deep, wide embankments will usually have to be built by constructing a narrow embankment first; this embankment is then widened either on one or both sides by further underfill blasting.

Considerable preliminary site investigation, as well as experience with explosives, is required if underfill blasting is to be carried out successfully. Nevertheless, it can be a most efficient and economical method of carrying out excavation to relatively great depths.

Ancillary methods. The following methods are normally used in conjunction with other more general methods of construction over peat. They are designed to provide additional support for the road, either by some form of stiffening or stabilizing the soft material itself, or by introducing external support.

Lateral support. Lateral support can be given to a road by installing sheet piling in the compressible soil or by constructing a berm of fill material along the side of the road. For instance, sheet piling is usually used to protect nearby buildings from any lateral disturbance that may occur when a road is built over a soft soil. At the same time additional support is provided to the roadway by the sheet piling. In a similar manner sand is often dumped alongside the road to form low, wide berms which will increase the resistance to sliding and/or bulging of underlying soft layers.

Drainage measures. It is common practice to lay fairly deep open ditches along the sides of roads constructed over peat, the intention being to drain the upper layers of peat and make them more firm. Although the shear strength of the peat is thus increased, the process is a slow one and there is a considerable volume of shrinkage of the peat which inevitably results in deformation of the pavement. Furthermore, the construction of ditches removes a certain amount of lateral support for the road and the underlying peat may be pushed into them. Other disadvantages are that the deep ditches require frequent maintenance and regrading. In addition, they may have to be extended considerable distances in order to obtain a suitable outfall, since peat often occurs in low hollows.

It is very likely that only very shallow drainage of peat should be carried out close to a road. General drainage other than for the removal of surface water should not normally be undertaken, although it may be satisfactory if it is carried out well in advance of the construction of a road. If this is done, however, care should be taken to see that the lateral support to the road is not reduced and that the peat foundation is not liable to undergo severe seasonal movements of moisture through being unduly exposed to atmospheric effects.

PREPARATION OF THE SUBGRADE

Clearing the site

The first step in preparing the subgrade for *either* a flexible or rigid pavement is to clear the site of all extraneous material. In rural areas this is usually a straight-forward problem involving primarily the removal of what might be termed the natural waste materials, i.e. clearing and grubbing. In urban and suburban locations, however, the problem is complicated by the need to remove footpaths, derelict buildings and other such artificial obstructions. In addition, the preparation of the subgrade in built-up areas is hampered by the necessity of locating and relocating extensive public utility mains and facilities.

Clearing and grubbing. The initial approach to clearing the site is to remove from the right-of-way all trees, tree stumps, underbrush, vegetation and rubbish. This means that the top-soil has to be removed to at least the depth controlled by the penetration of the plant roots. Ideally in clayey soils this should also mean the complete removal of the plastic B-horizon until the C-horizon is exposed, so that it can be used as the basement soil. Unfortunately this is only too often not economically practical, so that stripping is confined to a depth which will ensure the complete removal of vegetable matter. If this vegetable matter material is allowed to stay, it will decay with time and leave voids in the subgrade that may lead to eventual settlement of the pavement.

When the roadway is to pass through wooded areas, all tree stumps and matted roots should be removed to a depth of at least 500 mm below the stripped ground level on which the pavement is to be constructed. This is necessary also to avoid possible later settlement within the subgrade and, more important, to prevent interference with the compaction of the subgrade. All holes left by the removal of the stumps should be filled and thoroughly compacted prior to compacting the subgrade as a whole.

If the trees or shrubbery are not contained within the width of the roadway proper it is usually sufficient to cut the stumps level with the ground surface. Indeed where possible every effort should be made to preserve such vegetation unless it interferes with the road construction or the movement of traffic on the completed facility. In such cases it may be necessary to construct temporary fences about the protected trees and shrubs to ward off construction traffic.

Utilities. In clearing the site in urban areas it is usually necessary to tear up footpaths and old pavements and to knock down derelict buildings in order to prepare the subgrade. While these present certain initial problems, they are usually straightforward compared with those associated with the maintenance, relocation and installation of underground public utilities such as

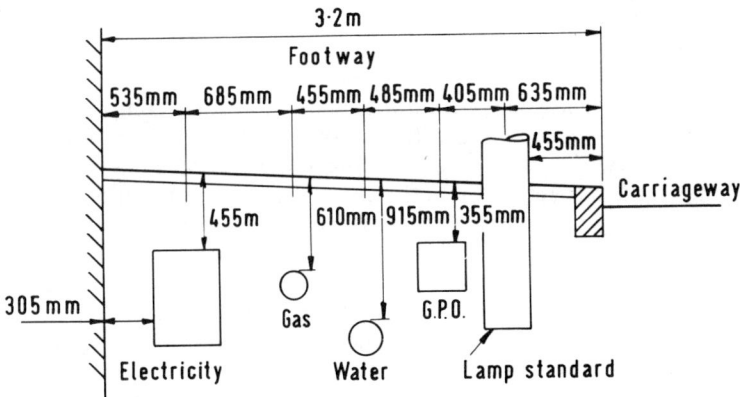

Fig. 6.31. Diagram of the relative locations of public utilities

sewers, drains, water and gas mains, and telephone and electricity mains. In addition, the problem of relocation is often complicated by the fact that these utilities are often the heritage of bygone days and the plans indicating their exact locations are no longer in existence.

This could become more of a problem in the future, both because of the increasing demand for these services and because of the increasing trend towards placing more and more of them below ground. Ideally, as many of these utilities as is possible should be placed under footways and verges instead of under the carriageways. In practice surface water and foul sewers, the lowest of all the utilities (in case of leakage), are usually laid under the carriageway; there is little risk of interference to traffic as sewers rarely need to be uncovered for repairs. Fig. 6.31 shows the locations of the remaining utilities under a wide footway as recommended in Britain. Note that the water mains should always be beneath the depth of frost penetration; however the actual depth of these and the other mains will vary, depending on the topography. All electric power lines should be above all water lines; they should also be horizontally away from them in order to avoid corrosion resulting from the water mains being within the electric field of the power lines. Communication mains should also be away from the electric power mains, and they should be above the water mains. The communication mains may contain ducts for telephone, telegraph, fire alarm and police lines, plus extra ducts to allow for future expansion.

Compaction

It is essential that the subgrade should be sufficiently dense and stable to withstand the stresses transmitted to it by the pavement. This means that the state of compaction attained in the subgrade must be essentially that anticipated in the design procedure.

There is no uniform agreement as to what depth the subgrade should be compacted to this design state. It is certain however that, if the basement soil is part of a high embankment, the top 300 mm at least, and preferably the top 750 mm, should be thoroughly compacted to this state. In cut sections the problem can be more complicated. If the basement soil is a heavy clay it may be found that a light roller can be used only to smooth the surface of the soil; undisturbed heavy clays cannot usually be further compacted beyond their natural state, and rolling with a heavy roller will only result in exudation of moisture with resultant remoulding and weakening of the soil. On the other hand, if the basement soil in the cutting is relatively coarse and loose, it may be quite easy to compact it to a substantial depth.

The principles and problems associated with soil compaction have been discussed in some detail in the chapters on Soil Engineering for Highways and Soil Stabilization so they will not be discussed here. A point to be mentioned, however, has to do with the compaction of the subgrade soil in highway cuttings. In many instances, in order to obtain the necessary depth of compacted material, it may be necessary first to remove a layer of soil, compact the exposed surface, and then replace and recompact the removed soil. The need for this method of compaction depends on the type of soil and the available compaction equipment at a particular site; in fact it can really only be determined by field trials at the site at the time of construction.

Detailed recommendations regarding the use of compaction equipment with soils and climatic conditions experienced in this country are available in the literature[29].

Transitions. When the subgrade passes from cut to fill, care should be taken that there is no abrupt change in the degree and uniformity of compaction. This is of particular importance when the change is from rock to soil.

In all such cases a soil transition-layer should be installed to allow any differential effects to occur gradually. At the 'point' of change the depth of the transition layer should be 1 m at the very least and it can be feathered down to about 150 mm about 14 m back on the rock. If the transition is from cut in soil to an embankment it may be necessary to excavate an exposed band of plastic B-horizon material and replace it with a more suitable soil. Again, this excavation should be carried to a depth of at least 1 m, as this can be a very moisture-susceptible material.

If the highway is being constructed through a rock cutting it will usually be necessary to interpose a 100–150 mm layer of suitable material between the pavement and the rock. This is necessary for two reasons. First of all it helps to provide a cushion for the transition between the rock and adjacent subgrade soil. Secondly, the layer levels the irregularities that occur in the rock surface during the rock-blasting. If necessary the larger depressions can be filled and brought up to the desired level by the addition of lean concrete. Before the cushion is laid, however, transverse grooves should be cut in the rock surface to allow lateral drainage of any moisture accumulations that may occur then or in the future.

Shaping

After the subgrade soil has been compacted, the surface will still be fairly rough. The next stage is to shape the rough surface to the final shape of carriageway. This ensures both that the pavement will be constructed to its proper shape and that any moisture intruding into the subgrade will drain away rapidly.

This final shaping can be easily carried out with a blade-grader. When it is completed, the loose material is removed from the surface and any irregularities that may still be apparent are removed by light rolling with smooth-wheeled or pneumatic-tyred rollers.

Once the final shape of the formation has been attained, only the most essential traffic should be allowed on it prior to the pavement being constructed. This is most important when the subgrade soil is fine-grained and easily subject to deformation. If it is known that the formation will be exposed for some time and/or that it is likely that weather conditions will be such as to cause changes in the moisture content to take place, then the prepared subgrade should be protected[30] by being sealed immediately.

ROADBASES AND SUB-BASES

As soon as the subgrade has been prepared, the base courses should be constructed as quickly as possible. These are usually composed of naturally or mechanically stabilized soil, chemically stabilized soil, crushed stone or gravel or slag, or, sometimes, waste materials such as hardcore and shale. Most of these materials have been described elsewhere in this text, so this discussion will be on the *manner* in which they are utilized in a pavement, particularly in roadbases.

Factors affecting the construction of stabilized soil pavements are of sufficient importance and so varied as to rate a separate discussion and have been included in the chapter on Soil Stabilization. A type of construction that will be discussed here, however, is that with lean concrete/ cement-bound granular materials. Although this is considered by some to be a form of rigid pavement, while others consider it but an advanced form of soil-cement, there is little doubt that in practice it is treated as a separate form of flexible construction.

One further point which must be mentioned is that, with some obvious exceptions, no attempt will be made in the following discussion to prove that any particular form of construction is 'best'. Indeed, it would be extremely foolish to do so since the choice in any particular situation is really a function of the amount of money available for the project, the volume and composition of the traffic, the availability of particular construction materials and equipment, the condition of the subgrade, the prevalent weather conditions, and the economic life of the pavement. All of these factors must be taken into account before the proper decision can be made as to what type of construction should be utilized in any particular situation.

Bases of waste material

Clinker, quarry waste, burnt colliery shale, spent-oil shale, hardcore, and such like waste materials have been used successfully in the past as pavement ingredients. In industrial areas, in particular, accumulations of waste materials are usually available in large quantities at low cost and there is often pressure for them to be utilized in highway construction.

It cannot be too strongly emphasized, nevertheless, that there is a very limited place for these materials in modern road construction. As a general rule it can be said that their usage as a pavement material should be confined to minor roads. At that they should only be used if it is certain that the subgrade soil is not likely to be affected by moisture intrusion or frost action.

Hardcore. If hardcore is used in road construction it is usually specified as any clean rock-like material, such as broken concrete and sound brickwork free from mortar, having few lumps greater than 100–150 mm. It is spread on a layer of clinker or fine granular material and then a layer of fine hard material is placed over it in order to fill the voids in the hardcore layer. The whole thickness is then compacted with a heavy smooth-wheeled roller.

Hardcore bases can be criticized on many accounts. First of all, the material is essentially debris collected from building sites, it is completely ungraded and little or no attention is paid to its cleanliness and quality. Its suitability as a construction material is left to the judgement of the site engineer and even though he may inspect every load it is very difficult for him to detect 'dirty' material.

It is useless to carry out any mechanical tests on hardcore material, since every lorry load may well be different, especially if the material is obtained from different sites. For this same reason there is no guarantee that uniformity of material is obtained when the hardcore is laid and this can lead to differential settlement and break-up of the pavement. As the hardcore is compacted it is more than likely to break beneath the roller, thereby adding to the fines already in the pavement and rendering it more susceptible to attack from moisture intrusion and frost action. This degradation and settlement is likely to be very much accelerated if traffic beyond that anticipated begins to use the roadway.

Shale. Shale used in pavements is usually specified as hard, well-burnt material free from ashy refuse and rubbish which may soften under the action of moisture. Normally material less than 76 mm maximum size is preferred for ease of laying and compaction.

Shale can be subject to the same objections as hardcore. The way in which it is produced in waste heaps means that the material is irregular in both quality and gradation. Again, it is left to the judgement of the site engineer to determine whether or not the material is acceptable. The colour of the shale is not a reliable guide to its quality, so that the decision can only be made by visual examination as the material is being removed from the tip. In addition it must be remembered that shale is essentially a clayey material; thus, although it may be apparently well-burnt, it is very likely to break down under conditions of moisture and frost[20] — and, hence, cannot normally be used in the pavement structure within the zone of freezing.

Nevertheless, since some 80 km^2 of land in the mining areas of England and Wales are covered with heaps of shale, it would be most desirable to use large quantities in road construction. Studies[31,32,33] have shown that shales are suitable for road fills and sub-bases below the zone of freezing. (Fears about unburnt shale igniting when used as fill are very low, if it is well compacted.) Preferably, however, the shale should not be placed within 0·5 m of bridge abutments, concrete pipes, or pavement layers containing cement-bound layers, because of the adverse effects that excess sulphates in the shale can have on cemented materials.

Hand-pitched bases

Prior to the second world war hand-pitching, similar to the Telford type described earlier in The Road in Perspective (Volume 1, Chapter 1), was the traditional method of roadbase construction in most countries. It consisted of hand-placing rough-cut, approximately rectangular, stones, each about 150–250 mm high by 100–150 mm wide, side by side on the prepared subgrade. Each stone was firmly settled in its seating and smaller stones were hand-wedged into the gaps between the larger stones. Projecting knobs of stone were then removed with a hammer, after which small stone was added to fill the surface voids. The layer was then compacted, extra material being added as necessary, until a roadbase having no surface depressions and true to the camber desired of the surfacing was obtained.

Hand-pitching is a very slow method of construction and requires a large labour force. For these reasons its usage in Britain has become extinct. However it still is, and will be for some time, an important method of pavement construction in underdeveloped countries where labour is cheap and modern construction equipment is not available.

Apart from the problem of high labour costs, the hand-pitched roadbase has other disadvantages which render its use undesirable in modern highways. For instance, it can normally be laid only in about 230 mm layers, irrespective of what thickness of roadbase is specified by the design method. Furthermore, because of the slowness of construction and the openness of the completed roadbase, the subgrade is usually open to the ingress of water for a considerable period of time, with the result that the subgrade soil can become softened and begin to work its way up through the roadbase.

In addition, experimental work has shown that layers composed of stone greater than about 100 mm in size are very difficult to compact sufficiently to meet modern traffic requirements. It has been found that excessive deformations have taken place *within* hand-pitched roadbases as a result of compaction under heavy traffic.

Modern macadam bases

The modern macadam-type of construction is an outgrowth of the early Macadam road (see Volume 1, Chapter 1). Present-day requirements in highway construction have favoured the macadam-type of roadbase, principally because of its suitability for laying by mechanical means. This has led to higher speeds of road-building with a small labour force, and therefore to a lower initial cost of construction. Furthermore, with the type of equipment that is now available, the macadam-type of construction can be brought to a highly stable state, with the result that little deformation will occur within the pavement under traffic. This latter factor permits bituminous surface courses to be laid concurrently with the roadbase instead of, as is customary with hand-pitching, laying the roadbase and then waiting for the traffic to complete compaction before regulating the surface and laying the final wearing course.

Several types of macadam pavements are used in highway construction today, the most common being dry-bound, crusher-run, water-bound, premixed water-bound, and bituminous coated macadams. (The coated macadams are discussed under Dense Bituminous Roadbases.) Two other types, which are essentially extinct as regards usage and thus merely of academic interest, are the traffic-bound and cement-bound macadams.

Aggregate materials used in macadam pavements are crushed stone, crushed gravel and crushed slag. All types of construction have a common stability mechanism; they primarily rely for their strength and resistance to deformation on the interlocking of the individual crushed particles and on the friction generated between the rough surfaces in contact. Some processes also utilize water or a bituminous binder to cement the particles together.

Basic macadam preparation. For economic reasons macadam bases are rarely greater than 600 mm deep, and are more usually 300 mm or less. The type of compaction to be used influences the method of construction. If heavy smooth-wheeled rollers are used, then it is advisable to lay macadam in layers not more than about 100 mm deep; a good rule-of-thumb states that the compacted layer thickness should not be more than 1·5 times the largest aggregate in the layer if good particle interlock is to be obtained. If, as is becoming more and more common, vibratory rollers are used, then single layers of up to about 200 mm compacted depth can be satisfactorily laid. In either case, the coarse macadam material is spread using a paving machine or spreader box operated with a mechanism which levels off the material to an even depth; this is usually about 125 per cent of the compacted layer thickness.

Before the first macadam layer can be laid some essential preliminary construction procedures must be carried out to ensure the stability of the finished pavement. These are basic to all types of macadam construction.

First of all, the underlying material, on which the macadam construction will rest, must be thoroughly compacted so that it presents an unyielding formation. If this is a silty or clayey soil it is likely to soften under the influence of moisture and penetrate the macadam layer; this can disrupt the mechanical interlock and friction and may cause distortion of the pavement. Therefore it is advisable to lay a 50–100 mm insulating layer of dry stone screenings or sand on top of the fine-grained soil. The gradation of this material will depend on whether it is intended to act as a drainage layer as well. The macadam construction is then laid on top of this granular sub-base.

Secondly, because of the dependence for stability that is placed on particle interlock and friction, it is desirable to prepare the edge of the roadway to resist the lateral thrust of the aggregate as it is being compacted. If the aggregate is not given stalwart support along the edges, it will give way laterally during compaction and much of the interlock strength will be lost.

Lateral support for the pavement can be provided by long steel or timber side forms laid alongside the edges. On the other hand, it may be more economical to construct and compact small earth embankments on either side and then to use a blade-grader to cut back the soil to form the required vertical side supports. After the pavement is constructed, the earth embankments can be shaped to the desired contours.

Irrespective of the type of compacter used, macadam compaction is normally begun at the edges in a direction parallel to the centre line in order to lock the outer stones, and is extended progressively toward the centre of the road. In the case of a superelevated curve, however, compaction is initiated at the lower edge and continued progressively toward the higher edge. In either case, rolling is continued until there is no obvious creep of the aggregate ahead of the roller.

Dry-bound macadam. The term dry-bound, when used with reference to a macadam pavement, is concerned with the manner in which the crushed aggregate is laid. In Britain the constituent aggregates are transported to the site in two separate sizes: a coarse material that is normally either 37·5 or 50 mm nominal single-size, and fine screenings graded from 5 mm to dust.

The first layer of coarse aggregate is spread on the prepared and compacted underlying material in an even layer 75–100 mm deep; the loose thickness should never be less than one third greater than the desired compacted thickness as otherwise it will be difficult to obtain the required compaction and the desired interlocking of the particles will not take place. Before compaction is initiated the depth of the loose layer is checked very carefully in order to avoid any undulations of the compacted layer that may later be reflected in the surface. Preliminary rolling and shaping of the layer is then begun, after which the surface is carefully examined and any local projections or depressions corrected. After this the fine aggregate is spread over the compacted layer. The reason for adding the fine aggregate is so that the voids between the larger particles can be filled (choked), thereby maintaining the interlock and increasing the internal friction. The fine aggregate is spread by some form of gritting attachment on the back of a lorry. Normally a loose layer of 25–50 mm thick, or from 20 to 40 per cent by weight of the coarse aggregate, is required. The exact amount will vary with the size and grading of the coarse material, and can really only be determined by experience on the site.

At this stage it is necessary to carry out a certain amount of brooming, either mechanically or by hand, in order to spread the fine choke material uniformly. This material is then compacted into the voids with a plate vibrator, a vibratory roller and/or an ordinary heavy steel-wheeled roller. Any open areas which develop during compaction are filled by brushing-in additional fines as compaction is continued. When it is clear that all the interstices are filled and there are no hungry patches, a final rolling is carried out with a heavy smooth-wheel roller until the rolls cease to mark the surface. By then a dense compact surface should be obtained.

When rolling is complete, the surface is broomed to remove excess fines and leave the coarse aggregate standing up to 10 mm proud, after which the side supports are repaired and built-up as necessary. Further layers are then laid as before until the macadam roadbase is fully constructed. When construction is finished the roadbase surface should contain no irregularities greater than 12·5 mm as determined by laying a 3 m straight-edge parallel to the centre line in any position on the surface.

(If the uppermost layer is to be utilized as a wearing course on which traffic will run—this is not common practice in Britain—it should contain more choke material than the lower layers so that a more smooth, dense, compact and waterproof layer is obtained. In this instance the maximum gap beneath a 3 m straight-edge laid on the surface should be less than 6·2 mm in order to ensure a high standard of riding quality.)

Dry-bound macadam has been shown to give excellent results in heavily-trafficked highway pavements. By using two separate materials of different sizes, segregation of the aggregate during stockpiling and transportation is kept to a minimum, and a uniform stable pavement is obtained at a relatively moderate cost. Stability is ensured by proper compaction; the unit weight of the compacted dry-bound layer is normally about 85 to 90 per cent of the calculated solid unit weight of the ingredients.

The main drawback to dry-bound macadam construction arises during wet, or even damp, weather. While the moisture has relatively little effect on the compaction of the coarse aggregate, it is practically impossible to ensure that the voids are properly filled by vibration if the fine material is in a wet state before compaction. Often, when attempting to shift the wet fines, the layer may receive an excessive amount of vibration; as a result the coarse material may ride upon a layer of fines with a resultant severe loss of stability.

Water-bound macadam. This process is one that is used extensively in the United States but not to any great extent in Britain. In essence it is a type of construction very similar to dry-bound macadam and so it will be only briefly discussed here.

As with the dry-bound macadam, the dry coarse aggregate layer is laid, compacted and checked for surface irregularities. A layer of fine choke material is then laid on the coarse layer surface, broomed, and vibrated and/or rolled so that the interstices between the coarse aggregate are filled. When it appears that a dense mixture is obtained, the loose material is swept from the surface, and it is then checked for irregularities.

With the wet-bound process the next step requires the compacted layer surface to be sprayed with water, after which rolling is recommenced. Spraying of the surface is usually done by means of a mechanical pressure-distributor vehicle. As rolling is being carried out wet aggregate is continuously broomed into any voids that may become apparent; if necessary new filler material is added to fill any local depressions. Moisture spraying and rolling are carried out continuously until a wave of grout appears ahead of the roller. This wave will appear when all the voids are full, indicating that compaction should cease.

In water-bound macadam construction the moisture is sprayed on to the surface of the layer for two reasons. Firstly, it acts as a lubrication for the aggregate particles, allowing them to slip over each other more easily. Provided that an excess of moisture is not added this will result in higher unit weights. Secondly, the dusts of certain aggregates have binding properties that are only activated when the dusts are brought into contact with moisture. Particularly noticeable in this respect is limestone dust, which becomes a weak cementing material when mixed with water.

The roadbase aggregate gradings recommended by the American Association of State Highway Officials for use in water-bound macadam pavements tend to give compacted coarse layers that are more open than their British counterparts for dry-bound macadam, and thus are more susceptible to subgrade intrusion. However, the greatest drawback to the use of water-bound macadam is that each compacted layer must be given time for the moisture to escape and to dry thoroughly before the next layer is laid. Particular care must be taken to allow the uppermost layer to cure for the required length of time before a bituminous covering is placed over it. While these curing times can be anticipated and taken into account in the construction schedule in dry climates, they can play havoc with pavement construction schedules in Britain's unsettled climate.

Crusher-run macadam. Macadam pavements depend for their stability on the interlocking and frictional characteristics of the ingredient particles. The extent to which these characteristics are utilized is dependent on the unit weight of the compacted layer. While high unit weights are achieved by both the dry and water-bound processes, it is obvious that higher, more economical and more predictable stabilities might be obtained if, instead of using two separate sizes of materials, just one material meeting high requirements regarding gradation and stability was used in each compacted layer. To achieve high stability the grading of the aggregate should be such that the finer particles completely fill the voids between the coarser particles, thus providing a condition for achieving a high unit weight after compaction.

This concept gave rise to the use of what is known in Britain as crusher-run macadam. It consists of natural or artificial aggregate which is generally passed through two stages of crushing to give a graded mixture of about 75 mm maximum size. This material is transported to the site, moistened and compacted in about 100–150 mm layers.

This type of construction is normally relatively cheap since only one material has to be handled. It is spread easily and the required compaction can be obtained relatively quickly and with the minimum of hand labour.

The main drawback to crusher-run macadam, and the one that has most limited its use, has to do with its most important characteristic, its gradation. An attempt is made to ensure the presence of particles of the various sizes between the maximum and minimum sizes in order to ensure high unit weight; in theory this is an excellent concept, but in practice it means that while the material is being carried from the crushing plant to the construction site considerable segregation can occur, especially over long-distance carriage. Thus when the material is spread it may consist principally of fines in some areas, whereas in others considerable amounts of choke material may be required. In addition during laying and compacting the finer material will tend to fall to the bottom of the layer, resulting in a loose surface that can be most difficult to compact.

Pre-mixed water-bound macadam (wet-mix). This consists of crushed natural or artificial aggregate, usually of 50 mm maximum size, which is graded down to filler so that a high unit weight is obtained when the material is compacted. Gradings are generally derived from a combination of individual sizes, but crusher-run material of suitable grading is also used. The graded aggregate is mechanically mixed with a carefully controlled amount of water, usually within the range of 2 to 5 per cent by weight. Great care is taken to ensure that a thorough mixing takes place and up-to-date machines have been specially designed for this purpose; where special plant is not available, the usual practice is to use a bituminous macadam plant.

Pre-mixed water-bound macadam, or 'wet-mix' as it is known in industry, has a basic cost that is slightly higher than the other macadams hitherto mentioned because of the processing of the material which is necessary before it is transported to the site. Its principal advantage is that because of the pre-mixing with water it can be transported over relatively long distances without significant segregation taking place. Similarly, little segregation takes place during laying and compaction of the wet-mix because of the cohesion generated between the fines and the larger particles.

The addition of moisture facilitates site work considerably and the amount of handwork required is negligible, since the pre-mixed graded aggregate can be laid speedily to the desired shape by automatic pavers. Compaction is carried out with the minimum of delay since there is little necessity to add choke material to harsh areas. High unit weights are also more easily obtained because the moisture films allow the particles to slide more easily over each other into their interlocking positions as compaction is carried out. The amount of water which should be added is very critical; just enough must be added to provide the cohesive films about the particles. If too much is added the cohesive bonds will not be formed and the particles will segregate out.

Less mixing moisture is normally used in winter than in summer, and care has to be taken that neither rain nor drying winds affect the moisture condition between the time of mixing and time of compaction. On warm days or on wet days it is necessary to sheet the wet-mix if it has to be transported any distance by lorry. For the same reason laying in rain must be avoided.

Laying of the wet-mix must on no account be carried out on foundations that are inadequately drained, since the free mix water must be allowed to drain away during and after compaction. This is of particular importance if the macadam construction is being laid in a trench and there is danger that the water will become ponded between the embankments. As soon as possible after compaction is complete, the uppermost layer should be sealed with a layer of bituminous coated grit or surface dressing. Assuming that the wet-mix is forming the roadbase construction, the desired bituminous surfacing is then placed on top of this sealing coat.

Dense bituminous roadbases

Since about the mid-1960's, premixed bituminous mixtures have come to be widely used in the roadbases of high quality roadways in Britain. The three mixtures most well-known in this capacity are *rolled asphalt* and the dense-coated macadams, i.e. *tarmacadam* and *bitumen macadam*. (These types of mixtures are described in some detail in the chapter on Bituminous Surfacings—this discussion will be therefore concerned only with their basic functions in roadbases.)

As the term 'dense coated macadam' implies, dense tarmacadams and bitumen macadams utilize relatively well-graded aggregates[29] which are premixed with either tar or bitumen before being placed in the pavement. Their main features are that 38 per cent of the aggregate materials pass the 3·18 mm sieve, and they utilize fairly high viscosity binders, i.e. 54°C evt tar or 100 pen bitumen for well trafficked roads, and the less viscous 50°C evt tar and 200 pen bitumen on the less heavily travelled ones. Rolled asphalt is the highest quality, as well as the oldest established, bituminous material used for roadbase construction; its roadbase composition is the same as that specified in a normal rolled asphalt basecourse mixture, containing 65 per cent coarse aggregate and, normally, a 50 pen or a 70 pen bitumen.

A major feature of the dense bituminous roadbases is their improved load spreading properties as compared with other non-coated materials used in pavements (see Fig. 6.12). Note that this is reflected in the design curves given in Fig. 6.21, in that the use of bituminous-bound roadbases permit a reduction in thickness as compared to what is required with other materials. Notwithstanding the increased stability resulting from the use of dense mixtures, the tar and bitumen binders also provide an elasticity component which gives to these mixtures a certain degree of flexibility without cracking; furthermore the binders possess sufficient plasticity to enable the material, without cracking, to accommodate a certain amount of pavement settlement. This non-cracking attribute is particularly desirable in areas subject to subsidence, e.g. mining areas.

Another advantage of these roadbases is that the application of the first layer of the bituminous mixture early in construction affords immediate protection to the sub-base and subgrade in bad weather. If desired, the next layer need not be superimposed immediately; indeed, the first layer may be used by traffic for short periods—a factor of particular importance in reducing traffic delays in, particularly, urban streets. The material is normally laid by machine in layers of up to 100 mm thick to very accurate levels; this also means that the unsurfaced complete roadbase can be very often used to carry construction traffic, without undue damage, prior to the thin surface course being added upon the completion of the construction.

Cement-bound bases

The grading requirements for cement-bound materials shown in Fig. 5.3 show that the coarser-gradations are given, in Britain, the titles of *Cement-bound Granular Materials* (*C.B.G.M.*) and *Lean Concrete*. Non-soil materials, e.g. rock aggregate, which meet the gradation and other governing criteria also fall into this category of roadbase and sub-base material.

Natural gravels used in lean concrete must meet the criterion that not more than 3 per cent shall pass the 0·075 mm sieve; to achieve this it is usually necessary to wash the material so that it is virtually free from silt and clay—which means that mixing can take place in stationary plant drum-type concrete mixers. With C.B.G.M., on the other hand, up to 10 per cent fines is permitted—which means that free-fall mixers cannot be used, and that power-driven pan or paddle mixers with a positive mixing action are required. The mixing process is one of the important aspects differentiating

C.B.G.M. from lean concrete, and failures have occurred with so-called lean concrete mixed in drum mixers when the material had a grading outside the specified limits.

As can be seen from Fig. 6.21, cement-bound materials are permitted materials in British pavement construction practice. In practice, lean concrete is quite widely used on the more heavily trafficked roads, C.B.G.M. is used to a certain extent on intermediate roads, while the 'remaining' soil cement is, in fact, relatively rarely used. More often than not, the coarse-grained materials are used only in the sub-bases of major roads; engineers tend to regard them as being useful as a means of reducing total depth or increasing pavement life, or else as a construction expedient which can speed other stages of construction or extend the working season. In minor roads, they are fairly widely used in roadbases.

Cement-bound materials are considered to belong in flexible pavement categorization because of their proneness to cracking. This cracking is initiated upon completion of setting and hardening of the cement after construction as a result of these factors in combination with the effects of temperature stresses set up with the soil cement slab. Much of the crack pattern in the roadbase may be sufficiently developed to be visible within a few days of its construction, and it may appear later through the surfacing; the worst situation is when, as often happens, most of these cracks are 2·5–6 m apart. The growth of these cracks is accelerated by the amount of traffic using the roadway. In fact, it is only in the stronger cement-bound materials in pavements of substantial thickness that general cracking of this nature can possibly be avoided.[34] Adequate extended curing also helps to control cracking in cement-bound roadbases[35].

The current British specifications for C.B.G.M. require a 7-day unconfined compressive strength in excess of $3·45 \text{ MN/m}^2$ for cubical test specimens; for lean concrete the requirement is a 28-day strength in excess of $9·65 \text{ MN/m}^2$. Most of the early post-war lean concrete bases were laid using cement contents of about 148 kg/m^3 (aggregate:cement, 14:1) and occasionally as much as 160 kg/m^3 (12:1) but recently mixes leaner than these have come into favour. The generally recommended practice is now to use a cement content of $101–136 \text{ kg/m}^3$ (20:1 to 15:1) regardless of the type of aggregate used.

Contrary to the conventional practice with concrete, the water content of a lean concrete or C.B.G.M. mixture is decided on the basis of the amount required to give the maximum compaction. Fig. 6.32 gives typical curves obtained when using an irregular gravel aggregate and shows the influence of moisture content on dry unit weight and on the compressive strength of cubes compacted to refusal. As can be seen, the moisture content is quite critical. If the mix is too dry the surface of the lean concrete may shear under the roller during compaction and appear loose after compaction is complete. If the mix is too wet, it will be picked up on the wheels of the roller or the roller may tend to sink into the lean concrete.

Assuming that the moisture content of the mix is suitable, the maximum thickness of lean concrete which can be compacted satisfactorily in one layer depends on the performance of the compaction plant, and on the quality of the subgrade or sub-base on which the lean concrete is being laid.

When the total thickness of lean concrete required exceeds the maximum recommended for good compaction, it should be laid in layers, the top layer being made as thick as is possible and consistent with obtaining good compaction. Thus, for example, in the case of a 300 mm thickness it is best to make the top layer 200 mm and the lower layer 100 mm thick. If the foundation soil is weak, a useful expedient is to delay laying the upper layer for about 3 days until the lower layer has hardened and better compaction can be obtained.

The compaction of a lean concrete or C.B.G.M. mixture must be complete within, at most, $1\frac{1}{2}$ hours of moisture being added to the dry cement

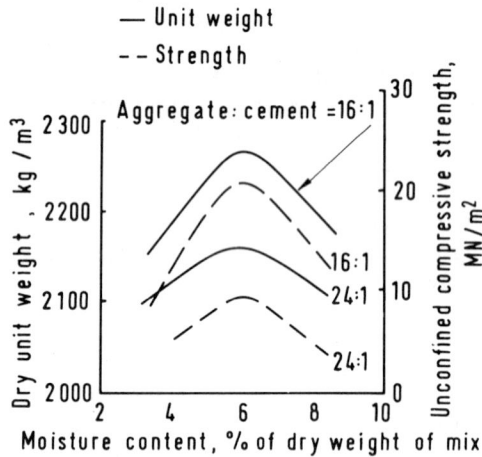

Fig. 6.32. Typical curves showing the influence of mix water content on the maximum dry unit weights and strengths obtainable from lean concrete cubes

and aggregate. If it is not, then the desired high unit weight will not be obtained and the strength will be considerably reduced. This means therefore that the length of haul may have to be limited in order to comply with this time requirement. Experience indicates that when the ready-mixed concrete is delivered in agitators, or truck mixers are used as agitators, an allowance of up to one hour for travelling time can be added to the recommended time limit of $1\frac{1}{2}$ hours for completion of the work.

It is preferable to use transporting vehicles with steel-lined bodies, and on long hauls cover should be available to protect the mixed material from wetting by rain or evaporation during dry weather. To minimize the possibility of segregation, a high level of discharge from the mixer to the lorry at the loading point should be avoided and the lean concrete should not be deposited in conical heaps. If segregation does occur, either at the mixer discharge or during transporting, it should be corrected by remixing during spreading.

Spreading can either be by hand, by blade grader or bulldozer, or by an automatic paving machine. It is not essential to use forms in laying the material, but it is advisable to do so. If forms are used they should be firmly fixed to resist lateral and vertical displacement, which will affect the regularity of the surface and the roadbase, as well as resulting in lack of compaction at the edges. Normally, the forms can be removed immediately after compaction, precautions being taken to avoid subsequent damage to the edges.

Most cement-bound roadbases are laid without joints, except for day-work joints, and at present there is no evidence to show that expansion joints are necessary. Particular attention, however, should be paid to compaction at the joints, which should all be vertical butt-joints. Poor compaction will result in weakness, which may cause failure when expansion takes place in the base. Transverse day-work joints can be formed by feathering off at the end of the work. If this is done it is most important to ensure that the material is cut back to a sound and vertical face before work is resumed against it; otherwise expansion in the roadbase may cause vertical displacement between the two days' work. A preferable way of forming day-work joints is to lay heavy baulks of timber, or well-supported forms, against which the roller can work. If necessary there should be further compaction by ramming to ensure dense material at the joint.

Curing. Lean concrete bases require curing to prevent evaporation of moisture. Experience has shown that the omission of curing can lead to scaling of the surface of the layers as well as cracking. The most convenient method of curing the top layer is to spray the surface with a bitumen emulsion, or cut-back bitumen; this membrane can then form part of a tack course on top of which the surfacing can be laid. When there is delay between laying the courses other methods of curing should be adopted for the lower layers, such as a light water spraying of the surface twice a day, or covering with wet hessian, waterproof paper blankets or plastic sheeting.

In frosty weather newly laid material should be protected for 7 days by a 76 mm layer of straw, covered by tarpaulins or some other waterproof sheeting. Lean concrete roadbases should not normally be laid when the air temperature falls below about 3 °C.

SELECTED BIBLIOGRAPHY

1. The WASHO Road Test—Part 2: Test data, analysis, findings, *Highway Research Board Special Report* No. 22, Washington, D.C., 1955.
2. The AASHO Road Test, *Highway Research Board Special Reports* 61A to 61E, Washington, D.C., 1962.
3. CRONEY, D. and J. A. LOE. Full-scale pavement design experiment on the A-1 at Alconbury Hill, Huntingdonshire, *Proc. Inst. Civ. Engrs*, 1965, **30**, 225–270.
4. LEE, A. R. and D. CRONEY. Research on the design of flexible road pavements, *Proc. Australian Road Research Board*, 1962. **1**, pt. 2, 622–642.
5. THOMPSON, P. D., D. CRONEY, and E. W. H. CURRER. The Alconbury Hill experiment and its relation to flexible pavement design, *Proceedings of the 3rd International Conference on the Structural Design of Asphalt Pavements*, 1972, **1**, 920–937.

6. WHIFFIN, A. C. and N. W. LISTER. The applications of elastic theory to flexible pavements, *Proc. International Conference on the Structural Design of Asphalt Pavements*, Ann Arbor, Michigan, 1962, 499–521.

7. DAVIS, E. H. Pavement design for roads and airfields, *Road Research Technical Paper* No. 20. London, H.M.S.O., 1951.

8. MALLOCK, H. R. A. Construction and wear of roads, *Proc. Inst. Civ. Engrs*, 1909, **178**, 111–181.

9. LISTER, N. W. The transient and long-term performance of pavements in relation to temperature, *Proceedings of the 3rd International Conference on the Structural Design of Asphalt Pavements*. 1972, **1**, 94–100.

10. CRONEY, D. and J. N. BULMAN. The influence of climatic factors on the structural design of flexible pavements. *Proceedings of the 3rd International Conference on the Structural Design of Asphalt Pavements*, 1972, **1**, 67–71.

11. STEEL, D. J. Discussion to classification of highway subgrade materials, *Proc. Highway Research Board*, 1945, **25**, 388–392.

12. MICHIGAN STATE HIGHWAY DEPARTMENT. *Field Manual of Soil Engineering.* Lansing, Michigan, 1952.

13. PORTER, O. J. The preparation of subgrades. *Proc. Highway Research Board*, 1938, **18**, pt. 2, 324–331.

14. Symposium on Development of the C.B.R. flexible pavement design method for airfields, *Trans. Amer. Soc. Civ. Engrs*, 1950, **115**, 453–589.

15. CRONEY, D., COLEMAN, J. D. and BLACK, W. P. M. Movement and distribution of water in soil in relation to highway design and performance, *Highway Research Board Special Report* No. 40, 226–252, 1958.

16. LEIGH, J. V. and D. CRONEY. The current design procedure for flexible pavements in Britain, *Proceedings of the 3rd International Conference on the Structural Design of Asphalt Pavements*, **1**, 1039–1048, 1972.

17. ROAD RESEARCH LABORATORY. *A Guide to the Structural Design of Pavements for New Roads*. London, H.M.S.O., 1970.

18. CURRER, E. W. H. and P. D. THOMPSON. The classification of traffic for pavement design purposes, *Proceedings of the 3rd International Conference on the Structural Design of Asphalt Pavements*, **1**, 72–81, 1972.

19. SHOOK, J. F., L. J. PAINTER and T. Y. LEPP. *Use of loadmeter data in designing pavements for mixed traffic. Highway Research Record* **42**, 41–56, 1963.

20. CRONEY, D. and J. C. JACOBS. The Frost Susceptibility of Soils and Road Materials. *RRL Report* LR 90, Crowthorne, Berks., The Road Research Laboratory, 1968.

21. HVEEM, F. N. and R. M. CARMANY. The factors underlying the rational design of pavements, *Proc. Highway Research Board*, 1948, **28**, 101–136.

22. MATERIALS AND RESEARCH DEPARTMENT. *Materials Manual of Testing and Control Procedures*. Sacramento, California Division of Highways, 1960.

23. PROKSCH, H. Thickness design methods for asphalt roads in the USSR, GDR, Poland and the CSSR, *Bitumen*, 1972, **34**, No. 6, 168–177.

24. BURT, M. E. Research on Pavement Design. *RRL Report* LR 243, Crowthorne, Berks., The Road Research Laboratory, 1969.

25. LINELL, K. A., F. B. HENNION and E. F. LOBOCZ. Current Corps of Engineers practice related to the design of pavements in areas of seasonal frost. *Paper presented at the 42nd Annual General Meeting of the Highway Research Board*, Washington, D.C., Jan. 1963.

26. HANRAHAN, E. T., *et al.* A series of three papers dealing with the problems of roads on peat and organic subsoils. In *Journal of the Institution of Highway Engineers*, 1967, **14**, No. 4, 8–38.

27. TRESSIDER, J. O. A review of existing methods of road construction over peat, *Road Research Technical Paper* No. 40. London, H.M.S.O., 1958.
28. CLARKE, A. C. and JOHNSON, B. W. Comparative designs, performance, and costs for a road on peat, *Journal of the Institution of Highway Engineers*, 1971, **18,** No. 3, 25–30.
29. MINISTRY OF TRANSPORT. *Specification for Road and Bridge Works*. London, H.M.S.O., 1969.
30. ROAD RESEARCH LABORATORY. The protection of subgrades and granular sub-bases and bases. *Road Note No.* 17, London, H.M.S.O., 1968.
31. LAKE, J. R., FRASER, C. K., and BURNS, J. A Laboratory Investigation of the Physical and Chemical Properties of Spent Oil Shale. *Roads and Road Construction*, 1966, **44,** No. 522, 155–159.
32. FRASER, C. K. and LAKE, J. R. A Laboratory Investigation of the Physical and Chemical Properties of Burnt Colliery Shale. *RRL Report* LR 125, Crowthorne, Berks., The Road Research Laboratory, 1967.
33. SHERWOOD, P. T. and RYLEY, M. D. The Effect of Sulphates in Colliery Shale on Its Use for Roadmaking. *RRL Report* LR 324, Crowthorne, Berks., The Road Research Laboratory, 1970.
34. LISTER, N. W. Design and performance of cement-bound bases, *Journal of the Institution of Highway Engineers*, 1972, **19,** No. 2, 21–33.
35. GEORGE, K. P. Shrinkage cracking of soil-cement base: Theoretical and model studies, *Highway Research Record* 351, 115–132, 1971.

7   Bituminous surfacings

About 95 per cent of the roads in Great Britain have flexible pavements within which are included bituminous surfacings of various types and thicknesses. Where properly used and constructed, these surfacings have long and economic lives while performing satisfactorily the following functions:

1. They provide smooth, relatively quiet running surfaces for vehicles.
2. They are highly resistant to surface wear and deformation.
3. They have high resistance to skidding, while having local surface drainage channels which also cause splash and spray to be minimized.
4. They help to transmit the applied wheel loads in such a way that the layers beneath are not overstressed.
5. They act as covers protecting the underlying pavement and subgrade from the detrimental actions of moisture.

For the purpose of discussion, bituminous surfacings can be divided according to whether they are close-textured (or dense), medium-textured, or open-textured surfacings, or surface treatments. Generally speaking this division makes it possible to treat surfacings in relation to their ability to meet the five criteria listed above. Thus close-textured mixtures will normally meet all five criteria, while the other two surfacings meet the first four. Although surface treatments are not in themselves surfacings, they help surfacings to meet all but the fourth criterion.

All bituminous surfacings are composed of a mixture of bitumen and/or tar and mineral aggregates. In close-textured ones, the binder is always a very viscous material, e.g. a penetration-grade bitumen or a high-viscosity tar. In the medium- and open-graded mixtures, the binders range from very viscous materials down to very fluid ones such as cut-back bitumens or low-viscosity tars. The surface treatments primarily use binders with relatively low viscosities. The aggregate content of a bituminous mixture may be composed of coarse, fine and filler material. By coarse aggregate is generally meant the material retained on the 2·36 mm sieve; this may be crushed stone, slag or gravel. The fine material is that which passes the 2·36 mm sieve and is retained on the 75 μm sieve; it is usually composed of angular natural sand particles. The filler is the dust ingredient, all of which passes the 75 μm sieve; it is most often composed of crushed limestone dust, but may also be quartz dust, lime, cement, asbestos or fly ash. The most commonly used dense mixtures contain proportioned amounts of all three types of aggregate, while the less dense ones are not as well-graded and are, on the whole, low in fines contents.

352

All close-textured surfacings are composed of hot-mix, hot-laid mixtures of aggregate and a high-viscosity binder. These are manufactured at central plants where both aggregates and binders are heated to very high temperatures, properly proportioned and mixed while hot, then transported rapidly in covered lorries to the sites and, while still hot, laid by mechanical spreaders and compacted by rollers. The commonly used medium- and open-textured mixtures are also mixed at central mixing plants, but mixing can take place at lower temperatures since lower viscosity binders are used. When transported to the site these mixtures may be laid hot, warm or cold, depending on the viscosities of the materials involved and the circumstances of mixing and laying. In surface treatments the aggregates are usually laid cold in the field while the binders may or may not be heated before application.

HOW BITUMINOUS SURFACINGS FAIL

Before discussing the various types of bituminous surfacing, and the methods of design which are used for dense bituminous mixtures, it is worthwhile reviewing the principal factors which influence the performance of these surfacings. While this may seem at first sight to be a negative approach to the problem of surface-course design, it is nevertheless a most useful one, because the state of the art at this time is still essentially one of design on the basis of experience. This is particularly so in Britain, where, instead of using mechanical procedures in surfacing design, reliance is placed on British Standard specifications for bituminous surfacings which are based on accumulated years of engineering experience with particular binder-aggregate mixtures.

Figure 7.1 summarizes the ways in which bitumen surfacings may fail and lists possible causes for each type of failure. While this diagram refers to surfacings prepared with a bitumen binder, it should be noted that the failure categories of instability, disintegration and fracture also apply to surfacings prepared with tar binders. Much of the following brief summary of the factors contributing to these failures is based on an excellent discussion available in the literature.[1]

It is appropriate to note here that the usual criterion of structural failure on experimental flexible pavements in Britain is the occurrence of about 25 mm permanent vertical deformation in the nearside wheel track measured from the original level[2]. This corresponds roughly to a rut depth of about 13 mm under a 1·2 m straight edge placed across the carriageway. Deformation of this magnitude is often accompanied by cracking of the wearing course, possibly with further cracking in bound layers at lower levels.

Instability

The stability—or, for that matter, the instability—of a bituminous mixture is a most difficult term to define, and it can mean different things to different authorities. There is general agreement, however, that stability refers to the resistance to deformation under traffic loading. It is perhaps the most important property of a surfacing, as reflected by the fact that unstable surfaces are very easily recognizable in the field; they are often referred to as having 'flowed', 'pushed', 'shoved', 'grooved', 'distorted', 'rutted', 'deformed', 'rippled' 'corrugated', or simply as 'washboarded'. These types of failure are most noticeable at intersections where the inertia resistance to slow-moving or stationary traffic cannot be developed to the extent that is possible elsewhere. Thus, for instance, 'shoving' is one of the terms used to describe the transversely-ridged deformation which occurs at areas subject to severe accelerating stresses, while 'rutting' is the result of the traffic channelization.

A resistance to deformation in which the engineer in, for example, an urban area is most interested is that exhibited under static or near-static loading conditions, since inertial resistance is only manifested under dynamic (fast-moving traffic) loading. This non-dynamic resistance is a function of both the interparticle and the binder frictions. Interparticle friction is dependent on the roughness of the surfaces of the particles and the intergranular contact pressure; it is lowered if too much binder is present and the particles are kept apart. By 'liquid friction' is meant the contribution to the resistance to sliding which is provided by the binder because of its viscosity characteristics. This means that sufficient binder must be present to coat all the particles, and that its viscosity must be high enough to give as great a liquid friction as is necessary. This latter consideration is influenced by the initial consistency of the binder, its ageing characteristics and the prevailing temperature conditions.

Disintegration

A bituminous surfacing which does not disintegrate as a result of weathering or the abrasive forces of traffic is described as being durable. Descriptive terms for surface failures which are primarily a result of the abrasive actions of traffic are 'ravelling', 'pitting' and 'pot-holing', all of which are variations of the same defect. 'Spalling' and 'stripping' are terms which are used to describe failures resulting primarily from the effects of water, frost and sunlight.

If disintegration is to be prevented, the binder must have a high degree of adhesion for the aggregate surfaces and adequate cohesive bonds between the aggregate particles to hold them together. This requires a combination of the proper selection of binder viscosity to fit the aggregate gradation, the addition of sufficient binder to coat all the particles thoroughly, the use of hydrophilic aggregates, and adequate compaction to produce an impervious surfacing. Disintegration is generally minimized by high binder contents, since they not only better waterproof the surfacings and thoroughly bind the particles, but in addition the films formed are more resistant to hardening and thus less liable to become brittle at an early age; care must be taken, however, to ensure that the binder content is not so great as to cause a lowering of stability. If the binder content is too low, or if the binder has become brittle, the aggregate can be pulled from the surfacing by the abrasive action of the vehicle wheels.

Fracture

This can be broadly defined as the cracking of a bituminous surfacing into sections. When this breaking develops into a regular pattern, it is sometimes described as 'haircracking', 'chicken-wiring' or 'alligatoring'. Major factors causing cracking are shrinkage, brittleness and slippage.

Shrinkage cracks are a result of volume change in the final roller-compacted surfacing. As illustrated in Fig. 7.1, shrinkage is a result of change in the volume of the binder due to loss of volatiles, change in temperature, absorption of the binder by porous aggregate material, or perhaps some internal structural readjustment in the binder such as that described by the term thixotropy. Shrinkage cracks are most often found in old bituminous surfacings which have very high binder contents and are little used by traffic. In present-day surfacings excess binder contents are rarely associated with this form of cracking, and even where they are, shrinkage cracking may be very limited since the kneading action of rubber-tyred motor vehicles tends to 'heal' the cracks.

The brittleness or inflexibility of a bituminous surfacing can be defined as its inability to conform to long-term changes in the shapes of the roadbase and sub-base, and its inability to bend repeatedly without cracking. Inflexibility or brittleness is a relative term, the degree to which it can exist in a surfacing being heavily dependent on the type, number and magnitude of the deflections involved and the thickness of the surfacing; in addition it is a function of the binder content, aggregate grading, and, especially, the final condition of the binder in the surfacing. In general, the more binder in a surfacing the less brittle it is likely to be, i.e. provided that the binder itself is a durable material. As the thickness of a bituminous surfacing increases, its flexibility decreases; on the other hand, thick dense surfacings resting on given roadbases will deform less and have greater radii of deformation curvatures than thin ones. Other things being equal, relatively open-textured surfacings are less brittle than dense-graded ones and can withstand larger deflections. Temperature is also a factor, and a bituminous surfacing is considerably less brittle at high temperatures than at low ones.

Slippage is a fairly rare characteristic of a bituminous surfacing which may occur at locations where sudden braking of heavily loaded vehicles is carried out. It is usually manifested in the form of a lateral displacement of the surfacing with respect to the roadbase or by the movement of the bituminous wearing course with respect to the basecourse. It results in a very characteristic cracking of the surface in the shape of a crescent or V with the apex of the V opposite to the direction of travel. Slippage cracks can be prevented by developing a proper bond between the sliding layers in the course of construction; this entails the removal of all loose debris from the lower layer, and perhaps the application of a light tack course prior to the laying of the upper layer. Normally the thicker the superimposed layer, the less likely it is to slide; this is due to its inherent inertial resistance to movement.

Objective	Types of Problems Encountered	Causes	Function of
IMPROVING PERFORMANCE OF ASPHALT PAVING MIXTURES	INSTABILITY	Low interparticle friction	Particle surface texture Quantity of bitumen (excess) Particle contact pressure (compaction)
		Low liquid friction (or 'mass' viscosity)	Gradation or surface area (i.e. points of contact) Type and amount of bitumen Density
		Lack of inertive resistance	Speed of vehicle Weight of vehicle Mass of pavement affected
	DISINTEGRATION	Low tensile strength (or cohesion)	Surface area (i.e. points of contact) Type and amount of bitumen Density
		Stripping (water action)	Adhesiveness, viscosity and amount (film thickness) of bitumen Wettability characteristic of aggregate
		Abrading (traffic action)	Amount of bitumen (insufficient) Relative hardness of aggregate Brittleness of binder
		Shrinkage (volume change)	Loss of volatiles and temperature changes Absorption of bitumen by aggregate Internal structural adjustments of bitumen
	FRACTURE (OR CRACKING)	Brittleness (inflexibility)	Amount, viscosity and brittleness of binder Grading of aggregate and thickness of bituminous paving Resiliency ('rebounding') characteristics of roadbase or sub-base
		Slippage	Bond between courses (insufficient) Tensile strength of pavement Thickness of slipping layer

Fig. 7.1.   An analysis of types and causes of failure in bituminous paving mixtures[1]

Loss of skidding resistance

Although not noted in Fig. 7.1, it is appropriate to discuss here another form of surfacing failure—that due to a reduction in skidding resistance. Many factors influence the frictional properties of a road surface, but in most instances the basic mechanism by which the road becomes slippery is the loss of microtexture brought about by tyres polishing the exposed surface. In the case of bituminous surfacings, the contact area of the exposed surface is composed largely of the mosaic of chippings—which is why in determining the skidding resistance of a road surface the type of aggregate, as characterized by its resistance to polishing (see also Chapter 3), is the single most important property of the surfacing material.

The macrotexture of the surfacing also contributes to skidding resistance, particularly at high speeds. It facilitates the drainage of water from the area of contact with the tyre, and it enables energy losses which occur in the tread rubber when it is deformed by projections in the road surface to be used in absorbing vehicle kinetic energy. Loss of macrotexture can result from the progressive embedment of the exposed aggregate/chippings in the bituminous surfacing, or from the failure of the binder to weather at an appropriate rate.

Many other factors affect the skidding resistance, as measured in terms of sideways force coefficient; they include type of binder, road temperature, road layout (e.g. excessive polishing is associated with braking or cornering forces), and size of stone (i.e. different sizes of the same stone may give different results under otherwise similar conditions). Overall, however, the most important external factor influencing the rate of resistance deterioration is undoubtedly the volume-speed-weight of traffic. This deterioration is particularly noticeable on traffic lanes carrying heavy flows of commercial vehicles. Nevertheless, it should be also pointed out that at the same time as traffic is tending to polish the surfacing, complex physio-chemical phenomena described as 'weathering' are tending to act in the opposite way to restore the microtexture of the exposed aggregate. Thus the resultant resistance to skidding measured at any given time is in effect a 'trade-off' between the effects of these naturally occurring phenomena and the density of traffic.

Recommendations regarding suitable aggregates for the wearing courses of high-quality bituminous surfacings, and for surface dressings, used at the locations indicated are given in Table 3.13. Also shown in this table are desirable sideways force coefficient values. One recent important study[3] carried out for British conditions has shown that the skidding resistance of a high quality bituminous surfacing at an average or easy site (see Table 3.13) can be predicted (correlation coefficient = 0·91) from the following equation:

$$sfc = 0.24 - \frac{0.663 q_{cv}}{10^4} + \frac{1}{10^2 PSV}$$

where sfc = sideways force coefficient,

q_{cv} = no. of comm veh/lane per day,

and PSV = polished stone value of the chippings.

The importance of this correlation is that it makes it possible to predict the level of skidding resistance at any reasonable site from a knowledge of the polishing stone value of the surface aggregate and an estimate of the traffic carried. The results should also serve as a guide in the selection of aggregates which are able to provide desired standards of skidding resistance. (Reference 4 summarizes the limited number of sources of British arenaceous rocks which can be used to give good skid-resistant surfacings.)

Recommendations for design

From information obtained relative to the types and causes of failures in bituminous surfacings the following general conclusions regarding design can be drawn:

1. Select a good quality, relatively hard and hydrophobic, rough-textured aggregate of a grading which will meet the requirements of workability, permeability, economy and (for a wearing course) with a high polishing stone value.

2. For the aggregate selected, choose a binder whose consistency is sufficiently soft to ensure good workability and long life, but is still hard enough to provide adequate tensile strength and resistance to moisture.

3. Use as much binder as possible in the surfacing without causing a loss of stability.

THE DESIGN AND SELECTION OF BITUMINOUS SURFACINGS

Empirical design procedures

From the previous discussion it is clear that the design of bituminous surfacing mixtures presents a most complex problem; indeed it is a problem which has yet to be solved, and the reader should have no doubt about this. While various 'design' procedures have been proposed, none of them can be considered as fully satisfactory at this time.

Before discussing procedures in use, it is worthwhile seeing how the rules given above relate to the design of high-quality (dense) aggregate-bituminous mixtures.

1. *Selecting the aggregate.* Since dense surfacings are used at locations which are subject to high and heavy traffic stresses, the aggregates chosen for use should have angular, bulky shapes and rough surfaces so that high stabilities can be achieved. Although hydrophobic aggregates are to be preferred, relatively hydrophilic ones may often be included in well-graded, densely compacted surfacings, since they are heated before being used with binder, and protected when compacted. For safety against skidding, the aggregates must have a considerable resistance to polishing. The selection of an appropriate gradation is very often based on practicalities. For instance, from both the stability and economic aspects it may be desirable to use a particular maximum size of aggregate in the surfacing, but practicalities of

compaction will restrict the maximum size to not more than 2/3 the thickness of the compacted layers. Again, ideally, the gradation should be chosen so as to give as close-textured a mixture as possible, but this may not always be economically possible with the locally available aggregates. Extra care must be taken if gap-gradations are used as these are especially sensitive to minor changes in binder contents.

All mix-design procedures in use at this time rely on mechanical tests to predetermine the qualities of the aggregates which it is proposed to use (see the chapter on Pavement Materials) and rigid specifications control their acceptance. Specifications also place limits on the range of gradations which may be used under particular circumstances.

2. *Choosing the binder.* Because of the locations at which dense bituminous surfacings are normally placed, it is desirable that binders composed of penetration-grade bitumens or high-viscosity road tars should be used, with the harder materials being utilized where the most severe conditions are encountered. Again there is no rational procedure for determining which grade of material should be used in any specific situation, and so at the present time selection is based on experience and governed by specifications.

3. *Amount of binder.* It is towards determining the amount of binder which need be incorporated in the surfacing that most efforts have been directed in terms of designing dense bituminous mixtures. Many empirical and semi-empirical design procedures have been devised which first attempt to evaluate various properties of bituminous mixtures and then base the binder-content determination on these evaluations. Some of the more widely known of these tests are as follows: (a) Hubbard-Field extrusion test[5,6] (b) OTL bearing index test[7] (c) Unconfined compression test[8] (d) Smith tri-axial test[6,9] (e) Wheel-tracking simulative test[10] (f) Marshall test[6,11] (g) Hveem test[6,12] (h) Splitting test[13] (i) Lee-Rigden test[14]

It is not possible to present here the procedures involved and the relative merits of all these tests, and the reader is referred to the literature for further information regarding them. However, just one of these tests, the Marshall test, and its associated design method will be discussed in detail here, as it is very likely that it is the test which has the greatest international acceptance. Before discussing the test and design method, it is necessary, however, to review the fundamentals underlying the theory of bituminous mix design.

Theory of bituminous mix design. The fundamental principles underlying the design of bituminous-aggregate mixtures are (a) that as much binder as possible should be incorporated in each mixture so that the voids are plugged, the aggregates thoroughly coated, and a durable impermeable mixture is obtained, but (b) excess binder should not be added which will lower the stability of the mixture below some minimum value which is controlled by the traffic and environmental conditions.

These principles indicate that for each mixture there is an 'optimum' binder content. There are two main approaches to determining this optimum content. These may be described as the *Surface-area Concept* and the *Voids Concept*.

Surface-area concept. As the name implies, this visualizes the optimum binder content as being a function of the surface area of the aggregate which is to be covered. This in turn is dependent on the size, shape and surface texture of the ingredient mineral particles, their absorptive capabilities, gradation and specific gravity, and the type of binder. Typical of the many formulae which have been devised to relate surface area and binder content is the Nebraskan Formula:

$$P = AG(0·02a) + 0·06b + 0·10c + Sd$$

where $P = \%$ by weight of bitumen residue in the bituminous mixture prior to laying,

   $A =$ absorption modifying factor for aggregate retained on the A.S.T.M. No. 50 (300 μm) sieve,

   $G =$ Specific gravity correction factor for aggregate retained on the A.S.T.M. No. 50 (300 μm) sieve; thus,

$$G = \frac{2·62}{\text{Apparent specific gravity of aggregate}}$$

   $a = \%$ by weight of aggregate retained on the A.S.T.M. No. 50 (300 μm) sieve,

   $b = \%$ by weight of aggregate passing the A.S.T.M. No. 50 (300 μm) and retained on the A.S.T.M. No. 100 (150 μm) sieve,

   $c = \%$ by weight of the aggregate passing the No. 100 (150 μm) and retained on the No. 200 (75 μm) sieve,

   $d = \%$ by weight of the aggregate passing the No. 200 (75 μm) sieve,

and $S =$ an experimental factor depending on the fineness and absorptive characteristics of the material passing the No. 200 (75 μm) sieve.

   At the present time, usage of surface-area formulae is confined (*a*) to determining the amount of binder to be used in low-cost roads and (*b*) as a first estimating of the quantities of binder to be later used in laboratory design procedures based on the voids concept.

Voids concept. According to this, the amount of binder which may be added is controlled by the void space in the aggregate framework, and by the desired volume of voids in the final compacted mixture. Obviously the amount of binder to be added is going to be influenced by the availability of the air voids in the aggregate framework—this in turn is predetermined by the aggregate gradation and method of compaction—and desirably as much binder as possible should be used to fill them. On the other hand, the compacted mixture must have a residue of air voids to allow for expansion of the binder and entrained air under hot weather conditions, to provide the mixture with a certain amount of elasticity, and to provide 'safety' space for further compaction of the surfacing under heavy traffic; against that, however, the residue of air voids should not be so great as to allow air and moisture to enter the surfacing and encourage disintegration. While it is generally agreed that maximum and minimum limits should be set on the amount of residual voids, there is considerable dissension as to what these limits should be. Recommended values have ranged from 2 to 3 per cent[15] to as high as 7 per cent maximum.[16]

Two points are to be emphasized about the voids concept and its relation to bituminous mixture design. First of all, it can only be applied to dense bituminous mixtures and not to open-graded ones. Where to draw the line between the two types of mixes, it is not possible to say; as a guide, however, it can be accepted that the concept is generally applied to mixtures whose gradations lie about those obtained by applying Fuller's gradation requirements for maximum unit weight. If more open-graded mixtures are used in conjunction with the voids concept, the binder content determined may be so great as to significantly reduce stability. The second factor which should be clearly understood is that the voids concept is not used on its own to any great extent as a means of determining the optimum binder content. As will be shown in the discussion on the Marshall design procedure, it is used primarily as a means of ensuring that durability considerations are taken into account in a test in which the determination of binder content for maximum stability is the prime object.

Marshall test and design procedure. The Marshall stability test itself is a type of unconfined compressive strength test in which a cylindrical specimen, 101·5 mm diameter by approximately 63·5 mm high, is compressed radially at a constant rate of strain of 50·8 mm/min. The maximum load in Newtons sustained by the specimen is recorded as the Marshall stability value, and the deformation at failure, in millimetres, is recorded as the Marshall flow value. Prior to the actual stability tests, however, void and unit weight determinations are carried out on the test specimens. The 'optimum' binder content then selected for design is essentially a compromise value which meets specified requirements for stability, deformation and voids content. The manner (American practice) in which this value is obtained can be outlined in a series of steps.

Step 1.   This involves the preparation of a series of test specimens for a range of different binder contents so that the stability test data show a well-defined optimum value at some binder content. To establish the binder contents to be used in these tests, the optimum content must first be estimated, either on the basis of experience or by means of a surface-area equation, to ensure that stability values above and below the optimum are obtained.

In preparing the individual Marshall specimens, each pre-heated mixture of aggregate and binder is placed in a heated mould and compacted with a specified number of blows of a 4·54 kg compaction hammer falling through 457 mm onto the top and bottom of each specimen.

Step 2. The bulk unit weight of each specimen is next determined. This determination is usually made by one of two methods, depending on the surface texture of the specimen being measured.

If the specimen has a compact, smooth surface, the determination is made by weighing the specimen in air and in water and then calculating as follows:

$$d = \frac{W_A}{V} = \frac{W_A}{W_A - W_W}$$

where d = bulk unit weight, g/cm^3

W_A = weight of specimen in air, g

V = volume of specimen, cm^3

and W_W = weight of specimen in water, g

If the specimen has an open and porous surface, it must be covered with a paraffin coating before being placed in the water. Then

$$d = \frac{W_A}{V} = \frac{W_A}{W_{PA} - W_{PW} - \dfrac{(W_{PA} - W_A)}{G_P}}$$

where W_{PA} = weight of specimen plus paraffin coating in air, g,

W_{PW} = weight of specimen plus paraffin coating in water, g,

G_P = specific gravity of paraffin,

and W_A and V are as defined above.

Step 3.   Calculate the percentage of air voids in each compacted specimen. This is done by first calculating the maximum theoretical unit weight and then expressing the difference between it and the actual unit weight as a percentage of the total volume of the specimen.

The maximum theoretical unit weight can be visualized as the unit weight which would result if the specimen had been compacted so that there were no voids in the aggregate-binder mixture. Thus,

$$\psi_t = \frac{W_A}{v_b + v_c + v_f + v_{mf}}$$

$$= \frac{W_A}{\dfrac{w_b}{G_b} + \dfrac{w_c}{G_c} + \dfrac{w_f}{G_f} + \dfrac{w_{mf}}{G_{mf}}}$$

where ψ_t = maximum theoretical unit weight, g/cm^3,

W_A = weight of the specimen, g,

v_b = volume of binder in the specimen, cm^3,

v_c, v_f and v_{mf} = volumes of coarse, fine and mineral filler fractions, respectively, of the aggregates in the specimen, cm^3,

w_b = weight of binder in the specimen, g,

w_c, w_f and w_{mf} = weights of coarse, fine and mineral filler fractions, respectively, of the aggregates in the specimen, g,

G_b = specific gravity of the binder, and

G_c, G_f and G_{mf} = *apparent* specific gravities of coarse, fine and mineral filler fractions, respectively, of the aggregates in the specimen.

Two points are to be noted with respect to this calculation. First of all, the weight of each component fraction of the specimen (i.e. w_b, w_c, w_f and w_{mf}) is determined by assuming each to be a proportional amount of the total weight of the specimen. Secondly, the specific gravity values of the aggregates used in this calculation are the apparent specific gravities and not the bulk specific gravities. In this way, the volumes of the permeable voids in the aggregates are excluded from the calculation, i.e. it is assumed that the voids are filled with binder material. While this is not necessarily true, it probably gives results which are closer to the true condition than if the bulk specific gravities were used.

The percentage of air voids in the compacted specimen is obtained by subtracting the actual from the theoretical unit weight and expressing the difference as a percentage of the theoretical unit weight. Thus

$$\%\text{V.T.M.} = \frac{\psi_t - d}{\psi_t} \times 100$$

where   %V.T.M. = voids in the total mixture, i.e. in the specimen, %

ψ_t = theoretical maximum unit weight, g/cm^3

and d = bulk unit weight, g/cm^3.

Step 4.   For each specimen calculate the percentage of voids in the compacted mineral aggregate framework which is filled with binder. This involves first determining the amount of voids in the aggregate framework (V.M.A.) and then calculating the percentage filled with binder material.

The V.M.A. is obtained by subtracting the volume occupied by the aggregate in the compacted specimen from the bulk volume of the compacted specimen, i.e. this is the volume of voids which in theory is available for filling with binder. Thus,

$$\text{V.M.A.} = V - v_c - v_f - v_{mf}$$
$$= \frac{W}{d} - \frac{w_c}{G_c} - \frac{w_f}{G_f} - \frac{w_{mf}}{G_{mf}}$$

where   V.M.A. = voids in the mineral aggregate framework, cm^3, and the remaining symbols are as defined in steps 2 and 3.

The voids in the mineral aggregate framework are often expressed as a percentage of the total volume of the specimen. Thus,

$$\%\text{V.M.A.} = \frac{\text{V.M.A.}}{V} \times 100$$

where   %V.M.A. = voids in the mineral aggregate framework,
                  % of total mix

         V.M.A. = voids in the mineral aggregate framework, cm^3

and   V = total volume of specimen, cm^3.

The percentage of voids in the aggregate framework which is filled with binder is determined from

$$\%\text{Voids filled with binder} = \frac{v_b \times 100}{\text{V.M.A.}}$$

where v_b and V.M.A. are as described above.

The same percentage may also be calculated from the following equation:

$$\%\text{Voids filled with binder} = \frac{\%\text{V.M.A.} - \%\text{V.T.M.}}{\%\text{V.M.A.}}$$

where %V.M.A. and %V.T.M. are the percentages of voids in the mineral aggregate framework and in the compacted mixture, respectively.

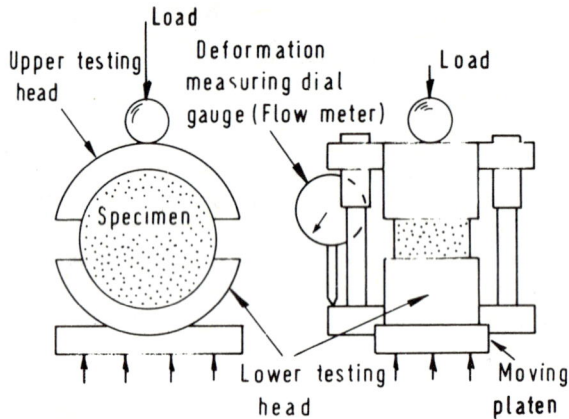

Fig. 7.2.   Illustration of the Marshall stability test

Step 5.   Determine the Marshall stability and flow of each specimen. The Marshall stability of a test specimen is the maximum load, in Newtons, required to produce failure when the specimen is preheated to a prescribed temperature, placed in the special testing head shown in Fig. 7.2 and the load applied at a constant rate of strain of 50·8 mm/min. While the stability test is in progress a dial gauge is used to measure the vertical deformation of the specimen; the deformation read at the load failure point is the flow value of the specimen.

Step 6.   The measured stability values are next corrected to those which would have been obtained if the specimens had been exactly 63·5 mm high. This is done by multiplying each measured stability value by an appropriate correlation ratio. The correlation values shown in Table 7.1 were derived by the U.S. Corps of Engineers in the course of their investigations into the use of the Marshall test in the design of bituminous surfacings for airport pavements.

TABLE 13.1.   *Marshall stability correlation values*[6]

Volume of specimen, cm³	Approximate thickness of specimen, mm	Correlation ratio	Volume of specimen, cm³	Approximate thickness of specimen, mm	Correlation ratio
200–213	25·4	5·56	421–431	52·4	1·39
214–225	27·0	5·00	432–443	54·0	1·32
226–237	28·6	4·55	444–456	55·6	1·25
238–250	30·2	4·17	457–470	57·2	1·19
251–264	31·8	3·85	471–482	58·8	1·14
265–276	33·3	3·57	483–495	60·3	1·09
277–289	34·9	3·33	496–508	61·9	1·04
290–301	36·5	3·03	509–522	63·5	1·00
302–316	38·1	2·78	523–535	65·1	0·96
317–328	39·7	2·50	536–546	66·7	0·93
329–340	41·3	2·27	547–559	68·3	0·89
341–353	42·9	2·08	560–573	69·9	0·86
354–367	44·5	1·92	574–585	71·5	0·83
368–379	46·0	1·79	586–596	73·0	0·81
380–392	47·6	1·67	599–610	74·6	0·78
393–405	49·2	1·56	611–625	76·2	0·76
406–420	50·8	1·47			

Step 7.   Prepare separate graphical plots for each of the following:
   (*a*) Binder content versus corrected Marshall stability.
   (*b*) Binder content versus Marshall flow.
   (*c*) Binder content versus percentage of voids in the total mix.
   (*d*) Binder content versus percentage of voids in the mineral aggregate framework filled with binder.
   (*e*) Binder content versus unit weight.
For illustrative purposes the data in Table 7.2, obtained from actual experiments, are plotted in Fig. 7.3, and the points joined with smooth curves.

Step 8.   Determine the optimum binder content. Before doing this, the design specifications must first be examined. Table 7.3 summarizes the criteria used by the British Ministry of Public Buildings and Works in the design of rolled asphalt for use in airport runway surfacings[11]. To determine the optimum binder content, the contents corresponding to maxi-

TABLE 7.2. *Typical Marshall test data*

Binder content %	Stability kN	Flow units	Percentage of air voids in the total mixture	Percentage of filled voids in the aggregate framework	Unit weight kg/m³
3	4·89	8·9	12·5	34	2169
4	7·06	9·4	7·2	65	2207
5	8·06	11·8	3·9	84	2255
6	7·52	14·6	2·4	91	2227
7	6·49	19·2	1·9	93	2190

* Binder content based on total weight of mixture.

mum stability, maximum unit weight, and to appropriate percentages of voids in total mix and aggregate voids filled with binder, are determined from Fig. 7.3. Thus,

Property	Unit weight	Stability	4% V.T.M.	80% Aggregate voids filled
Binder content, %	5·0	5·0	4·9	4·8

The optimum binder content is then taken as the average of these four binder contents; in this case it is 4·9 per cent.

TABLE 7.3. *Marshall design criteria used for airfield surfacings in Britain*

Test property	Wearing course	Basecourse
Stability, N	>8000	>8000
Flow, mm	0–4·1	0–4·1
Voids in total mix, %	3–4	3–5
Aggregate voids filled, %	76–82	67–77
Unit weight, kg/m³	None	None

Step 9. Check that the optimum binder content gives a mixture which will still meet the wearing course criteria in Table 7.3. Thus, re-entering the curves shown in Fig. 7.3 with the optimum binder content, the following values are determined:

Property	Stability	Flow	% V.T.M.	% Aggregate voids filled
Value	8060	5·08	4·4	83

In this instance the stability value is just barely adequate, the flow value is significantly in excess of the maximum and the permitted %V.T.M. is also exceeded, as is the limiting requirement *re* %V.M.A. Therefore this mix requires redesigning, and the tests performed again, before it can be used in the pavement.

Discussion. The Marshall stability tests and design procedure, like most other laboratory-based pavement design methods, has many advocates and many adversaries. The pros and cons of the method are too numerous to detail here; suffice it to say that, where it is now used as a design procedure, the results have been extensively checked against performance in the field. At the present time, however, the Marshall method of design is not commonly used for road design purposes in this country. It is now clear that a Marshall-type design method will be introduced within the near future. It can be anticipated[17] that this method will use two criteria for determining

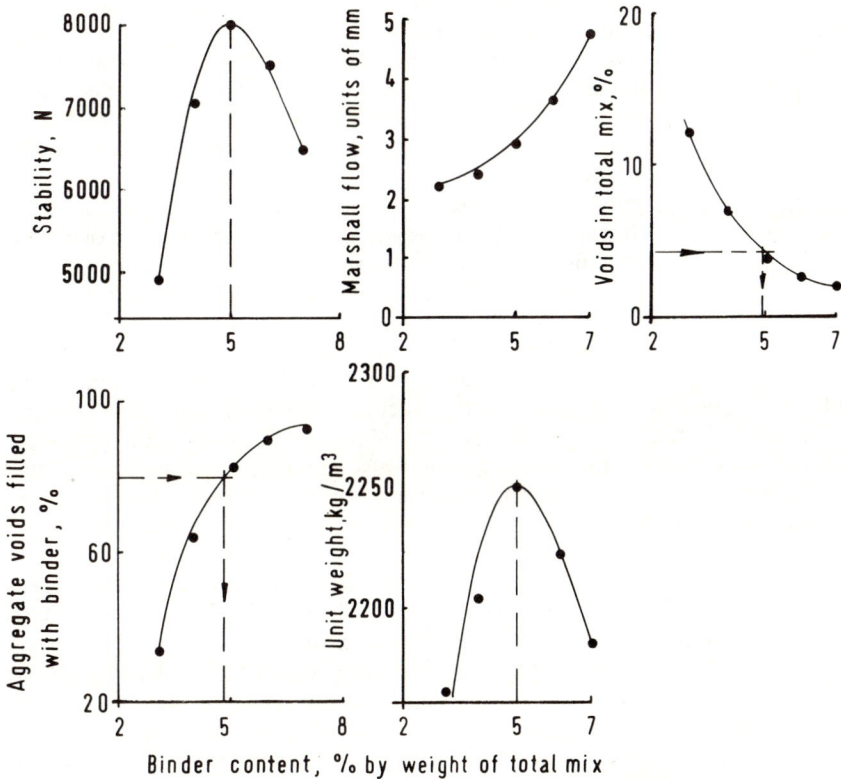

Fig. 7.3. Typical Marshall test data

the optimum binder content of sand-filler asphalts; these are the bitumen content that enables the maximum unit weight of the compacted asphalt to be obtained, and the bitumen content that enables the closest packing of the sand particles to be obtained. The mean binder content will then be the design content.

The Marshall test results shown in Fig. 7.3 are of particular interest in that they reflect some trends which are common to most laboratory tests for optimum binder content and, indeed, to most studies of actual road surfacing behaviour. These may be outlined as follows:

1. The unit weight of the mixture increases with increasing binder content until a maximum value is obtained, after which the unit weight decreases. This characteristic is similar to that observed when moisture is added to a natural soil. At first the binder acts as a lubricant and helps the aggregate particles to slide over each other. Once an optimum amount of binder has been added, however, it acts only to displace the particles and so the 'wet' unit weight decreases. If therefore a dense mixture is to be obtained, it is important that the amount of binder should not exceed the optimum for the compactive effort applied.

2. The stability value of the mixture also increases with increasing binder content until a maximum value is obtained, after which stability decreases. Generally the optimum binder content for stability is close to the optimum value for unit weight but on the dry side of it. The optimum content for stability can be explained by noting that the Marshall test is actually a type of 'unconfined' compressive strength test in which some degree of lateral support is given to the specimen; thus the maximum stability value occurs at the binder content at which the combination of the internal friction component of stability, provided by the interlocking aggregates, and the cohesive component, provided by the thin viscous bituminous films coating the particles, is a maximum under the conditions of test. Of the two, the test is by far a more true measure of the cohesive component of stability than of the internal friction one. Whether it is an exact measure of the change in stability which would occur in the roadway itself is, of course, very doubtful; there is no doubt, however, that each bituminous-aggregate mixture placed in a pavement has an optimum binder content for stability under the prevailing environmental conditions.

3. The flow value increases as the binder content increases. In addition, the rate of deformation change is slow at low binder contents but increases rapidly as high binder contents are reached. Again this is what perhaps might be expected in a roadway. For instance, surfacings with low flows and high stabilities will not deform easily but are likely to be brittle, while those with low stabilities and high flows deform easily under traffic.

4. The percentage of voids in the total mix decreases with increasing binder content until a value is reached at which it begins to level off. The %V.T.M. is critical as regards durability, since the greater the air voids content, the more easily air and moisture can attack the binder and the binder-aggregate bond. When the air voids content is too low, perhaps approaching 1 per cent or less, the surfacing is likely to 'flush' or 'bleed' under traffic. Since the error in the air voids determination may be as much as one per cent, due to the limits of precision of the standard test methods, the minimum air voids value which is normally recommended is 3 per cent. A further point to note is that, if the stability is low while the air voids content is satisfactory, it is probable that the low stability is the result of poor quality aggregate being used in the mixture.

5. The percentage of aggregate voids filled with binder increases with increasing binder content until a maximum value is reached. Again the rate of increase is fastest at low binder contents, and levels off at high contents. It is important to note that, unless the %V.M.A. is large, the paving mixture will be deficient in binder content or air voids, or both. When surfacings are deficient in binder, they become brittle and crack in early life and may ravel seriously under traffic; the effects of a low air voids content is, as has been stated, to promote flushing. Thus the voids in the mineral aggregate should be large enough to ensure that there is sufficient room between the particles in the thoroughly compacted mixture to contain at least 2 to 3 per cent air voids plus the minimum amount of binder required for a durable surfacing. Consequent on this, the voids in the mineral aggregate framework must be as nearly filled with binder as is possible.

British practice with bituminous surfacings

As stated before, it is not yet common practice in this country to utilize laboratory tests such as the Marshall test in the design of bituminous surfacings. Instead, over the years, a considerable amount of information has been collected with respect to the performance of different types of surfacings under various traffic conditions, both on specially constructed test roads and on the normal roads of the country. The results of research on these data have been combined with commercial experience in formulating particular aggregate and binder compositions that ensure good durability while at the same time providing reasonable working tolerances for manufacture and laying. The number and apparent complexity of these recipe-type of surfacings can seem daunting, however, to the person introduced to them for the first time. The emphasis in this discussion will be therefore on those aspects which explain the system and illustrate how and why one surfacing is used as compared with another.

Figure 7.4 outlines the types and usages of the main types of bituminous surfacings used in Britain.[27] Table 7.4 summarizes the governmental recommendations regarding them.

Rolled Asphalt. Next to mastic asphalt, this hot-mix, hot-lay surfacing is probably the most stable and durable bituminous road mixture used in Britain today. In essence it is a mortar of fine aggregate and penetration-grade bitumen to which is added a quantity of coarse aggregate or stone. 'Low' coarse aggregate content mixtures (about 30% retained on the 2·36 mm sieve) are used solely for wearing courses, while 'high' stone content mixtures (about 60%) may also be used for wearing courses but are more usually limited to basecourses (and roadbases). It possesses considerable mechanical strength and this makes it more effective than most other types of surfacing in reducing the traffic stresses within the underlying parts of the pavement. Assessments of the relative economic value of various types

TABLE 7.4. *Bituminous surfacings recommended for newly constructed flexible pavements in Britain*[18]

Traffic (Cumulative no. of standard axles $\times 10^6$)			
>11	2·5–11	0·5–2·5	<0·5
Wearing course (min thickness = 40 mm)	Wearing course (min thickness = 40 mm)	Wearing course (min thickness = 20 mm)	Two course (a) Wearing course (min thickness = 20 mm)
1. *Rolled asphalt*[19] Crushed rock or slag coarse aggregate only. Pitch-bitumen binder may be also used	1. *Rolled asphalt*[19] Crushed rock or slag coarse aggregate only. Pitch-bitumen binder may be also used	1. *Rolled asphalt*[19] Pitch-bitumen binder may be also used 2. *Dense tar surfacing*[20] 3. *Cold asphalt*[21] 4. *Med-textured tarmacadam*[22] See note 4 5. *Dense bitumen macadam*[23] See note 4 6. *Open-textured bitumen macadam*[23] See note 4 Basecourse 7. *Rolled asphalt*[19] See note 2 8. *Dense bitumen macadam or tarmacadam*[24] 9. *Single-course tarmacadam*[22,25] See notes 2 and 5 10. *Single-course bitumen macadam*[23,26] See notes 2 and 5	1. *Cold asphalt*[21] See note 4 2. *Coated macadam*[22,23,25,26] See notes 2 and 4 (b) Basecourse 3. *Coated macadam*[22,23,25,26] See note 2 Single course 4. *Rolled asphalt*[19] Pitch-bitumen binder may be also used 5. *Dense tar surfacing*[20] 6. *Med-textured tarmacadam*[22] See note 4 7. *Dense bitumen macadam*[23] See note 4 8. *60 mm of single-course tarmacadam*[22,25] See note 4 9. *60 mm of single-course bitumen macadam*[23,26] See note 4

Note 1. The thicknesses of all layers of bituminous surfacings should be consistent with the appropriate B.S. specification

Note 2. When gravel, other than limestone, is used, 2% Portland cement should be added to the mix and the percentage of fine aggregate reduced accordingly

Note 3. Gravel tarmacadam is not recommended as a basecourse for roads designed to carry more than $2·5 \times 10^6$ std axles

Note 4. When the wearing course is neither rolled asphalt nor dense tar surfacing, and where it is not intended to apply a surface-dressing immediately to the wearing course, it is essential to seal the construction against the ingress of water by applying a surface dressing either to the roadbase or to the basecourse

Note 5. Under a wearing course of rolled asphalt or dense tar surfacing the basecourse should consist of rolled asphalt or of dense coated macadam

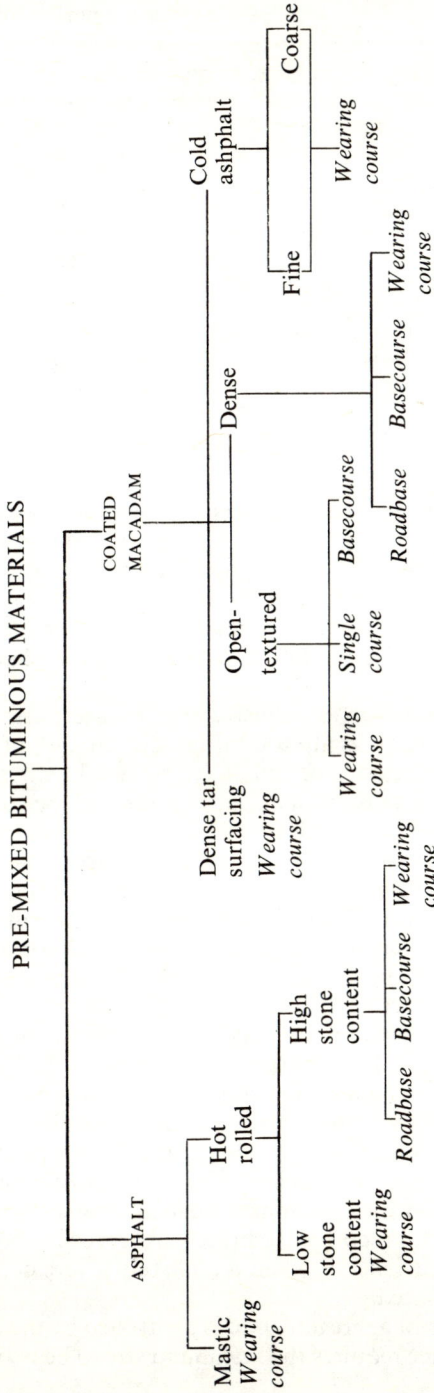

Fig. 7.4. The types and uses of pre-mixed bituminous materials

of bituminous surfacings have indicated that a properly constructed rolled asphalt surfacing can be expected to have a maintenance-free life of about 20 years provided the pavement otherwise remains sound. Life-comparisons between rolled asphalt and some other common types of surfacing which will be discussed are given in Table 7.5.

TABLE 7.5. *Lowest average lives to be expected for different types of surfacing*[28]

Surfacing	Traffic		
	Heavy	Medium	Light
Surface dressings	$4\frac{1}{2}$	5	6
Bitumen macadam carpets	4	$4\frac{1}{2}$	6
Fine cold asphalt	12	12	—
Hot-rolled asphalt	20	21	—

The compositions of rolled-asphalt surfacings are controlled by a British Standard[19] which at the moment lists 71 possible wearing-, base- and single-course combinations which may be used in highway pavements. The choice of which mixture to use in a given situation depends on economic considerations, the thickness to be laid, site and traffic conditions, availability of particular types of aggregate and grades of binder, etc. The following are the four types of construction in which rolled asphalt mixtures are normally used:

1. A two-course construction consisting of a basecourse and a wearing course. Two-course surfacings are used when the strength of the roadbase is considered to be inadequate for single-course work, where single-course construction might otherwise be used if the surface of the roadbase was not irregular.

2. A wearing course without a basecourse, called a single course. It is normally used when the surface thickness required is too small for two-course construction, but traffic conditions require a fairly thick high-quality surfacing. The cost of construction may also be a consideration, since normally the cost of a single-course is less than that of a two-course construction of the same overall thickness.

3. Thin wearing courses or 'carpets' which are sometimes used instead of single-course mixtures. These carpets are usually applied to already existing surfaces from which an uneven or otherwise defective wearing course has been wholly or partially removed beforehand by burning or planing.

4. Multiple-course construction. This is the description applied to the situation where use is made of successive layers of basecourse material. This form of construction, with or without a wearing course, is normally necessary when the desired total compacted thickness exceeds about 100 mm. An illustration of this occurrence is when a relatively severe curve on a motorway is being superelevated.

The maximum size of aggregate used is controlled by the thickness of the layer. The usual practice requires the maximum size to be between a half and a third of the thickness of the compacted layer. If too large a size is used, the graded aggregate will not compact satisfactorily under the rollers; if too small a size is used, there is a significant reduction in mechanical stability.

As mentioned before, rolled asphalt is essentially a mortar of sand and bitumen to which is added a quantity of coarse material. The choice of the fine aggregate and its gradings is the result of nearly seventy years' experience with this material; it permits a relatively wide tolerance in the bitumen content without detriment to the surfacing, while allowing maximum ease of spreading and compaction.[28] The resistance to deformation under traffic is controlled primarily by using a bitumen of suitable hardness and, to a lesser extent, by the incorporated coarse aggregate. Thus the use of low penetration bitumens is associated with locations where the foundation material is strong and a high resistance to deformation is required, e.g. on heavily travelled rural and city roads, particularly at severely stressed locations, such as bus stops. In contrast, high penetration bitumens are associated with thin surfacings or those with high coarse-aggregate contents or on less heavily trafficked routes. Rolled asphalt surfacings with high binder contents are necessary at locations which, when wetted, do not dry out easily; they are also used in roads carrying light traffic, since durability and not stability is the primary design factor here. Low bitumen contents are used where there are heavy volumes of moving traffic.

Rolled asphalt surfacings usually have a surface dressing of coated chippings rolled in during the laying operation; this is to increase their resistance to skidding. If a rolled asphalt wearing course has a high coarse-aggregate content, the addition of a surface dressing may not be possible; indeed if the coarse aggregate is such that it does not polish easily under traffic, the surface dressing will not be necessary. Since adequate resistance to skidding is dependent on the extent to which the chippings and/or aggregate are able to remain proud of the surface throughout its long life, the choice of binder is particularly important on very heavily travelled roads where binder contents tend to become more critical with, in particular, close-textured low-void content mixtures. The ability of chippings to stand proud is determined to a large extent by the nature of the binder and the way it permits the surface of the mortar to be slowly worn away by the combined action of traffic and weather (see Chapter 3).

Mastic asphalt. Manufactured mastic asphalt[29] consists of a mixture of limestone powder and asphaltic bitumen, in such proportions as to yield a plastic and voidless mass which, when applied hot, can be trowelled or 'floated' on a pavement to form a very dense impermeable surfacing. This is an 'old' surfacing; the records indicate that a number of the major streets in central London were surfaced with manufactured mastic asphalt as long ago as 1874. Although accepted as being perhaps the most stable and durable of the dense bituminous surfacings used in Britain, its usage is economically limited at this time since, unlike rolled asphalt, it does not lend itself to mechanization and so must be laid by hand in the pavement.

Unlike the other types of pre-mixed bituminous materials, manufactured mastic asphalt is not normally created in one process. Rather the mastic (or mortar) part is manufactured in special mixtures at a depot and cast into blocks. These are then taken to the site, reheated and mixed with up to 45 per cent of chippings (to improve stability) after which the material is spread, usually by hand floats. Its usage is limited to the construction of wearing courses which are normally laid on top of rolled asphalt basecourses.

Because of its complete imperviousness, high resistance to deformation and extreme durability, mastic asphalt is particularly suitable for roads and areas that have to carry iron-wheeled and tracked vehicles, as well as at street locations subject to very high stresses. It has a very fine texture which provides little resistance to skidding and so pre-coated chippings are normally applied to the running surface to improve its anti-skid characteristics. These chippings are rolled into the surface while the mastic material is still sufficiently plastic to allow them to become partially, but firmly, embedded. Here again refined lake asphalt has special wearing properties which render it most effective in maintaining the chippings proud of the surface. Although the specifications do not recommend it at present, there appears to be no reason why pitch-bitumen could not be used with advantage as the binding material in mastic asphalt surfacings also.

Compressed natural rock asphalt. As its name implies, compressed natural rock asphalt is a bitumen-impregnated limestone which is ground to a powder (it must entirely pass the 2·36 mm sieve and at least 15 per cent must pass the 75 μm sieve[30]), heated and, after it has been laid on the roadway and spread by hand, compressed to form a dense impermeable layer. The surfacing obtained has very high stability and durability properties which also make it an excellent material for use at highly stressed road locations.

On its own, compressed natural rock asphalt presents to traffic a smooth surface texture and hence it is necessary to cover it with a layer of chippings in order to produce a 'non-skid' surfacing. This is normally done by painting the surface with a tack-coat of a high-penetration bitumen, adding pre-coated hard igneous chippings, rolling the chippings into the surface, and then brushing either cold or warm natural rock asphalt powder into the interstices, taking care that no powder is left proud of the chippings.

Natural rock asphalt surfacings were first introduced into England in 1869 when they completely revolutionized the method of surfacing city streets, and, indeed, initiated the era of the durable hygienic pavement. Nowadays compressed natural rock asphalt is used only as a thick wearing course on top of a rolled asphalt basecourse or a concrete pavement. It is an expensive surfacing; not only does it require considerable experience to lay, but it cannot be laid or compacted by the normally used surfacing machines. As a result, its usage in Britain is now most rare.

Rock (non-skid) asphalt. This is a bituminous 'surfacing' in which the emphasis is placed on its use as a surface dressing rather than as a wearing course, and it should not be confused with the compressed natural rock asphalt described above. Also known as 'Natural-Rock-Asphalt Non-Skid Surfacing (Cold Process)', it consists of a dressing of 19–25 mm pre-coated chippings embedded in a 12–19 mm thick layer of cold rock-asphalt powder which is 'tacked' to the top of a rolled asphalt or mastic asphalt wearing course. This presents a very rough-textured surface which gradually closes up under the action of heavy traffic. While there is no British Standard for this construction, it is highly stable and durable.

Dense Tar Surfacing. Although this surfacing is included in Fig. 7.4 under the general heading of Coated Macadam, it should be appreciated that its strength depends to a large extent on its mortar—which is why DTS[20] is generally considered to be the tar macadam-type surfacing which corresponds most closely to rolled asphalt. It is a hot-mix, hot-laid material consisting of a mixture of coarse and fine aggregates, filler and high viscosity tar, the proportions of which are selected so that, when rolling is complete, a dense and impervious surfacing is obtained which is capable of carrying heavy and varied traffic flows over long periods of time.

A dense tar surfacing is normally laid in new construction in the form of a dense impervious wearing course on top of a dense tarmacadam basecourse, so that the combined thickness makes up a surfacing of the desired stability. On a sound roadbase, a single wearing course is normally adequate for resurfacing work; if the roadbase is excessively irregular or requires strengthening, two-course construction must be used.

The dense tar wearing course mixtures are of two types, one of which contains 35 per cent coarse aggregate and the other 50 per cent. Both types of mixture can be used on all types of road, but if the lower stone content mixture is used in heavily trafficked roads it is necessary to roll coated chippings into the running surface to improve its skid-resisting qualities. Chippings cannot easily be applied to the surface of the 50 per cent coarse aggregate mixture, but then again this surface has a rough texture which renders the addition of chippings unnecessary if the right quality of stone is used. The maximum size of aggregate is determined by the desired thickness of layer; for good compaction the maximum size should lie between one-half to one-third the layer thickness.

The tars used in dense tar surfacings have viscosities that are as high as can be practically used without fuming occuring during mixing and the tar hardening too rapidly; this limits the tars to those with viscosities below 60–65°C e.v.t. The same viscosity binders are used in both the surface and wearing courses so that the lower layer is not softened during the hot-laying process.

Coated macadams

These surfacings differ from the dense mastic-type asphalts in that they generally contain relatively little fine aggregate and filler material, and the intimate interlocking of the aggregate particles is a major factor in the strength of the surfacing. Furthermore, they are not initially impervious to water; in many instances also the aggregates have to be more resistant to abrasion and crushing, since a matrix of fine particles is no longer available to protect the coarse particles from the traffic. These surfacings are normally laid at locations where all-weather roads are required but the traffic conditions do not justify the construction of the more expensive dense surfacings.

The coated macadams contain coarsely graded mineral aggregates which are pre-coated with either tar or bitumen; the compacted mixtures have a substantial proportion of voids. They are prepared in large plants situated in quarries or pits which, as a result of the geological formations of Britain, are sufficiently uniformly spaced throughout the country to provide surfacing materials at most road sites at economical prices. They are easy to manufacture and lay and, when properly designed, they provide adequate stability for all roads but those carrying the heaviest traffic. At the present time the compositions of most coated macadam road surfacings are controlled by four British Standard specifications.[22,23,25,26]

Coated macadam materials are normally used in one of the four following forms of construction:

1. A two-course construction which consists of a wearing course on top of a basecourse. This type of construction is most often used in the reconstruction of existing roads and as the surfacing in new roads carrying medium to light traffic. In two-course construction the basecourse serves the dual function of supporting the wearing course and of acting as a regulating course between the wearing course and roadbase, thereby improving the riding quality of the wearing surface. Semi-dense graded bituminous macadams[24] which are not yet covered by a British Standard are also used as basecourse material beneath rolled asphalt and dense tar surfacings. The nominal size of aggregate used in any given layer depends on the compacted thickness, traffic conditions and, for wearing courses, on the running surface texture desired.

2. A single-course construction, because it has to serve the dual function of being both a wearing and basecourse, is more densely graded than a normal coated macadam basecourse which would be used in a two-course surfacing. Single-course surfacings are considerably cheaper to construct than two-course ones, and so they are in wide use as reconstruction surfacings on already existing roads. In new pavements they are confined to lightly trafficked roads where the uniformity of the final running surface is not a major design criterion.

3. A 'carpet' whose compacted thickness is less than about 50 mm. A coated macadam carpet is used only when it is desired to give a new riding surface to an already existing pavement, and its function is usually to remove minor irregularities in the running surface.

4. A multiple-course construction which consists of more than two layers of coated macadam material. This method of building a surfacing is used primarily to build up superelevations at curves and in widening existing road pavements.

Coated macadam wearing courses are most usually described as being open, medium or close-textured. Open-textured mixtures contain 15 per cent fines (i.e. material passing 3·18 mm sieve) or less, medium-textured about 25 per cent fines, and close-textured or dense macadams about 35 per cent fines.

The skid-resisting properties of open-textured wearing courses are determined by the nature of the aggregate; although these surfacings give the motorist a feeling of security in wet weather driving, this can be deceptive when the aggregate is one that polishes easily. Open-graded coated macadam mixtures have the disadvantage that not only are they very permeable, but they are also more open to the disintegration effects of air and water. To resist these effects they have available, unfortunately, less binder per unit volume of surfacing than any other pre-mixed material. Since the coarse aggregates are no longer protected by the filler-binder matrix, the aggregates used in open-textured surfacings must have high resistance to abrasion and crushing. Although the open mixtures have a wide tolerance in gradation with respect to stability, the tendency nowadays is to use the more closely textured mixtures in road surfacings. These have the dual advantages of providing more durable as well as more skid-resistant surfacings. Close-textured surfacings must be used if the underlying roadbase and/or subgrade are liable to be detrimentally affected by moisture percolating down through an open-textured layer. Close-textured basecourses must also be used under wearing courses formed from rolled asphalt and dense tar surfacing, since the open-textured ones utilize low viscosity binders which would soften when the hot wearing course is applied. This in turn could result in surface deformations taking place beneath the roller during compaction, thereby producing a poor running surface.

Coated macadams may also be divided according to the manner in which they are laid. *Hot-laid* macadams are those in which comparatively high-viscosity binders are hot-mixed with the aggregate and then transported to the site and laid at relatively high temperatures. They are used at locations where high stabilities are required. Since, however, these surfacings must be compacted while still hot, their use is limited to within short transportation distances of the mixing plant. The most widely used macadams are the *warm-laid* ones. The bitumens used in these mixtures range from medium-curing cutbacks to high penetration grade bitumens, and the road tars have medium viscosities. The materials in these aggregate-binder mixtures must be heated if an intimate mix is to be obtained; if properly sheeted they can be transported fairly long distances and still be laid warm the same day. The longer the journey or the colder the weather or the more lightly-trafficked the road, the lower the viscosity of the binder which is used. *Cold-laid* macadams are made with low viscosity tars or very fluid cutback bitumens and hence they are suitable for transporting over long distances or for storing for a period of time before laying. Full stability is not attained with these mixtures until the volatile oils present have evaporated, and so the surfacings cannot be sealed until this stage is reached. These surfacings are used only on lightly trafficked roads or for patching more heavily travelled ones.

While the British Standards designate fairly wide ranges of binder contents to be used with different aggregates under various conditions, it is left to the engineer's discretion as to what contents should be chosen within these ranges. In general, this choice is guided by knowledge of the prevailing traffic conditions. If the traffic is heavy, binder contents towards the lower ends of the ranges are selected in order to avoid the bleeding which would be likely to occur as the traffic further compacted the surfacing. On the other hand, if the traffic is light, then as much binder as possible is included, since in this case durability, and not stability, is the prime design consideration.

Cold asphalt.   This material is a special type of coated macadam which is usually manufactured with slag as the aggregate, although limestone aggregate is also used. A British Standard[21] which covers the composition of cold asphalt mixtures divides them into fine cold asphalt, which is made with an aggregate all of which passes a 6·35 mm sieve, and coarse cold asphalt, which is made with an aggregate substantially all of which passes the 9·52 mm sieve and only 65–85 per cent passes the 6·35 mm sieve. In both types the amount of material passing the 75 μm sieve is small.

Cold asphalt mixtures may contain bitumens of various grades, ranging from medium-curing cutbacks to refinery bitumens of high penetration. Cutbacks of low viscosity are used when it is necessary to stockpile the material or when it is desired to lay it cold in the roadway. If the penetration grade binders are used the mixtures must—in spite of the name 'cold asphalt'—be heated prior to laying.

Cold asphalts are used primarily as thin wearing coats or for patching purposes. In the latter case, the fine cold asphalt has the particular advantage that it can be premixed and stockpiled for several weeks prior to being laid. When used as a wearing course, fine cold asphalt is laid in thicknesses of 13–19 mm on lightly travelled roads and in thicknesses of 19–25 mm on the more heavily trafficked ones. Coarse cold asphalt mixtures are laid only in thicknesses of 19–38 mm. In both instances, cold asphalt must be laid either on top of a new basecourse or an already existing surface which has all defects repaired and excess binder burnt off.

When first laid, cold asphalt wearing courses are not impervious to water, and hence they are not used as surfacings over pavements which are likely to suffer damage as a result of the entry of water. In time, however, the surfacing does become quite impermeable under the kneading compaction of traffic.

When fine cold asphalt is laid and compacted it has a sandpaper texture which is slippery to fast traffic, which is why it is customary to apply a surface dressing of pre-coated chippings to the running surfaces of these mixtures. With the coarse cold asphalt surfacings, the surface dressing is not necessary.

SURFACE TREATMENTS

'Surface treatment' is the general term used to describe work carried out to alter the qualities of a wearing surface, but which does not add appreciable thickness to it. While by far the most well-known type of surface treatment is that of surface dressing, other treatments which are also of importance deal with the application of tack coats and sealing coats.

Surface dressings

A surface dressing normally consists of a layer of small aggregate such as chippings of stone, slag or gravel, which is superimposed on a thin layer of road tar or bitumen which is freshly applied to an existing road surface. They are used:

1. To seal new or old surfacings against the entrance of air or water.
2. To improve the non-skid properties of a surfacing.
3. To arrest any disintegration which may already be taking place.
4. To provide a clear demarcation between the carriageway and the shoulders.
5. To improve the night visibility of bituminous surfacings.

It must be emphasized that a surface dressing cannot restore the riding qualities of an already mis-shapen road and to obtain good results the surfacing being treated must be sound, clean and have a uniform texture. A surface dressing is simply a thin layer of aggregate held by a film of binder and has no structural strength of its own.

Surface dressings are used successfully on all types of road, from very lightly travelled country lanes to motorways carrying many thousands of vehicles per day. They are applied to surfacings which range from those which are soft and rich in binder to those that are hard and stony. In addition, if it is necessary to protect the subgrade or sub-base from the effects of inclement weather during the construction of a new pavement, a surface dressing may be applied immediately after compaction of the subgrade or laying of the sub-base.

When a surface dressing on a wearing course has a long and successful life, it is usually found that final failure occurs through the aggregate being worn away by the traffic; with a well-chosen aggregate this may take as long as 16 years, but it is more normally about 5 years. A surface dressing which fails prematurely does so either by 'scabbing'—i.e. early dislodgement of the stone from the dressing—or by 'bleeding' or 'fatting-up'—this occurs when the binder works its way up to the surface of the dressing. Thus the production of a good surface dressing consists primarily in designing to prevent scabbing or bleeding.

A considerable amount of research has been carried out in Britain into this problem and the results show that the factors determining whether scabbing or bleeding is likely to occur may be divided into the following three main groups:[31]

A. The properties of the materials used.
    (a) Pertaining to the binder,
        1. Type and quality.
        2. Viscosity.

(b) Pertaining to the aggregate chippings,
    1. Type and strength.
    2. Size and shape.
    3. Surface properties, i.e. texture, porosity, dryness and dustiness.
B. The method of application.
    1. Quantity or rate of spread of binder.
    2. Quantity or rate of spread of chippings.
    3. Weight and type of roller.
    4. Time interval before opening to road traffic.
C. External conditions.
    1. Condition of road surface.
    2. Traffic.
    3. Weather and time of year.

The following discussion relates to the influence which each of the above factors has upon the quality of surface dressings in Britain. The order in which they are taken into account in the design process[32] is summarized in Fig. 7.5.

Materials. The binders most easily available for surface dressing work in Britain are road tar, cutback bitumen, bitumen emulsions (cationic and anionic), and tar-bitumen blends containing not less than 30 per cent tar. Of these, the tars and cutback bitumens are by far the most widely used. The principal considerations governing the choice of binder are, first, that the adhesive when applied must be sufficiently fluid to 'wet' both the surface of the road and the aggregate and, secondly, it must then set sufficiently quickly to hold the chippings firmly in position against the displacing forces of traffic in the early stages of the life of the surface dressing. The first of these can be overcome by using heated tar or cutback bitumen, or an emulsion. If a heated binder is used, its viscosity must be such that after application it is still sufficiently fluid to coat the aggregate particles; since the binder temperature drops almost immediately to the prevailing road temperature, this means that the viscosity of the binder at road temperature must not exceed some maximum value. On the other hand, if the chippings are to be held in place against the early dislodging actions of traffic, then the viscosity of the binder at the road temperature must be above some minimum value. Thus the binder selected in any given situation must represent a compromise between these two requirements, and in Britain this has resulted in the recommendations given in Table 7.6. Note that for motorway-type traffic (>2000 comm veh/day per lane) the preferred binder is road tar as, in the event of the binder rising to cover the applied chippings because of embedment, tar will tend to weather and wear rapidly at the surface thereby re-exposing the aggregate and maintaining a non-skid surface. It can be also seen that the higher viscosity materials are used with higher temperatures as well as on the most heavily travelled roads.

Emulsions can also be used for surface dressing purposes; they have the particular advantage over the heated binders that no heating equipment is required. Emulsions have, however, the very significant disadvantage that

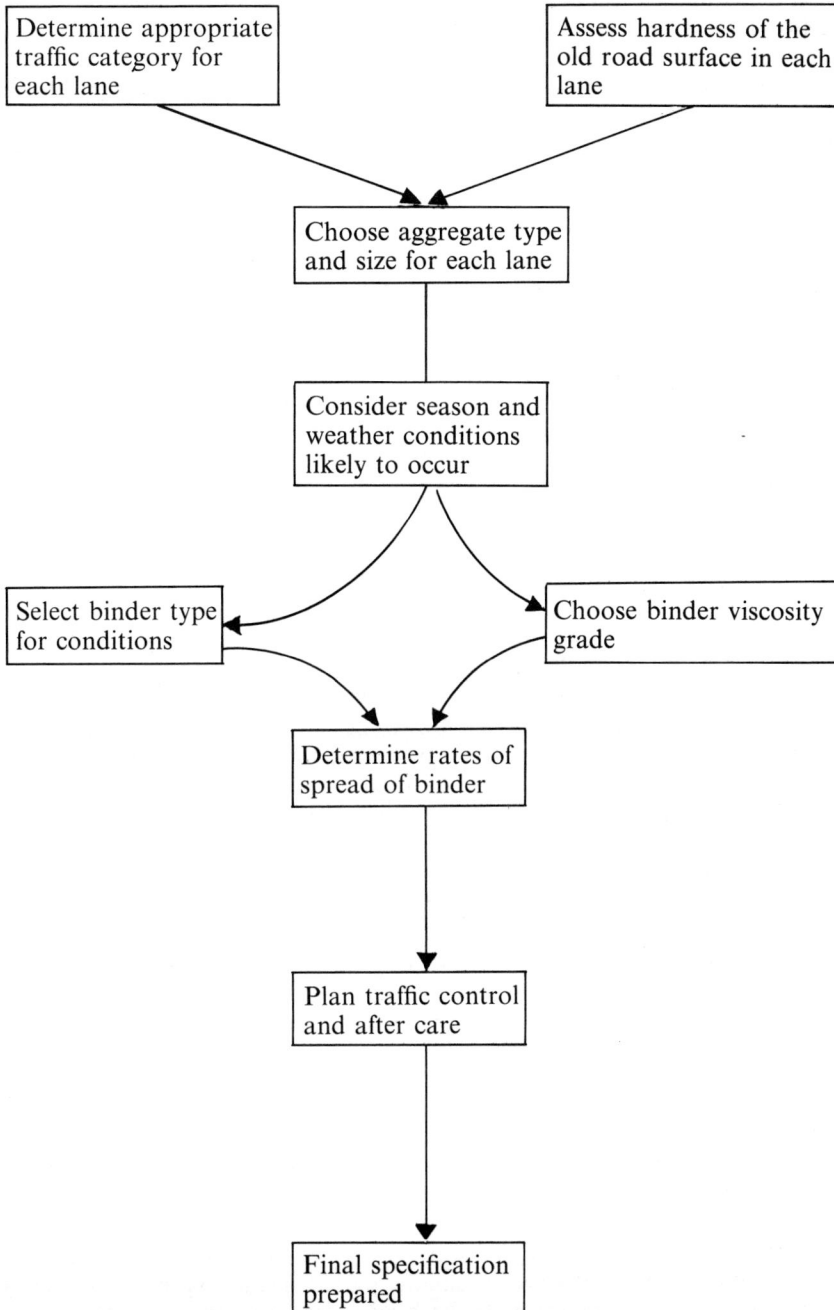

Fig. 7.5. Flow diagram for the surface dressing design process

their performance is much more affected by adverse weather conditions. In addition, the rate of spread of an emulsion is limited by its high fluidity, i.e. a heavy application is not possible on a road with a camber or slope while, because of the relatively thin film of residue binder which is left after breaking of the emulsion on the road, smaller chippings must be used with the emulsion than with the heated binders. For these reasons, use of emulsion binders is generally limited to locations where it is not practical to use heated binders.

While both slag and gravel aggregates are used to surface-dress pavements, it is found that hard igneous rock *chippings* are generally preferred because of their high resistance to crushing and polishing. High crushing

TABLE 7.6.   *Binders used for normal surface dressing purposes in Britain*[32]

No. of comm. veh/day per lane	Period of the year	Viscosity grade		
		Road tar (BS76, Table 1)	Cutback bitumen (BS3690, Table 2)	Tar bitumen blends
> 2000	Mid-May to mid-July	A46	—	—
1000–2000	Mid-May to mid-July	A46	200 sec (rubberized)	200 sec
< 1000	Apr, May, Sept	A34, A38	50, 100 sec	100 sec
	June, July, Aug	A38, A42	100, 200 sec	

strengths are necessary because of the considerable stress imposed by the roller used during the laying of the dressing as well as by the subsequent traffic. If an aggregate is used which polishes easily, then slippery surfaces will be produced and skidding accidents will very likely occur.

The size of chippings used at any particular site depends on the volume of traffic and the hardness of the existing road surface, i.e. on the forces causing embedment and the resistance to embedment. Recommended sizes of chippings which can be used with either tar or bitumen binders are given in Fig. 7.6. This shows that the smaller chippings are used for light traffic on hard surfacings while the largest chippings are used on the softer surfacings with the heavier traffic. As interpreted in this figure, a hard surface is one which has a close mosaic of aggregate that resists penetration by the chippings, while a soft surface is one which is usually rich in binder and into which the chippings can penetrate easily and deeply. Normal surfaces are those into which the chippings are embedded only slightly by the action of the traffic.

A further point to be noted about Fig. 7.6 is that only single-sized chippings are recommended. British experience indicates that if the chippings contain a range of sizes, the smaller particles tend to prevent the larger ones from making contact with the binder film. In support of the use of single-size aggregate, it might be pointed out that if the chippings have an appreciable size range and segregation takes place, then the small particles are likely to be almost entirely submerged in the binder layer, while the larger particles may only have a small depth of embedment when placed on the roadway.

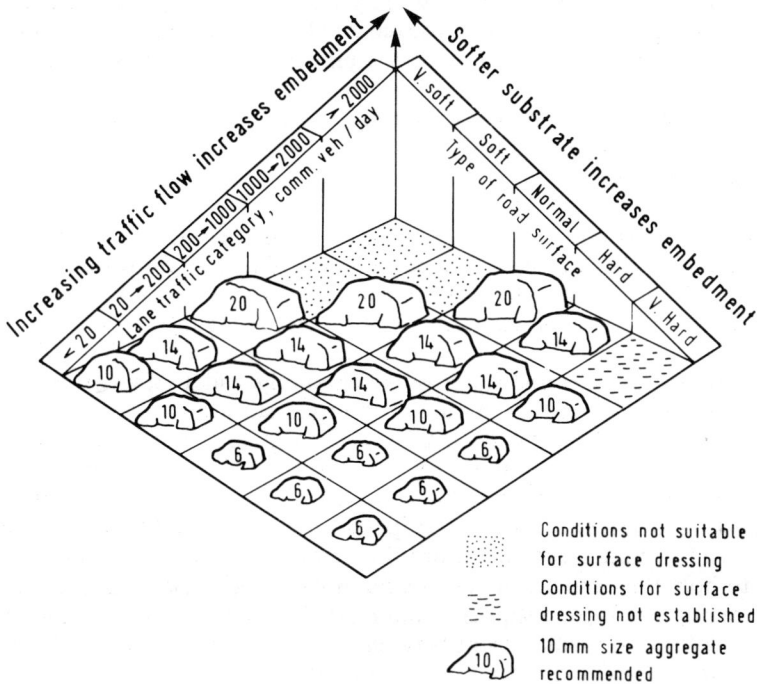

Fig. 7.6. Nominal sizes of chippings used for surface dressing purposes in Britain[32]

In all cases the aggregate particles should have cubical shapes as these are best for combining durability with a good non-skid property. This does not by any means preclude the use of gravel aggregates, especially if they are crushed beforehand and the correct amount and type of binder is used to secure good adhesion. Dust on the surface of the aggregate, especially if it is wet, will greatly delay the wetting of the stone by the binder and this increases the risk of premature failure in wet weather or conditions of high humidity; this is an additional reason for avoiding the use of soft stones. The risk of water displacing the binder from the aggregates is least with those chippings having rough surface textures, and greatest with those having glassy surfaces.

A special problem arises in the case of open-textured surfacings, e.g. coated macadams which are being surface-dressed for the first time. With these the chippings chosen must be of a size such that they can enter into and be held within the pores of the surfacing, thereby sealing it. The usual practice for porous surfacings is to determine the appropriate size of chipping from Fig. 7.6 and then use the next size smaller, although never less than 6 mm.

The 'dust' problem with chippings becomes most acute with the smaller sizes of stone. This is best overcome by using chippings that are pre-coated with thin films of tar or bitumen (usually 1·2–0·5% by weight). The use of as thin a coating as possible of hardened binder ensures that surface dust is eliminated while rapid adhesion to the binder film is easily achieved. Correctly coated chippings separate from each other easily and flow readily through mechanical spreaders; heavily coated or otherwise incorrectly treated chippings will tend to bind together in warm weather and cannot be applied mechanically.

Applications.   The rate of spread of the binder is the single most important item affecting the life of a surface dressing. There is no doubt that the majority of premature failures arise from using the wrong rate of spread. A considerable amount of data has been collected regarding rate of binder application, and this indicates that the correct rate of spread is primarily dependent on the size and shape of the chippings and, to a lesser extent, on the traffic intensity and the nature of the road surface.

The function of the binder is to bind the aggregate particles both to the road surface and to each other, while at the same time providing the chippings with the maximum possible support before the dislodging actions of traffic. This means that, since the chippings must be embedded in the binder, their size is a major consideration; the depth of embedment must be sufficient to hold the particles in position but not so much that the aggregate will not remain proud and a slippery running surface is produced. Because of the greater volume of voids present between rounded particles, gravel chippings require more binder to attain a given depth of embedment than cubical particles of the same sieve size. Similarly, flat or flaky particles require less binder than cubical ones. Once the aggregate particles are firmly embedded in the binder, the tendency is for them to be pushed by the traffic into the surface of the road; the heavier the traffic and the softer the old road surface, the more this is likely to occur.

It is difficult to give exact criteria as to how to select the correct rate of spread for any particular surface, since such an assessment is largely a matter of experience. For example, with old surfacings information regarding the rate of application of the binder may be best obtained by studying the old records for the road and assessing the performances of previous surface-dressing applications. Some suggested rates of spread of tar are summarized in Table 7.7; these target values, which apply to the chipping sizes given in Fig. 7.6, should only be used for general guidance, however, and the engineer must be prepared to vary them according to the needs of local conditions.

The quantity of chippings added to a surfacing must be sufficient to cover the entire film of applied binder after rolling. Since, however, it is rarely possible to lay the exact amount of aggregate which will do this, and since some of the aggregate is going to be whipped off the surface by the traffic anyway, the rate of spread of chippings which is applied is usually slightly in excess of that actually required to give shoulder to shoulder coverage.

Observations carried out in Britain have resulted in the recommendations shown in Table 7.8 regarding the rate of spread of chippings. It should be noted, however, that the rate at which the chippings are applied depends on their size, shape and specific gravity. Thus, while these data can be used as a guide in most instances, there will be occasions when it is necessary to depart from them.

TABLE 7.7. *Target rates of spread of tar at spraying temperature, litre/m^2*

| Type of surface | No. of comm veh/lane per day | | | | |
	2000	1000–2000	200–1000	20–200	< 20
Very hard	—	1·4	1·2	1·4	1·5
Hard	1·2	1·4	1·2	1·2	1·3
Normal	1·2	1·3	1·2	1·2	1·2
Soft	—	1·2	1·1	1·2	1·2
Very soft	—	—	1·1	1·1	1·1

A most interesting method of estimating the rate of application of the chippings is used in Australia.[34,35] The approach of this method is based on the assumption that each chipping in a true one-sized aggregate ultimately takes up a position on the road such that it lies with its least dimension vertical. The design then aims at applying just enough particles so that, when this stage is reached, they exactly cover the whole surface with each chipping having the same least dimension, thereby providing a dressing of uniform thickness. Since under actual conditions this means that a loose layer of aggregate having a thickness greater than one stone must be spread before compaction, the problem becomes one of determining the thickness of this loose layer. This is done by considering the following.

It is known that the voids in loose layers of 6–38 mm one-sized aggregates, when spread shoulder to shoulder on a firm surfacing, vary from

TABLE 7.8. *Recommended rates of spread of chippings*[32]

Size of chippings, mm	Rate of application, kg/m^2	Size of chippings, mm	Rate of application, kg/m^2
6	6– 8	14	12–15
10	9–12	20	15–20

(Before opening the road to traffic, however, any excess chippings should be swept up as they represent a hazard to safety, particularly on high speed roads.)

46 to 54 per cent of the total volume; thus the voids in a loose layer can be taken as an average of 50 per cent. If N is the average thickness of a loose layer—obtained by dividing the loose volume by the area of the surfacing covered—and since the layer contains 50 per cent voids, then the volume of

voids and the volume of stone can each be represented by a height of $0.5N$. If X is taken as the height of the final traffic-compacted layer, then the voids in this final layer can be represented by $X - 0.5N$. In practice it is found that the void space in a layer of one-sized aggregate which is compacted by a roller is approximately 30 per cent of its compacted depth, and that when this is further compacted by traffic the void content of the traffic-compacted layer is about 20 per cent of this final layer. Thus, equating the voids contents,

$$X - 0.5N = 0.2X$$

and
$$X = 0.625N$$

or
$$N = 1.6X$$

where N and X are the depths of the initial and final layers respectively. To allow for the traffic whip-off of chippings, an allowance of 9 per cent is made, so that the gross loose depth becomes

$$N = 1.75X$$

However, the final compacted depth X is also equal to the 'average least dimension' of the aggregate (A.L.D.); it is also found that the A.L.D. is essentially the sieve size, in millimetres, through which 50 per cent of the chippings will pass. Thus, if the average least dimension of a single-size aggregate is known, then the loose depth of chippings which must be laid to obtain complete coverage can be determined by multiplying this value by 1.75.

All the calculations and recommendations regarding the rates of application of the binder and the chippings are relatively valueless if these materials are not uniformly applied. British practice indicates that the most uniform application of the binder is carried out by mechanical tankers fitted with spray-bars through which the binder is made to circulate at constant pressure. These spray-bars have atomizing jets, such that each jet makes a hollow cone of spray which interpenetrates others so that any one spot on the road receives binder from four or five jets. There is a British Standard which helps to ensure the proper performance of these machines.[36] Best results with respect to the application of chippings are obtained by using mechanical spreaders fitted with metering devices.

Immediately after the chippings have been laid on the binder, they are rolled until a uniform compact running surface is obtained. The roller has the dual function of pushing the chippings into the binder and then flattening the layers so that it is less susceptible to dislocation by the traffic. Best results are obtained by using a slow-moving rubber-tyred roller, since with the steel-wheeled ones some detrimental degradation of the chippings will certainly take place. If a steel-tyred roller is used, it should be the lightest available, and in no case should its weight exceed 8t.

The success of a surface dressing, particularly if it is on a heavily travelled road, is very much dependent on the length of time before traffic is allowed on the dressing, and then on the control of traffic speeds over it during the initial phase of its life. It should be kept in mind that a surface dressing is at its most vulnerable when freshly laid and that the proper consideration of this fact will very often differentiate between its success or failure. The length of time which must elapse between roller compaction and allowing traffic can vary from as little as 15 minutes to as much as a couple of days, depending primarily on the binder used and the climatic conditions. When traffic is allowed, safety considerations require that vehicle speeds should be kept below 30 km/h, until it is certain that sufficient adhesion has been developed between the aggregate and the binder so that the chippings will not be appreciably disturbed.

External conditions. As can be discerned from the previous discussions the external and, only too often, uncontrollable factors affecting the design of a good surface dressing are the state of the existing road surface, the traffic using the road, and the prevailing weather conditions. Since one of the major functions of the binder is to stick the chippings to the road surface, this means that, for proper adhesion to take place, the road surface must be free of dust and standing water. It is not necessary for the surface to be absolutely dry when applying the binder since adhesion will still take place; it does so, however, at a relatively slower rate so that the dressing is exposed to possible injury for a longer period of time than if the binder was applied to a dry surface. When a new surface dressing is being placed over an accumulation of old ones there is also the possibility that bleeding may occur. To avoid this, and the slippery surface usually associated with it, the excess binder in the old surfacing should be removed by heating and planing prior to the application of the new dressing.

A particular problem arises when a concrete pavement is being surface-dressed as it may be necessary to apply two coats of surface dressing. Since the chippings cannot be embedded at all in the concrete surface, the smallest sizes of chipping must be used in the initial course. When this course has been well compacted by the traffic, a second surface dressing course is applied, but in this case a larger size of chipping is used. The interval between the application of the two dressings depends on the weather and the traffic, but normally it does not exceed a couple of weeks.

The effects resulting from allowing traffic directly on a new surface dressing are both good and bad. It is good in that the traffic acts as an additional roller and helps to embed the chippings in the binder; it is for this reason that roads carrying light traffic require more binder than those with heavy traffic. On the other hand, early heavy traffic sets up very severe stresses which the new surface dressing may not be able to withstand. While these can very often be overcome by using high viscosity binders, there are certain locations at which it is not advisable to lay surface dressings at all. Typical of these locations are very heavily travelled streets and intersections in urban areas where the stresses caused by stopping, turning and accelerating vehicles may be more than the surface dressing can take.

The final, but by far from being the least, consideration is the weather. British weather is notable for its inclemency and only too often have failures of newly laid surface dressings been caused by rain falling during or within a few days of laying, and displacing the binder from the chippings. Damage due to wet weather can be prevented by using any one of three simple processes which render the dressings immune to the action of water throughout even the most unsettled weather.[33] In all three processes, a cationic surface-active agent is used to waterproof the stone, thereby making the bond more resistant to moisture. In two of these processes, a solution of the cationic agent in creosote is applied to the interface between the binder and the chippings. The creosote solution can be applied either by pre-coating the chippings—this is the better but more expensive way—or by sprinkling the solution on to the spread binder prior to the application of the chippings. So far, the agents found effective for these two processes are some quaternary ammonium compounds, and certain long-chain amines and amides. In the third process, the cationic agent is incorporated in the binder before spraying begins. The quantity of agent required for this method is considerably greater than with the other processes and hence its cost is likely to be greater. It has, however, the advantage of requiring no extra equipment and thus is a more convenient procedure. So far only certain long chain amines have been found suitable as the surface-active agents in this process, but these have been shown to have the most effective results during extremely adverse weather conditions.

As mentioned previously, a development which has helped to make surface dressings a successful process under arduous conditions is the use of coating chippings. The binder pre-coatings not only give the chippings extra protection against wet weather, but also provide additional resistance to the dislocating forces of traffic, especially in the early days of the dressing.

Sealing coats

If a bituminous binder is applied as a thin film to a road surface in order to close the voids in the surface with the object of rendering it waterproof, the coating is called a sealing coat. Emulsions are particularly useful as sealing materials, since they provide a means of applying additional but thin quantities of binder to the voids of an open-textured surfacing with little danger of accumulating a layer of excess binder on the running surface which could lead to vehicles skidding. Very often the spread binder (usually an emulsion containing 30 to 35 per cent bitumen) is blinded with a very thin layer of bitumen-coated grit which lodges in the voids of the surface.

Tack coats

A tack coat is not a surface treatment in the true sense of the word since traffic does not move directly upon it. Rather it is a treatment which is given to the underside of a wearing or some other such course before those layers are laid. The purpose of a tack coat is to bind the previously prepared layer to the superimposed layer.

Normally, use of a tack coat involves simply the single application of a film of bituminous binder to a previously prepared surface. This binder is most often a bitumen emulsion, a cut-back bitumen, or a low viscosity tar. Tack coats are most commonly used when a new wearing surface is to be placed on an old bituminous surfacing, or on a new or old concrete pavement, or on a brick, stone or block pavement, or on a previously-primed granular roadbase. It is normally not necessary to apply a tack coat between two bituminous layers which are laid within a short time of each other; usually these are sufficiently plastic to interlock and there is sufficient binder at the interface to form a good bond. A tack coat can do more harm than good if applied where not necessary or in too great a quantity. For instance, if an extra amount of binder is applied to a hot bituminous layer having a binder content already at the optimum for the mix, the result may be such as to exceed the optimum content for the mix, with consequent loss of stability.

One particular form of tack coat is known as a *Prime Coat*. This is a film of binder which is applied to a previously untreated porous layer so that it penetrates the top of the layer and is completely absorbed. A prime coat has a number of functions of which by far the most important is to aid in binding the penetrated layer to the superimposed one; normally, however, the actual bond is provided by a secondary tack coat application. As well as helping to provide this bonding action, a prime coat also serves to make firm the surface of the layer which is primed and this may be a considerable help during construction. For instance, it is not at all uncommon practice to prepare the subgrade of a road and place over it a surface dressing to protect it from the weather until construction of the pavement at a later time.[37] In such instances, the prime coat may not only help to bind the top of the subgrade but, if the subgrade is granular, it prevents the surface dressing from scaling off under traffic. In addition it will act as a membrane preventing the rise of vapour or capillary moisture into the pavement from a moisture source beneath.

SELECTED BIBLIOGRAPHY

1. VALLERGA, B. A. On asphalt pavement performance, *Proc. Ass. Asphalt Paving Technologists*, 1955, **24**, 79–102.
2. BURT, M. E. and NICHOLAS, J. H. Future Research on Bituminous Materials at the Road Research Laboratory. *TRRL Report* LR 442, Crowthorne, Berks., The Transport and Road Research Laboratory, 1972.
3. SZATOWSKI, W. S. and HOSKING, J. R. The Effect of Traffic and Aggregate on the Skidding Resistance of Bituminous Surfacings. *TRRL Report* LR 504, Crowthorne, Berks., The Transport and Road Research Laboratory, 1972.
4. HAWKES, J. R. and HOSKING, J. R. British Arenaceous Rocks for Skid-Resistant Road Surfacings. *TRRL Report* LR 488, Crowthorne, Berks., The Transport and Road Research Laboratory, 1972.
5. HUBBARD, P. and J. FIELD. A practical method for determining the relative stabilities of fine aggregate-asphalt paving mixes, *Proc. A.S.T.M.*, 1925, **25**, Pt. 2, 335–348.
6. ASPHALT INSTITUTE. *Mix Design Methods for Asphalt Concrete.* (Manual Series No. 2). Washington, D.C., The Institute, 1962.
7. CAMPEN, W. H. and V. R. SMITH. A study of the Omaha Testing Laboratory's

bearing index test for bituminous mixtures. *Proc. Ass. Asphalt Paving Technologists*, 1950, **19**, 369–382.

8. MCLOUGHLIN, J. F. and W. H. GOETZ. Comparison of the unconfined and Marshall test results, *Proc. Ass. Asphalt Paving Technologists*, 1952, **21**, 203–236.

9. SMITH, V. R. Triaxial stability method for flexible pavement design, *Proc. Ass. Asphalt Paving Technologists*, 1949, **18**, 63–94.

10. SPEER, T. L. Progress report on laboratory traffic tests of miniature bituminous highways, *Proc. Ass. Asphalt Paving Technologists*, 1960, **29**, 316–361.

11. BRIEN, D. Design of bituminous mixes by the Marshall method. Lecture J in the *Proceedings of the Conference on Bituminous Materials for Flexible Pavements* held at the University of Nottingham (Department of Civil Engineering) on April 17–21, 1972.

12. CALIFORNIA DIVISION OF HIGHWAYS. *Materials Manual: Testing and Control Procedures*, Vol. 1. Sacramento, California, 1962.

13. LIVNEH, M. and E. SHKLARSKY. The splitting test for determination of bituminous concrete strength. *Proc. Ass. Asphalt Paving Technologists*, 1962, **31**, 457–476.

14. LEE, A. R. and P. J. RIGDEN. The use of mechanical tests in the design of bituminous road surfacing mixtures. Part I: Dense tar surfacings, *J. Soc. Chem. Ind.*, 1945, **64**, No. 6, 153–161.

15. CAMPEN, W. H., J. R. SMITH, L. G. ERICKSON and L. R. MERTZ. Factors that control asphalt content requirements of bituminous paving mixtures and a method for determining the proper asphalt content, *Proc. Ass. Asphalt Paving Technologists*, 1963, **32**, 530–552.

16. WEETMAN, B. and D. W. HURLBURT. The effect of asphalt viscosity on stability of asphalt paving mixtures, *Proc. Ass. Asphalt Paving Technologists*, 1947, **16**, 249–263.

17. LEE, A. R. Letter to the Editor on ʻA Design Method for Gap-Graded Asphalt Mixesʼ, *Roads and Road Construction*, 1972, **50**, No. 596–7, 240.

18. ROAD RESEARCH LABORATORY. A Guide to the Structural Design of Pavements for New Roads. *Road Note* No. 29, London, H.M.S.O., 1970.

19. BS 594:1961. *Rolled Asphalt (Hot Process)*. London, British Standards Institution, 1968.

20. DEPARTMENT OF THE ENVIRONMENT. Dense Tar Surfacing Wearing Course. *Technical Memorandum* H6/72, London, The Department of the Environment, June 29, 1972.

21. BS 1690:1962. *Cold Asphalt*. London, British Standards Institution, 1968.

22. BS 802:1967. *Tarmacadam with Crushed Rock or Slag Aggregate*. London, British Standards Institution, 1967.

23. BS 1621:1961. *Bitumen Macadam with Crushed Rock or Slag Aggregate*. London, British Standards Institution, 1961.

24. MINISTRY OF TRANSPORT. *Specification for Road and Bridge Works*, London, H.M.S.O., 1969.

25. BS1241:1959. *Tarmacadam and Tar Carpets (Gravel Aggregate)*. London, British Standards Institution, 1959.

26. BS 2040:1953. *Bitumen Macadam with Gravel Aggregate*. London, British Standards Institution, 1953.

27. *Modern flexible road construction*. London, The Asphalt and Coated Macadam Association, 1971.

28. LEE, A. R. Bituminous road surfacings in Great Britain, *Proc. Australian Road Research Board*, 1962, **1**, 681–690.

29. BS 1447:1962. *Mastic Asphalt (Limestone Aggregate) for Roads and Footways.* London, British Standards Institution, 1966.
30. BS 348:1948. *Compressed Natural Rock Asphalt.* London, British Standards Institution, 1948.
31. LEE, A. R. and G. H. FUIDGE. *The Technique of Surface Dressing with Tar.* London, British Road Tar Association, 1959.
32. TRANSPORT AND ROAD RESEARCH LABORATORY. Recommendations for Road Surface Dressing. *Road Note* 39, London, H.M.S.O., 1972.
33. ROAD RESEARCH LABORATORY. Prevention of wet-weather damage to surface dressings by the use of surface-active agents, *Road Note* No. 14. London, H.M.S.O., 1964.
34. SWAMI, S. A. Australian method of surface dressing, *J. Inst. Highway Engrs*, 1965, **12,** No. 1, 22–31.
35. HERRIN, M., MAREK, C. R., and MAJIDZADEH, K. Surface Treatments: Summary of the Existing Literature. *Special Report* 96, Washington, D.C., The Highway Research Board, 1968.
36. BS 1707:1970. *Hot Binder Distributors for Road Surface Dressings.* London, British Standards Institution, 1970.
37. ROAD RESEARCH LABORATORY. Protection of subgrades and granular sub-bases and bases, *Road Note* No. 17. London, H.M.S.O., 1968.

8 Rigid pavements

Rigid pavements are those which contain sufficient beam strength to be able to bridge over localized subgrade failures and areas of inadequate support. Thus, in contrast with flexible pavements, depressions which occur beneath properly designed and constructed rigid pavements are not reflected in their running surfaces.

While in theory it might be argued that pavements composed of many different types of material can be classified as 'rigid', in practice the only ones recognized as such are cement-concrete pavements, and it is perhaps true to say that no other type of pavement has received so much study. It is not possible in the space available here to attempt to cover all the investigative effort which has been carried out into this form of road-making, and hence the emphasis in this chapter is placed on the fundamental concepts underlying the design of concrete pavements.

ELEMENTS OF A RIGID PAVEMENT

Before discussing the various factors influencing the design and construction of a concrete road, it is necessary to mention some relevant terminology. As with flexible roads, the cross-section of a concrete road is composed of a pavement superimposed on the subgrade. With respect to the composition of the pavement, there are some differences in the literature in the nomenclature which is applied to the various layers.

To avoid any misunderstanding, the layer between the concrete slab and the subgrade will be called the *sub-base* throughout this text, irrespective of whether it is part of a two- or three-layer rigid pavement system. Similarly, the layer above this sub-base will be called the concrete slab, again irrespective of whether it is part of a two- or three-layer system. When the term surfacing is used, it will refer to the topmost layer of a three-layer pavement system.

SUBGRADES AND SUB-BASES

Subgrade

The most important property of the subgrade in relation to concrete road design is whether or not it will provide *uniform* support for the slab. If the subgrade can be relied on to provide uniform support throughout the life of the pavement, then the slab can probably be laid directly on the prepared

392

in situ soil and there will be no need to construct a sub-base. If the subgrade soil conditions are poor, then it will be necessary to interpose a sub-base between the slab and the subgrade.

The only information regarding the subgrade which is required for design purposes in Great Britain is its classification as given in Table 14.1, i.e. whether it is very stable, normal or weak. These subgrade classifications are, of course, very wide—although in practice they are quite adequate for design purposes. This can be explained by considering the manner in which a concrete slab distributes the wheel loads applied by the traffic.

TABLE 8.1. *Classification of subgrades for concrete roads and minimum thickness of sub-base required*

Type		Min. thickness of sub-base, mm
Weak	All subgrades of C.B.R. $\leqslant 2\%$ (see Fig. 6.17b)	150
Normal	Subgrades other than those defined by the other categories	80
Very stable	All subgrades of C.B.R. $\geqslant 15$ (see Fig. 6.17b). This category includes undisturbed foundations of old roads	0

Concrete used in road construction is normally required to have a crushing strength of more than $27 \cdot 6 \, \text{kN/m}^2$ after 28 days. This material usually has a flexural strength of more than $3 \cdot 1 \, \text{kN/m}^2$, which indicates that it has significant beam strength. Its modulus of elasticity is in the order of $34 \cdot 5 \, \text{MN/m}^2$; this means that the concrete slab has a high degree of rigidity. Because of these properties of rigidity and beam strength, the wheel loads applied to a concrete slab are distributed over a large area and so deflections are small and unit pressures on the subgrade very low.

It cannot be too strongly emphasized however that, notwithstanding the fact that the inherent strength of the subgrade plays a less important role in the design of a concrete road than it does in a flexible one, the old maxim 'a road is as good as its foundation' still holds for a rigid pavement. Thus adequate care must always be taken to ensure that the drainage and compaction conditions of the subgrade which are selected at the design stage are those which are likely to be experienced throughout the life of the pavement.

Sub-bases

The function of a sub-base in a rigid pavement is not so much to increase the structural stability of the pavement as to counteract or correct unsatisfactory subgrade conditions which could lead to non-uniform support for the slab. Particular instances where sub-bases are used to provide uniform, stable and permanent support for concrete slabs are when damage is anticipated from one or more of the following causes: 1. Frost action. 2. Poor drainage. 3. Mud-pumping. 4. Swell and shrinkage of high-volume-change soils. 5. Construction traffic.

Preventing or minimizing frost action. Frost action in a soil can result in very severe differential heaving and softening of the subgrade, and this in turn can result in the break-up of the pavement (see also Chapters 4 and 6). Specific recommendations on how to deal with the frost action problem with respect to concrete roads vary from country to country and from authority to authority. In general it can be said that the only guaranteed solution to the frost action problem is to remove the subgrade soil and replace it with a non-frost-susceptible sub-base material to a depth equal to that of the frost-penetration below the road surface. Practically, however, it has been found that, because of the structural abilities of the concrete slab, it is very often sufficient to carry out subgrade removal and replacement to a depth equal to only about half of the frost-penetration into the subgrade. British practice[1] is to ensure that the total thickness of slab and sub-base above a frost-susceptible subgrade is never less than 450 mm.

Improving drainage and minimizing the accumulation of water within the pavement. Adequate sub-surface drainage is another fundamental requirement for any road pavement (see also Chapters 2 and 6). A granular sub-base which acts as a drainage layer is absolutely necessary whenever there is a danger of water accumulating within a pavement.

Preventing mud-pumping at the joints, edges and cracks of pavements. By mud-pumping is meant the forceful ejection of a mixture of water and soil which can occur when concrete road slabs are laid directly on top of fine-grained plastic soils. If too much soil is removed, eventual failure of the pavement may occur through lack of subgrade support. Pumping occurs with most severity on soils with high clay contents. It generally does not occur, even under the heaviest wheel loads, on non-plastic granular soils when more than 55 per cent of the material is retained on the sieve.[2] Three factors must be present before pumping can occur. These are (a) free water, (b) a soil that will go into suspension, and (c) frequent passage of heavy wheel loads. If any one of these three is missing, pumping will not occur.

The third factor emphasizes why it is that on major concrete roads soil pumping can be expected, since it is a result of the frequent deflecting pumping action of the slab ends brought about by the heavy wheel loads which pass over the carriageway. The tendency of a slab to deflect and pumping to occur is minimized by limiting the number of expansion joints to as few as is absolutely necessary and substituting closely spaced and dowelled contraction joints.

Water which infiltrates through the pavement joints and cracks and along the edges of the pavement is invariably the source of the free water which causes pumping. If the subgrade is not free-draining the soil will go into suspension where pockets of water are collected at void spaces in the subgrade, and it is this suspension which is ejected when the slab is deflected. Continued pavement deflection soon leads to the ejection of enough subgrade soil to leave the slab ends unsupported and extensive cracking will then occur. Thus pumping resulting from water entry is best minimized by ensuring that all pavement cracks and joints are adequately sealed against the infiltration of water, and that surface water is drained away over the shoulders as quickly as possible.

Finally, pumping can occur only if the subgrade soil is able to go into suspension in the water. This can be prevented either by chemically stabilizing the subgrade to an adequate depth or by providing a sub-base composed of either dense-graded or open-graded materials. A dense-graded sub-base is usually constructed in an undrained trench which is somewhat wider than the concrete pavement; to be successful it must be graded and compacted so that it is practically impervious to water. If the gradation of the sub-base is poorly designed it will have a high moisture-holding capacity, while lack of proper compaction will lead to pavement cracking at joints as a result of the traffic causing further compaction of the sub-base. Open-graded sub-bases must also be properly compacted, but here the coarseness of the gradation must be such as to permit adequate drainage without itself becoming clogged by the intrusion of soil from the subgrade. An open-graded sub-base is more expensive to construct since it must be provided with a subsidiary drainage system which must be maintained and kept open throughout the life of the roadway. As described in the chapter on Flexible Pavements, this form of pumping-prevention entails constructing the sub-base in such a way that it extends through the shoulders into shoulder drains or longitudinal pipe underdrains. Only about 50 mm of sub-base thickness is normally necessary to prevent mud-pumping, but between 75 and 150 mm is usual in practice, so as to make sure of obtaining the depth necessary to prevent pumping.

Minimizing volume changes. As has been described in other chapters, certain soils are highly susceptible to volume changes due to taking up moisture in wet weather and giving up moisture in the dry season. If, for instance, construction takes place when the soil is very dry, subsequent absorption of moisture may result in warping of the concrete pavement and differential uplift of adjacent slabs at joints. If the soil is very wet and in a highly expanded state, then subsequent drying-out of the subgrade will take place from the shoulders towards the centre of the pavement, with the result that the soil at the edges shrinks and leaves the slabs unsupported at critical locations.

Partial design for combating a swelling subgrade may require chemical stabilization of the subgrade and/or the construction of a granular sub-base of 75–300 mm thick, so that a sufficient thickness (and weight) of pavement is obtained which will both hold down the subgrade and act as a ' blanket' to absorb some of the expansion. Fortunately the soil expansions and contractions which result in the development of rough riding surfaces are not common occurrences in Britain. They are most prevalent in the semi-arid regions of the world.

Forming a working surface. A sub-base is often utilized as a working surface by vehicles during the construction of a road. The construction used for this purpose should be a sufficient depth of either dense, well-compacted granular material or a chemically stabilized soil which is capable of withstanding the applied wheel stresses and protecting the subgrade from the detrimental actions of the weather. The usual practice is to increase the minimum subbase thickness by an additional 150 mm if the subgrade has a C.B.R. $\geqslant 4\%$, and by 80 mm on other 'normal' subgrades.

CONCRETE

Cement-concrete is a mixture of cement, coarse aggregate, fine aggregate and water which is combined into a solid mass as a result of the chemical reaction which takes place between the water and the cement. In many ways concrete is an excellent material for a highway pavement. Because of its high strength qualities, it can withstand heavy wheel loads and high tyre pressures. Well-designed and well-made concrete roads are known to last for at least 40 years and need relatively little maintenance. The principal disadvantage associated with concrete construction is that it can be a relatively high initial cost as compared to the flexible type of construction (see Table 8.2).

Materials

The ingredients of a concrete mixture are cement, aggregate and water. These materials, and the chemical reaction which occurs when cement is hydrated, have been described elsewhere in this text, and so this discussion is confined to emphasizing factors basic to the design and construction of a concrete pavement.

Cement. Ordinary Portland cement is the binding material most often used in concrete road slabs, but where a high early strength is required, so that a road can be opened to traffic earlier than is usual, a rapid-hardening cement may be used. In cold weather or when very rapid hardening is required, high alumina cement may be utilized. In certain parts of Britain, Portland blast-furnace cement is also used for roadmaking purposes. In certain instances, as for example where it is desired to define a particular traffic lane, coloured cements are utilized: these are made by mixing pigments with a white cement made from kaolin and chalk.

As has been shown previously, most cements are essentially mixtures of clay and limestone which are burnt to very high temperatures and the resultant clinkers then ground to a powder. Because of inherent variations in the raw materials and the manufacturing processes, it can be expected that there will be variations in the quality of cements obtained at different times and from different sources. Various tests are carried out, usually by the manufacturers, to ensure that when a cement reaches a road scheme it can be relied upon to meet certain minimum requirements. Two of the most important of these tests are the setting time and strength tests.

TABLE 8.2. *Estimated initial construction costs for 1·6 km of pavement*[3]

Type of road	Form of construction	Type of construction	Mean cost £	Range of costs, £	
				Lower limit	Upper limit
1. Rural motorway	Flexible	Composite base (lean concrete)	134 000	126 000	169 000
		Fully flexible base	153 000	138 000	190 000
	Concrete	Hot-poured joint sealants	151 400	117 400	180 900
		Cold-poured joint sealants	153 500	119 500	183 000
2. Peri-urban (dual carriageway) road	Flexible	Composite base (wet-mix)	81 600	76 000	93 300
		Composite base (lean concrete)	92 500	85 600	107 000
	Concrete	Fully flexible base			
		Hot-poured joint sealants	86 700	69 600	102 000
		Cold-poured joint sealants	88 100	71 000	103 400
3. Rural secondary road	Flexible	Wet-mix base	25 000	20 900	31 500
		Lean concrete base			
		Fully flexible	31 500	26 200	37 800
	Concrete	Hot-poured joint sealant	36 300	29 100	42 500
		Cold-poured joint sealant	37 200	29 900	43 400
4. Housing estate road	Flexible	Wet-mix base	15 700	13 400	19 700
		Lean concrete base			
		Fully flexible base	18 100	15 500	22 200
	Concrete	Reinforced slabs	21 000	16 800	24 800
		Unreinforced slabs	22 900	18 000	26 000

Setting time. The term 'setting' is used to describe the stiffening of the paste which begins to occur some time after the water is added to the cement, i.e. the change from a fluid to a rigid state. The start of this process is arbitrarily called the 'initial set' while its completion is called the 'final set'. Hardening of a cement paste is not considered to begin until after setting has 'ended'. The setting properties of a cement are most important in the construction of a concrete road and excellent results are normally obtained when mixing, transporting, placing and compacting of the concrete takes place before the initial set has occurred. On the other hand, concrete which is placed after the initial set can have higher strengths provided that mixtures are sufficiently workable to be compacted. Freshly laid concrete is particularly susceptible to frost action which occurs prior to the onset of the final set of the cement paste.

A purely arbitrary method of test is used to establish the setting times of a cement. Utilizing what is known as the Vicat apparatus, the test is essentially a penetration test which measures the time required by a loaded needle with a 1 mm square cross-section to penetrate into a standardized patty of neat cement and water. The setting times are expressed as the time which elapses from the instant when water is added until, for initial set, the needle fails to penetrate to the bottom of the test sample by about 5 mm, and for final set it just fails to penetrate to a distance of 0·5 mm from the surface of the test sample. The British Standard requirements for cements commonly used in road construction are given in the literature[4, 5, 6, 7].

Crushing strength. The strength of the hardened concrete is the property which is of primary importance in the construction of a rigid pavement. It is perhaps not surprising therefore that strength tests are required in all cement specifications. While there are several forms of strength test, the one on which most reliance is normally placed is the unconfined compression test. The British Standards specify two methods for evaluating the compressive strength of a cement. In the first method cubes having 70·6 mm square sides—each side has an area of 50 sq. cm.—are mechanically prepared in a specified way from a 1:3 cement-sand mortar. The weight of water used in the standard mix is 10 per cent by weight of the dry materials. According to the second method, 101 mm concrete cubes are prepared by hand from a workable mixture of cement, coarse and fine aggregates and water which has a slump of between 12·5 and 50 mm. The minimum strengths which must be attained by the cubes are specified in the relevant British Standards.[4,5,6,7] The concrete cube test is said to result in a smaller cube-to-cube variation than the mortar cube test.

Aggregates. Since mineral aggregates form roughly 80 per cent of the weight of a normal concrete slab, it follows that aggregate quality is of considerable importance in concrete road-making. Fortunately, however, good concrete can be made with most aggregates, even though they be of different mineralogical composition, and results are less dependent on the type of aggregate than on the grading, size and shape of the particles. Good quality aggregates are free from coatings of clay or organic impurities which may prevent the full bond from being developed between the particles and the cement paste. In addition, good aggregates are essentially non-absorbent when in water and do not contain ingredients that are liable to decompose or change in volume from exposure to the atmosphere.

While most aggregates used in concrete roads are composed of crushed stone, gravel or sand, blastfurnace slag is another material which is sometimes used. Irrespective of whether they are natural or artificial materials, aggregates used in concrete road-making are usually classified as coarse and fine. Coarse aggregate is composed of particles which are retained on the 4·76 mm sieve. Natural sand, crushed gravel or crushed stone which passes the 4·76 mm sieve is called fine aggregate or, simply, sand. The ratio of

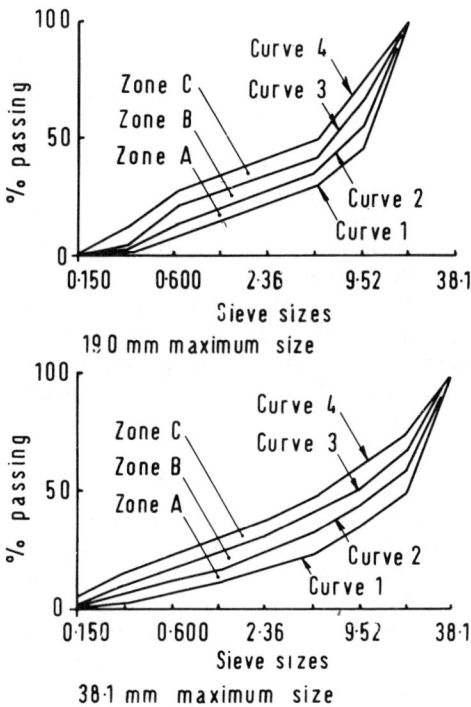

Fig. 8.1.   Concrete aggregate grading curves[8]

fine to coarse aggregate which is used in any mixture is dependent on their individual gradations and on the desired gradation of the combined materials.

As a rough rule the largest size of aggregate used in a mix should not exceed one quarter of the thickness of the slab if the best combination of strength and workability is to be obtained. In practice the maximum sizes which are most often used in concrete mixes in Britain are 38 and 19 mm. While larger sizes can result in a reduction in the cement content required to attain a prescribed strength, they give rise to construction problems at joints in the pavement.

For a given maximum size of aggregate there is no such thing as an 'ideal' gradation which will satisfy all desirable criteria. In general, the best gradation is a function of the sizes and shapes of the particles, and of the workability required of the plastic concrete mixture. Various suitable gradations have been recommended by different authorities, but the ones which are perhaps the most widely known in Britain are those which are shown in Fig. 8.1. In practice, the gradations actually used are more likely to lie in the zones between these curves than directly along them. A summary of the suitability of gradings which fall into the different zones for use in various cement: aggregate mixtures is given in Table 8.3.

Water. The final ingredient of a concrete mixture is water. In most instances, the specifications simply prescribe that the water should be potable. This requirement is normally sufficient to ensure that the water does not contain any impurities that will be detrimental to the quality of the cement paste.

Additives. In certain instances a fourth ingredient may be added to a concrete mix for one of the following reasons:

1. To improve the workability of a relatively dry concrete mixture. This is rare usage in concrete roadworks.

2. To increase the heat of hydration during setting so that the concrete can be mixed and placed at lower temperatures than are normally possible. The accelerating agent most used for this purpose is calcium chloride.

3. To decrease the susceptibility of the hardened concrete to frost action, de-icing salts, and/or bleeding. This involves the addition of an air-entraining agent to the concrete.

Air-entraining agents. If a wet concrete mixture is placed in the roadway, the coarse particles are liable to settle, with the result that there is a tendency for a layer of mortar to accumulate at the surface during compaction. This process, known as 'bleeding', can result in the surface-layer-mixture having a high water:cement ratio which has poor resistance to abrasion. If freezing temperatures occur, the surface-layer is particularly liable to be attacked by frost action. This is made worse by the deleterious actions of the de-icing agents used in cold weather to keep the road surface safe for traffic.

If any one of many air-entraining agents, e.g. rosin or vinsol resin, are incorporated in the concrete, either directly or indirectly in the form of an air-entraining cement, these deleterious effects can be largely prevented. The addition of the air-entraining agent results in the concrete being intermixed with countless numbers of uniformly distributed tiny air-bubbles which literally provide buoyancy to retard the settling of the coarse particles. After the concrete has hardened it retains these tiny bubbles, with the result that

TABLE 8.3(a). *Suitability of aggregate gradings for different mixes*[8]

Degree of workability required

Grading zone (see Fig. 8.1(a))	Low			Medium			High		
	Rounded aggregate	River aggregate	Crushed aggregate	Rounded aggregate	River aggregate	Crushed aggregate	Rounded aggregate	River aggregate	Crushed aggregate
Zone A Coarse	Suitable for all mixes	Suitable for all mixes	Suitable for all mixes	Suitable for 1:6 mixes and richer. Upper limit suitable for 1:7½ mixes	Suitable for 1:4½ mixes and richer. Upper limit suitable for 1:6 mixes			Suitable for 1:3 mixes only	Suitable for 1:3 mixes only
Zone B Medium	Suitable for all mixes	Suitable for all mixes	Upper limit requires high water content with 1:6 and 1:7½ mixes	Suitable for all mixes	Suitable for 1:6 mixes and richer. Upper limit suitable for 1:7½ mixes	Upper limit requires high water content with 1:6 mixes	Suitable for all mixes	Suitable for 1:4½ mixes and richer. Upper limit suitable for 1:6 mixes	Suitable for 1:3 mixes only. Upper limit suitable for 1:4½ and 1:6 mixes
Zone C Fine	Upper limit requires high water content with 1:4½, 1:6 and 1:7½ mixes	Requires high water content with 1:7½ mixes. Upper limit requires high water content with 1:4½ and 1:6 mixes	Requires high water content with 1:6 and 1:7½ mixes. Upper limit requires high water content with 1:3 and 1:4½ mixes	Upper limit requires high water content with 1:6 and 1:7½ mixes	Upper limit requires high water content with 1:4½ and 1:6 mixes	Requires high water content with 1:6 mixes. Upper limit requires high water content with 1:4½ mixes	Suitable for all mixes. Upper limit requires high water content with 1:6 and 1:7½ mixes	Suitable for 1:6 mixes and richer. Upper limit suitable for 1:7½ mixes	

(a) 19mm maximum size

TABLE 8.3(b). *Suitability of aggregate gradings for different mixes*[8]

Grading zone (*see Fig. 8.1(b)*)	Degree of workability required		
	Low	*Medium*	*High*
Zone A Coarse	Suitable for all mixes	Suitable for all mixes	Suitable for 1:4½ mixes and richer. Upper limit suitable for 1:6 mixes
Zone B Medium	Suitable for all mixes	Suitable for all mixes	Suitable for 1:6 mixes and richer. Upper limit suitable for 1:7½ mixes
Zone C Fine	Suitable for all mixes. Upper limit requires high water content with 1:6 and 1:7½ mixes	Suitable for all mixes. Upper limit requires high water content with 1:6 and 1:7½ mixes	Suitable for all mixes

(b) 38 mm maximum size

they can act as 'expansion chambers' which help to relieve the hydraulic pressure developed in capillaries in the concrete in the initial stages of freezing. As freezing proceeds the cavities limit the growth of microscopic bodies of ice, and thus protect the thin 'shells' of concrete surrounding them. The thickness of the shells between adjacent air voids (bubble spacing is critical) should be at least less than 0·25 mm and possibly as low as 0·05 mm. Bubbles should be as small as possible—they are usually 0·05–1·25 mm—so that the total volume of entrained air is low and any consequent strength losses are minimized.

While the advantages associated with air-entraining by far outweigh the disadvantages, nevertheless the inclusion of air can and does cause a reduction in strength (see Table 8.4).

In summary[9], it can be said that, for a given dosage of air-entraining agent, the amount of air entrained in concrete is *increased* by increasing the sand content, alkali content of cement to 1·5 per cent (expressed as $Na_2 0$), and workability; it is not appreciably affected by sand grading but is *decreased* by increasing cement content, specific suface cement, organic impurities, concrete temperature, and handling and vibration. The percentage loss in concrete strength for each 1 per cent of entrained air is greatest for high air contents, high cement contents, and large maximum sizes of aggregate. Air-entrainment produces smaller changes in the flexural strength of concrete than in the compressive strength.

While the cost of the air-entraining agent and the additional amount of

TABLE 8.4. *Effect of entrained air on the strengths of medium-workability concretes*[9]

Cement content, kg/m³	Max aggregate size, mm	Change in flexural strength, %		Change in compressive strength, %	
		Air content, %		Air content, %	
		3	6	3	6
225	38	−3·0	−12·6	−1·2	−24·6
	19	+4·5	0	+6·6	0
310	38	−12·0	−24·0	−15·9	−36·0
	19	−8·1	−16·2	−10·8	−23·4
390	38	−10·5	—	−12·3	—
	19	−5·7	−14·4	−9·9	−27·0

cement required is partly offset by the increased yield of the concrete due to the air content, it can generally be said that air-entrained concrete costs about 2 per cent more per cu. metre. This is roughly equivalent to increasing the cost of the concrete slab by about 1 per cent.

Design of mixes

Before construction of a concrete road can begin, it is necessary therefore to determine what proportions of cement, water, and aggregate should be present in the mixture. This is known as the mix design.

There are a number of different methods for 'designing' mixes, all of which have either directly or indirectly the ultimate aim of obtaining a concrete of the desired strength and workability at the lowest cost. Another characteristic which all the methods have in common is that, even though the selection of the initial combination may be based on well-tried data obtained from previous experience, the ultimate design is always a result of extensive trial mixes which are tested at the site.

Before describing the procedure which is most often utilized when constructing concrete roads in Britain, it is necessary to discuss in some detail what is probably the most important concept in concrete engineering, viz. the relationship between the water:cement ratio and strength.

The water:cement ratio 'law' was formulated by Dr. Duff Abrams over fifty years ago.[10] Essentially Abrams stated that for a combination of cement and conventional aggregates in *workable* mixtures, under similar conditions of placement, curing and test, the strength of concrete is solely a function of the ratio of cement to the free water in the plastic mixture. When the water:cement ratio is too low, then the mixture is too dry to be compactable and, as shown in Fig. 8.2(a), this can result in low strengths. The peak-point on the curve can be visualized as occurring where the concrete mix is just sufficiently workable to obtain near-complete compaction with the maximum energy which can be applied. As the water:cement ratio is increased beyond the peak-value, the strength begins to fall, but the rate of decrease is less than that experienced for uncompacted conditions. It is here that the water:cement law begins to operate and it is seen that, as the ratio is increased, the strengths obtained are decreased. Note, however, that because

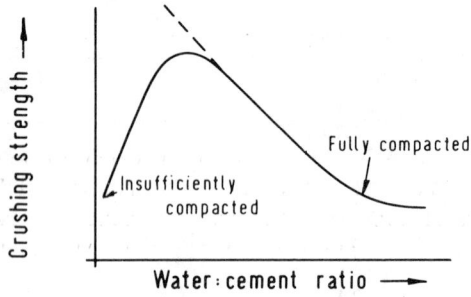

(a) Basic concept

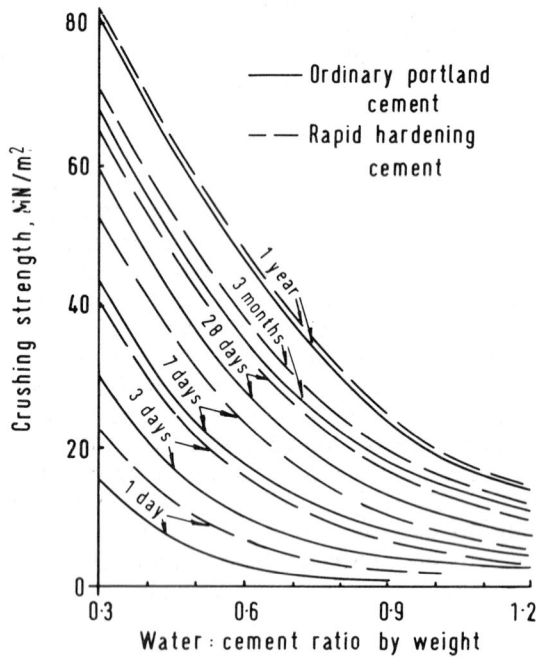

(b) Actual relationships established
with 100 mm fully compacted cubes

Fig. 8.2.   Water:cement ratios versus strengths

of the excess water the mixtures are very workable and full compaction can
always be obtained.

It is the right-hand side of the curve, i.e. the part which follows Abrams'
law, that the concrete engineer is most interested. Ideally he wants to utilize
water:cement ratios which are neither too wet nor too dry but which result
in strength values as close to the peak as is consistent with workability
requirements. A very wet mix, while very easy to compact, results not only in

a substantial reduction in crushing strength but also has a higher shrinkage factor, is subject to greater moisture movement in the slab, and is more susceptible to attack by frost. On the other hand, if too low a water:cement ratio is used, then the mixture will be very difficult to compact and a smooth sealed running surface may not be obtained. Figure 8.2(b) shows a family of curves relating water:cement ratio to strength which are based on normal concrete measurements carried out by the Transport and Road Research Laboratory. The water content of the water:cement ratio is the free water content of the mixture, i.e. the moisture over and above that absorbed by the aggregate.

In the discussion so far free use has been made of the term *workability*. This is a term which is easily understood but is relatively hard to define. Perhaps the easiest way to do so is to say that the more workable a mix, the less work is required to compact it fully. The degree of workability required of the concrete on any road scheme is dependent on the nature of the construction and the method of compaction. Thus, a road slab in which mesh reinforcement is utilized requires a mixture with greater workability than a plain mass concrete slab, while a concrete mix which is to be compacted by mechanical vibration can be less workable than one which is to be tamped by hand.

Factors which affect the workability of a mixture are as follows:

1. *Water:cement ratio*. Assuming constant cement and aggregate proportions, the workability of a mix increases as the water content is increased.

2. *Cement:aggregate ratio*. If the water:cement ratio is kept constant, a rich mix, i.e. one having a high cement content, is more workable than a lean mix.

3. *Shape, size and grading of the aggregate*. The effects of these factors can be observed by noting the results of adding aggregate to a neat cement paste of a particular water:cement ratio. It is found that the smaller the surface area of the added material in relation to the given quantity of cement, the greater is the workability, since more water is available to wet the surfaces of the particles; thus it is that a concrete mix having smooth and round aggregate particles is more workable than one having rough angular particles of a similar grading. Flaky aggregates result in the harshest and least workable concrete mixes. Workability also increases as the maximum size of the aggregate is increased. A corollary to this is that for a given degree of workability the fines content of a mixture may be reduced as the maximum size of the aggregate is increased. In general the better-graded a mix, the more workable it is; this effect is more noticeable as the mix gets leaner.

Unfortunately, at this time there is no single test which is universally accepted as truly measuring workability. Two tests, however, which are of particular interest are the Slump and the Compacting Factor tests.

The *Slump Test* is the workability test which is most widely used throughout the world today. There are minor differences in the test procedures carried out in different countries, but essentially all the tests can be said to be the same as that carried out in this country and which is specified by BS 1881:1952. In the test the newly mixed concrete is placed in a specified manner in the frustum of a cone having a bottom diameter of 203 mm, a top diameter of 101·5 mm and a height of 305 mm. When the container has been filled with concrete and levelled at the top, the mould is carefully removed by raising it vertically and the freed concrete is allowed to subside. The difference between the height of the mould and the final height of the concrete is called the slump.

The slump test is only sensitive to fairly great changes in workability, and even then it gives variable results. Its use is very much limited when the concrete is fairly dry and harsh, as is often so in modern road construction.

The *Compacting Factor Test* results in an indirect measure of workability, since in essence it measures the degree of compaction which is achieved by a standard amount of work. In the test the concrete is gently placed in an inverted frustum of a cone having a bottom diameter of 127 mm, a top diameter of 254 mm and a height of 279 mm, and which is fitted with a flap door at the bottom. When this is opened the concrete falls 203 mm into a lower frustum which, being slightly smaller, fills to overflowing; the concrete is now considered to be in a standard condition. The flap door on this frustum is opened and the concrete allowed to fall another 203 mm into a 152·5 mm diameter by 305 mm high cylinder which is also filled to overflowing. The surplus concrete is carefully struck off and the weight in the cylinder is then determined. The weight of the same volume of fully-compacted concrete is also determined, either by filling the cylinder with a series of very shallow layers and tamping or vibrating until all the air has been extracted, or else by calculation from the specific gravities and proportions of the constituents. The ratio of the observed weight of the concrete in the cylinder to the weight of the fully compacted concrete in the same cylinder is called the compacting factor. As might be anticipated, the more workable a concrete mix, the higher is its compacting factor.

The Slump test and the Compaction Factor test, as well as other commonly used tests to evaluate concrete, are fully described in BS 1881:1952, *Methods of Testing Concrete.*

From the previous discussion it will be seen that good mix design is not only concerned with obtaining a concrete in which a water:cement ratio is used that will give a required strength, but also the mix selected must be sufficiently workable to be properly, and economically, compacted with the means available. In designing a mix, these factors can be considered separately, the water:cement ratio being chosen to give the desired strength, while the ratio of cement to aggregate and the grading of the aggregate are selected to give the desired workability. This is best illustrated by carrying out an illustrative sample design.

EXAMPLE. A mix design is required for a conventional concrete pavement which is to be part of an urban motorway. The concrete slab is to be continuously reinforced and will be covered with a bituminous surfacing. Aggregate can be economically supplied at the site in two sizes, i.e. 19 mm to 4·76 mm gravel, and river sand.

Step 1. Determine the mix design strength. One specification for this type of road specifies that the concrete mix should be designed on the basis of a minimum crushing strength of 27·6 kN/m² at 28 days, using ordinary Portland cement. For mix design purposes this minimum figure must be increased to allow for the normal variation in strength of works test cubes. This is a motorway project and therefore strict control can be expected.

TABLE 8.5. *Estimated relation between the minimum and average crushing strengths of works cubes for different works conditions*[8]

Conditions	Minimum strength as percentage of average strength
Very good control with weigh batching, use of graded aggregates, moisture determinations on aggregates, etc. Constant supervision.	75
Fair control with weigh batching. Use of two sizes of aggregate only. Water content left to mixer-driver's judgement. Occasional supervision.	60
Poor control; inaccurate volume batching of all-in aggregates. No supervision.	40

By reference to Table 8.5, it is seen that the minimum strength may be expected to be about 75 per cent of the average strength. Thus the average strength to be aimed at in the mix design procedure is

$$27·6 \times \frac{100}{75} = 36·8 \text{ MN/m}^2.$$

Step 2. Determine the water:cement ratio. Figure 8.2(b) shows that a water:cement ratio of 0·5 for ordinary Portland cement is required to attain a strength of 36·8 MN/m².

Step 3. Determine the degree of workability required. This is a motorway scheme and the modern conventional equipment will be used on the construction. This suggests that a 'very low' workability is suitable for initial design purposes.

Step 4. Determine the cement content. British studies have resulted in the data given in Table 8.6, which show the proportions of cement required to give each of four degrees of workability with various water:cement ratios, taking into account the grading and type of aggregate. The aggregate shapes referred to in Table 8.6 are (a) rounded, e.g. beach and other well-worn gravels, (b) irregular, e.g. water-worn river gravels, and (c) angular, e.g. crushed rocks. In the experiments from which the tables were derived, the fine aggregate used was in each instance of the same shape as the larger sizes; it is therefore important to remember that interpolation must take place if the fine and coarse aggregates are of different shapes.

TABLE 8.6. *Aggregate: cement ratio required to give four degrees of workability with different gradings and shapes of 19 mm aggregate*[8]

Degree of workability	Very low				Low				Medium				High			
Grading of aggregate (curve no. on Fig. 8.1(a))	1	2	3	4	1	2	3	4	1	2	3	4	1	2	3	4
Water:cement ratio by weight 0·35	4·5	4·5	3·5	3·2	3·8	3·6	3·2	3·1	3·1	3·0	2·8	2·7	2·8	2·8	2·6	2·5
0·40	6·6	6·3	5·3	4·5	5·3	5·1	4·5	4·1	4·2	4·2	3·9	3·7	3·6	3·7	3·5	3·3
0·45	8·0	7·7	6·7	5·8	6·9	6·6	5·9	5·1	5·3	5·3	5·0	4·5	4·6	4·8	4·5	4·1
0·50	—	—	8·0	7·0	8·2	8·0	7·0	6·0	6·3	6·3	5·9	5·4	5·5	5·7	5·3	4·8
0·55	—	—	—	8·1	—	—	8·2	6·9	7·3	7·3	7·4	6·4	6·3	6·5	6·1	5·5
0·60	—	—	—	—	—	—	—	7·7	—	—	8·0	7·2	×	7·2	6·8	6·1
0·65	—	—	—	—	—	—	—	8·5	—	—	—	7·8	×	7·7	7·4	6·6
0·70	—	—	—	—	—	—	—	—	—	—	—	—	×	—	7·9	7·2
0·75	—	—	—	—	—	—	—	—	—	—	—	—	×	—	—	7·6
0·80	—	—	—	—	—	—	—	—	—	—	—	—	×	—	—	—
0·85	—	—	—	—	—	—	—	—	—	—	—	—	×	—	—	—

(a) Rounded aggregate, e.g. beach gravels

TABLE 8.6. (cont.)

Degree of workability	Very low				Low				Medium				High			
Grading of aggregate (curve no. on Fig. 8.1(a))	1	2	3	4	1	2	3	4	1	2	3	4	1	2	3	4
Water:cement ratio by weight																
0.35	3.7	3.7	3.5	3.0	3.0	3.0	3.0	2.7	2.6	2.6	2.7	2.4	2.4	2.5	2.5	2.2
0.40	4.8	4.7	4.7	4.0	3.9	3.9	3.8	3.5	3.3	3.4	3.5	3.2	3.1	3.2	3.2	2.9
0.45	6.0	5.8	5.7	5.0	4.8	4.8	4.6	4.3	4.0	4.1	4.2	3.9	×	3.9	3.9	3.5
0.50	7.2	6.8	6.5	5.9	5.5	5.5	5.4	5.0	4.6	4.8	4.8	4.5	×	4.4	4.4	4.1
0.55	8.3	7.8	7.3	6.7	6.2	6.2	6.0	5.7	×	5.4	5.4	5.1	×	4.8	4.9	4.7
0.60	9.4	8.6	8.0	7.4	6.8	6.9	6.7	6.2	×	6.0	6.0	5.6	×	×	5.4	5.2
0.65	—	—	—	8.0	7.4	7.5	7.3	6.8	×	×	6.4	6.1	×	×	5.8	5.6
0.70	—	—	—	—	8.0	8.0	7.7	7.4	×	×	6.8	6.6	×	×	6.2	6.1
0.75					—	—	—	7.9	×	×	7.2	7.0	×	×	6.6	6.5
0.80					—	—	—	—	×	×	7.5	7.4	×	×	×	7.0
0.85									×	×	7.8	7.8	×	×	×	7.4
0.90									×	×	×	8.1	×	×	×	7.7
0.95									×	×	×	—	×	×	×	8.0
1.00													×	×	×	×

(b) Irregular aggregate e.g. river gravels

TABLE 8.6. (cont.)

Degree of workability	Very low				Low				Medium				High			
Grading of aggregate (curve no. on Fig. 8.1(a))	1	2	3	4	1	2	3	4	1	2	3	4	1	2	3	4
Water:cement ratio by weight																
0·35	3·2	3·0	2·9	2·7	2·7	2·7	2·5	2·4	2·4	2·4	2·3	2·2	2·2	2·3	2·1	2·1
0·40	4·5	4·2	3·7	3·5	3·5	3·5	3·2	3·0	3·1	3·1	2·9	2·7	2·9	2·9	2·8	2·6
0·45	5·5	5·0	4·6	4·3	4·3	4·2	3·9	3·7	3·7	3·7	3·4	3·3	3·5	3·5	3·2	3·1
0·50	6·5	5·8	5·4	5·0	5·0	4·9	4·5	4·3	4·2	4·2	3·9	3·8	×	3·9	3·8	3·5
0·55	7·2	6·6	6·0	5·6	5·7	5·4	5·0	4·8	4·7	4·7	4·5	4·3	×	×	4·3	4·0
0·60	7·8	7·2	6·6	6·3	6·3	6·0	5·6	5·3	×	5·2	4·9	4·8	×	×	4·7	4·4
0·65	8·3	7·8	7·2	6·9	6·9	6·5	6·1	5·8	×	5·7	5·4	5·2	×	×	5·1	4·9
0·70	8·7	8·3	7·7	7·5	7·4	7·0	6·5	6·3	×	6·2	5·8	5·7	×	×	5·5	5·3
0·75	—	—	8·2	8·0	7·9	7·5	7·0	6·8	×	×	6·2	6·1	×	×	5·8	5·7
0·80	—	—	—	—	—	—	7·4	7·2	×	×	6·6	6·5	×	×	6·1	6·0
0·85					—	—	7·8	7·6	×	×	7·1	6·9	×	×	6·4	6·3
0·90									×	×	7·8	7·3	×	×	×	6·7
0·95									×	×	8·0	—	×	×	×	7·0
1·00									×	×	—	—	×	×	×	7·3

(c) Angular aggregate e.g. crushed rock

— Indicates that the mix was outside the range tested.
× Indicates that the mix would segregate.

Here both aggregates may be classed as irregular and so the data in Table 8.6(b) can be used directly. Entering this table and following the line marked water:cement ratio = 0·50 across to the columns headed 'very low' workability, it is seen that the aggregate:cement ratios required for each of the gradings shown in Fig. 8.1(a) are as follows:

Grading curve no. 1 7·2:1

Grading curve no. 2 6·8:1

Grading curve no. 3 6·5:1

Grading curve no. 4 5·9:1

It will be most economical therefore if an aggregate grading approximating to curve no. 1 can be utilized in the design.

Step 5. Determine the gradation to be used. This is selected on the basis of how closely the available coarse and fine aggregates can be blended to meet the gradings shown in Fig. 8.1. If, for instance, the blending results in a grading which is midway between curves 1 and 2, then an aggregate:cement ratio of 7·0:1 should be used for design purposes.

Step 6. Make up a trial mix at the site. This mix should contain the exact amounts of coarse and fine aggregates, cement and water. If it is then found not to be sufficiently workable, it should be made richer while keeping the water:cement ratio constant. If there is workability to spare, the mix can be made leaner.

Comment. In the above discussion, extensive reference has been made to one British publication, *Road Note* No. 4[8], which was published in 1950. The reader should be aware that this publication is now being reconsidered, and a new *Note* could well be issued in the near future. It should be pointed out, nevertheless, that the *approach* to mix design illustrated in the above discussion will be essentially the same; what could change are the actual numbers used or method of test specification. For example, one change that can be certainly expected is that the strength of pavement quality concrete shall be assessed using the *Indirect Tensile Test*. As is discussed later in some detail, the performance of a concrete road slab is critically affected by its ability to withstand the tensile stresses imposed on it; it is not illogical therefore that the quality of pavement concrete should be specified in terms of its tensile strength. (Data are available[12] which can assist in the design of mixes on the basis of indirect-tensile strength for both plain and air-entrained concretes.)

STRESS CONSIDERATIONS IN CONCRETE SLABS

Stresses in concrete pavements stem from a variety of sources, of which the applied wheel loads, changes in the temperature and moisture content of the concrete, and the volumetric changes in the foundation soil are by far the most important. These factors tend to result in deformations of the concrete slab which cause tensile, compressive and flexural stresses of varying magnitude to be developed. Analysis of these stresses presents a most complex problem, and in general it can be said that they cannot be truly evaluated at this time. As a result, most American methods of concrete pavement design[14] in use today combine a theoretical determination of the load stresses with an experimental or theoretical approach to the fatigue of concrete, although in some cases the theoretical analysis is supplemented or modified by experience gained from field observations of the performance of in-service pavements. In Britain, it is not usual practice to carry out a theoretical stress analysis prior to designing a concrete pavement. Instead it is considered better to rely on the very considerable experience which has accumulated over many years of concrete road construction. Nevertheless, before undertaking to make use of the practical method of pavement design used in this country, the highway engineer should have a basic understanding of these stresses and how they are produced.

Wheel load stresses

When a vehicle travels over a concrete carriageway it obviously causes pavement stresses which vary with the position of the wheels at any given time. In theory therefore the stresses in each slab should be analysed when the wheel loads are at all points on the slab, and the most severe stresses then evaluated and used for design purposes. In this respect it should be noted

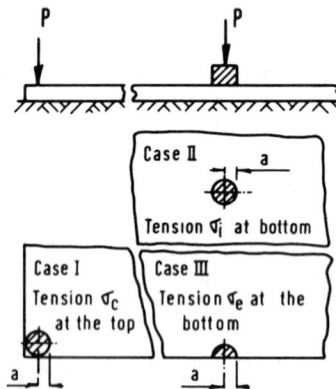

Fig. 8.3.   Three cases of loading considered by Westergaard in his original analysis

that, with the characteristics of concrete being what they are, a rigid pavement will fail under load when the load or bending moment is so great that the developed flexural stress exceeds the modulus of rupture of the concrete.

The most commonly used methods of theoretical/semi-theoretical analysis are based on that derived by Dr. H. M. Westergaard of the University of Illinois, which was first presented by him in 1925.[13] In his original analysis, Westergaard assumed that:

1. The concrete slab acts as a homogeneous isotropic elastic solid in equilibrium.
2. The reactions of the subgrade are vertical only and they are proportional to the deflections of the slab.

3. The reaction of the subgrade per unit of area at any given point is equal to a constant k multiplied by the deflection at the point. The constant k is termed 'the modulus of subgrade reaction', and is assumed to be constant at each point, independent of the deflection, and to be the same at all points within the area of consideration.

4. The thickness of the slab is uniform.

5. The load at the interior and at the corner of a slab is distributed uniformly over a circular area of contact; for the corner loading the circumference of this circular area is tangent to the edge of the slab.

6. The load at the edge of a slab is distributed uniformly over a semi-circular area of contact, the diameter of the semi-circle being at the edge of the slab.

Westergaard then examined three critical conditions of loading; as is illustrated in Fig. 8.3, these are at the corners, edges and interior of the slab. For these cases, Westergaard developed the following *original* equations[14]:

$$\sigma_i = 0.31625 \frac{P}{h^2} \left[4 \log_{10}\left(\frac{l}{b}\right) + 1.0693 \right]$$

$$\sigma_e = 0.57185 \frac{P}{h^2} \left[4 \log_{10}\left(\frac{l}{h}\right) + 0.3593 \right]$$

$$\sigma_c = \frac{3P}{h^2} \left[1 - \left(\frac{a\sqrt{2}}{l}\right)^{0.6} \right]$$

where P = point load, lbf,

σ_i = max. tensile stress, lbf/in^2 at the bottom of the slab directly under the load, when the load is applied at a point in the interior of the slab at a considerable distance from the edges,

σ_e = max tensile stress, lbf/in^2, at the bottom of the slab directly under the load at the edge, and in a direction parallel to the edge,

σ_c = max tensile stress, lbf/in^2 at the top of the slab, in a direction parallel to the bisector of the corner angle, due to a load applied at the corner,

h = slab thickness, in,

μ = Poisson's ratio for concrete ($= 0.15$ in these equations),

E = modulus of elasticity of the concrete, lbf/in^2,

k = subgrade modulus, lbf/in^2/in,

a = radius of area of load contact, in (the area is circular for corner and interior loads, and semi-circular for edge loads),

b = radius of equivalent distribution of pressure at the bottom of the slab = $(1.2a^2 + b^2)^{1/2} - 0.675h$,

and l = radius of relative stiffness, in

$$= \left[\frac{Eh^3}{12(1 - \mu^2)k} \right]^{1/4}$$

The most critical situation illustrated by the above equations—which were later modified—refers to the corner loading where, due to local depressions of the subgrade or warping of the slab, the corner portion may become unsupported and, in extreme circumstances, behave as a cantilever. An important variable in the corner formula is the modulus of subgrade reaction, k.

This reaction coefficient, which is a measure of the stiffness of the supporting soil—this may be either the sub-base or the subgrade—is evaluated for the unit weight and moisture content conditions which it is expected will be approximated beneath the slab under service conditions. The k-value is usually obtained by loading a 762 mm diameter plate which is placed on compacted soil sections and then calculating

$$k = \frac{p}{\delta}$$

where p = unit load on the plate,

  δ = deflection of the plate,

and k = modulus of subgrade reaction.

Values for k vary very widely, depending upon the type of soil, its unit weight and its moisture condition. A relationship established between California Bearing Ratio values and modulus of subgrade reaction values is shown in Fig. 8.4.

It is to be noted that the influence of the modulus of subgrade reaction is such that a large change in its value has but a relatively small influence on the calculated stress in a slab. This is one of the reasons why it is usually considered unnecessary to build up the supporting capacity of the subgrade beneath a rigid pavement with sub-base material.

The least critical situation considered by Westergaard was when the load is applied at the interior of the slab. By assuming the pressure to be distributed uniformly over the contact area of a small circle, Westergaard determined that the critical stress is the tensile stress which is produced at the bottom of the slab under the centre of the circle. This stress is the critical stress except when the radius of the circle is so small that some of the vertical stresses near the top become more important; this latter exception need not be considered, however, in the case of a wheel load which is applied through a rubber tyre. The initial formula derived by Westergaard for this condition was later modified by him to allow for the reactions of the supporting soil closely concentrated about the load being greater than if the reaction is assumed to be proportional to the deflection.[15]

The third situation considered by Westergaard was when the wheel load is at the edge of the slab, but at a considerable distance from any corner. When the load is applied, the edge deflects downward immediately under the load and upward at a distance away. The critical tensile stress is therefore immediately beneath the centre of the circle on the underside of the slab; the

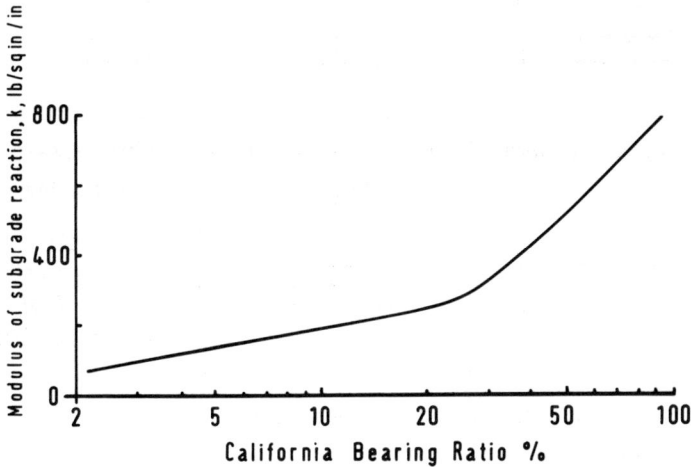

Fig. 8.4.   Relationship between California Bearing Ratio and modulus of subgrade reactions

tensile stresses at the upper surface of the edge at a distance away are considerably smaller than that at the bottom of the slab beneath the centre of the semi-circle. Westergaard's original stress formula for edge loading was later modified following further investigation which took into account the situation where the slab was in a warped condition.[16]

Other loading considerations. The stress relationships developed for the three critical conditions which have just been described only refer, in general, to loading in the form of a single wheel load at rest. Obviously therefore the values given by the equations have to be modified to take into account the effects brought about by multiple wheels and by moving traffic loads.

While the effects of multiple wheel loads are not yet fully understood, it is known that they result in increases in the stresses produced at the corners and interior of a slab and a reduction in the stresses caused at the edges of the slab. The multiple loading conditions which cause the greatest influence on the maximum stresses are given in Fig. 8.5. Analyses which have been carried out suggest that the effect of multiple wheel load distributions is to increase the stresses caused by a single wheel load by about 25 per cent for corner loading and 50 per cent for interior loading; for edge loading the effect can be considered as unimportant.

It is well known that the stresses caused by a statically applied load can be substantially increased by having it impact upon the surface. As has been described in a previous chapter, the impact effect of a moving wheel upon a *smooth* road surface is either negligible or it results in a reduction in the maximum pavement stress; this is due to the mass and elasticity of the pavement preventing the full deflection from occurring before the wheel has moved on to another position. On a concrete road, however, impact effects

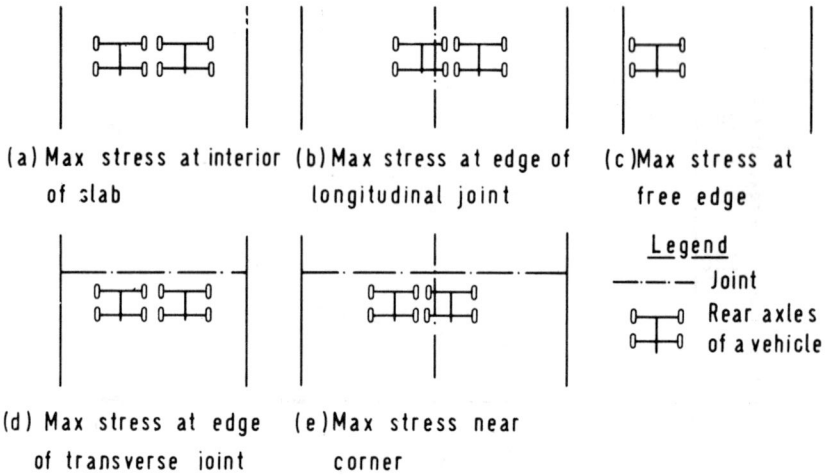

(a) Max stress at interior of slab   (b) Max stress at edge of longitudinal joint   (c) Max stress at free edge

(d) Max stress at edge of transverse joint   (e) Max stress near corner

Legend

— · — Joint

Rear axles of a vehicle

Fig. 8.5.   Loading positions for maximum stresses in concrete road slabs[17]

may be quite considerable depending on whether the adjacent slabs composing the pavement are 'flush' with each other. This in turn depends on whether or not the joints (and cracks) are properly protected. One American recommendation on how to handle the impact effect suggests that, where transverse joints are adequately provided with effective load transfer devices, such as dowels, the static load selected for design purposes should be increased by 25 per cent, and it should be increased by 50 per cent if the transverse joints are not fitted with load transfer devices or if there are cracks in an existing unreinforced pavement.[18]

The effect of frequency of loading, i.e. intensity of traffic, is another factor which must be considered in any analysis of stresses within a concrete pavement. As is well known, considerable research has been carried out, particularly in the past fifteen years, on the fatigue behaviour of concrete. Unfortunately, however, most of this has been based on regular repetitions of stress at quite high frequencies and this is very much in contrast to what actually does occur on a roadway. On a highway, for instance, there is invariably a sufficiently long period of time between applications of the same successive heavy loads to allow time for recovery of the concrete. The fact that these recovery times are available is very important, since it is known that they reduce the magnitude of the detrimental effects brought about by repeated loading.

Temperature-warping stresses

If a thin sheet of cellophane is placed on the (warm) palm of the hand, it will be found to curl upwards as a result of the thermal expansion of its lower surface. Placing it on a cool surface will cause curling in the opposite direction. A long narrow sheet will curl either up or down about its major longitudinal axis, but as the length to breadth ratio of the sheet approaches unity, the major curling or warping will shift to a diagonal direction.

The action of the cellophane is a very useful analogy which helps in understanding the warping stresses produced in a concrete pavement by temperature differentials throughout its thickness. When a road slab is maintained at a constant temperature it will rest flat on the supporting soil, whether it be a subgrade or a sub-base. If, however—as is usual in the evenings—the air temperature is lowered, then the initial reaction of the top surface of the concrete is to attempt to contract. Since the thermal conductivity of concrete is relatively low, the bottom of the slab remains at the same initial temperature and so the corners and edges of the slab tend to curl upwards. If the temperature conditions are reversed then the tendency will be for the slab to warp downwards. If, as with the cellophane sheet, warping of the slab were allowed without restraint, then the stresses produced in the slab would be negligible and no stress problem would arise. In practice, however, the tendency to curl is resisted both by the weight of the slab itself and by load-transfer devices, or friction, at the pavement joints, and these induce warping stresses in the pavement.

The problem of temperature warping was also considered by Westergaard.[19] Using the assumption that the temperature gradient from top to bottom of a concrete road slab is in the form of a straight line, Westergaard developed equations for three different cases. In the simplest one of these, he assumed that the slab was infinitely large and derived the expression

$$\sigma_0 = \frac{E\varepsilon t_1}{2(1-\mu)}$$

where σ_0 = tensile stress developed, lb/in^2
E = modulus of elasticity of concrete, lb/in^2
ε = coefficient of linear thermal expansion of concrete per degree F,
μ = Poisson's ratio for concrete,
and t_1 = temperature difference between the top and the bottom of the slab.

In Britain, it has been shown that a temperature gradient of 5°C can occur from top to bottom of a 150 mm concrete slab, so that on the basis of Westergaard's analysis the stresses caused by restrained temperature warping can be quite large even in this country. In general, the temperature stress results which he gives tend to be on the high side when compared to what actually does occur. To account for this it has been shown that the temperature gradient is in fact closer to a curved line, and that the use of this latter form of gradient results in calculated stress values which are considerably lower than those given by Westergaard's analysis.

In practice the stresses induced by temperature warping are not as detrimental as might otherwise be expected. There are a number of reasons for this:

1. At slab corners, where the load stresses are actually the greatest, the warping stresses are negligible since the tendency of a slab to curl at these locations is resisted by only a very small amount of concrete.

2. At the interior of a slab and along its edges, significant curling stresses may be developed which, under certain circumstances, are additive to load stresses. However, since concrete slabs are normally designed to have a uniform thickness based on the corner-load needs, the margin of strength present in the interior and edges of a slab is usually sufficient to offset the warping stresses which are produced at these locations.

3. Long-term temperature record studies which were carried out in America have indicated that, at least in that country, the temperature at the bottom of the slab exceeds that of the top more often than the reverse, so that curling stresses in the interior and at the edges are more frequently subtractive than additive.

Moisture-warping stresses

Differences in moisture content between the top and bottom of a slab also cause warping stresses. This is due to the ability of concrete to shrink when its moisture content is decreased and to swell when the moisture content is increased. Very little is known about the extent to which this type of warping occurs, and as yet it has not been possible to develop a method of analysis which enables the stress produced by this phenomenon to be calculated.

Warping stresses caused by moisture differences are not considered by most engineers utilizing calculations to obtain a thickness design since, in fact, the omission of this calculation generally results in a solution which errs on the side of safety and reduces the possibility of failure of the pavement. The reason for this is that the slab normally lies on a moist subgrade and therefore the bottom will have a moisture content which, on average, is significantly higher than that at the top. Furthermore, because of its higher water content, the bottom of the slab is in a more expanded condition and so there is a tendency for the slab to curl upward. The net result is that the stresses produced by the moisture differential act to resist the stresses caused by the wheel loads and, very often, the warping stresses caused by the temperature differential within the slab.

Temperature-frictional stresses

Stresses are also produced in concrete pavements as a result of the changes in the 'average' temperature of the slab, which cause it to expand and contract. In other words, as the pavement temperature increases or decreases, each end of the slab tries to move away from or towards the slab centre. If cooling takes place uniformly, a crack may occur about the centre of the slab. If expansion is excessive, and adequate joints are not provided, then 'blow-ups' can result in adjacent slabs being quite dramatically jack-knifed into the air. (Blow-outs more usually occur, however, at joints, including dowelled joints, in the late afternoon when the air temperature in the shade is greater than $32°C^{(21)}$.)

Assuming that adequate widths of joint were provided, the stresses due to 'average' temperature changes could be considered negligible provided that there was no friction between the slab and the supporting soil. In fact, however, considerable friction may be developed between the slab and the soil, with the result that when the slab attempts to expand it is restrained from doing so and compressive stresses are produced at its underside. As the slab contracts the same type of restraint is exerted but this time it results in tensile stresses in the bottom of the slab. It is to be noted that stresses resulting from restraint of this type are only important when the slabs are quite long (say, over 30 m). They are critical only when conditions allow them to be applied when the combined loading and warping stresses from other sources are at their maximum. Since the maximum tensile stress due to frictional restraint only occurs when a slab is contracting and since the warping stresses resulting from temperature gradients are not at their maximum at this time, the net result is that in practice these restraint stresses can usually be neglected when calculating the maximum tensile stresses in a concrete road slab for thickness design purposes.

One further point which might be mentioned is that the magnitude of the restraint tensile stresses which are developed is heavily dependent on the temperature conditions prevailing at the time the slab is laid. For instance, if a pavement is laid at a temperature of about 32°C in Britain, it can expect to spend by far the greater part of its life in a state of permanent contraction, with consequent development of permanent tensile stresses. If, on the other hand, it is laid at a much lower temperature, then much of its life will be spent in an expanded state and the stresses developed will be compressive ones which the concrete is usually well capable of withstanding.

Moisture-friction stresses

A concrete slab will also expand and contract as its 'average' moisture content changes, throughout its life. Little is actually known about the stresses developed as a result of these actions, but it can be assumed that they are developed in direct opposition to the temperature stresses, e.g. when the temperature is raised (and the slab tends to expand), the moisture content is lowered and this results in a tendency for the slab to contract. Since tests have suggested that the frictional stresses resulting from moisture movements are always less than those resulting from temperature changes, it is usually considered that they can be omitted from concrete road stress calculations.

REINFORCEMENT IN CONCRETE ROAD SLABS

The previous discussion on stress considerations has dealt only with plain road slabs, and no mention has been made so far of the effects of the introduction of steel reinforcement. The reason for this omission is that at this time there is no fully acceptable theory on which such a discussion could be based.

Reinforcing steel is, and has been for many years, used in concrete road construction. It should be clear, however, that the term 'reinforcing' is used very loosely indeed in reference to concrete road slabs, as both the amount of steel used and its function are very different from those utilized in 'normal' reinforced concrete building construction. When a bridge, a building or some other such structure is being constructed, the concrete is normally expected to carry only the compressive stresses which are produced, and reinforcing steel is incorporated in the structure to withstand the tensile stresses. The reason for this is that if the relatively low tensile strength of the concrete is exceeded, the result could well be the complete collapse of the structure, with perhaps the loss of many lives. With a concrete road, on the other hand, slab 'failure' does not result in any such dramatic happenings since the slab will still be supported. Indeed, if the pavement is on a good foundation the riding quality of the road may be so little affected that the only obvious damage is the aesthetic one of unsightly cracks in the surface. As a result the generally accepted design approach has been primarily to include only sufficient reinforcing steel to minimize the development of cracks and only secondly to regard the steel as taking any of the induced tensile stresses. A practical by-product of this form of usage is that it allows a considerably greater spacing to be used between transverse joints.

Examination of the literature on concrete roads shows that in the past there have been considerable differences of opinion as to where and how much of the reinforcing steel should be placed in a concrete pavement. This was primarily due to a lack of understanding of its function in the slab. If tensile stress resistance is the principal criterion, then it is of course logical that two layers of steel should be used, so that one layer is at the top and the other at the bottom of the slab. The function of the top layer of steel is to resist the load and warping tensile stress at the edges and corners at the top surface, while the bottom layer resists the tensile stresses set up at the interior of the underside of the slab. If only one layer of reinforcement is used and the governing criterion is again resistance to tensile stress, then the layer should be placed near the top of the slab, since it is there that the more critical stresses are induced. Even if it were possible to analyse these stresses exactly, it is doubtful whether in normal road construction the inclusion of sufficient steel in a road slab to completely resist the stresses would be justified; not only is it not necessary from a safety aspect, but it is probably not an economical proposition either, since with modern construction methods additional load-carrying capacity could probably be secured at less cost by using additional thicknesses of concrete than by using the relatively large amounts of steel needed to increase the pavement's flexural strength to the same extent. Hence the decision, as has been stated before, to add only the relatively small amounts of steel which are necessary to resist crack development at the top of the slab.

Amount, type and position of reinforcement

Even when an unreinforced concrete slab is placed on an excellent foundation it can be expected to crack. If the slab is well designed and constructed, these cracks will only be of the 'hairline' variety, and so are of no dangerous consequence whatsoever. If, however, say, the joints between slabs are too far apart, then the cracks may be larger and more numerous so that foreign material can enter them and water can percolate easily to the subgrade. If traffic is heavy the broken edges may become spalled, so that the cracks become wider, there is a loss in shearing resistance and eventually the pavement fails. Early maintenance in the form of cleaning out the cracks and then filling them with a plastic sealing compound can, of course, prevent too great a deterioration taking place. This however is very expensive—not only because of the cost of the actual repairs but also due to the cost of the hold-ups to traffic—and it also results in an unsightly surface appearance about which the general public are likely to be very critical. In the past this has resulted in otherwise structurally sound concrete pavements being 'damned' and in an unfair bias against this form of construction.

It can be seen, therefore, that the principal function of steel in concrete pavements is to *control* the opening of cracks in the slab, so that the interlocking faces of the concrete remain in tight contact; thus slabs act together as the vehicle approaches. The aim is to furnish sufficient steel area to resist the forces tending to pull the crack faces apart. These forces develop when the slab tends to shorten as a result of a drop in temperature, concrete shrinkage, and/or a reduction in moisture content. As the slab contracts, the movements are resisted by the friction between the slab and the underlying subgrade or sub-base. The resistance to movement produces a direct tensile stress which may cause the concrete to crack—which is when the tensile stress is transferred to the steel reinforcement. This tensile stress is greatest at the middle of the slab and so the steel is usually designed to withstand the stresses at this location. A simple expression which is used in U.S. practice to calculate the minimum amount of longitudinal steel is as follows:

$$A_s = \frac{LCW}{2f_s}$$

where A_s = area of steel per unit width of slab,
L = distance between free end joints,
C = coefficient of friction between the slab and the underlying soil. This value is considered to range between 1 and 2, 1·5 being the most commonly used figure.

W = downward force per unit width resulting from the weight of the slab,

and f_s = allowable working stress in the steel.

Experience has shown that small bars or mesh are more effective reinforcing agents than the same area of larger bars since they can be distributed more uniformly in a concrete slab. This, combined with the fact that higher working stresses can be used with welded wire mesh, has resulted in this material being the most widely used form of steel in reinforced concrete road construction.

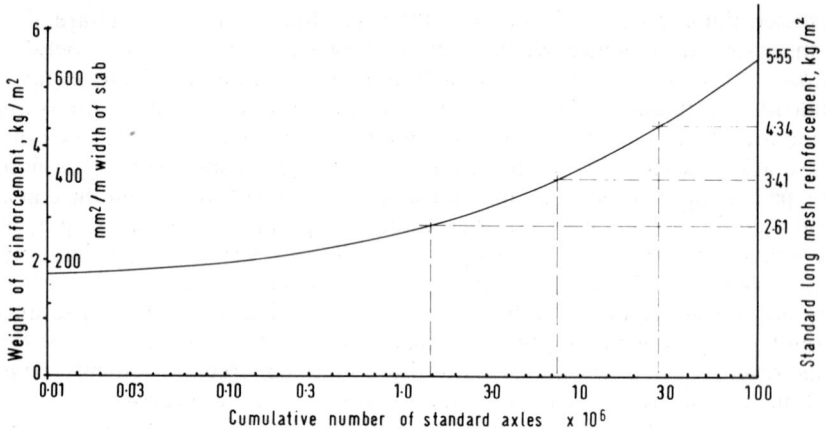

Fig. 8.6.   Recommended minimum weight of reinforcement for concrete slabs[22]

As has been stated, the primary function of the steel is to limit the maximum width of crack that may develop and only incidently is it expected to resist the induced flexural stresses. This means that the width of a reinforced crack is dependent on the elongation properties of the steel and not on its strength properties. Since both mild and high-tensile steel have approximately the same modulus of elasticity, it follows that the choice of steel to be used for crack control can be considered to be independent of the grade of steel utilized.

Since reinforcing steel is not designed to act in flexure, its exact location within the slab is not particularly important as long as it is well bonded to the concrete and adequately protected from corrosion. Common practice in Britain is to place it about 60 mm below the slab surface. British practice is also to use air-entrained concrete either for the full depth of the slab or, at least, for the full depth above the reinforcement.

British governmental recommendations for the use of reinforcement in reinforced concrete pavements are given in Fig. 8.6. Note that the minimum weight of reinforcement is given in relation to the cumulative number of standard axles to be carried in terms of both weight of long mesh reinforcement and the area of steel per unit width of pavement. Reinforcement fabric must be in accordance with BS 4483. Deformed bar reinforcement must be in accordance with BS 4449 or BS 4461. Detailed information regarding the arrangement of the reinforcement, e.g. joint and bar spacing, lengths of mesh sheets, etc. are available in the literature.[23]

Continuous reinforcement

Concrete roads can also be built with no transverse joints if sufficient reinforcing steel is employed in their design. Roads which are so constructed and which contain enough steel to keep closed the many more cracks which can be expected are said to have continuously reinforced pavements.

Experience to-date has indicated that minute cracks will occur more frequently in continuously reinforced pavements than in normally reinforced slab-type roadways. In fact the greater the amount of steel used, the closer the distance between cracks. In this case, however, because of the large amount of steel which is used, the cracks remain intact and do not increase in size. If too little steel is used the cracks will be formed farther apart but they are liable to widen.

Performance studies indicate[14] that (a) the numerous fine transverse cracks are usually not visible to the driver or passenger, and crack widths remain insignificant under all conditions and sealing is not necessary, (b) although the edges of the cracks will be under traffic, this is only a surface condition which does not affect the load-carrying capacity or the riding qualities, (c) there is no spalling, faulting or pumping at the tightly closed cracks, and (d) the cracks are held tightly closed, which prevents the passage of water and the infiltration of solid material. An important finding from the various studies is that longitudinal movements are limited to the end sections (60–150 m) and that the long central portion is, for all practical purposes, fully restrained.

Continuously reinforced concrete pavements cost more to construct, as well as requiring more steel, than normally reinforced slab-type pavements. Thus their usage at this time is generally limited to very heavily travelled roadways on which it is desired to keep traffic delays to a minimum, e.g. on busy city streets, especially on subgrades of doubtful quality or those extensively disturbed by excavations, or where there are shallow service trenches. This form of construction also minimizes the risk of uneven settlement where the pavement crosses areas formerly occupied by buildings.

In Britain, continuously reinforced concrete pavements are also not normally used unless they are likely to carry in excess of 2·5 million standard axles during the design life. The main factor controlling thickness design is the stability of the subgrade (see Fig. 8.7(b)). Note that British practice is to always cover the continuously reinforced slab with a bituminous surfacing not less than 90 mm thick and to place it on an appropriate sub-base conforming the Table 8.1. Long mesh reinforcement not lighter than 5·5 kg/m^2 or longitudinal deformed bar reinforcement of cross sectional area not less than 650 mm^2/m width of road is recommended in view of the absence of transverse joints.

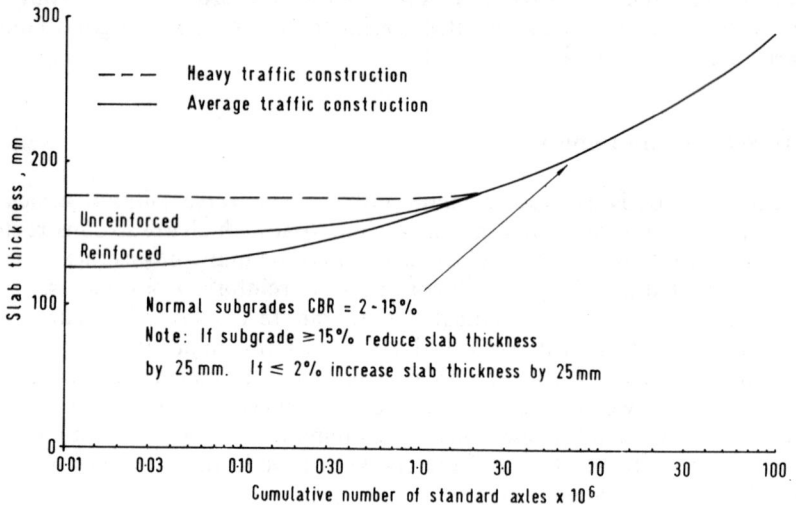

(a) Reinforced and unreinforced slabs

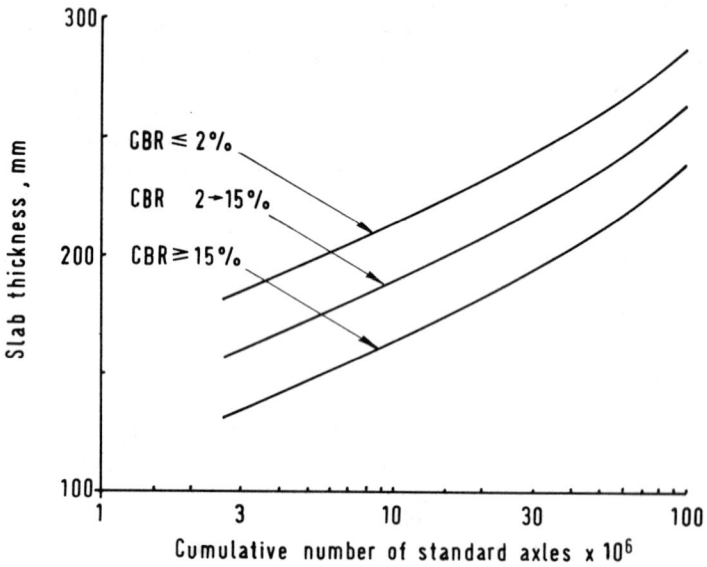

(b) Continuously reinforced with bituminous surfacing

Fig. 8.7.   Thickness design curves for concrete pavements[22]

Prestressed concrete pavements

Although the principle of prestressing has been widely used in other structural fields, its application to road construction has been relatively limited to-date. Several prestressed roads have been built in this country and abroad and, on the basis of the available evidence, it appears that prestressed concrete roads may be economically comparable with other forms of construction. In one excellent discussion it was suggested that as constructional techniques improve, prestressed pavements could become cheaper to build than normal reinforced concrete pavements.[24]

The purpose of prestressing a concrete structure is to augment its tensile strength by inducing an initial compressive stress in the concrete before it is subjected to loading. Prestressing is carried out by two main methods. In the first of these, steel tendons are incorporated in the concrete, stretched and subsequently fixed to the concrete in such a way that the induced tension in the steel is balanced by compression in the concrete. Stretching may take place before the concrete has hardened (pretensioning by wires) or after the concrete has hardened (post-tensioning by cables or bars placed in ducts). The second method of prestressing is by casting the concrete between suitable rigid abutments which are of sufficient strength to receive the prestressing thrust. Jacks are then placed between the abutments and the concrete and operated to compress the concrete.

Prestressed concrete pavements can be classified according to whether they are of the individual slab type or of the continuous type. With the exception of one experimental section of roadway designed by the Transport and Road Research Laboratory, it is probable that all of the prestressed concrete pavements built in Britain to date have been of the individual slab type. The method of construction employed is to cast the slabs so that they are separated only at expansion joints, and to prestress the slabs by internal tendons or cables. This constructional procedure allows each slab to expand or contract relatively freely and, since the coefficients of thermal expansion for steel and concrete are similar, the prestress in each slab remains essentially constant, except for the effects of subgrade friction. In addition to the customary construction equipment and processes, a prestressed slab pavement requires (a) special jointing, since transverse joints may be 100–350 m apart, (b) sleeper slabs under the joints, (c) friction-reducing layers between the pavement and the subgrade or sub-base, (d) steel tendons and conduits distributed throughout the area of the pavement, (e) filling the conduits with grout after the tensioning is completed, and (f) jacks for stressing the tendons. Prestressed slabs are especially suitable for use in roads having inferior subgrade conditions and curves. In contrast, the continuous type of construction is more suited to fairly straight level lengths of roadway over good subgrades. The continuous method utilizes jacks to apply the thrust and then, after the slab has been prestressed, the gaps in which the jacks are accommodated are filled, so that the strain remains in the slab. This means that, since the prestressed slab cannot expand between the abutments, the stress in it does not remain constant but varies with temperature and moisture changes.

The main advantages attributed to prestressing are reductions in the slab thickness, in the number of joints and in the amount of cracking. Reductions in road slab thickness can result in some saving in concrete, and this is important in areas where good construction materials are not readily available. Fewer joints expedite and simplify construction of the road and result in the carriageway having a better riding quality. Fewer cracks result in a longer life and lower the maintenance cost for the roadway.

The principal disadvantages associated with the use of prestressing are the difficulties encountered in applying prestress on vertical and horizontal curves and, in urban areas, the difficulty of repairing services located beneath the road since the prestress is destroyed by cutting trenches. At the present time the application of prestressing to road construction is still only in the development stage and hence, prestressed pavements are still more expensive to construct than normal reinforced concrete pavements. A further disadvantage is that strict safety precautions must be enforced during the actual prestressing process.

JOINTS

As described previously, concrete is subject to changes in dimensions due to variations in temperature and moisture which cause it to warp and expand and contract. If these changes are entirely resisted such high stresses may be induced in the pavement as to cause the concrete to develop tension cracks or buckling under compressive stresses to occur. It is to keep these stresses within safe limits, thereby conserving the strength of the concrete to resist the stresses induced by the traffic loads, that joints are provided in concrete pavements.

Necessary though joints may be, their installation is the most prolific cause of defects in concrete roads. Joints are, in effect, deliberate planes of weakness inset into the pavement, and so the design engineer should take care that only those which are absolutely necessary are included in his design. On his part, the site engineer should ensure that the joints receive the highest standard of workmanship when they are being made. The cost of the extra attention to detail during construction is easily repaid by the increase to the life of the road, the reduction in maintenance costs, and the improvement in the riding qualities of the carriageway.

Requirements of joints

The divisions between slabs can be divided into longitudinal joints, which are parallel to the centre line of the road, and transverse joints, which are at right-angles to the centre line. Irrespective of the positions of the joints, the importance of the highest standards of care in their design and construction must be emphasized. The following summarizes the chief functional requirements of joints if they are to be satisfactory constituents of a concrete pavement:

1. *A joint must be waterproof at all times.* This is perhaps the single most

important criterion. If surface water is allowed to enter through a joint, softening of the subgrade will take place and the uniformity of its supporting power may be significantly reduced. If the underlying soil is clayey then pumping may be promoted. This criterion emphasizes the importance of selecting and placing properly long-life sealing materials in the joints.

2. *Free movement of the slabs must be permitted at all times.* Both the filler and the sealing materials must be capable of withstanding repeated expansion and contraction of the concrete. There is a particular danger with expansion joints that if grit and other such foreign matter gets between the slabs, then expansion of the concrete will be prevented, with resultant increased stresses in the concrete and subsequent spalling of the edges. This danger can only be overcome by taking care in the selection and pouring of the sealing compound.

3. *Joints should not detract from the riding quality of a roadway.* Improper design and/or construction can result in excessive relative deflections of adjacent concrete slabs. Not only is this structurally undesirable but it also results in a most uncomfortable riding surface. Similarly, if improper care is taken when selecting and/or placing the joint sealing compound, it may mean that the motorist experiences an irritating and continuous series of impacts as the vehicle is driven over transverse ridges of sealing material which has been pushed above the level of the surface of the carriageway.

4. *A joint should not be the cause of an unexpected undesigned structural weakness in a pavement.* For instance, transverse joints on either side of a longitudinal joint should not be staggered from each other, as transverse cracks will be induced in the slabs in line with the staggered joints. Furthermore, no joint should be constructed at an angle of less than 90° to an adjacent joint or edge of the slab unless it is an intentional part of the design. The importance of not having acute-angled corners can be gathered by considering a corner which is completely unsupported, i.e. cantilevered, and then subjected to wheel loading. In this case the estimated increase in stress as compared to a right-angled corner is 20 per cent for a corner angle of 80°, 75 per cent for an angle of 60°, and 175 per cent for an angle of 40°.[25]

5. *Joints should interfere as little as possible with the placing of the concrete pavement.* This is best accomplished by having as few joints as possible. Where they must be used, then easy and economical construction is facilitated by using the simplest types of joint consistent with the structural design of the pavement.

Functions of joints

Joints may be classified according to the primary purpose for which they are installed in pavements. Thus there are expansion joints, contraction joints, warping joints and construction joints.

Expansion joints. These joints are primarily designed to provide space into which expansion of concrete slabs can take place when the temperature rises above that at which the concrete was laid. Their provision prevents the development of compressive stresses of damaging magnitude, as well as buckling or 'blow-up' of slabs. As well as being used between slabs, expansion joints are necessary at locations where pavements join fixed structures such as bridge abutments or at intersections with other pavements. Expansion joints also serve as contraction joints, warping joints and construction joints.

Expansion joints are normally used transversely in a pavement. As indicated by Fig. 8.8(a), expansion joints are formed by embedding in the concrete a vertical layer of non-extruding compressible material shaped so that there is a complete division between abutting slabs. The wider space above the filler material is sealed with a different compound to prevent the entry of water and grit. The thickness of an expansion joint at the time of construction is usually 25 mm, and only rarely is it less than 12·5 mm thick. The narrower thicknesses are used only when the expansion joints are closely spaced and/or construction takes place during hot weather.

Mechanical load-transfer devices are desirable at all expansion joints, even though the edges and corners are designed to carry the wheel loads without overstressing the concrete. Loads which pass, over an unprotected joint may cause excessive deflections and consequent damage to the

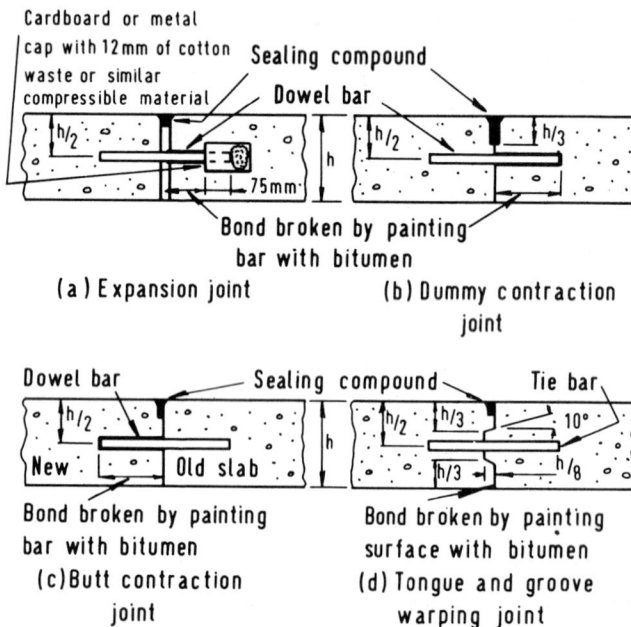

Fig. 8.8.   Illustrations of typical joint details

subgrade. This in turn can result in excessive vertical displacements of the slab ends and eventual faulting at the joints, even though the stresses induced in the concrete were not originally excessive. The load-transfer devices used at expansion joints are called *dowels*. They are placed across openings so as to allow the joint to expand and contract freely while holding the slab ends on either side at essentially the same elevation. The deflection of either slab under load is then resisted, through the dowels, by the other slab, which in turn is also caused to deflect and carry its share of the load imposed upon the first slab.

The varieties of suitable load-transfer devices on the market are literally too numerous to describe here, and the reader is referred to an excellent Transport and Road Research Laboratory publication in which detailed and referenced information on the various types is given.[25] The most commonly used device is also illustrated in Fig. 8.8(a). This is simply a round steel bar, one-half of which is anchored in either slab while the other half is free to move longitudinally within the adjacent slab. Longitudinal movement is economically ensured by coating the free-end half of the dowel with a rapid-curing bituminous material and having it terminate in an expansion cup containing a readily compressible material.

It is most important to note that at a given transverse joint the dowels must be placed parallel to each other and to the surface and centre line of the pavement. If this is not done then free movement of the slabs may be restricted and very high pressures set up between the bars and the concrete, which result in over-stressing and cracking of the pavement.

Very many complex factors must be taken into account in any attempt to design by analysis a dowel system at a joint. The load-transfer capacity of such a system is not only dependent on the bending and bearing capacities of the individual dowels, but also on their spacing, the pavement thickness, the supporting qualities of the subgrade, the means by which the loads are applied, and the location of the loads with respect to the edge of the pavement. Thus, as with the slab thickness, it is not possible to confidently utilize theoretical methods only when designing dowels. Table 8.7 summarizes the British recommendations regarding the dimensions of dowel bars for both expansion and contraction joints. Fig. 8.9 indicates the recommended maximum spacing of joints in reinforced concrete roads; every third joint should be an expansion joint, the remainder being contraction joints. The recommended maximum spacing of expansion joints in unreinforced concrete pavements is 60 m for slabs of 200 mm or greater thickness, and 40 m for slabs of lesser thickness. For both reinforced and unreinforced concrete slabs, the expansion joints may be replaced by contraction joints if construction takes place in Summer.

Contraction joints. The purpose of a contraction joint is to limit tensile stresses induced in the pavement due to contraction or shrinkage of the concrete and to prevent or control cracking. As with expansion joints, contraction joints are normally constructed at right-angles to the centre-line of the roadway, as it is seldom considered necessary to allow for transverse expansion and contraction of the pavement. Again also, as with expansion joints, contraction joints permit some angular movement between adjacent

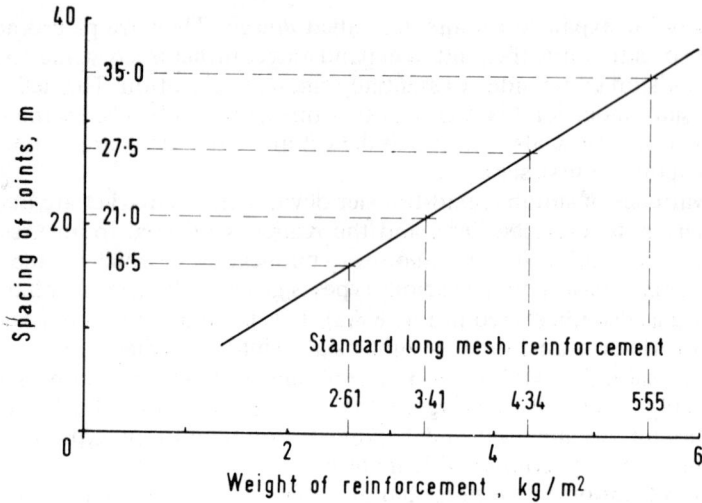

Fig. 8.9.   Recommended maximum spacing of joints in reinforced concrete roads

TABLE 8.7.   *Dimensions of dowel bars for expansion and contraction joints*

Slab thickness, mm	Expansion joints		Contraction joints	
	Dia. mm	*Length, mm*	*Dia. mm*	*Length, mm*
150–180*	20	550	12	400
190–230	25	650	20	500
240 and over	32	750	25	600

* Dowel bars are not recommended for slabs thinner than 150 mm.

slab sections, thereby providing some extra relief to the warping stresses developed within a concrete pavement.

There are two main types of contraction joint; these are *butt* joints and *dummy* joints. On large concrete road schemes, the most commonly used type by far is the dummy joint.

As illustrated in Fig. 8.8(b), a dummy joint consists essentially of a deliberate groove which is placed in a slab in order to form a vertical plane of weakness and thus induce a controlled crack. Three main varieties of dummy joints are in use; these utilize either a groove in the top of the slab (as illustrated), or an insert in the bottom of the slab, or a combination of groove at the top and an insert at the bottom. Use of a bottom insert alone results in the formation of an unsightly irregular crack at the surface, which is not only liable to spall but is also difficult to seal as there is no deliberate surface groove; hence this type of contraction joint is not to be recommended. By far the most widely used is the surface-groove type. The groove is formed in the surface of the slab as soon as the finishing process is complete. In Britain, it is usually formed to a depth of between $\frac{1}{3}$ and $\frac{2}{5}$ the thickness of the slab. As soon as the concrete has hardened the groove is filled and sealed to prevent water from percolating through to the subgrade. In relatively recent years, the use of sawed dummy joints has become very popular. This is formed by using special parallel saws to cut into the surface of the pavement after it has set. While this is a very simple yet efficient method of construction, care must be taken to ensure that sawing takes place at exactly the right time. If sawing takes place too soon, aggregate may be pulled out of the concrete and the surface torn; if too late, uncontrolled cracking will take place in the pavement.

When contraction joints are less than 4·5–6·5 m apart it is usually not necessary to provide load-transfer devices at dummy joints. Normally the crack-opening will be small enough for the interlocking of the aggregate particles at the faces of the joint to be sufficient to provide an adequate amount of load transference without the need for dowels. If, however, the grooves are more widely spaced, all contraction joints should be provided with dowel-bars. These bars are similar to those in expansion joints except that a receiving cap is not usually a necessity.

The butt-type of contraction joint is normally used only on small road schemes which utilize the alternate-bay method of concrete construction. (As illustrated in Fig. 8.8(c), this is also a construction joint since it runs from top to bottom of the pavement.) Irrespective of the distance between these joints, they should normally be provided with load-transfer devices. When the pavement is reinforced, the mesh is not carried through the joints; in other words, the reinforcing is used to control cracking *between* the contraction joints.

Warping joints. Known also as hinge joints, warping joints are simply breaks in the continuity of the concrete which allow a small amount of angular movement to occur between adjacent slabs. By so doing, these joints prevent excessively high stresses due to restrained warping from being developed within the pavement. While warping joints are also used as transverse joints (particularly in unreinforced concrete pavements), their main usage is for longitudinal jointing purposes. In contrast to expansion and contraction joints, appreciable changes in joint width are prevented at warping joints. This is done either by continuing the reinforcing steel through each joint or by utilizing tie bars to draw the sides of the joint together.

Dummy warping joints are essentially the same as dummy contraction joints except that the tie bars are not lubricated to allow sliding to occur. In contrast to dowel bars, tie bars are not normally intended to act as load-transfer devices, but are designed to bond both slabs together and to withstand the tensile forces which pull them apart. Load-transference at dummy joints is provided by the interlocking of the aggregate particles at the faces of the joints. The *tongue-and-groove* type of warping joint is illustrated in Fig. 8.8(d). In this instance the slabs are held close together by the tie bars and load-transference is provided by the tongue-and-groove interlock. Tongue-

and-groove construction is generally not considered suitable for slabs of thickness less than 200 mm.

Construction joints. As the name implies, construction joints are those joints other than deliberately designed expansion, contraction or warping joints which are formed in the course of construction of the pavement. They are formed when construction work is unexpectedly interrupted, as for instance by mechanical breakdown or by the onset of bad weather, at points where joints are not normally required by the design. The structural integrity of the pavement is best maintained by treating construction joints as warping joints and utilizing tie bars and/or overlapping mesh reinforcement to bind the old and the new slabs together.

Seals and Fillers

Regardless of their function, all joints in a concrete pavement should be made waterproof at the time of construction and maintained in that way through the life of the pavement. Furthermore, any cracks which develop in the slabs throughout their life should also be made waterproof as soon as they are detected. Failure to do so can result in the infiltration of grit and water with consequent spalling of joint edges and the creation of non-uniform foundation conditions. The materials which are placed in joints to prevent these occurrences are called fillers and seals. Both materials are used in expansion joints, while only seals are used in the other types of joint.

Fillers. These are used to provide the gaps for expansion joints at the time of construction and to provide support for the sealing compound. Materials which are to be used as fillers should satisfy a number of criteria. The most important of these are that a filler should be capable of being compressed without extrusion, sufficiently elastic to recover its original thickness when the compressive force is released, and able to retain these properties throughout the design life of the pavement. Filler materials used in joints which satisfy these criteria are softwood, fibreboard impregnated with a light cut-back bitumen or tar distillate, and cork. Materials which have been used as fillers, but which are not to be recommended because they extrude when compressed in hot weather, are pre-formed sheet bitumen and bitumen-bound cork granules.

Seals. Expansion joints which are formed with filler materials only are neither watertight nor do they prevent the ingress of grit. Thus not only are contraction, warping and construction joints sealed with a watertight compound, but so also are expansion joints. The requirements of a good sealing material have been enunciated as follows:[17]

1. It should be capable of adhering firmly to concrete under all weather conditions.

2. It should be sufficiently ductile in cold weather to accommodate itself to the widening of the joint without cracking and be capable of withstanding

repeated expansion and contraction over long periods without disintegrating.

3. It should not flow either along or down the joint in hot weather.

4. It should, under all weather conditions, effectively resist the ingress of grit and water.

5. It should be capable of being poured at temperatures up to 165°C.

6. It should be durable and not harden or soften with time.

7. Its colour, when used on transverse joints, should not unduly contrast with the road surface.

Some of these requirements contradict each other as, for instance, the fact that a hard material is required to prevent the ingress of grit while a soft material is needed to provide good adhesion. It is therefore perhaps not surprising that no sealing compound has been found yet which is considered satisfactory in all respects and which does not require regular maintenance. Best results so far have been obtained with a hot-poured rubber-bitumen compound composed of a soft bitumen containing a high natural or synthetic rubber content.[26]

The amount of material needed to seal the joints is dependent on the function of the joints. Since sealing compounds are incompressible, the grooves should never be filled to a height greater than 3·6 mm below the surface of the concrete. If the grooves are made deeper than is required for the sealing compound they should be caulked to an appropriate depth with a soft, compressible filling material.

THICKNESS DESIGN

Notwithstanding the many erudite analyses of stresses within concrete pavements which have been carried out, the problem is so complex that, at this time, there is no highway agency in the world which attempts a completely theoretical analysis when designing a concrete road pavement. In particular, it may be stressed that all theoretical design procedures, of necessity, make use of a number of assumptions and approximations, some of which—especially those relating to the magnitude of thermal stresses and of subgrade restraint—can have a profound effect on the conclusions reached relating to thickness requirements for concrete slabs. The result is that, in Britain concrete pavement design procedures have tended to rely heavily on the results of full-scale road experiments conducted under controlled conditions[27] in order to gain design information re slab thickness, degree of reinforcement, spacing of joints, strength of concrete and sub-base thickness. The performances of these concrete roads have been evaluated almost exclusively on the rate of crack development. The influencing factors about which information was gathered at these test sites are subgrade strength and traffic.

The British design procedures for unreinforced and reinforced rigid pavements makes use of the data given previously in this chapter and in the chapter on Flexible Pavements. They are best explained by giving some examples from the governmental recommendations.[22]

Unreinforced concrete

EXAMPLE DESIGN. An unreinforced rigid pavement (with gravel aggregate) is to be constructed in a residential area. The carriageway width will be 7 m and it is not expected to carry any public service vehicles during its 40-year design life. The finished pavement will, however, have to carry heavy construction traffic bringing in construction materials to a new housing estate of several hundred houses. The subgrade soil is a sandy clay with L.L. = 35%, and P.L. = 20%. The water table is 3 m below the final road surface.

Solution. Reference to the table on p. 302 shows that the traffic will likely be 10 comm. veh/day in each direction at the time of construction. Over 40 years this will amount to 0.4×10^6 comm. veh/lane (see Fig. 6.22(c)). Table 6.4 recommends a conversion factor of 0·45 to obtain the number of design standard axles from the cumulative number of commercial vehicles carried by each lane, i.e. 0.4×0.45 million = 0·18 million.

The Plastic Index of the soil (L.L. − P.L.) = 15%. Fig. 6.17(b) shows that for a deep water table, the subgrade will likely have a C.B.R. = 6–7%. From Table 8.1, this subgrade can be classified as ' Normal '.

From Table 8.1, the minimum requirement for *sub-base* thickness = 80 mm. (Note: This will be structurally adequate as heavy construction traffic will not be using the sub-base.)

From Fig. 8.7(a), a *slab-thickness* of 16 mm would be normally required on a normal subgrade. Because, however, the completed rigid pavement will be used by heavy construction traffic (because of the housing estate), the minimum thickness should be increased to 180 mm.

As discussed elsewhere in this chapter, the recommended maximum spacing of *expansion joints* (winter construction) for slabs less than 200 mm thick is 40 m, with intermediate *contraction joints* at 5 m intervals. The omission of expansion joints is not to be recommended in this instance as the rigid pavement will not butt against flexible construction.

The *total pavement thickness* is $80 + 180 = 260$ mm. This is considerably less than the 450 mm within which it is recommended that no frost susceptible material should be contained. The frost susceptibility of the subgrade soil should be therefore carefully examined in the light of local experience.

Reinforced concrete

EXAMPLE DESIGN. A reinforced rigid pavement is to be constructed to carry 2200 comm. veh/day (both directions) at the time of construction, with a growth rate of 5% per annum over a design life of 20 years. It will be necessary for heavy vehicles to travel on the sub-base during construction. The subgrade soil is a clay with L.L. = 65%, and P.L. = 25%. The water table is 500 mm below the finished road surface.

Solution. Figure 6.22(b) shows that for initial traffic of 1100 comm. veh/day (one direction) and a growth rate of 5%, each slow lane will carry about 12·5 million commercial vehicles during its 20-year design life. Table

6.4 recommends a conversion factor of 1·08 to obtain the number of design standard axles from the cumulative number of commercial vehicles, i.e. $1·08 \times 12·5 = 13·5$ million.

The P.I. of the subgrade soil $= 40\%$. Figure 6.17(b) shows that when the water table is less than 600 mm below the formation, the subgrade C.B.R. $= 2\%$. From Table 8.1, this subgrade can be classified as weak.

From Table 8.1, the minimum requirement for *sub-base* thickness $=$ 150 mm. This will be continuously traversed by heavy construction traffic, however, so the minimum thickness should be increased by an additional 150 mm so as to minimize sub-base/subgrade damage. This gives a sub-base thickness $= 300$ mm. (The material used in the lower half of the sub-base may be of a lower quality than the upper material, provided both meet minimum requirements.)

From Fig. 8.7(a) a *slab thickness* of 250 mm is required for traffic corresponding to 13·5 million standard axles on a weak subgrade, i.e. $222 + 25 = 247$ mm, rounded up to 250 mm.

From Fig. 8.6 it can be seen that the minimum weight of *reinforcement* required in the slab is $3·8 \text{ kg/m}^2$.

The next *standard weight of reinforcement* fabric above the minimum of $3·8 \text{ kg/m}^2$ (which is what will be used) is $4·34 \text{ kg/m}^2$. The *joint spacing* appropriate to this latter weight is 27·5 m. Since every third joint should be for expansion purposes, this means that *expansion joints* will be 82·5 m apart, and there will be two *contraction joints* in between at 27·5 m spacing. (If the road is constructed during Summer, the expansion joints may be omitted, and contraction joints only used at 27·5 m spacing.)

The *total pavement thickness* is $300 + 250 = 550$ mm. Since this exceeds the minimum value of 450 mm, the frost susceptibility of the subgrade need not be considered.

CONSTRUCTION

Prior to 1935, all concrete road construction in Britain was carried out by hand. The only instances where mechanization was employed were in the batching and transporting of materials, the mixing of the concrete and, in some instances, its distribution from the mixer. About 1935 interest in the use of machines for spreading, compacting and finishing became aroused in Britain and this was accelerated during the Second World War. Now it is true to say that all major concrete road construction in this country is to all intents and purposes completely mechanized. Mechanized construction results in less manpower costs, an increase in the speed of construction, and— since it allows the use of leaner concrete mixes with low water:cement ratios—an effective saving in the cost of materials.

The construction stages involved in conventional concrete pavement construction can be summarized as follows: 1. Subgrade and sub-base preparation. 2. Form laying. 3. Mixing and placing the concrete. 4. Compacting and finishing. 5. Curing.

Preparing the subgrade and sub-base

The preparation of the subgrade and sub-base represents one of the most critical stages in the conventional construction process. Thus considerable care must be taken to ensure that each of these foundation layers is graded and compacted to a smooth hard surface. In particular, the subgrade should be dressed so that it conforms to the required line, profile and cross-section. If the pavement is to be constructed on top of an existing roadbed, it should be scarified to a depth of not less than 150 mm, and then recompacted and shaped to the correct line and grade in order to provide uniform support. The type of equipment used to compact the subgrade and sub-base is dependent on the nature of the material being compacted. Thorough compaction is absolutely essential, and particular care should be taken during the backfilling of drainage excavations and other such openings. Generally the subgrade is trimmed to the correct elevations to a width of at least 0·6 m beyond each edge of the pavement to be constructed. This allows the sideforms to be placed on foundations which are hard and true to grade, which considerably facilitates the high-quality construction necessary to obtain a pavement with smooth riding qualities.

Laying the forms

After the subgrade and sub-base have been compacted and graded, the next step is to lay the side-forms for the concrete. It used to be that timber forms were used exclusively in concrete roadworks; it was found, however, that even the best quality timber formwork warped after contact with wet concrete on only very few occasions, and so today they have been supplanted entirely by purpose-made steel forms that are suitable for continual use with concrete thicknesses of up to 300 mm.

Forms often serve dual purposes. Not only do they shape the sides of the concrete slab and define its top surface, but in addition many of the machines used in its construction run on rails supported by the forms. Where the machinery is expected to be supported in this way, particular care should be taken to ensure that the forms are rigidly supported so that they will not be displaced in line or level by the lateral stresses and vibration caused by the processes of spreading and compaction of the concrete. These forms are usually bedded on lean concrete foundations at least 75 mm thick which are laid in advance of the concreting operations.

Upon completion of form laying and aligning, the subgrade (or sub-base) is usually further prepared so that the elevations and shape of the surface are exactly as desired. At this stage it may be necessary to treat the foundation soil to prevent water from being absorbed from the concrete when it is placed. One way of doing this is to wet the surface just before the concrete is placed. A more desirable method, however, is to cover the surface with polythene sheeting or waterproof bitumen-bonded paper. Use of a waterproof covering has the additional advantage that adhesion between the concrete slab and its foundation is prevented and so friction stresses are reduced.

Sometimes instead the foundation surface is sealed with a bituminous material. In this latter case the bituminous covering should be blinded with sand and not chippings as otherwise the resultant friction will tend to restrict the free movement of the slabs.

Mixing

There are essentially three methods of organizing the mixing and placing of concrete on large conventional concrete schemes, and all produce satisfactory results if properly controlled. These methods are as follows:

1. Batching at a central plant, and transporting the dry batched materials to mixing and spreading machines at the site.

This is the method most often utilized in American concrete roadmaking. The aggregates and cement are batched by weight at a centrally located batching plant at a quarry or gravel pit away from the site. The dry materials are then transported by sheeted lorry to the site for wet-mixing and laying. At the site the mixer may be a crawler-mounted machine travelling on the foundation soil or on the shoulder or, as has been mentioned before, one which is mounted on a mobile framework in the form of a bridge running on rails and spanning the foundation. Whichever method is used, the mixer travels forward in conjunction with the spreading and compacting machines so that the freshly-mixed concrete is always adjacent to the location where it has to be placed.

This method of organization has the particular advantage that there is continual and excellent control over the mixing time of the concrete, since setting cannot begin until water is added at the site just prior to spreading. It is very beneficial from the point of view of control of consistency and the proportioning of ingredients, since if any defect is found in the concrete during construction it can be rectified immediately before another batch is started. Furthermore, if there is a breakdown in equipment at the site, the unmixed materials can be saved and used at another time.

2. Batching and wet mixing at a central plant, and transporting the concrete to the spreading machines at the site.

This also requires the batching of the materials to take place at a central plant removed from the construction site. In this instance, however, the mixing machines are also located at the central plant and so wet-mixing takes place there. The wet concrete is then hauled to the spreader at the site by sheeted lorries or dumpers.

This is the method most commonly used in Britain. It has the obvious advantage of obviating the necessity to move the mixer continuously in the course of construction. In addition, the permanency of the plant encourages the usage of more elaborate storage facilities and batching and mixing equipment, with consequent beneficial influence on the degree of quality control which can be attained. This method is particularly valuable when the central plant is capable of being used for several other concrete schemes in the same locality.

The disadvantages associated with this method are all concerned with the fact that the wet concrete has to be transported to the road site. If travelling takes too long, say more than about 20 or 30 min, the concrete will begin to stiffen and its consistency and workability on arrival at the site may be detrimentally affected. Furthermore, if the haul road is bumpy, segregation may take place in the wet mixture; this will result in difficulty in discharging the material as well as defects in the compacted concrete. Finally, because the mixing and placing activities are separated, there is an inevitable loss in efficiency. Thus, for example, if a major piece of equipment at the site breaks down or if some defect is found in the concrete, all material in transit between the mixer and the site may have to be wasted if it cannot be diverted for other purposes.

3. Batching at a central plant, and then mixing and transporting the materials at the same time in a transit-mixer vehicle on its way to the site.

This can be considered a compromise between the previous two methods, since batching takes place at the central plant, and then wet-mixing is initiated either just before the mixture leaves for the site or en route during the trip. This method is most useful when the central plant is located a long distance from the road and working space at the site is limited. If the water is added and the mixing begun just prior to starting the journey, bulking is minimized and more material can be carried per unit volume of the mixer. If the journey is long and/or evaporation is likely to take place, water need not be added and mixing begun until the transit mixer is actually approaching the site. In either case, the fact that the concrete mixture can be agitated in the course of travel means that segregation can be practically eliminated.

The main disadvantage associated with this method of operation is that of cost. The operating cost of the transit-mixing equipment is quite high and as a result the unit cost of the concrete may be higher than with the other methods. In addition, the sizes of most transit-mixer drums in relation to their batch capacities are smaller than the central plant or on-site mixers and so the mixing operation tends to be less efficient.

Spreading

The placing of concrete should be carefully carried out to obtain an even depth of spread with the minimum of segregation. The aim is to ensure that the concrete is in a state of uniform precompaction. The machines used to spread the concrete are as follows:

1. Box-hopper machines into which the concrete is first transferred from the mixing machines, and which then move longitudinally on the side rails and transversely across the subgrade and sub-base in order to deposit the concrete.

These spreading machines are by far the most widely used in conventional concreting in Britain. The usual procedure is for the power-operated hopper—this is a V-shaped box which can be opened on the bottom, and which is mounted on a wheeled chassis—to be filled to a uniform height from side-tipping or end-tipping lorries, and it then travels transversely across the width of pavement and deposits the wet concrete to the desired height; the height of the hopper is adjustable to obtain the required depth of slab. The machine on which the transverse-moving hopper is mounted can be moved longitudinally on the side-forms so that a new layer can be spread.

The amount by which the depth of spread concrete exceeds the ultimate slab thickness is called the surcharge, and its height primarily depends on the workability of the concrete. For workabilities within the usual range of compaction factor values, i.e. 0·80–0·85, the concrete is usually spread to a depth between 25 and 18 per cent greater than that of the compacted slab. The amount of surcharge required decreases as the workability increases, as is illustrated by the following empirical formula which has been found to give heights close to those used in practice:[31]

$$\text{Surcharge} = \text{depth of slab required} \times \left[\frac{1}{\text{compacting factor}} - 1 \right]$$

In practice the surcharge height is obtained by trial at the beginning of the work so that when fully compacted later it is level with the form elevations.

Usually concrete is spread in two layers, and the reinforcement mat placed on the top of the lower layer.

2. Machines which use a moving blade or rotating screw to distribute the concrete from heaps of the material deposited on the foundation material.

With this type of operation the freshly-mixed concrete is dumped in small well-distributed heaps on the subgrade or sub-base, as the case may be, and is then distributed uniformly across the pavement width by transversely-moving blade or screw spreaders. A typical blade spreader has on its front a heavy reciprocating blade with a face inclined at 45° in plan, which turns at the end of the stroke; thus it pushes the concrete forward and sideways at the same time. With the screw-spreader the blade is replaced by a revolving screw which is similar to that used in a screw conveyor. This screw revolves alternately in each direction and distributes the concrete in front of striking-off paddles which, in a manner similar to that for the blade spreader, strike off the spread concrete at a predetermined height.

As with the hopper-spreading machines, the blade and screw spreaders are part of wheel-mounted machines which can move longitudinally along the side-forms. Unlike the hopper machines, the blade and screw machines are rarely used in Britain, but they are the spreaders most commonly used in the United States.

Compaction and finishing

After the concrete has been distributed to the desired depth, the next stages are its compaction and finishing. The sequence of operations is for the concrete to be compacted and then the finisher obtains the exact surface contour desired in the completed slab.

The aim when compacting is to get a dense homogeneous slab of concrete which is free from voids, honeycombing and surface irregularities.

Compaction is accomplished either by surface vibration, internal vibration or tamping of the wet concrete. Surface vibration, either by means of power-propelled vibrating machines or by means of hand-operated vibrating screeds, is the process most widely used. The vibrators are usually supported by the side-forms and are moved longitudinally as compaction is carried out.

The finishing operation is carried out immediately compaction is complete. The object is to obtain a running surface which is free from depressions and other irregularities, is highly skid-resistant, gives a pleasing appearance and has good light reflectance properties. Experience has shown that when normal surface vibratory compaction is complete the surface may still be irregular. These irregularities are normally removed by a heavy beam which oscillates transversely across the width of slab as it follows immediately behind the compactor. Final finishing may then be necessary to remove the transverse ridges left by this finishing. This is best obtained by using a longitudinal beam or float which is set parallel to the centre line of the road and then moved from one side of the slab to the other with a sawing or wiping motion which eliminates irregularities along the direction of travel. To obtain a highly skid-resistant surface texture, the floated surface is very often lightly brushed or dragged to obtain very shallow corrugations. Although these corrugations will wear with time under the action of traffic, the surface of the concrete road need not necessarily lose its skid-resistant qualities. If a hard sharp fine aggregate is used in the concrete, then as the corrugation marks wear the sand particles will be left proud of the surface and its skid-resistant qualities maintained.

Slip-form pavers.   Before leaving the manner in which pavements are laid, mention should be made of what is probably the most exciting development in concrete road construction in recent years. This is the developing use of the slip-form paver.[28,29] This is a machine which replaces all the machines in the conventional paving train except the mixer, lays the pavement in a single pass, and eliminates the need for laying and using side-forms.

In slip-form paving, the preparation of the subgrade and sub-base is similar to that for the conventional construction method, only now it does not have to be followed by the laying of side-forms. Instead wet concrete is deposited in front of the moving paver—it is self-propelled and runs on caterpillar tracks which reduce the pressure on the sub-base or subgrade—between slip forms attached to the machine, and is then struck off by a paddle or ram, leaving an approximate surcharge of suitable height to form a slab of (in Britain) up to 13 m wide. The paver then forces a row of either poker vibrators or motorized angle plates through the concrete, which drive out the air and render the concrete fluid by intense vibration. As the machine moves forward at the rate of 6 m/min, it engulfs the concrete (between the moving forms on each side and a conforming plate bearing on its surface) while it is still maintained in a fluid condition by a transverse vibrator placed in front of the conforming plate. By the time the concrete emerges behind the machine, it is beyond the effective range of vibration and is thus able to retain its shape. This is aided by the use of a concrete with a low compacting factor and air-entrained cement, i.e. wet concrete made with air entrained cement is sticky, cohesive, and quite plastic.

Fig. 8.10 shows the main features of a machine recently developed in Great Britain[30] to plain concrete slabs in one operation. Note that with this machine all the transverse contraction joint dowels and the longitudinal joint tie bars are automatically inserted by the paver as the slab is slip-formed. Another important principle is the use of a plastic strip to create a vertical plane of weakness in the slab, thus cutting out the need for a longitudinal joint crack inducer (non-tongue-and-groove). As is indicated in Fig. 8.10 the strips are towed on steel keels under the paver and under a joint seal inserter through the green concrete while it is still being vibrated; the strip's length is such as to ensure that it leaves the concrete in a fully compacted state outside the influence of further vibration. Another improvement is the fitting of polythene underlay rolls in front of the hopper, so that the sheets are unrolled directly under the paver as it moves forward.

A major advantage of the machine illustrated in Fig. 8.10 is that it lends itself to the construction of minor roads such as in housing estates. One reason for this is that it does not require the great lengths of train associated with conventional pavers.

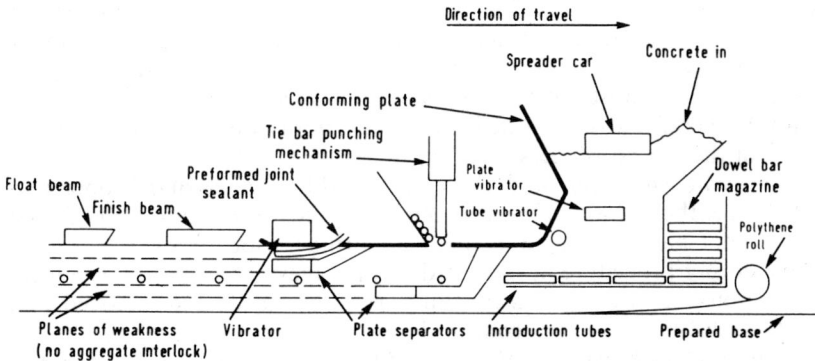

Fig. 8.10 Gunter and Zimmerman/McGregor modified slip form paver

Curing

The primary reason for curing concrete is to prevent the evaporation of water so that the chemical reaction in which the cement combines with the water can occur. It has also been found that the amount of cracking in concrete pavements after many years of traffic is a function of the efficiency of the curing and heat insulation given to the slabs immediately after finishing is complete. This is believed to be the result of shrinkage and/or temperature cracks occurring in the hardening concrete during hot or dry weather but which did not open sufficiently to be detected for many years. Thus there is need not only to cover the concrete in order to prevent water from evaporating during curing, but also to provide for thermal insulation in hot weather, particularly during periods of large temperature fluctuation.

The curing of concrete should begin as soon as possible after the mixture has been laid between the side-forms. In warm weather, concrete containing ordinary Portland cement is usually cured for about fourteen days, while rapid-hardening cement-concrete needs only to be cured for seven days. In cold, but not frosty, weather these curing periods normally have to be extended by about 50 per cent. In frosty weather, the curing periods are usually further increased by the number of days of frost.

The main means by which concrete pavements are cured are as follows:

1. *Spraying waterproof membranes.* Probably the most widely used procedure is to spray a white-pigmented insoluble resinous compound on to the concrete immediately after finishing. Thus, not only is the moisture prevented from evaporating by the impervious cover, but the white pigment reflects the radiant heat of the sun and so detrimental temperature effects are reduced. The membrane has the further advantage that it needs no additional treatment after spraying and can be left to be worn off by the traffic once it is allowed on the roadway.

2. *Covering with damp fabric covers.* This is a method which is still widely used in conventional concreting. It consists of placing damp fabric coverings, i.e. cotton matting, hessian, jute and cotton, directly on top of the concrete. The coverings are sprinkled and kept constantly damp throughout the curing period. The principal disadvantages of this method are, firstly, that the coverings must not only be placed but also removed, and, secondly, they must be constantly sprinkled with water; both of these factors increase the cost of the curing process.

3. *Covering with damp sand.* Once the initial hardening of the concrete is complete, i.e. the initial curing, it is often covered with about 5 cm of sand which is kept moist throughout the final curing period. While it is a quite effective curing procedure, it has the disadvantages of being both expensive and laborious, i.e. two separate coverings must be placed and removed, and again the sand must be kept constantly wet.

4. *Covering with waterproof paper.* With this method a layer of waterproof paper is laid on top of the concrete, weighted down, and kept in close contact with the surface so that the wind does not penetrate beneath and free circulation of air is prevented. This allows droplets of condensed water to collect on the underside of the paper and the concrete is cured under moist conditions.

5. *Water ponding.* This is a method of curing which was fairly widely used in the past but is rarely used now. Essentially it consisted of building small clay dams about the slab and then flooding it with water. While this is certainly a most effective way of curing a concrete slab, the disadvantages associated with it are obvious.

SELECTED BIBLIOGRAPHY

1. ROAD RESEARCH LABORATORY. A guide to the structural design of flexible and rigid pavements, *Road Note* No. 29. London, H.M.S.O., 1970.
2. HIGHWAY RESEARCH BOARD. Final report of Committee on maintenance of concrete pavements as related to the pumping action of slabs, *Proc. Highway Research Board*, 1948, **28**, 281–310.
3. The Cost of Constructing and Maintaining Flexible and Concrete Pavements over 50 years. *RRL Report* LR 256, Crowthorne, Berks., The Road Research Laboratory, 1969.
4. BS 12:Pt. 2:1971. *Portland Cement (Ordinary and Rapid Hardening)*. London, British Standards Institution, 1971.
5. BS 146:1958. *Portland-Blast-furnace Cement*. London, British Standards Institution, 1967.
6. BS 915:1947. *High Alumina Cement*. London, British Standards Institution, 1962.
7. BS 4027:1972. *Sulphate Resisting Portland Cement*. London, British Standards Institution, 1972.
8. ROAD RESEARCH LABORATORY. Design of concrete mixes, *Road Note* No. 4. London, H.M.S.O., 1950.
9. CORNELIUS, D. F. Air Entrained Concretes: A Survey of Factors affecting Air Content and a Study of Concrete Workability. *RRL Report* LR 363, Crowthorne, Berks., The Road Research Laboratory, 1970.
10. ABRAHAMS, D. A. Design of concrete mixtures, *Bull.* No. 1, Structural Materials Research Laboratory. Chicago, Illinois, Lewis Institute, Dec. 1918.
11. MINISTRY OF TRANSPORT. *Specification for Roads and Bridge Works*, London, H.M.S.O., 1969.
12. FRANKLIN, R. E. and KING, T. M. J. Relations between Compressive and Indirect-Tensile Strengths of Concrete. *RRL Report* LR 412, Crowthorne, Berks., The Road Research Laboratory, 1971.
13. WESTERGAARD, H. M. Stresses in concrete pavements computed by theoretical analysis, *Public Roads*, 1926, **7**, No. 2, 25–35.
14. COMMITTEE ON RIGID PAVEMENT DESIGN. State of the Art of Rigid Pavement Design. Special Report 95, pp. 1–33, Washington, D.C., The Highway Research Board, 1968.
15. WESTERGAARD, H. M. Analytical tools for judging results of structural tests of concrete pavements, *Public Roads*, 1933, **14**, No. 10, 185–188.
16. TELLER, L. W. and E. C. SUTHERLAND. The structural design of concrete pavements, Pt. 5: An experimental study of the Westergaard analysis of stress conditions in concrete pavement slabs of uniform thickness, *Public Roads*, 1943, **28**, No. 8, 167–212.
17. SPARKES, F. N. and A. F. SMITH. *Concrete Roads*. London, Arnold, 1962.
18. BRADBURY, R. D. *Reinforced Concrete Pavements*. Washington, D.C., Wire Reinforcement Institute, 1938.
19. WESTERGAARD, H. M. Analysis of stresses in concrete pavements caused by variations in temperature, *Public Roads*, 1928, **8**, No. 3, 54–60.
20. THOMLINSON, J. Temperature variations and consequent stresses produced by daily and seasonal temperature cycles in concrete slabs, *Concrete and Constructional Engineering*, 1940, **35**, No. 6, 298–307; No. 7, 352–360.
21. STOTT, J. P. and BROOK, K. M. Report on a Visit to the U.S.A. to Study Blow-Ups in Concrete Roads. *RRL Report* LR 128, Crowthorne, Berks., The Road Research Laboratory, 1968.

22. ROAD RESEARCH LABORATORY. A Guide to the Structural Design of Pavements for New Roads. *Road Note* No. 29, London, H.M.S.O., 1970.
23. GREGORY, J. M. The Effect of the Revision of Road Note No. 29 on Reinforcement Requirements for Reinforced Concrete Pavements. *TRRL Report* LR 460, Crowthorne, Berks., The Transport and Road Research Laboratory, 1972.
24. STOTT, J. P. Prestressed concrete roads, *Proc. Inst. Civ. Engrs*, Pt. II, 1955, **4**, 491–538.
25. ROAD RESEARCH LABORATORY. *Concrete Roads*. London, H.M.S.O., 1955.
26. WRIGHT, P. J. F. Full scale tests of materials for sealing expansion joints in concrete roads, *Roads and Road Construction*, 1963, **41**, No. 485, 138–146.
27. CRONEY, D. The Design of Concrete Road Pavements. Paper presented at a *1-day Meeting on Concrete Roads* on Sept. 20, 1967, London, The Concrete Society, 1967.
28. BURKS, A. E. Application of the slip-form paver to British road construction, *Cement, Lime and Gravel*, 1965, **40**, No. 5, 173–178.
29. BURKS, A. E. and MAGGS, M. F. Rigid pavements and the advent of the slip-form paver, *Journal of the Institution of Municipal Engineers*, 1968, **95**, No. 8, 237–239.
30. ANON. Concrete road breakthrough: McGregor slips fully dowelled slabs in Essex, *New Civil Engineer*, p. 20, Oct. 12, 1972.
31. BLAKE, L. S. and K. M. BROOK. The construction of major concrete roads in Great Britain, 1955–1960, *Technical Report* TRA/363. London, Cement and Concrete Association, 1962.

Index

Index

Index

A.A.S.H.O. soil classification system, *see* Highway Research Board system of soil classification
Abram's water:cement law, 403
Abrasion tests on aggregates, 138, 140
Accelerated polishing test, 138, 140
Additives for concrete, 400–1
Adsorption, in soils, 176–7
Aeolian soils
  dune sands, 161
  loess, 161
Aerial photography, vertical
  advantages, 27
  criteria governing usage, 7, 20
  disadvantages, 28
  flight line, height and overlap, 19–21
  ground control for, 16–17
  interpretation of, 23–6
  mosaics of, 17–18
  photographic scale used in, 15–16, 20
  planimetric and topographic maps developed from, 18–19
  stereoscopic vision, 21–3
Aerial surveys, *see* Aerial photography
Aggregate gradation calculations
  blending soils and aggregates, 233–6
  combining aggregates, 147–51
  comparing specifications, 143–7
Aggregate tests and their significance
  descriptive tests, 132–3
  durability, 135–41
  non-destructive quality, 132–5
  specific gravity, 141–3
Aggregates
  artificial, 131
  in concrete, 398–400, 405
  natural, 125–31
  tests and their significance, 132–43
  usage in pavements, 124–5
Aggregation of soil particles, due to the addition of lime, 254–5

A-horizon in soil, 190–1
Air-entrained concrete, 400, 402–403
Air photo interpretation, 23–6
Alluvial soils
  deltas, 160
  fans, 160
  flood-plains, 159–60
  See also Glacial soils
Alumina cement, high, 122–4
American Association of State Highway Officials soil classification system, *see* Highway Research Board system system of soil classification
Anionic emulsions, 100–3
Apparent cohesion, 178
Apparent specific gravity, of aggregates, 142
Ash content test on bituminous binders, *see* Composition tests
Asphalt
  cold, 378
  compressed natural rock, 93–4, 374
  mastic, 373–4
  rock (non-skid), 374–5
  rolled, 369–73
  Trinidad Lake, 92–3
Asphaltic-based petroleum crude, components of, 94
Atterberg's, for soils
  consistency limits, 207–8
  size fractions, 165–6
Auger borings in soil exploration, 34–5

Basecourse, definition of, 271
  See also Bituminous surfacings
B-horizon in soil, 190–1
Bilham formula for selecting design rainfall intensity, modified, 60
Bitumen
  comparison with tar, 107
  cut-back, 95–9

Bitumen (*cont.*)
definition, 91–2
emulsions, 99–103
natural rock asphalt, 93–4
penetration-grade, 95
refinery, 94–9
rubber in, 107–9
specifications for
cut-back, 100
penetration-grade, 97
Trinidad Lake asphalt, 92–3
Bituminous binder tests and their significance
composition, 116–20
consistency, 110–16
Bituminous binders
materials, 91–120
terminology, 90–1
See also Bitumen *and* Tar
Bituminous mix design
Marshall test and design procedure, 361–9
surface-area concept, 360
voids concept, 360–1
Bituminous roadbases, dense, 345–6
Bituminous stabilization
mechanism of, 258–9
oiled earth, 263–4
sand-bitumen, 259–60
sand-gravel-bitumen, 260–1
soil-bitumen, 261–3
Bituminous surfacings
design and selection of, 358–78
function of, 352–3
how they fail, 353–8
recommendations for design, 358
See also British practice with bituminous surfacings
Blending of soils and aggregates, *see* Aggregate gradation calculations
Borings, *see* Subsurface exploration
Boussinesq theory related to stress distribution in soils, 272–5
Bridge surveys, 11–12
British design procedure
for flexible pavements, 300–9
for rigid pavements, 433–5
British practice with bituminous surfacings
compressed natural rock asphalt, 374
coated macadam, 375–8
dense tar, 375

mastic asphalt, 373–4
rolled asphalt, 369–73
Bulk specific gravity of aggregates, 142
Burmister theory of stress distribution in highway pavements, 276, 277

C.B.G.M., 346–9
Calcium acrylate, for soil stabilization, 266–7
Calcium chloride, for soil stabilization, 236–8
Calcium sulphate, effects in soil, 244
California Bearing Ratio
factors affecting results, 297–300
relationship with modulus of subgrade reaction, 414–15
test, 294–7
use in British design procedure, 300–9
values for British soils, 296
California stabilometer design method, *see* Stabilometer pavement design method
Capillary fringe, 197
Capillary water, 197–8
Carbonation of lime, 256–7
Casagrande system of soil classification
extended system, 179–83, 194
related to Manning's roughness coefficient, 69–70
related to maximum permissible velocity of flow in open channels, 69–70
Cation exchange capacity, 172–4
Cationic emulsion, 100–2
Cement
high alumina, 122–4
portland, 120–2, 123–4
setting time of, 398
strength of, crushing, 398
Cement-bound bases
C.B.G.M., 346–9
curing, 349
lean concrete, 346–9
Cement stabilization
construction, 248–50
definition of, 221
factors affecting stability and usage, 243–8
mechanism of, 242–3
types of, 240–2
Cement-treated soils, *see* Cement stabilization

Chézy formula, 67
Chloride stabilization
  mechanism of, 236–8
  types of chloride used for, 236
  usage, 238
C-horizon in soil, 190–1
Ciment Fondu, 122
Classification of soils, *see* Soil
    classification
Clay
  minerals, 168–72
  physical and mineralogical properties
    of, 167–8
  size limits of, 165
Clearing the site
  clearing and grubbing, 334–5
  utilities, 335–6
Coated macadam surfacings, 375–8
Coefficient of consolidation, 218–19
Cohesiometer test, 313
Cohesion in soils
  explanation of, 177–8
  typical values of, 217
Cold asphalt, 378
Colloid mill, used to manufacture emul-
    sions, 100–1
Colluvial soils, 157
Compacting factor test for concrete
    mixes, 406
Compacting the subgrade, 336–7
Compaction
  by impact, 228
  definition of, 224
  degree of, 164
  effect of chlorides on, 237
  factors affecting compaction in the
    field, 224–8
  of cement-treated soil mixtures, 246–7
  of subgrade, 336–7
  relative, 164
  specification and control of, 228–9
  with pneumatic-tyred rollers, 226–7
  with sheepsfoot rollers, 225–6
  with smooth-wheeled rollers, 225
  with vibratory compactors, 227–8
Composition tests on bituminous
    binders
  ash content, 118
  distillation, 117
  flash and fire point, 120
  loss-on-heating, 118
  solubility, 118–19

specific gravity, 119–120
  water content, 117–18
Compressed natural rock asphalt, 93–4,
    374
Compressibility of a soil, explanation of,
    326–7
Compressible soils, construction of pave-
    ments on, 325–34
Compressive index, 218–19
Concentration, time of, 58–9
Concrete
  additives in, 400–4
  aggregates in, 398–400
  air-entrained, 400, 402–3
  crushing strength, 398, 407
  lean, 346–9
  mix design, 403–7, 411
  pavements, stresses in, 411, 412–19
  setting time, 398
  See also Reinforcement in concrete
    pavements *and* Construction of
    concrete pavements
Concrete pavements, *see* Rigid
    pavements
Consistency tests for bituminous binders
  definition, 110
  ductility, 116
  penetration, 111
  softening point, 115–16
  viscosity, 111–15
Consistency tests for soils
  liquid limit, 207–8
  plastic limit, 209
Consolidation of soils
  methods of, 229–30, 328–30
  test, 218–19
Construction of concrete pavements
  compacting and finishing, 439–40
  curing, 441–2
  joints, 432
  laying the forms, 436–7
  mixing the concrete, 437–8
  preparing the subgrade and sub-base,
    334–7, 436
  spreading the concrete, 438–9
  use of slip-form pavers, 440–1
Construction of flexible pavements
  conventional roadbases and sub-
    bases, in Britain, 337–49
  in frost areas, 320–25
  on compressible soils, 325–34
  preparing the subgrade, 334–7

Construction on compressible soils, methods of
  ancillary methods, 333–4
  direct embankment, 328–30
  direct lightweight, 327–8
  removal and replacement, 331–3
Continuously reinforced concrete pavements, 423
Contour lines, characteristics of, 18–19
Contraction joints in concrete pavements, 429–431
Creosote oils, 104
Crude tar
  components of, 103–4
  production from British coal, 105
Crushing strength of concrete, 398, 407
Crushing value test, aggregate, 140–1
Culverts
  hydraulic design of, 74–6
  location of, 72–3
  types of, 71–2
Cumulous soils, 157
Curing of
  cement-treated soil mixtures, 248
  concrete pavements, 441–2
  lean concrete pavements, 349
Cut-back bitumens, 95–9
Cut-back tars, 107

Degree of compaction of soils, 164
Degree of saturation of soils, 163–4
Deliquescent properties of chlorides, 236–7
Delta soils, 160
Dense bituminous, British practice
  roadbases, 345–6
  surfacings, 369–75
Dense tar surfacing, 375
Descriptive tests on aggregates, 132–3
Design and selection of bituminous surfacings
  British practice, 369–78
  empirical design procedures, 358–9
  Marshall test and design procedure, 361–9
  theory of mix design, 359–61
Design storm of surface drainage determination, 59
Digital ground models, 19
Direct shear test, 216–18
Disintegration of bituminous surfacings, 354

Distillation tests on bituminous binders, see Composition tests
Ditches
  side, 66
  open-channel hydraulic design of, 66–71
Dowels for expansion and contraction joints, 428–31
Drainage
  of compressible soils, 334
  patterns on air photos, 25
  role of sub-base in improving, 394
  subsurface, 78–88
  surface, 56–77
Drumlin soils, 158
Dry density of soil, see Moisture-unit weight test
Dry unit weight of soils
  determination of, 211–15
  typical test results, 213–15
Ductility test on bituminous binders, see Consistency tests
Dune sands, 161
Durability tests on road aggregates
  abrasion, 138, 140
  toughness, 140–1
Dust palliatives
  chlorides, 236–7
  lignin, 239
  molasses, 239–40
  oiled-earth treatment, 263–4

Earthwork quantities, estimating and distributing, 45–53
Economic haul, 50–1
Elastic soils, 325
Elastic theories of stress distribution, 272–8
Electrical resistivity method of subsurface exploration, 37–8, 42
Emulsifiers, anionic and cationic, 100
Emulsions
  bitumen, 99–102
  mechanism of action, 102–3
  tar, 107
Equi-viscous temperature system, and test, for bituminous binders, 113
Esker soils, 158
Evaporation of moisture, see Dust palliatives
Exchange capacity of a clay soil, 172–4

Expansion joints in concrete pavements, 427–9
Expansion-pressure test, California, 311
Extended Casagrande system of soil classification, 179–83, 194
Exudation-pressure test, California, 311

Fatigue of flexible pavement, 284–5
Fillers and seals for joints in concrete pavements, 432–3
Final location survey
    for bridges, 12
    for roads in rural areas, 9–11
    for roads in urban areas, 12–13
    using aerial techniques, 27
    See also Urban location controls
Flash and fire point tests, 120
Flexible pavements
    construction elements, 269–70
    effect of edge loading on, 287
    effect of number of axles, 283
    effect of pavement thickness and material on, 285–6
    effect of repeated loads on, 284–5
    effect of static and moving loads on, 283–4
    effect of tyre pressure on, 278–81
    effect of wheel configuration on, 281–3
    effect of wheel load on, 281
    cement bound bases of, 346–9
    in frost areas, 320–5
    on compressible soils, 325–34
    preparing the subgrade for, 334–7
    roadbases and sub-bases of, 271, 337–49
    theoretical stress distribution in, 272–8
    See also Frost action and Methods of flexible pavement design
Flight height and line for aerial surveys, 19–21
Flocculent soil structure, 175–6
Flood-plain soils, 159–60
Fluff-point of a soil-bitumen mix, 262–3
Fluxed bitumens, 98
Fly ash, see Lime and lime-pozzolan stabilization
Forms for concrete pavements, 436–7, 440–1
Fracture of bituminous surfacings, 355
Free-haul, 50–1

Frost action
    detrimental effects of, 321
    in concrete, 394
    pavement design for protection against, 322–5
    role of sub-base in preventing, 394
    using additives to resist, 324–5
Frost heaving, 199–201, 321
Frost penetration, 201–2
Frost-susceptible soils, 202–4
Fuller's gradation law, 232
Fully stable anionic emulsions, 101–2

Geological Survey maps, 4
Geophysical methods of subsurface exploration, 36–42
Glacial soils
    drumlins, 158
    eskers, 158
    kames, 158
    moraines, 157–8
    outwash deposits, 159
Gradation tests
    for aggregates, 132, 134
    for soils, 205–7
    See also Aggregate gradation calculations
Graft rubber in bitumen/tar, 108
Gravitational water in soil, 196
Ground control for aerial surveying, 16–17
Group index
    method of pavement design, 290–3
    use of, in classifying soils, 188–9
Gulleys on urban roads, 76–8

Hand-pitched roadbases, 339–40
Hardcore in road construction, 338–9
Haul, free-haul and overhaul, in earthworks, 50–1
    See also Economic-haul and Mass-haul
Highway Research Board system of soil classification, 183–9, 194
Hinge joints, 431–2
Honeycomb soil structure, 175
Hot-rolled asphalt
    average life of, 372
    surface course construction, 369–73
    surface dressings on, 373
Hveem stabilometer, see Stabilometer method of pavement design

Hydrated lime
   in soil stabilization, 252, 254–7
   process, for bituminous surfacings,
      260
Hydraulic design
   of culverts, 71–6
   of ditches, 66–71
   of gulleys, 76–7
Hydraulic lime, 251
Hydrological study
   precipitation, 57–60
   run-off, 61–5
Hygroscopic moisture, 198
Hygroscopic properties of chlorides,
   236–8

Ice lensing, see Frost heaving
Igneous rocks, 125–6
Illite clay mineral, 172
Impact compaction, 228
Instability value test for aggregates,
   140–1
Instability of bituminous surfacings,
   353–4
Intensity of rainfall, 58–65
Internal friction in soils
   explanation of, 178
   typical values of, 217
International Society of Soil Science's
   size-fractions, 165

Joints in concrete pavements
   construction, 432
   contraction, 429–31
   expansion, 427–9
   requirements of, 426–7
   seals and fillers for, 432–3
   warping, 431–2

Kame soils, 158
Kaolinite clay mineral, 170–1

Labile anionic emulsions, 101
Landslides, 25–6, 30
Latex rubber in bitumen/tar, 107–9
Lean concrete roadbases, 346–9
Lignin stabilization, 239
Lime and lime-pozzolan stabilization
   mechanism of, 254–7
   pozzolans in, 252–4
   types of lime, 251–2
   usage, 257

Liquid limit
   description of, 207–8
   usage in soil classification, 183, 188–9
Lloyd-Davis run-off formula, 63–5
Location of highways
   controls influencing the, in urban
      areas, 13
   principles of, 1–3
   surveys in rural areas, 3–11
   surveys in urban areas, 12–14
   See also Aerial photography; Bridge
      surveys and Subsurface explora-
      tion
Loess soils, 161
Loss-on-heating test for bituminous binders, see Composition tests

Macadam roadbases
   basic preparation, 341
   coated, dense, 345–6
   crusher-run, 344
   dry-bound, 341–3
   pre-mixed water bound, 344–5
   water-bound, 343–5
   wet-mix, 344–5
Manning formula, 67
Manning roughness coefficient, 69–70
Marine soils, 160–1
Marshall method, and test, of bituminous mix design, 361–9
Mass-haul diagram, 51–3
Mastic asphalt, 373–4
Maximum dry unit weight of soils
   typical test values, 213–14
   See also Moisture-unit weight test
Mechanical analysis, see Gradation tests
Mechanical stabilization
   by altering soil gradation, 231–6
   by compaction, 223–4
   by consolidation, 229–30
   by electrical and thermal methods, 230
   definitions of, 230
   usage, 222
   with chlorides, 236–8
   with lignin, 239
   with molasses, 239–40
Methods of flexible pavement design
   based on empirical methods, using a
      dispersion angle for load distribu-
      tion, 289–90
   using a soil classification test, 290–3
   using a soil strength test, 293–309

based on precedent, 288
based partly on theory and partly on experience, 309–15
based primarily on theory, 315–18
British practice, 300–9
research on, current and future, 318–20
See also Flexible pavements
Medium-curing cut-backs, 98–9
Metamorphic rocks, 127
Michigan State Highway pedological method of pavement design, 293
See also Pedological classification system
Ministry of Health rainfall intensity formulae, 59–60
Modulus of subgrade reaction, *k*, 414–15
Moisture
 and frost action, 202
 content, of cement-treated soil mixtures, 246
  of soil bituminous mixtures, 262–3
 control in soils, subsurface, 78–88
 movements in soils, 78–82, 196–8
 -unit weight test for soils, 211–15
 See also Surface moisture control
Moisture-density test and relationship in soils, *see* Moisture
Moisture-friction stresses in concrete pavements, 419
Moisture-unit weight test and relationship in soils, *see* Moisture
Moisture-warping stresses in concrete pavements, 418
Molasses stabilization of soil, 240
Monmorillonite clay mineral, 171–2
Moraine soils, 157–8
Mosaic, *see* Aerial photography
Mud-pumping of concrete pavements, 198, 394–5

Natural rock asphalt, 93–4
Non-destructive quality tests, aggregate
 gradation, 132, 134
 shape, 134–5
 water absorption, 135

Oiled-earth treatments, 263–4
Open-channel design of ditches, 66–71
Optimum moisture content of soils
 determination of, 211–15
 typical test results, 213–15

Ordinary Portland cement, *see* Portland cement
Ordnance Standard Datum Plane, 7–8
Ordnance Survey maps and plans, 4
Organic cationic stabilization, 265–6
Organic matter, detrimental effects of, on cement-stabilized soils, 244

Paraffinic crudes, for bitumen, 94
Particle size analysis, *see* Gradation tests
Pedological classification system for soils, 189–193
Penetration-grade bitumens, 95, 97
Penetration test on bitumens, *see* Consistency tests
Per cent air voids in soil, 163
Per cent water voids in soil, 163
Photogrammetry, *see* Aerial photography
Piling construction on compressible soils, 330
Pitch-bitumen mixtures, 109–10
Plans, preparation of, 42–53
Plastic index and plastic limit of soils, 209
Plastic soil-cement, 242
Plasticity chart for Extended Casagrande soil classification system, 179, 183
Plasticity index
 definition of, 209
 blending two soils to meet a specified, 233–4
Pneumatic-tyred rollers, *see* Rollers
Poisson's ratio, for concrete, 274, 413
Porosity of soils, 162–3
Portland cement
 blast furnace, 121–2
 chemical composition of, 123–4
 ordinary, 121
 rapid-hardening, 121
 sulphate resisting, 122
Pozzolans and pozzolanic reactions, *see* Lime and lime pozzolan stabilization
Precipitation, intensity and duration of, in surface moisture control and drainage, 57–60
Preliminary surveys
 for bridges, 11–12
 for roads in rural areas, 6–9
 for roads in urban areas, 12–13

Preliminary surveys (*cont.*)
  using aerial techniques, 27
  See also Urban location controls
Preliminary work for subsurface investigations, 29
Prestressed concrete pavements, 425–6
Primary structure of soil, 174–6
Proctor test, see Moisture-unit weight test
Public Roads classification systems, *see* Highway Research Board system of soil classification
Pulverized fuel ash, properties of, 253–4
Pumping of fine-grained soils, see Mud-pumping

Quantities, preparation of, 53
Quarries and gravel pits in Britain, 130
Quicklime, 251–2

Rapid-curing cut-back bitumen, 99
Rapid-hardening Portland cement, 121
Reconnaissance surveys
  for bridges, 11
  for roads in rural areas, 4–6
  for roads in urban areas, 13
  using aerial techniques, 26
  See also Urban location controls
Refined Trinidad Lake asphalt, 92–93
Refinery bitumens, 94–99
Reinforcement in concrete pavements, 420–430
Removal and replacement method of road construction on compressible soils
  ancillary methods of, 333–4
  displacement-by-blasting method of, 332–3
  displacement-by-gravity method of, 331–2
  mechanical methods of, 331
Research into pavement design, some comments on, 318–20
Residual soils, 157
Resin stabilization of soils, 266–7
Retread process for bituminous surfacings, 101
Rigid pavements
  concrete in, 396–8
  construction of, 435–41
  curing of, 441–2

  definition of, 392
  joints in, 426–33
  reinforcement in, 420–6
  stresses in, 411–19
  sub-bases and subgrades for, 392–6
  thickness design of, 433–5
Roadbases and sub-bases, types of
  cement-bound, 346–9
  dense bituminous, 345–6
  function, 271
  hand-pitched, 339–40
  macadam, 340–5
  soil-aggregate, 233
  waste-material, 338–9
Rock core drillings, 35–6
Rock (non-skid) asphalt, 373–4
Rolled asphalt, 369–73
Rosin soil stabilization, 266–7
Rollers, for soil compaction
  pneumatic-tyred, 226–7
  sheepsfoot, 225–6
  smooth-wheeled, 225
  vibratory, 227–8
Rotary drilling in soils, 35
Rubber-bitumen/tar, 107–9
Run-off from catchment area, 61–5

Sand-bitumen mixtures, 259–60
Sand characteristics, 166
Sand-drains, 329–30
Sand-gravel-bitumen mixtures, 260–1
Sealing coats, bituminous, 388
Seals and fillers for joints in concrete pavements, 432–3
Secondary structure of soil, 176
Sedimentary rocks, 126–7
Seepage from high ground, 79
Seismic refraction method of subsurface exploration, 39–42
Semi-stable anionic emulsions, 101
Setting time of cement, 398
Shale in road construction, 339
Shape tests on aggregates, *see* Non-destructive quality tests
Shear tests for soils, 216–18
Sheepsfoot rollers, *see* Rollers
Sheet rubber in bitumen, 108
Shrinkage of soil, 48–50
Side-drains, hydraulic design of, *see* Open-channel design of ditches
Sieve analysis, *see* Gradation tests

Silt characteristics, 166
Single-grained soil structure, 174–5
Skidding resistance of bituminous surfacings, loss of, 357–8
Slip-form pavers for concrete road construction. 440–1
Slow curing cut-back bitumens, 96, 98
Slump test for concrete, 405–6
Smooth-wheeled rollers, *see* Rollers
Sodium chloride as a soil stabilizing agent, 236–8
Sodium silicate as a soil stabilizing agent, 264–5
Softening point test, *see* Consistency tests
Soil
  classification, 178–95
  colloids, 168
  definition of, 154–5
  formation of, 155–7
  frost action in, 198–204
  frost-susceptible, 202–4
  methods of testing, 204–19
  moisture, 195–8
  phases, 161–4
  physical and mineralogical characteristics of, 166–74
  series, 191–2
  solid constituents of, 164–74
  stabilization, *see* Bituminous stabilization; Cement stabilization; Lime and lime-pozzolan stabilization *and* Mechanical stabilization
  structure, 174–6
  Survey of Great Britain, 29
  surveys, *see* Subsurface exploration
  systems, forces in, 176–8
  types of, 157–61
Soil-aggregate stabilization
  blending soils and aggregates, 233–6
  general requirements for surfaces and roadbases, 231–2
  usage, 232–3
Soil auger borings, 34
Soil-bituminous mixtures, *see* Bituminous stabilization
Soil cement, *see* Cement stabilization
Soil classification
  extended Casagrande system, 179–83, 184–5
  Highway Research Board system, 183, 186–9

illustrative examples of, 194–5
  pedological system, 189–93
  textural system, 193
Soil-lime mixtures, *see* Lime and lime pozzolan stabilization
Soil surveys, *see* Subsurface explorations
SOJUSDORNII method of pavement design, 316–18
Solubility tests on bituminous binders, *see* Composition tests
Soundings, as used in subsurface exploration, 36
Specific gravity tests
  for aggregates, 141–3
  for bituminous binders, 119–20
  for soils, 210–11
Specifications, for highway construction, 53–4
Stabilometer method of pavement design, California, 310–15
Standard Tar Viscometer test, 113
Stereoscopic vision in aerial photography, 21–3
Sticky limit for soils, 207
Strength tests
  shear, 216–18
  unconfined compressive, 215–16
Stress distribution
  in concrete slabs, 411–16
  in flexible pavements, *see* Flexible pavements
Strip map, usage of, in highway location, 8
Structure of the highway location process, 3–4
Structure, soil
  flocculent, 175–6
  single-grained, 174–5
Sub-bases, function of
  for flexible pavements, 271
  for rigid pavements, 393–6
  See also Roadbases and sub-bases
Subgrades, beneath rigid pavements, classification of, 392–3
Subgrade, preparing the
  clearing the site, 334–6
  compaction of, 336–7
  shaping, 337
Subsurface exploration
  determining the soil profile, 30–2
  locating the water table, 33
  methods used to carry out, 33–42

Subsurface exploration (*cont.*)
preliminary work for, 29
site examination for, 29–30
usage, 28–9
Subsurface moisture control
methods of control, 78–82
sub-drain design, 83–88
Sulphate-resisting cement, 122
Sulphates, detrimental effects of, on cement-stabilized soil, 244–5
Surcharge
depth of, in concrete road construction, 439
method of construction on compressible soils, 328–9
Surface course, function of, 250–1
Surface drainage, *see* Surface moisture control
Surface dressings, bituminous
application of materials, 384–7
average life of, 372
external conditions affecting the success of, 387–8
functions of, 379
materials used in, 380–4
on rolled asphalt, 373
Surface moisture control
culverts, 71–6
design storm for, 59
gulleys used for, 76–8
hydraulic design for, 66–76
hydrological study for, 57–65
Surface tension in soils, 178
Surface treatments, bituminous
sealing coats, 388
surface dressings, 379–88
tack coats, 388–9
Surfacings
bituminous, 352–79
function of, in flexible pavements, 270–1
soil-aggregate, 231–2
Surveys, for highway location
aerial, 7, 14–28
bridge, 11–12
final location, 9–11, 12–13
preliminary, 6–9, 12–13
reconnaissance, 4–6, 13
See also Subsurface explorations
Swell of soil, 50

Tack coats, bituminous, 388–9

Tar
-bitumen mixtures, 109
comparison with bitumen, 107
cut-back, 107
definition of, 103
emulsions, 107
preparation of crude, 103–4
production from various coals, 105
road, used in Britain, 104–9
-rubber mixtures, 107–9
specifications, 106
See also Binder tests and their significance *and* Pitch bitumen
Temperature-frictional stresses in concrete pavements, 418–19
Temperature-warping stresses in concrete pavements, 416–18
Ten per cent fines test, 140–4
Test pits, for subsurface exploration, 34
Textural classification, 193, 195
Theories of bituminous mix design, 360–1
Thickness design of concrete pavements, *see* Rigid pavements
Thickness design of flexible pavements, *see* Methods of flexible pavement design
Toughness tests for aggregates, 140–1
Trade groups of rock aggregates, 127–9
Traditional rock aggregate names, 131
Triaxial shear test for soil, 217–18
Trinidad Lake asphalt, 92–3
Tyre pressure, effects on flexible pavements, *see* Flexible pavements

Unconfined compressive strength test for soils, 215–16
United States Department of Agriculture, Bureau of Soils, size fractions used by the, 165
Unvulcanized rubber powder in bitumens, 108
Urban location controls, 13–14
Utilities
beneath pavements in urban areas, 335–6
effect on highway location of, 14

Val de Travers region, asphaltic bitumen mined in the, 93
Vapour movement in soils, 82

Vibratory compactors, *see* Rollers

Vinsol resin as a soil stabilizing agent, 266

Viscosity tests on bituminous binders, *see* Consistency tests

Void ratio of soil, 162

Vulcanized rubber powder in bitumen, 108

Warping joints in concrete pavements, 431–2

Wash borings in soil exploration, 35

Water
absorption test on aggregates, 135
:cement ratio in concrete mix design, 403–5
content test for bituminous binders, 117–18
table, 33, 79–80, 196

Westergaard analysis of stresses in concrete pavements, 412–15

Wet-aggregate (hydrated lime) process, for bituminous mixes, 260

Wet-mix (pre-mixed water bound) macadam construction, 344–5

Wheel load stresses in pavements, *see* Flexible pavements *and* Rigid pavements

Workability of concrete, 405–11